SEAWEED BIOTECHNOLOGY

Biodiversity and Biotechnology of Seaweeds and Their Applications

Innovations in Biotechnology

SEAWEED BIOTECHNOLOGY

Biodiversity and Biotechnology of Seaweeds and Their Applications

Edited by

Jeyabalan Sangeetha, PhD
Devarajan Thangadurai, PhD

AAP | APPLE ACADEMIC PRESS

First edition published 2023

Apple Academic Press Inc.
1265 Goldenrod Circle, NE,
Palm Bay, FL 32905 USA

760 Laurentian Drive, Unit 19,
Burlington, ON L7N 0A4, CANADA

CRC Press
6000 Broken Sound Parkway NW,
Suite 300, Boca Raton, FL 33487-2742 USA

2 Park Square, Milton Park,
Abingdon, Oxon, OX14 4RN UK

Library and Archives Canada Cataloguing in Publication

Title: Seaweed biotechnology : biodiversity and biotechnology of seaweeds and their applications / edited by Jeyabalan Sangeetha, PhD, Devarajan Thangadurai, PhD.

Names: Sangeetha, Jeyabalan, editor. | Thangadurai, D. (Devarajan), 1976- editor.

Description: First edition. | Series statement: Innovations in biotechnology | Includes bibliographical references and index.

Identifiers: Canadiana (print) 20220175357 | Canadiana (ebook) 20220175381 | ISBN 9781774910900 (hardcover) | ISBN 9781774910917 (softcover) | ISBN 9781003300854 (ebook)

Subjects: LCSH: Marine algae—Biotechnology. | LCSH: Marine algae.

Classification: LCC QK570.2 .S43 2023 | DDC 579.8/177—dc23

Library of Congress Cataloging-in-Publication Data

Names: Sangeetha, Jeyabalan, editor. | Thangadurai, D. (Devarajan), 1976- editor.

Title: Seaweed biotechnology : biodiversity and biotechnology of seaweeds and their applications / edited by Jeyabalan Sangeetha, PhD, Devarajan Thangadurai, PhD.

Description: First edition. | Palm Bay, FL, USA : AAP/Apple Academic Press, [2023] | Series: Innovations in biotechnology | Includes bibliographical references and index. | Summary: "Seaweeds are known for their rich bioactive compounds, which promote health in human beings and are a good for the ecosystem as well. They are also natural resources that are a major source of raw material for different industries. There are still undiscovered and unexploited compounds synthesized by seaweeds that may have potential applications in the pharmaceutical, nutraceutical, food, and cosmetics industries. This book serves as a comprehensive knowledge source for the predominant roles of seaweeds in various sectors, particularly in the areas of health, environment, and agriculture. It explores the diverse biodiversity aspects of seaweeds and their derivatives. The book critically reviews the present industrial challenges to investigate the novel compounds synthesized by seaweeds and their unique characteristics and benefits. The volume covers the various biodiversity attributes of tropical seaweeds, their cultivation and bioactive compounds, and the diverse agricultural and biomedical applications of new seaweed derivatives. The authors also discuss the current challenges, emerging markets, and latest developments in extracting the useful biomolecules from seaweeds as well as the role of seaweeds in food security and environmental mitigation. With chapters written by experts and professionals in the field, this volume, Seaweed Biotechnology: Biodiversity and Biotechnology of Seaweeds and Their Applications, provides a deep understanding of the biodiversity of seaweeds around the world and their industrial, biomedical, and environmental applications"-- Provided by publisher.

Identifiers: LCCN 2022007971 (print) | LCCN 2022007972 (ebook) | ISBN 9781774910900 (hbk) | ISBN 9781774910917 (pbk) | ISBN 9781003300854 (ebk)

Subjects: LCSH: Marine algae--Biotechnology. | Marine algae industry. | Algae products.

Classification: LCC QK570.2 .S39 2023 (print) | LCC QK570.2 (ebook) | DDC 579.8/177--dc23/eng/20220321

LC record available at https://lccn.loc.gov/2022007971

LC ebook record available at https://lccn.loc.gov/2022007972

ISBN: 978-1-77491-090-0 (hbk)
ISBN: 978-1-77491-091-7 (pbk)
ISBN: 978-1-00330-085-4 (ebk)

INNOVATIONS IN BIOTECHNOLOGY BOOK SERIES

SERIES EDITOR

Devarajan Thangadurai, PhD

Associate Professor, Karnatak University, Dharwad, South India

BOOKS IN THE SERIES

Fundamentals of Molecular Mycology

Devarajan Thangadurai, PhD, Jeyabalan Sangeetha, PhD, and Muniswamy David, PhD

Biotechnology of Microorganisms: Diversity, Improvement, and Application of Microbes for Food Processing, Healthcare, Environmental Safety, and Agriculture

Jeyabalan Sangeetha, PhD, Devarajan Thangadurai, PhD, Somboon Tanasupawat, PhD, and Pradnya Pralhad Kanekar, PhD

Phycobiotechnology: Biodiversity and Biotechnology of Algae and Algal Products for Food, Feed, and Fuel

Jeyabalan Sangeetha, PhD, Devarajan Thangadurai, PhD, Sanyasi Elumalai, PhD, and Shivasharana Chandrabanda Thimmappa, PhD

Biogenic Nanomaterials: Structural Properties and Functional Applications

Devarajan Thangadurai, PhD, Saher Islam, PhD, Jeyabalan Sangeetha, PhD, Natália Cruz-Martins, PhD

Seaweed Biotechnology: Biodiversity and Biotechnology of Seaweeds and Their Applications

Jeyabalan Sangeetha, PhD, and Devarajan Thangadurai, PhD

Microalgal Biotechnology: Bioprospecting Microalgae for Functional Metabolites towards Commercial and Sustainable Applications

Jeyabalan Sangeetha, PhD, Svetlana Codreanu, PhD, and Devarajan Thangadurai, PhD

Algal Metabolites: Biotechnological, Commercial, and Industrial Applications

Jeyabalan Sangeetha, PhD, Devarajan Thangadurai, PhD, Saher Islam, PhD, and Ravichandra Hospet, PhD

ABOUT THE EDITORS

Jeyabalan Sangeetha, PhD
Assistant Professor, Central University of Kerala, Kasaragod, India

Jeyabalan Sangeetha, PhD, is an Assistant Professor at Central University of Kerala, Kasaragod, South India. She has edited/co-edited several books in her research areas, which include environmental toxicology, environmental microbiology, environmental biotechnology, and environmental nanotechnology. She earned her BSc in Microbiology and PhD in Environmental Science from Bharathidasan University, Tiruchirappalli, Tamil Nadu, India. She holds an MSc in Environmental Science from Bharathiar University, Coimbatore, Tamil Nadu, India. She is the recipient of a Tamil Nadu Government Scholarship and a Rajiv Gandhi National Fellowship of the University Grants Commission, Government of India, for her doctoral studies. She served as Dr. D.S. Kothari Postdoctoral Fellow and UGC Postdoctoral Fellow at Karnatak University, Dharwad, South India, during 2012–2016 with funding from the University Grants Commission, Government of India, New Delhi.

Devarajan Thangadurai, PhD
Associate Professor, Karnatak University, Dharwad, India

Devarajan Thangadurai, PhD, is an Associate Professor at Karnatak University in South India and Editor-in-Chief of the international journals *Biotechnology, Bioinformatics, and Bioengineering* and *Acta Biologica Indica.* He has authored/edited over 30 books with national and international publishers and has visited 24 countries in Asia, Europe, Africa, and the Middle East for academic visits, scientific meetings, and international collaborations. He received his PhD in Botany from Sri Krishnadevaraya University in South India as a CSIR Senior Research Fellow with funding from the Ministry of Science and Technology, Government of India. He served as a Postdoctoral Fellow at the University of Madeira, Portugal; University of Delhi, India; and ICAR National Research Centre for Banana, India. He is the recipient of a Best Young Scientist Award with a Gold Medal from Acharya Nagarjuna University, India, and the VGST-SMYSR Young Scientist Award of the Government of Karnataka, Republic of India.

CONTENTS

Contributors .... *xi*

Abbreviations .... *xv*

Symbols .... *xix*

Preface .... *xxi*

1. **Biodiversity of Tropical Seaweeds 1**
Yosmina Tapilatu

2. **Seaweed Cultivation for Community Welfare in Coastal Areas and Small Islands 15**
Joeharnani Tresnati, Ujiie Kiyokazu, Inayah Yasir, Hakiki Melanie, Risal Aprianto, and Ambo Tuwo

3. **Bioactive Compounds from Egyptian Seaweeds 47**
Madelyn Nabil Moawad and Soha Shabaka

4. **Recent Advances and Research Challenges in Bioprospecting of Brown Seaweeds 81**
Sreekala Kannikulathel Gopidas and Nagaraj Subramani

5. **Brown Seaweed Polysaccharides: A Roadmap as Biomolecules 97**
Diana Pacheco, João Cotas, Adriana Leandro, Sara García-Poza, Ana Marta Mendes Gonçalves, and Leonel Pereira

6. **The Global Seaweed Industry: Applications and Emerging Markets 153**
Zahira Belattmania, Abdeltif Reani, and Brahim Sabour

7. **Seaweed as Plant Biostimulants 183**
Tiago Sousa, João Pedro Nunes, José Lopes, João Cotas, Ana Marta Mendes Gonçalves, Kiril Bahcevandziev, and Leonel Pereira

8. **Seaweeds and Their Derivates as a Multirole Tool in Agriculture 201**
Paulo Cesar De Melo, Tiago Sousa, Rita Teixeira, João Cotas, Diana Pacheco, Ana Marta Mendes Gonçalves, Kiril Bahcevandziev, and Leonel Pereira

9. **Seaweeds as a New Source of Bioactive Compounds and Potent Biotechnological Applications 229**
Fatima El Khalloufi and Brahim Oudra

10. Roles of the Ingredients Contained in Edible Seaweeds on Particulate Matter (PM)-Induced Skin Damages 289
Kalu Kapuge Asanka Sanjeewa and You-Jin Jeon

11. Seaweed Biotechnology: An Answer to Environmental Issues and Human Health Problems 311
Elhan Khan and Iffat Zareen Ahmad

12. Seaweeds for Food Security and Environmental Mitigation 335
Ambo Tuwo, Mitsutoshi Nakajima, Joeharnani Tresnati, Inayah Yasir, Hakiki Melanie, and Risal Aprianto

Index 369

CONTRIBUTORS

Iffat Zareen Ahmad
Natural Products Laboratory, Department of Bioengineering, Integral University, Lucknow – 226026, Uttar Pradesh, India

Risal Aprianto
Multitrophic Research Group, University of Hasanuddin, Kampus Tamalanrea KM. 10, Makassar – 90245, Sulawesi Selatan, Indonesia

Kiril Bahcevandziev
Agricultural College of Coimbra (ESAC/IPC), Research Centre for Natural Resources Environment and Society (CERNAS), Institute of Applied Research (IIA), Coimbra – 3045-601, Portugal

Zahira Belattmania
Phycology Research Unit, Department of Biology, Faculty of Sciences, University Chouaib Doukkali, PO Box 20, El Jadida – 24000, Morocco

João Cotas
Department of Life Sciences, University of Coimbra, MARE-Marine and Environmental Sciences Centre, Coimbra – 3001-456, Portugal

Sara García-Poza
Department of Life Sciences, University of Coimbra, MARE-Marine and Environmental Sciences Centre, Coimbra – 3001-456, Portugal

Ana Marta Mendes Gonçalves
Department of Life Sciences, University of Coimbra, MARE-Marine and Environmental Sciences Centre, Coimbra – 3001-456, Portugal; Department of Biology and CESAM, University of Aveiro, Aveiro – 3810-193, Portugal

Sreekala Kannikulathel Gopidas
Centre for Advanced Studies in Botany, University of Madras, Guindy Campus, Chennai –600025, Tamil Nadu, India

You-Jin Jeon
Department of Biosystem Technology, Faculty of Technology, University of Sri Jayewardenepura, Pitipana (10206), Sri Lanka; Marine Science Institute, Jeju National University, Jeju Self-Governing Province – 63333, Republic of Korea; Laboratory of Marine Bioresource Technology, Department of Marine Life Science, Jeju National University, Jeju 63243, Republic of Korea

Fatima El Khalloufi
Department of Geography, Biology, and Geology, Polydisciplinary Faculty of Khouribga (FPK), Sultan Moulay Slimane University of Beni-Mellal, Khouribga – 25000, Morocco

Elhan Khan
Natural Products Laboratory, Department of Bioengineering, Integral University, Lucknow – 226026, Uttar Pradesh, India

Ujiie Kiyokazu
Doctoral Program in Appropriate Technology and Sciences for Sustainable Development, University of Tsukuba, 1-1-1 Tennoudai, Tsukuba Ibaraki – 305-8572, Japan; Master's Program in Agro-bioresources Science and Technology, University of Tsukuba, 1-1-1 Tennoudai, Tsukuba Ibaraki – 305-8572, Japan; College of Agro-Biological Resource Sciences, University of Tsukuba, 1-1-1 Tennoudai, Tsukuba Ibaraki – 305-8572, Japan

Adriana Leandro
Department of Life Sciences, University of Coimbra, MARE-Marine and Environmental Sciences Centre, Coimbra – 3001-456, Portugal

José Lopes
Agricultural College of Coimbra (ESAC/IPC), Research Centre for Natural Resources Environment and Society (CERNAS), Institute of Applied Research (IIA), Coimbra – 3045-601, Portugal

Hakiki Melanie
Research Center for Chemistry, Indonesian Institute of Sciences, Kawasan Puspiptek Serpong, Tangerang Selatan – 15314, Indonesia

Paulo Cesar De Melo
Department of Agriculture, Federal University of Lavras/UFLA, Lavras, Brazil; University of Coimbra, MARE-Marine and Environmental Sciences Centre, Department of Life Sciences, Coimbra, Portugal

Madelyn Nabil Moawad
National Institute of Oceanography and Fisheries (NIOF), Alexandria, Egypt; Marine Chemistry Laboratory, Marine Environment Division, National Institute of Oceanography and Fisheries, Alexandria, Egypt

Mitsutoshi Nakajima
Tsukuba Life Science Innovation Program (T-LSI), School of Integrative and Global Majors (SIGMA), University of Tsukuba, 1-1-1 Tennoudai, Tsukuba Ibaraki – 305-8572, Japan; Graduate School of Life and Environmental Sciences, University of Tsukuba, 1-1-1 Tennoudai, Tsukuba Ibaraki – 305-8572, Japan

João Pedro Nunes
Agricultural College of Coimbra (ESAC/IPC), Research Centre for Natural Resources Environment and Society (CERNAS), Institute of Applied Research (IIA), Coimbra – 3045-601, Portugal

Brahim Oudra
Department of Biology, Faculty of Sciences Semlalia Marrakech, Cadi Ayyad University, Marrakech – 40000, Morocco

Diana Pacheco
Department of Life Sciences, University of Coimbra, MARE-Marine and Environmental Sciences Centre, Coimbra – 3001-456, Portugal

Leonel Pereira
Department of Life Sciences, University of Coimbra, MARE-Marine and Environmental Sciences Centre, Coimbra – 3001-456, Portugal

Abdeltif Reani
Phycology Research Unit, Department of Biology, Faculty of Sciences, University Chouaib Doukkali, PO Box 20, El Jadida – 24000, Morocco

Brahim Sabour
Phycology Research Unit, Department of Biology, Faculty of Sciences, University Chouaib Doukkali, PO Box 20, El Jadida – 24000, Morocco

Kalu Kapuge Asanka Sanjeewa
Department of Biosystem Technology, Faculty of Technology, University of Sri Jayewardenepura, Pitipana – 10206, Sri Lanka

Soha Shabaka
National Institute of Oceanography and Fisheries (NIOF), Alexandria, Egypt; Hydrobiology Laboratory, Marine Environment Division, National Institute of Oceanography and Fisheries, Alexandria, Egypt

Tiago Sousa
University of Coimbra, MARE - Marine and Environmental Sciences Centre, Department of Life Sciences, Coimbra – 3001-456, Portugal

Nagaraj Subramani
Centre for Advanced Studies in Botany, University of Madras, Guindy Campus, Chennai 600025, Tamil Nadu, India

Yosmina Tapilatu
Centre for Deep-Sea Research, The National Research and Innovation Agency (PRLD-BRIN), Jl. Y. Syaranamual Guru-guru Poka, Ambon – 97233, Indonesia

Rita Teixeira
Agricultural College of Coimbra (ESAC/IPC), Research Centre for Natural Resources Environment and Society (CERNAS), Institute of Applied Research (IIA), Coimbra, Portugal

Joeharnani Tresnati
Department of Fisheries, Faculty of Marine Science and Fisheries, University of Hasanuddin, Kampus Tamalanrea KM. 10, Makassar – 90245, Sulawesi Selatan, Indonesia; Multitrophic Research Group, University of Hasanuddin, Kampus Tamalanrea KM. 10, Makassar – 90245, Sulawesi Selatan, Indonesia; Center of Excellence for Development and Utilization of Seaweed, University of Hasanuddin, Kampus Tamalanrea KM. 10, Makassar – 90245, Sulawesi Selatan, Indonesia

Ambo Tuwo
Department of Marine Science, Faculty of Marine Science and Fisheries, University of Hasanuddin, Kampus Tamalanrea KM. 10, Makassar – 90245, Sulawesi Selatan, Indonesia; Multitrophic Research Group, University of Hasanuddin, Kampus Tamalanrea KM. 10, Makassar – 90245, Sulawesi Selatan, Indonesia; Center of Excellence for Development and Utilization of Seaweed, University of Hasanuddin, Kampus Tamalanrea KM. 10, Makassar – 90245, Sulawesi Selatan, Indonesia

Inayah Yasir
Department of Marine Science, Faculty of Marine Science and Fisheries, University of Hasanuddin, Kampus Tamalanrea KM. 10, Makassar – 90245, Sulawesi Selatan, Indonesia; Multitrophic Research Group, University of Hasanuddin, Kampus Tamalanrea KM. 10, Makassar – 90245, Sulawesi Selatan, Indonesia; Center of Excellence for Development and Utilization of Seaweed, University of Hasanuddin, Kampus Tamalanrea KM. 10, Makassar – 90245, Sulawesi Selatan, Indonesia

ABBREVIATIONS

3,6-AG	3,6-anhydro-L-galactose
AD	atopic dermatitis
AECS	acetone extract of *C. spongiosus*
AEP	aqueous extract processed
AgNPs	silver nanoparticles
Ahr	aryl hydrocarbon receptor
AOM	azoxymethane
ARE	antioxidant responsive element
BNF	British Nutrition Foundation
CAT	catalase
CH_4	methane
CME	chemically modified electrode
COX	cyclooxygenase
CQ-r	chloroquine-resistant
CQ-s	chloroquine-sensitive
CQ-s	CQ-susceptible
DHA	docosahexaenoic acid
DPHC	diphlorethohydroxycarmalol
DPPH	2,2-diphenyl-1-picrylhydrazyl
DW	dry weight
EAE	enzyme-assisted extraction
EC	European Commission
ECM	extracellular matrix components
ED_{50}	effective dose
EFSA	European Food Safety Authority
EPA	eicosapentaenoic acid
ERK	extracellular signal-regulated kinases
ETM	extracellular tissue matrix
EU	European Union
FAO	Food and Agricultural Organization
FCRs	feed conversion ratios
FDA	Food and Drug Administration
FTIR	Fourier transform infra-red

GAE	gallic acid equivalent
GC	gas chromatography
GC-MS	gas chromatography-mass spectroscopy
GHG	greenhouse gas
GPx	glutathione peroxidase
GSH	glutathione
H_2O_2	hydrogen peroxide
Ha	hectares
HABs	harmful algal blooms
HAV	hepatitis A virus
HCMV	human cytomegalovirus
HDF	human dermal fibroblasts
HDL	high-density lipoprotein
HELA	cervix cancer cell line
HeLa	Henrietta lacks
HIV	human immunodeficiency virus
HPLC	high-performance liquid chromatography
HRBA	human rights-based approach
HSV	herpes simplex virus
HSV-1	herpes simplex virus types-1
HSV-2	herpes simplex virus types-2
HTT	high-temperature-tolerance
I%	inhibition %
IAA	indole-3-acetic acid
ICZM	Integrated Management of Coastal Zones of the Kingdom of Morocco
IKK	IκB kinase complex
IL-6	interleukin-6
IMTA	integrated multitrophic aquaculture
IZ	inhibition zone
JNKs	Jun amino-terminal kinases
LDL	low-density lipoprotein
LIFDC	low-income food-deficit countries
LMWF	low molecular weight fucoidan
LUC	land-use change
MAA	mycosporine-like amino acid
MAE	microwave-assisted extraction
MAPK	mitogen-activated protein kinases
MDA	malondialdehyde

MDP	marine-derived polysaccharide
MeV	measles virus
MIC	minimum inhibitory concentration
MMPs	matrix metalloproteinases
MRI	magnetic resonance imaging
MRSA	methicillin-resistant *Staphylococcus aureus*
MT	metric tons
N_2O	nitrous oxide
NADA	National Agency for the Development of Aquaculture
NDV	Newcastle disease virus
NF-κB	nuclear factor kappa B
NK	natural killer
NMR	nuclear magnetic resonance
NO	nitric oxide
NPs	nanoparticles
OH	hydroxyl radicals
OMRI	Organic Materials Review Institute
PAHs	polycyclic aromatic hydrocarbons
PBP2a	penicillin-binding protein 2a
PC3	prostate cancer cell line
PEEs	polysaccharide enriched extracts
PEG	polyethylene glycol
PGA	propylene glycol alginate
PGE2	prostaglandin E2
PLS	sulfated polysaccharides
PM	particulate matter
PPARγ	peroxisome proliferator-activated receptor gamma
PPs	polyphenol-rich extracts
PUFA	polyunsaturated fatty acids
QE	quercetin equivalent
RDA	recommended dietary allowances
ROS	reactive oxygen species
RSV	respiratory syncytial virus
SC	stratum corneum
SCWE	subcritical water extraction
SDG	sustainable development goals
SFV	Semliki forest virus
SLF	seaweed liquid fertilizer
SOD	superoxide dismutase

S-O-M	subject, object, and method
SREBP1	sterol regulatory element-binding protein 1
TGF-α	transforming growth factor-alpha
TNF-α	tumor-necrosis factor-alpha
TPA	tetradecanoylphorbol-13-acetate
UAE	ultrasound-assisted extraction
UFA	unsaturated fatty acid
UN	United Nations
UPLC Q-TOF	ultra-performance liquid chromatography quadruple time-of-flight
US$	United States Dollar
USA	United States of America
UV	ultraviolet
UVA	ultraviolet A
VAP	virus-associated protein
ZnO NPs	zinc oxide nanoparticles

SYMBOLS

%	percent
°C	degrees Celsius
^{13}C	carbon-13
^{1}H	proton
Ba	barium
C_2	second carbon position
C_4	fourth carbon position
Ca	calcium
Co	cobalt
CO_2	carbon dioxide
Cu	copper
Fe	iron
K	potassium
kg	kilogram
m	meter
Mg	magnesium
Mn	manganese
N	nitrogen
Na	sodium
P	phosphorus
pH	hydrogenic potential
Rp	Indonesian rupiah
Se	selenium
Sr	strontium
TJ	terajoule
Zn	zinc
α	glycosidic bond between two carbons of same stereochemistry
$\alpha(1\rightarrow2)$	glycosidic bond between carbon 1 of one monomer and carbon 2 of another
$\alpha(1\rightarrow3)$	glycosidic bond between carbon 1 of one monomer and carbon 3 of another

α(1→4)	glycosidic bond between carbon 1 of one monomer and carbon 4 of another
α-L	levo; hydroxyl group of chiral carbon on the left
β	glycosidic bond between two carbons of different stereochemistry
β(1→3)	glycosidic bond between carbon 1 of one monomer and carbon 3 of another
β(1→6)	glycosidic bond between carbon 1 of one monomer and carbon 6 of another
β-D	dextro; hydroxyl group on chiral carbon on the right

PREFACE

Seaweeds comprise enormous marine species such as plants and algae growing up to 180 meters in depth along rocky shorelines of the marine ecosystems, including backwaters and estuaries. The greatest seaweed species diversity is found in the temperate ecosystem. Seaweeds actually do not have true roots and stems, but for protection from corrosive salty ocean water, they have thick, rubbery stems. Their roots have a simpler structure and leave to hold it in place and absorb sunlight. Seaweeds often form dense growth on the solid surface of the plants and dead corals, pebbles, and rocks in the different regions of the marine ecosystem. Based on the different pigment compositions, seaweeds are divided into four taxonomic groups: Chlorophyta (green), Rhodophyta (red), Ochrophyta (brown), and Cyanophyta (blue-green). The world is comprised of 372,000 miles of coastal line, enriched with the different types of marine ecosystems. A recent study on the estimation of seaweed diversity reported that around 72,500 taxa have been discovered in the global ocean. Seaweed (macroalgae) is used almost globally for the filtration of unwanted toxins from water; they also provide shelter to the sea creatures. Their richness in diversity is a promising source for food and feed free from pollutants.

The complicated species have specialized bladders, which allow them to float. The ocean-associated seaweeds are flavorsful due to their richness in a variety of minerals, vitamins, and fiber. These nutritious seaweeds are used as the raw material for many industries, including food, cosmetics, pharmaceuticals, paints, and other chemical and gelatin industries. Seaweeds are also known for their antimicrobial, antioxidant, and anti-inflammatory potential, so the Romans used them to cure skin burns, open wounds, and rashes. Human consumption of seaweeds rich in polyunsaturated fatty acids, metabolites like phlorotannin, catechin, hesperidin, and other phenolic metabolites, proteins, minerals, and vitamins are responsible for a healthy life due to their deep potential of pharmaceutical and nutraceutical properties that prevent various diseases like cancer, diabetics, cardiovascular-related issues, rheumatoid arthritis, and hypertension.

Recently various seaweed applications have been developed with respect to plant growth promotion through the improvement of seed germination, shoot, and root formation and health simulation through increasing the uptake of nutrients from the soil and developing resistance. The increase in demand for nutritious food is making seaweed a sustainable food in the near future. In solving the fuel shortage problem, which is at its peak, seaweed has turned out to be a blessing because of its potential to grow fast-forming biomass, which can be used for the production of biofuel. These macroalgae are also used as bioindicators for environmental pollutants in water, and some of the species are involved in pollution remediation.

The present book provides a deep understanding of the biodiversity of seaweeds around the world and their industrial, biomedical, and environmental applications. Chapter 1 discusses the various biodiversity attributes of tropical seaweeds. Cultivation of seaweeds and preparation of bioactive compounds from seaweeds are detailed in Chapters 2 and 3. Current challenges, emerging markets, and the latest developments in extracting the useful biomolecules from seaweeds are elaborated in Chapters 4 to 6. Chapters 7 to 9 review the diverse agricultural applications of novel seaweed derivatives. Biomedical applications of seaweeds are covered in Chapters 10 and 11. The last chapter deals with the prominence of seaweeds as an influential indicator in food and environmental safety.

—Jeyabalan Sangeetha, PhD
Devarajan Thangadurai, PhD

CHAPTER 1

BIODIVERSITY OF TROPICAL SEAWEEDS

YOSMINA TAPILATU

Centre for Deep-Sea Research, The National Research and Innovation Agency (PRLD-BRIN), Jl. Y. Syaranamual Guru-guru Poka, Ambon – 97233, Indonesia

ABSTRACT

This chapter offers an updated insight on the tropical status of seaweeds biodiversity in tropical areas as of July 2020. Aspects of their ecology and biogeography, as well as their taxonomy, diversity, and distribution, are discussed. The climate change impact on tropical seaweeds is subsequently highlighted. Underutilized seaweed species from Indonesia, one of the major producers of tropical seaweeds, is presented at the end of this chapter. Increasing efforts are required to study seaweed biodiversity, especially in tropical countries and geographical areas where relevant data are scarce or practically nonexistent. Other than that, it is necessary to identify underutilized seaweeds with potential for human consumption and biotechnological applications. As 40% of the world population living in tropical areas, finding new seaweeds that could be an alternative source of food would be of our high interest in ensuring food security in a world with a changing climate and increasing anthropogenic pressure.

1.1 INTRODUCTION

Seaweeds are members of the macroalgae adapted to the marine environment and are individually visible to the naked eye. This chapter refers to those that are multicellular. Published books have already discussed various aspects of seaweeds globally (Buchholz et al., 2012; Hurd et al., 2014). However, few if any treats specifically about the biodiversity of seaweeds in the tropical marine environment.

Tropical seaweeds remain an important marine bioresource that are harvested continuously, either in the uncultivated coastal area or in aquaculture farms in South East Asian countries. This harvest is to fulfill demands not only in its producing countries but also at a global scale. However, they are essentially an ignored marine resource in all countries of tropical Central America, the Caribbean, and most of Africa (Radulovich et al., 2014). A thorough review to highlight French tropical territories seaweeds was done pretty recently (Stiger-Pouvreau and Zubia, 2020). Whilst literatures are abundant on its applications, notably on some that are widely used in various industries, not many on the updated information regarding the biodiversity status of tropical seaweed. Frequently encountered around the globe, marine macroalgae are used since centuries ago until now by humans mainly as a source of food, but also for other purposes such as in pharmaceutical, agricultural, and cosmetic fields.

This chapter provides a review of the current knowledge of seaweeds biodiversity in the tropical area. It highlights the progress achieved so far concerning this important marine primary producer, and how seaweeds fare against climate change impact in the tropical area. The case of underutilized seaweeds in Indonesia, the major producer among tropical countries will be briefly discussed at the end of this chapter.

1.2 ECOLOGY AND BIOGEOGRAPHY

The tropical term used throughout this chapter refers to earth's surface influenced by the tropical climate. Seaweeds in the tropical seascape prefer growing on reef flats, rocky shores, although some are also found in the sandy coastal areas. It is argued that in terms of species number, contrary to other marine organisms, no clear evidence indicates an increase in the tropical area compared to the temperate ones (Garbary et al., 1990; Bolton, 1994). The environmental parameters are some of the various factors contributing to the habitat complexity that influenced species decrease (Keith et al., 2014). Specifically, Bolton (1994) suggested that the absence of a peak in species numbers in tropical latitudes compared to the temperate ones is caused by: (i) the small size of the said region; (ii) seawater temperature patterns and other aspects of marine climates; and (iii) stability of habitat condition.

In the tropical area, seaweeds distribution and density depend on seabed profile, seasonal hydrography, and species competition (Soegiarto, 1977). In terms of ecological function, like their counterparts in other regions, tropical seaweeds are one of the primary producers, notably in the coastal area. This leads to their contribution also in the blue carbon system (Queirös et al., 2019). Because of their presence in the mangrove, coral reef, and seagrass ecosystems, seaweeds in the tropics are food sources to other marine organisms (e.g., mollusks, crustacea, Echinodermata, and various fishes). Tropical seaweed beds and farms function as spawning grounds and habitat of several commercially important fishes and invertebrates, as well as safe and healthy nursery grounds for young fish and crustaceans (Waters et al., 2019).

1.3 TAXONOMY, DIVERSITY, AND DISTRIBUTION

1.3.1 TAXONOMY

Seaweeds are polyphyletic, which means they derive their morphological, physiological, and metabolic diversity from various phylogenetic lineages (Stengel et al., 2011). They are members of three phyla, i.e., chlorophytes (green algae), ochrophytes (brown algae), and rhodophytes (red algae) (Figure 1.1).

As of 2019, AlgaeBase recorded 11,017 macroalgae species distributed in temperate, polar, and tropical regions. These species are divided into 1,901 green algae (class: Ulvophyceae), 7,083 red algae (classes: Cyanidiophyceae, Stylonematophyceae, Compsopogonophyceae, Rhodellophyceae, Porphyridiophyceae, Bangiophyceae, and Florideophyceae), and 2,033 brown algae (class: Phaeophyceae) (Guiry and Guiry, 2020). Red multicellular macroalgae found in the tropical region are only from the following four classes: Compsopogonophyceae, Rhodellophyceae, Bangiophyceae, and Florideophyceae.

Seaweeds identification in the past is highly dependent on their morphological characteristics, notably on their pigmentation and photosynthetic properties. However, seaweeds physical characteristics tend to vary as they are directly influenced by environmental factors (Norton et al., 1981).

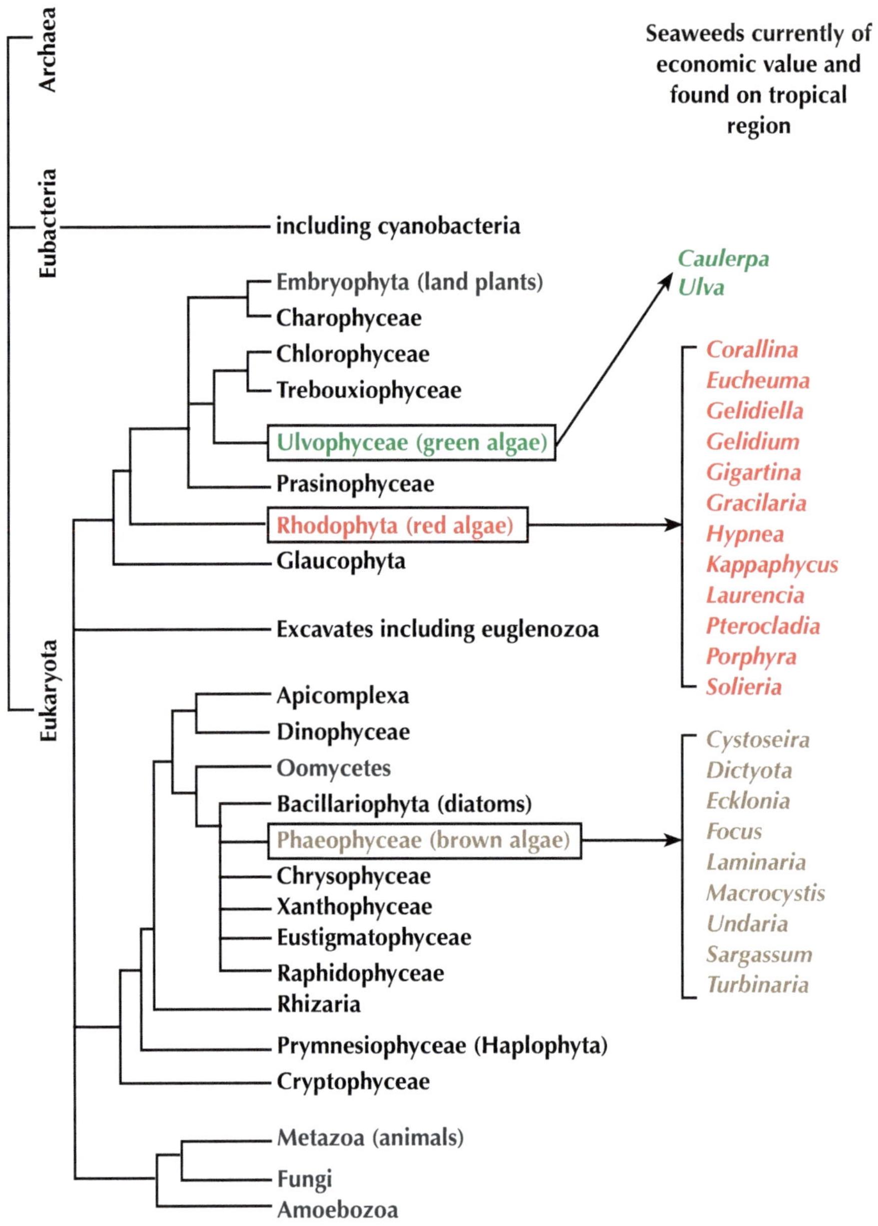

FIGURE 1.1 Phylogenetic tree showing the polyphyletic group of seaweeds positions in the different lineages in regards with micro- and other types of algae and their positions in regards with the other major lineages of life (Archaea and Eubacteria) (left). Redrawn with major modifications from Stengel et al. (2011).

1.3.2 GENETIC DIVERSITY

The genetic diversity of a given population of seaweeds inhabiting tropical regions is influenced by three sources of variation: mutation, recombination, and immigration of genes (Griffiths et al., 2000), as in other organisms belonging to the Eukaryotic lineage of life. The development of knowledge in seaweeds genetic diversity using molecular phylogeographic approaches was started in the mid-1990s. An overview done by Hu and Fraser (2016) of this development from 1994 to 2014 indicated that the seaweeds phylogeographic studies were taxonomically biased, in that more data were available on brown and red seaweeds than those of green seaweeds. Besides, the phylogeographic of seaweeds growing in the tropical region was understudied during those decades, compared to those found in the sub-tropic or temperate regions. This is unfortunate as lack of information about seaweeds from the tropical area will hinder the investigation of the mechanism and process contributing to seaweeds diversification and evolution at a global scale.

A very recent attempt to review the status of coral reef-associated seaweeds diversity in the French tropical overseas territories (Stiger-Pouvreau and Zubia, 2020) reveals that not much of a change was observed after 2014. The incomparable nature of the sampling effort between territories contributed to the still existing gap of information.

However, the limited information available indicated the existence of cryptic species in tropical seaweeds, which was undetected when identification was done using only morphological properties. For example, 10 new species of the brown seaweed *Lobophora* sampled in New Caledonia were revealed when molecular tools were used (Vieira et al., 2014). This was previously unsuspected since traditional identification using the morphology properties included only 6 taxonomically accepted species until 2014. On the other hand, molecular approaches allowed the taxonomic revision of tropical seaweeds as well, as in the case of *Sargassum* sampled in French Polynesia tropical and sub-tropical areas. The combination of morphology and molecular approaches, in this case, led to the reclassification proposition of 18 epithets into only 3 valid species of *Sargassum* from the said area (Mattio et al., 2008).

Knowledge of genetic diversity has helped to avoid the misidentification of cultivated seaweed strains with important economic value, something that has caused productivity decline and the quality of cultured seaweeds (Yong et al., 2016). Lim et al. (2014), for example, used a genetic approach to identify known haplotypes for *Kappaphycus alvarezii* (seven), *K. striatus* (seven), and *Eucheuma denticulatum* (two) from highly commercialized species sampled in Malaysia, the Philippines, Indonesia, and Vietnam. This knowledge is also important, among others, to identify new strains, trait studies (Tan et al., 2017), and climate resilience (Waters et al., 2019).

1.3.3 DISTRIBUTION AND SPECIES DIVERSITY

Heatmap (Figure 1.2) done with the data retrieved and compiled from Guiry and Guiry (2020) in July 2020 indicated that despite its highly dynamic nature, the database provides a basic insight into seaweeds hotspot in the tropical region. From 84 countries and geographical areas with coastal surfaces in this region, Mexico overall has the highest number (2,021) of species recorded, followed by the Philippines (1,206) and Indonesia (1,078). Continent-wise, Brazil has the second-highest number (1,077 species recorded) after Mexico, followed by Cuba (700) in America. India was the third country with the highest number of species recorded in Asia after the Philippines and Indonesia. In the African continent and surrounding islands, Mauritius has the highest number of species recorded (666), followed by Tanzania (517) and Kenya (453). The number of seaweed species recorded was higher in tropical regions of America and Asia continents coastal waters, compared to those in Africa.

In terms of division (phylum), from the three phyla that are included in seaweeds in this chapter, Rhodophyta (classes: Bangiophyceae, Compsopogonophyceae, Florideophyceae, and Rhodellophyceae) has the highest number of species recorded on all top three countries with the highest species records, from all three continents concerned. Green seaweeds (class: Ulvophyceae) are in second place, followed by brown seaweeds (class: Phaeophyceae).

The highest number of red seaweeds species recorded from the coastal area of Mexico, the Philippines, and Mauritius has the same pattern, in that it was predominantly members of the class Florideophyceae. The most numerous genera recorded from this class are *Gracilaria*, followed by *Ceramium*. This interesting trend of these two as the genera with the most numerous species recorded was probably due to their nature as the producer of high-quality agar. Thus, the inventory effort that took place was possibly based on the economic motives, of selecting the best raw materials to produce agar. Seaweeds' specific diversity variations are arguably influenced not only by the habitat diversity, each territory surface but also by the sampling effort carried out (Stiger-Pouvreau and Zubia, 2020).

Country/Geographic area	Number of species recorded: Rhodophyta (Red algae)	Ochrophyta (Brown algae)	Chlorophyta (Green algae)	Total species recorded
Africa and surrounding islands				
Angola	85	21	24	130
Benin	7	12	6	25
Cameroon	74	20	24	118
Comoros	0	0	0	0
Congo	16	1	2	19
Democratic Republic of Congo	0	0	0	0
Djibouti	22	11	11	44
Equatorial Guinea	43	15	18	76
Eritrea	82	42	41	165
Gabon	69	32	14	115
Gambia	71	35	17	123
Guinea	20	2	5	27
Guinea-Bissau	14	5	4	23
Ivory Coast	61	26	19	106
Kenya	243	68	142	453
Liberia	65	30	19	114
Madagascar	211	70	100	381
Mauritania	171	67	26	264
Mauritius	398	91	177	666
Mayotte	0	0	0	0
Mozambique	109	27	85	221
Nigeria	35	17	15	67
Reunion	0	0	0	0
Saint Helena	0	0	0	0
Sao Tome and Principe	87	33	30	150
Seychelles	242	65	127	434
Senegal	217	61	52	330
Sierra Leone	79	31	32	142
Somalia	83	44	76	203
Sudan	0	0	0	0
Tanzania	286	81	150	517
Togo	20	18	10	48
America and surrounding islands				
Anguilla	0	0	0	0
Antigua and Barbuda	4	0	1	5
Aruba	0	0	0	0
Bahamas	202	44	113	359
Barbados	142	49	71	262
Belize	155	41	102	298
Brazil	666	164	247	1077
Cayman Islands	22	13	20	55
Colombia	324	63	50	437
Costa Rica	207	25	50	282
Cuba	379	96	225	700
Dominica	0	0	0	0
Dominican Republic	5	3	1	9
Ecuador (incl. Galapagos Islands)	302	67	62	431
El Salvador	94	4	33	131
Grenada	3	0	1	4
Guatemala	11	0	0	11
Guyana	15	2	8	25
French Guiana	0	0	0	0
Haiti	6	0	5	11
Honduras	1	0	5	6
Jamaica	160	46	93	299
Martinique	0	0	0	0
Mexico	1457	282	282	2021
Montserrat	0	0	0	0
Netherlands Antilles	81	32	0	113
Nicaragua	0	0	0	0
Panama	176	48	140	364
Peru	200	37	40	277
Puerto Rico	204	30	83	317
Saint Barthelemy	0	0	1	1
Saint Kitts & Nevis	0	0	0	0
Saint Lucia	0	0	0	0
Saint Martin	1	0	0	1
Saint Vincent & the Grenadines	0	0	0	0
Suriname	4	0	0	4
Trinidad & Tobago	141	48	64	253
Turks & Caicos Islands	56	17	29	102
Venezuela	308	85	131	524
British Virgin Islands	197	47	86	330
United States Virgin Islands	0	0	0	0
Asia				
Brunei Darussalam	0	0	0	0
Cambodia	0	0	0	0
East Timor	0	0	0	0
India	510	247	272	1029
Indonesia	625	185	268	1078
Malaysia	104	54	49	207
Myanmar (Burma)	0	0	0	0
Philippines	725	210	271	1206
Singapore	140	109	114	363
Thailand	94	41	91	226
Viet Nam	558	191	227	976

FIGURE 1.2 Seaweed distribution heatmap in tropical countries or geographic area with coastal area according to the division (phylum): Rhodophyta (classes: Compsogonopycheae, Rhodellopyceae, Bangiopyceae, and Florideophyceae), Ochrophyta (class: Phaeophyceae) and Chlorophyta (class: Ulvophyceae).

Source: Guiry and Guiry (2020), retrieved on July 2020.

Interestingly, some tropical countries/geographical areas do not have any published data on seaweeds species. This could be caused by various reasons. The first possible explanation is that the area's marine environment does not facilitate any seaweeds species growth (e.g., Aruba). Second, seaweeds in some areas could already be extinct due to anthropogenic pressure as possibly was the case with the St. Helena coastal area. Another possible reason is simply the lack of seaweeds inventory done in the area at all due to conflicts (e.g., Sudan) or to lack of competent human resources and adequate scientific infrastructures available. The inventory task is thus still far from complete, particularly in the inaccessible areas to give more updated information regarding the seaweeds distribution in the tropical region.

1.4 CLIMATE CHANGE IMPACT

Climate change multiple components could affect biodiversity at all levels, from single organisms to biome levels (Bellard et al., 2012). The separate impacts of global warming and ocean acidification (e.g., pH decrease) on seaweed communities were studied in 2012 (Harley et al., 2012). It was indicated that seaweeds physiological responses are still poorly understood, with different responses to be expected, in function of region, but also on algal communities. More studies are thus required to obtain more information on continuous warming and acidification impact in today's specific climate change settings, notably in the tropical region.

As observed by Harley et al. (2006) in the French overseas territories. natural and anthropogenic stresses on these macroalgae both contribute to the loss of tropical coastal diversity. Many seaweeds communities with predominant large thalli in numerous coasts around the world may disappear or suffer from anthropogenic pressure and global climate change. If there is a lot of grazing, the vacant place will be occupied by turf algae, coral debris or even become barren (Harley et al., 2006; Johnson, 2011; Filbee-Dexter and Wernberg, 2018), which in turn threaten the normal algal diversity of coral reefs (Stiger-Pouvreau and Zubia, 2020).

1.5 UNDERUTILIZED SEAWEEDS SPECIES: THE CASE OF INDONESIA

Indonesia is the major producer of seaweeds among tropical countries since 2010. In 2018, 9,320,3000 tons of seaweeds were reported from this country alone (FAO, 2020). This massive number concerned mainly four out of more than 1,078 species currently recorded, i.e., *Eucheuma*, *Kappaphycus*, *Gracilaria,* and *Gelidium*. Most of them are widely cultivated throughout the country to supply the industrial demand for their carrageenan and agar contents. However, many other existing species are also potential to be developed further. In this chapter section, three genera will be highlighted, i.e., *Sargassum*, *Porphyra,* and *Caulerpa* (Figure 1.3). On a much smaller scale, members of these three genera have already been harvested for food by locals (Kadi, 2004). Most of them grow on small islands that some are remote and thus need adequate transportation facilities.

FIGURE 1.3 Example of under-utilized seaweeds in Indonesia from the three phyla: Rhodophyta (*Porphyra* sp.) (A), Chlorophyta (*Caulerpa* sp.) (B) and Ochrophyta (*Sargassum* sp.). All seaweeds pictured here were sampled from Maluku Province in eastern Indonesia. Scale bar: 1 cm. Picture credit: Yosmina Tapilatu (A) and Ferdinand Pattipeilohy (B and C); used with permission.

The first one, *Porphyra* sp. is a member of the class Bangiophyceae. It is found mainly in the eastern part of the Indonesian archipelago, notably in certain coastal areas of Maluku (local name: *sayur lao*) and West Papua Provinces. Only two species were recorded from Indonesia as of July 2020 (Guiry and Guiry, 2020). *Porphyra*'s presence in Indonesia is seasonal, meaning it is currently harvested annually from the uncultivated coastal area during the southeast monsoon period (June to August). Subtropical species of *Porphyra* are extensively used in other parts of the world, mainly in Japan, China, and Korea as preferred raw materials for Nori production and other purposes. *Porphyra* harvest on a global level was valued at ~1.3 billion dollar US per year in 2011 (Blouin et al., 2011). It is potentially useful as antihypertensive peptides and other bioactive compounds producers (Qu et al., 2010). However, its usage in Indonesia is limited in its fresh form as seasonal food for the locals in the eastern part of Indonesia. Unsuccessful multiple attempts of cultivating *Porphyra* sp., either in coastal areas and in the laboratory or through tissue culture especially in Maluku Province, led to the abandonment of its exploration as an emerging seaweed resource to be benefitted from.

Conversely with *Porphyra*, *Caulerpa* sp. is found in coastal areas other than in eastern Indonesia. It is a member of green seaweeds (class: Ulvophyceae) and reportedly found in Java, Sulawesi, Bali, and Kei islands coastal areas. As of July 2020, there are 95 species (including variety and forms) recorded from

Indonesia (Guiry and Guiry, 2020). Long time consumed as traditional sea vegetables, *Caulerpa* production in Indonesia has increased slowly annually during the last fourteen years to meet the industrial demand, nationally, and on a global level as raw material for nutraceutical and fertilizer industries, other than human food. However, the appropriate method to cultivate industrial level quality *Caulerpa* in Indonesian coastal waters setting to reach optimal yield is still under development. This explains its small-scale production despite its potential utilization.

The third genus, *Sargassum*, a member of brown seaweed (class: Phaeophyceae). It has an even larger distribution area in Indonesia, as it is practically reported to be found from the western to the eastern part of the archipelago. Data from Guiry and Guiry (2020) showed 117 species including variants of *Sargassum* recorded from Indonesia. Various studies have been done on its potential application, ranging from its properties to produce alginate, to its nutritional value and its bioactive compounds. Certain species of *Sargassum* growing in Indonesia so far seemed to meet the quality requirement as raw material for the alginate industry in China (minimal alginate content of 17%, viscosity 47 cP, and gel strength 139 $g.cm^{-2}$). The alginate content of various *Sargassum* species found in Indonesia is ranging from 17.07 to 26.96% (Tambunan et al., 2013). However, Indonesian alginate industry expansion is still limited by low continuity of raw material provision (Subaryono, 2011).

1.6 CONCLUSION

This chapter reviewed the current status of seaweeds found in the tropical regions. More efforts are pressingly needed to study seaweeds biodiversity, especially in countries and geographical areas in the tropical region where there is scarce or nonexistent data. Studies including taxonomy, genetic diversity, spatial, and temporal variability of seaweeds are indispensable to have a more complete picture of this resource in the said region. As anthropogenic disturbances such as pollution, harvesting, invasive species introduction or global warming will eventually alter seaweeds biodiversity patterns in the tropics, new approaches for monitoring their biodiversity and sustainable management should be applied in these studies.

It would be helpful also to identify underutilized seaweeds with potential for human consumption and biotechnological applications. As 40% of the world population living in the tropical area, finding new seaweeds that could be an alternative source of food would be of our high interest in ensuring food security in a world with a changing climate and increasing anthropogenic pressure.

ACKNOWLEDGMENTS

The author wishes to thank Ferdinand Pattipeilohy for *Caulerpa* sp. and *Sargassum* sp. picture preparation. This chapter is contribution number 0013 from Marine Microbiology and Biotechnology Laboratory, Center for Deep-Sea Research, Indonesian Institute of Sciences (LIPI).

KEYWORDS

- **anthropogenic pressure**
- **Bangiophyceae**
- **Compsopogonophyceae**
- **Florideophyceae**
- **Phaeophyceae**
- **Rhodophyceae**
- **Ulvophyceae**
- **underutilized macroalgae**

REFERENCES

Bellard, C., Bertelsmeier, C., Leadley, P., Thuiller, W., & Courchamp, F., (2012). Impacts of climate change on the future of biodiversity. *Ecol. Lett., 15,* 365–377. https://doi.org/10.1111/j.1461-0248.2011.01736.x.

Blouin, N. A., Brodie, J. A., Grossman, A. C., Xu, P., & Brawley, S. H., (2011). *Porphyra*: A marine crop shaped by stress. *Trends Plant Sci., 16*(1), 29–37.

Bolton, J. J., (1994). Global seaweed diversity: Patterns and anomalies. *Botanica Marina, 37,* 241–245. https://doi.org/10.1515/botm.1994.37.3.241.

Buchholz, C. M., Krause, G., & Buck, B. H., (2012). Seaweed and man. In: Wiencke, C., & Bischof, K., (eds.), *Seaweed Biology: Ecological Studies (Analysis and Synthesis)* (pp. 471–493). Springer, Berlin, Heidelberg. https://doi.org/10.1007/978-3-642-28451-9_22.

FAO, (2020). *The State of World Fisheries and Aquaculture*. Rome.

Filbee-Dexter, K., & Wernberg, T., (2018). Rise of turfs: A new battlefront for globally declining Kelp forests. *BioScience, 68*(2), 64–76. https://doi.org/10.1093/biosci/bix147.

Garbary, D. J., (1990). Biogeography of marine algae. In: *Encyclopedia of Life Sciences.* John Wiley & Sons, Ltd. https://doi.org/10.1038/npg.els.000312.

Griffiths, A. J. F., Miller, J. H., Suzuki, D. T., Lewontin, R. C., & Gelbart, W. M., (2000). *An Introduction to Genetic Analysis* (7th edn.). New York: W.H. Freeman.

Guiry, M. D., & Guiry, G. M., (2020). *Algae Base*. World-Wide Electronic Publication, National University of Ireland, Galway. https://www.algaebase.org (accessed on 13 October 2021).

Harley, C. D. G., Anderson, K. M., Demes, K. W., Jorve, J. P., Kordas, R. L., Coyle, T. A., & Graham, M. H., (2012). Effects of climate change on global seaweed communities. *J. Phycol., 48*(5), 1064–1078. https://doi.org/10.1111/j.1529-8817.2012.01224.x.

Harley, C. D. G., Hughes, A. R., Hultgren, K. M., Miner, B. G., Sorte, C. J. B., Thornber, C. S., Rodriguez, L. F., et al., (2006). The impacts of climate change in coastal marine systems. *Ecol. Lett., 9*(2), 228–241. https://doi.org/10.1111/j.1461-0248.2005.00871.x.

Hu, Z. M., & Fraser, C., (2016). Seaweed phylogeography from 1994 to 2014: An overview. In: Hu, Z. M., & Fraser, C., (eds.), *Seaweed Phylogeography: Adaptation and Evolution of Seaweeds under Environmental Change* (pp. 3–22). Springer, Netherlands. https://doi.org/10.1007/978-94-017-7534-2.

Hurd, C. L., Harrison, P. J., Bischof, K., & Lobban, C. S., (2014). *Seaweed Ecology and Physiology* (2nd edn.). Cambridge University Press.

Johnson, M. T. J., (2011). Evolutionary ecology of plant defenses against herbivores. *Funct. Ecol., 25*(2), 305–311. https://doi.org/10.1111/j.1365-2435.2011.01838.x.

Kadi, A., (2004). Potensi rumput laut di beberapa perairan pantai Indonesia. *Oseana, 29*(4), 25–36.

Keith, S. A., Kerswell, A. P., & Connolly, S. R., (2014). Global diversity of marine macroalgae: Environmental conditions explain less variation in the tropics. *Global Ecol. Biogeogr., 23*(5), 517–529. https://doi.org/doi:10.1111/geb.12132.

Lim, P. E., Tan, J., Phang, S. M., Nikmatullah, A., Hong, D. D., Sunarpi, H., & Hurtado, A. Q., (2014). Genetic diversity of *Kappaphycus* Doty and *Eucheuma* J. Agardh (Solieriaceae, Rhodophyta) in Southeast Asia. *J. Appl. Phycol., 26*(2), 1253–1272. https://doi.org/10.1007/s10811-013-0197-y.

Mattio, L., Payri, C. E., & Stiger-Pouvreau, V., (2008). Taxonomic revision of *Sargassum* (Fucales, Phaeophyceae) from French Polynesia based on morphological and molecular analyses. *J. Phycol., 44*(6), 1541–1555. https://doi.org/10.1111/j.1529-8817.2008.00597.x.

Norton, T. A., Mathieson, A. C., & Neushul, M., (1981). Morphology and environment. In: Lobban, C. S., & Wynne, M. J., (eds.), *The Biology of Seaweeds*, (pp. 421–451). Berkeley: University of California Press.

Qu, W., Ma, H., Pan, Z., Luo, L., Wang, Z., & He, R., (2010). Preparation and antihypertensive activity of peptides from *Porphyra yezoensis*. *Food Chem., 123*(1),14–20. https://doi.org/https://doi.org/10.1016/j.foodchem.2010.03.091.

Queirós, A. M., Stephens, N., Widdicombe, S., Tait, K., McCoy, S. J., Ingels, J., Rühl, S., et al., (2019). Connected macroalgal-sediment systems: Blue carbon and food webs in the deep coastal ocean. *Ecol. Monogr., 89*(3), 1–21. https://doi.org/10.1002/ecm.1366.

Radulovich, R., Umanzor, S., Cabrera, R., & Mata, R., (2014). Tropical seaweeds for human food, their cultivation and its effect on biodiversity enrichment. *Aquaculture, 436*, 40–46. https://doi.org/10.1016/j.aquaculture.2014.10.032.

Soegiarto, A., (1977). Indonesia seaweed resources, their utilization and management. In: *9th International Seaweed Symposium* (p. 15). Santa Barbara, California, USA.

Stengel, D. B., Connan, S., & Popper, Z. A., (2011). Algal chemodiversity and bioactivity: Sources of natural variability and implications for commercial application. *Biotechnol. Adv., 29*(5), 483–501. https://doi.org/10.1016/j.biotechadv.2011.05.016.

Stiger-Pouvreau, V., & Zubia, M., (2020). Macroalgal diversity for sustainable biotechnological development in French tropical overseas territories. *Botanica Marina, 63*(1), 17–41. https://doi.org/10.1515/bot-2019-0032.

Subaryono, S., (2011). Distribution of brown seaweed producing alginate in Indonesia and the potential utilization. *Squalen Bulletin of Marine and Fisheries Postharvest and Biotechnology, 6*(2), 55–62. https://doi.org/10.15578/squalen.v6i2.61.

Tambunan, A. P. M., Rudiyansyah, & Harlia, (2013). Pengaruh konsentrasi Na_2CO_3 terhadap rendemen natrium alginat Dari *Sargassum cristaefolium* asal perairan lemukutan. *Jurnal Kimia Khatulistiwa, 2*(2), 112–117. https://jurnal.untan.ac.id/index.php/jkkmipa/article/view/3943/3927 (accessed on 13 October 2021).

Tan, J., Lim, P. E., Phang, S. M., & Hurtado, A. Q., (2017). Biodiversity, biogeography and molecular genetics of the commercially important genera *Kappaphycus* and *Eucheuma*. In: Hurtado, A. Q., Critchley, A. T., & Neish, I. C., (eds), *Tropical Seaweed Farming: Trend, Problems and Opportunities. Developments in Applied Phycology,* (Vol. 9, pp. 29–43), Springer, Cham. https://doi.org/10.1007/978-3-319-63498-2_2.

Vieira, C., D'hondt, S., De Clerck, O., & Payri, C. E., (2014). Toward an inordinate fondness for stars, beetles and *Lobophora*? species diversity of the genus *Lobophora* (Dictyotales, Phaeophyceae) in New Caledonia. *J. Phycol., 50*(6), 1101–1119. https://doi.org/10.1111/jpy.12243.

Waters, T. J., Lionata, H., Wibowo, T. P., Jones, R., Theuerkauf, S., Usman, S., Amin, I., & Ilman, M., (2019). *Coastal Conservation and Sustainable Livelihoods through Seaweed Aquaculture in Indonesia: A Guide for Buyers, Conservation Practitioners, and Farmers.* Arlington, VA, USA and Jakarta, Indonesia: The Nature Conservancy.

Yong, W. T. L., Chin, G. J. W. L., & Rodrigues, K. F., (2016). Genetic identification and mass propagation of economically important seaweeds. In: Thajuddin, N., & Dhanasekaran, D., (eds.), *Algae – Organisms for Imminent Biotechnology*, (pp. 278–305). IntechOpen. https://doi.org/10.5772/62802.

CHAPTER 2

SEAWEED CULTIVATION FOR COMMUNITY WELFARE IN COASTAL AREAS AND SMALL ISLANDS

JOEHARNANI TRESNATI,[1,6,7] UJIIE KIYOKAZU,[2–4] INAYAH YASIR,[5–7] HAKIKI MELANIE,[8] RISAL APRIANTO,[6] and AMBO TUWO[5–7]

[1]*Department of Fisheries, Faculty of Marine Science and Fisheries, University of Hasanuddin, Kampus Tamalanrea KM. 10, Makassar – 90245, Sulawesi Selatan, Indonesia*

[2]*Doctoral Program in Appropriate Technology and Sciences for Sustainable Development, University of Tsukuba, 1-1-1 Tennoudai, Tsukuba Ibaraki – 305-8572, Japan*

[3]*Master's Program in Agro-bioresources Science and Technology, University of Tsukuba, 1–1-1 Tennoudai, Tsukuba Ibaraki – 305-8572, Japan*

[4]*College of Agro-Biological Resource Sciences, University of Tsukuba, 1-1-1 Tennoudai, Tsukuba Ibaraki – 305-8572, Japan*

[5]*Department of Marine Science, Faculty of Marine Science and Fisheries, University of Hasanuddin, Kampus Tamalanrea KM. 10, Makassar – 90245, Sulawesi Selatan, Indonesia*

[6]*Multitrophic Research Group, University of Hasanuddin, Kampus Tamalanrea KM. 10, Makassar – 90245, Sulawesi Selatan, Indonesia*

[7]*Center of Excellence for Development and Utilization of Seaweed, University of Hasanuddin, Kampus Tamalanrea KM. 10, Makassar – 90245, Sulawesi Selatan, Indonesia*

[8]*Research Center for Chemistry, Indonesian Institute of Sciences, Kawasan Puspiptek Serpong, Tangerang Selatan – 15314, Indonesia*

ABSTRACT

Seaweed is a high economic commodity, but it does not contribute much to community welfare; therefore, this chapter outlines how seaweed cultivation could improve community welfare in coastal areas and small islands using the holistic-integral approach S-O-M (Subject, Object, and Method) Analysis. The actual seaweed cultivation condition is a low contribution to the community welfare in the coastal area and small islands due to limited domesticated species, low carrying capacity, brackish water ponds, low technology input, low-quality seaweed products, and low-capacity building. The related paradigm of seaweed cultivation is sustainable aquaculture. The associated issues of seaweed cultivation are sustainable development, equity, gender, livelihood, and poverty. The alternative solutions to improve community welfare are the domestication of wild species, multitrophic cultivation development, improvement support system technology, and socio-techno-economic aspects. The expected condition of seaweed cultivation is a high contribution to community welfare. There are 28 concepts of implementation to achieve the expected conditions (high contribution of seaweed cultivation to the community welfare), outputs (sustainable development), and outcomes (national resilience increase).

2.1 INTRODUCTION

Globally, fish production was 171 million tons, with an aquaculture contribution of 47% of total fisheries production and 53% if non-food use was excluded. For a long time, aquaculture has played an essential role in fulfilling fish consumption. Fish contributes 2% to meeting human calorie needs, and 15% of human protein consumption comes from the sea, both marines caught and marine farmed. In 2015, the global seafood production put the total at 138 million metric tons (MT), of which 40% aquaculture production (Anonymous, 2018a). Indonesia is the second-largest aquaculture country in the world after Tiongkok. Tiongkok is the largest producer of aquaculture in Asian countries, while Indonesia is the largest producer of aquaculture in South East Asian countries.

Aquaculture contributed 96.5% by the total 32 million tons of aquatic plants production. The aquatic plant production dominated by seaweeds, grew from 14 million tons in 1995 to more than 30 million tons in 2016.

Rapid cultivation growth occurred in Indonesia in two species of tropical seaweed, namely *Kappaphycus alvarezii* and *Eucheuma* spp. (Saunders and Lindsay, 1978; Tuwo et al., 2020). Aquaculture is a hope for the future because catches in the sea are already flat, sometimes decreasing. World capture decreased from 81 million tons in 2015 to 79 million tons in 2016 (Anonymous, 2018a).

In 2016, the aquaculture production was 110 million tons, with an estimated value was US$ 244 billion. The total production included eighty million tons of food fish (value US$ 232 billion), 30 million tons of aquatic plants (value US$ 12 billion), as well as 38 tons of non-food products (value estimated US$ 215 million). Seaweed dominates Indonesia's aquaculture production, with a share of 61% of the total aquaculture production in Indonesia. In 2016, Indonesia was one of the world's major seaweed producers, with wet seaweed production reaching 12 million tons or 39% of total world seaweed production (Hendartyo, 2020).

Seaweed cultivation development is very relevant to millennium development goals (MDGs). There are nine of the seventeen MDGs goals that are directly and not directly related to seaweed development, which is: (i) end of all forms of poverty; (ii) improve gender equality and women empower; (iii) improve water availability and sanitation; (iv) ensure access to sustainable energy; (v) promoting economic growth and productive employment; (vi) reduce inequality; (vii) mitigate the climate change and its impacts; (viii) sustainable use of marine resources; and (ix) revitalize the global partnership for sustainable development.

Seaweed cultivation for community welfare in the coastal area and small islands was described using a holistic-integral approach. In this approach, an actual condition (i) that is not expected due to the problems encountered (ii) can be overcome if the developing paradigm (iii) and the existing strategic environmental conditions (iv) are considered in the formulation of alternative solutions (v). A formulated alternative solution will lead to the expected condition (goal) (vi). The expected condition can be realized if alternative solutions are implemented successfully. The implementation concept is formulated by using S-O-M Analysis (Subject, Object, and Method Analysis) (vii). In the medium term, five years, success in implementing the implementation concept will deliver the expected output (viii); and in the long term, ten years will deliver the expected outcome (ix) (Figure 2.1).

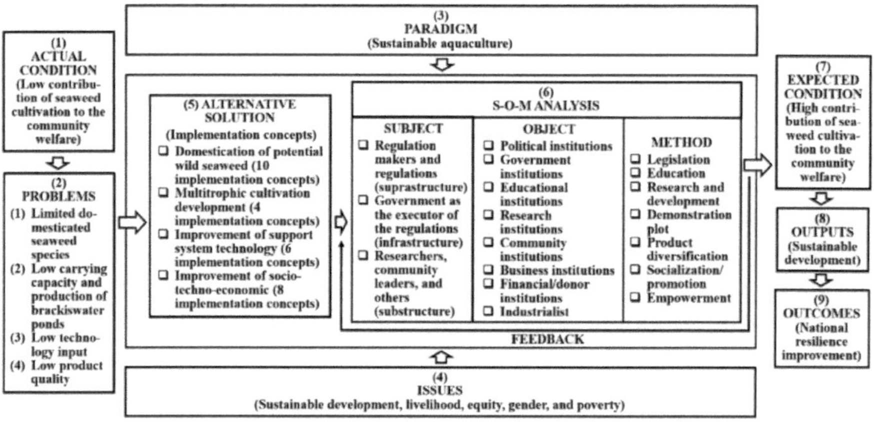

FIGURE 2.1 Holistic-integral approach used for seaweeds cultivation for community welfare in coastal areas and small islands.

2.2 ACTUAL CONDITION AND PROBLEMS

The actual condition of seaweed cultivation is a low contribution to the community welfare in the coastal area and small islands due to limited domesticated species, low carrying capacity and production of brackish water ponds, low technology input, and low product quality.

2.2.1 *LIMITED DOMESTICATED SPECIES*

Domestication is considered a long and complex process in which domesticators choose and modify organisms that can thrive in the human ecological environment and express traits of interest to human use (Tanno and Willcox 2006; Larson et al. 2014). There are 9,800 macroalgae species, but only four species have been domesticated as aquaculture species. The most cultivated species are Gulaman (*Gracilaria* sp.), gusô (*Eucheuma* sp.), Japanese kelp (*Laminaria japonica*), elkhorn sea moss (*K. alvarezii*), and wakame (*Undaria pinnatifida*). Intensive seaweed domestication is needed to support the diversification of cultivated seaweed species. The alternative for macroalgae domestication is enormous; the target macroalgae domestication can be selected in three major groups of macroalgae, i.e., brown algae (Phaeophyceae), green algae (Chlorophyta), and red algae (Rhodophyta). Of these three major groups, it is estimated 1,800 species of brown macroalgae, 6,200 are red macroalgae, and 1,800 are green macroalgae (Mouritsen, 2013).

2.2.2 LOW CARRYING CAPACITY AND PRODUCTION OF BRACKISH WATER PONDS

Brackish water pond is the principal area for seaweed cultivation, especially *Gracilaria* sp. (Arbit et al., 2019). For the Indonesian case, Brackish water pond production is only around 50 kg per hectare (Anonymous, 2018b). This production is lower than extensive brackish water pond production, which can produce 400–500 kg ha^{-1} each growing season (Kordi, 2010), and brackish-water pond production with intermediate technology (semi-intensive), which can produce 1200–3000 kg ha^{-1} each growing season (Amri, 2003). Brackish water ponds with low and intermediate technology can be developed with a multitrophic cultivation system, where *Gracilaria* sp. functions as an autotroph and shrimp as heterotroph (Tuwo et al., 2019a).

Gracilaria sp. is generally cultivated with shrimps and fish in brackish water ponds, which have many problems, including low carrying capacity. In the last twenty years, the actual problems of brackish water ponds, for Indonesian case, were (i) carrying capacity of brackishwater ponds is low due to environmental degradation; (ii) high mortality rate of shrimp culture, (iii) more than 95% of coastal ponds are small scale business; (iv) more than 95% of shrimp farmers do not have the capital to apply intensive and supra intensive shrimp cultivation practices; (v) difficulty in accessing the production facilities and infrastructure, especially electrical energy; (vi) most shrimp farmers do not have the knowledge and skills required for intensive and supra intensive shrimp cultivation practices; (vii) shrimp farmers are challenging to master the success factor of intensive shrimp cultivation; and (viii) shrimp farms are often considered unsustainable and less environmentally-friendly (Tresnati and Tuwo, 2017). The seaweed multitrophic cultivation could be an alternative to increase the carrying capacity and productivity of brackish-water ponds.

2.2.3 LOW TECHNOLOGY INPUT

Cultivation in brackish water ponds and coastal areas in developing countries, such as Indonesia, is generally carried out by farmers with financial limitations. The brackish water pond production with low technology (extensive) is practically no input technology; water supplied relying on the tidal cycle, the ponds are filled with seawater high tides. The input provided is only inorganic fertilizer and fish or shrimp juvenile from hatcheries. The brackish water pond production with intermediate technology (semi-intensive), the input technology is the fertilizer, fish or shrimp juvenile, water pumps, and artificial feed. Low technology inputs like this caused the difficulty to produce as expected because brackish water ponds that have reached decades have high organic matter and pyrite content, so it requires much oxygen to oxidize these organic materials and pyrite. Therefore, to increase productivity, technology inputs such as aerators or water wheels are needed to increase the water oxygen content and circulate the brackish water pond (Tresnati and Tuwo, 2017).

The problem faced by fish farmers is the high operational costs when using water wheels (Anonymous, 2018c). To develop semi-intensive ponds, it needs an electricity supply. The electricity supply will increase operational costs, burdening the farmers because they have limited financial capacity. Therefore, to increase the success, a supply of appropriate input technology to increase dissolved oxygen at an affordable cost is needed (Tuwo et al., 2018).

2.2.4 LOW PRODUCT QUALITY

The leading cause of low product quality is low carrying capacity and low economic capacity. Low carrying capacity causes a low growth rate so that the size of fish, shrimp, or seaweed produced is small. Small-sized fish, shrimp, and seaweed are categorized as low product quality, and of course, the selling price is also low. While low economic capacity, besides causing low technological input, also causes brackish water pond farmers to harvest fish, shrimp, and seaweed before harvest time. Seaweed that is harvested prematurely will be much lost during the seaweed industry processing because seaweed that is harvested prematurely is easily broken or destroyed during processing.

The low economic capacity of brackish water pond farmers causes the farmers not to do post-harvest processing or other value-added processes; all of their products are sold as raw materials. Indonesian seaweed export is in the form of raw materials. Therefore, even though Indonesia is a large producer of fisheries resources globally, the fisheries contribution has not been significant in national economic development. Indonesian fisheries products are generally still exported in raw materials, so the selling value is low. Based on September 2018 data, the selling value per kg was US$8.85 for shrimp, and US$1.38 for seaweed (Anonymous, 2019).

2.3 PARADIGM

The aquaculture paradigm is that sustainable aquaculture. The 2030 sustainable development agenda offers a vision of a more just and peaceful world where nothing is left behind. The 2030 agenda sets out fisheries and aquaculture development objectives towards food security and the rational use of the resources to ensure the socio-economic improvement and environmental sustainability (Anonymous, 1995). Sustainable aquaculture is needed to support food security and improve the welfare of rural communities. Globally, in 2016, the peak of fish production was 171 million tons. The aquaculture was 47% of the total fish production; and 53% if non-food use is excluded. Sustainable aquaculture will be increasingly expected in the future because marine fisheries resources decline continuously. The fraction of marine fish stocks captured at a biologically sustainable level has shown a declining trend, from 90.0% in 1974 to 66.9% in 2015. The cultivation is the future hope because catches in the sea are already flat, sometimes decreasing. World capture decreased from 81.2 million tons in 2015 to 79.3 million tons in 2016, representing a decrease of almost 2.1 million tons (Anonymous, 2018a).

Ecologically, aquaculture is a friendlier and efficient way of producing protein than terrestrial livestock due mainly to the pond fertility and good feed conversion ratios (FCR) of fish. Nevertheless, even aquaculture contributes directly to food security by increasing availability and accessibility; it is essential to understand the aquaculture contribution to global greenhouse gas (GHG) emissions and how to mitigate it. In 2013, global aquaculture produced 0.45% of anthropogenic GHG emissions, which can have significant non-GHG impacts, such as water quality and marine biodiversity. Therefore, it is essential to improve aquaculture's efficiency to make an essential contribution to food security. Technical innovation is a great scope for improving aquaculture's efficiency (MacLeod et al., 2019).

The major GHGs produced by aquaculture are: (1) nitrous oxide (N_2O) as a result of the microbial activities in ponds that excreted nitrogen (N); (2) carbon dioxide (CO_2) arising during pre-farm, post-farm, distribution, and processing due to energy use. CO_2 emissions also could be generated by the changes in land reserves caused by land use and land-use change (LUC); and (3) methane (CH_4) arising arises during fish waste management (MacLeod et al., 2019).

Seaweed cultivation is more friendly than another aquaculture. The total GHG emissions of nine species-groups (Bivalves, Catfishes, Cyprinids,

General Freshwater Fishes, Indian Major Carps, Marine Cereals, Salmonids, Shrimps, and Tilapias) were 201 metric tons (MT) of CO_2. It is representing 92% of total shellfish and finfish production, but it is considerably lower than livestock emissions that are estimated at 14.5% (Gerber et al. 2013). Globally, aquaculture emissions are lower than livestock because there was more livestock production. Aquatic animals produced only about 7% of global protein consumption, approximately half of which comes from aquaculture, compared to 33% of the protein consumption came from livestock products. Emissions correlate closely with production. The cyprinids account for 34% of emissions and 32% of production. There are exceptions; shrimp account for 18% of emissions but only 9% of production, while bivalves produce 9% of emissions but represent 23% of production. Production of feed materials accounts for 40% of total aquaculture emissions (MacLeod et al., 2019).

Aquaculture is relatively young compared with terrestrial livestock sectors, but it offers excellent technical innovation scope to increase resource efficiency (Waite et al., 2014). There are four technological approaches to reduce the aquaculture impact, i.e., (i) improve breeding and genetics, (ii) promote disease control, (iii) improve nutrition and feeding quality, and (iv) promote low-impact and sustainable production systems. It is essential to reduce emissions while increasing the affordable protein supply (MacLeod et al., 2019).

2.4 ISSUES

Many issues that develop in the fisheries and aquaculture sector are related to sustainable development, equity, gender, livelihood, and poverty.

2.4.1 SUSTAINABLE DEVELOPMENT

Climate change is one issue that can threaten the sustainable use of fisheries and aquaculture resources. Climate change can impact aquatic organisms, both catch species in the sea and aquaculture species. Climate change can affect the mobility and connection of fish habitats in the sea; fish species can adapt to climate change by shifting their distribution patterns to cooler regions or deeper waters. Climate change can affect the oceans' primary productivity, which is the food web of marine organisms. Climate change also affects the fisheries production. Fisheries production is expected to decrease by 6% by 2100 and by 11% in the tropical zone (Kwiatkowski et al., 2017).

Climate change can increase the storms and rising sea levels, directly, and indirectly affecting the fishing and aquaculture industry (Seggel et al., 2016). Various models predict worse conditions whereby 2050, the total global fishing potential is estimated to decrease by more than 10% (Cheung et al., 2010). Although the impact will be negative for fisheries and aquaculture in the tropics, opportunities will arise in temperate climates (Barange et al., 2014).

CO_2 emissions are another issue that can affect aquatic organisms. CO_2 can affect water acidity so that which can threaten the sustainable use of fisheries and aquaculture resources. Increased CO_2 uptake by the oceans will increase water acidity. Seawater will disturb the calcifying organisms in the sea. High acidity will also cause marine culture or cultivation facilities to be damaged more quickly.

In addition to natural impacts, activities that utilize fisheries resources that are not environmentally friendly also exacerbate natural impacts. The overfishing symptoms present for a long time; the fraction of fish stocks within the biologically sustainable levels had exhibited a decreasing trend from 90% to 66.9% in 41 years (1974–2015). The biologically unsustainable stocks fished increased from 10% to 33.1%, with the most significant was in the late 1970s and 1980s. In 2015, the maximum sustainably fished stocks was 59.9% and underfished stocks was only 7% of the total valued stocks (Anonymous, 2018a).

The use of aquaculture resources that are not environmentally friendly contributes to the natural impacts. Intensive aquaculture activities worldwide have raised concerns about the dangers of environmental degradation (Pillay, 2004). The degradation due to large amounts of artificial feed, so it is necessary to find alternatives for sustainable cultivation using natural feeds, for example, multitrophic aquaculture. Multitrophic cultivation is similar to polyculture (Syahid et al., 2006). At the multitrophic cultivation, producers act as aquaculture products and source of energy or food (Tacon, 1987), the producer or seaweed also acts as an absorber of the organic matter and functions as a nutrient biofilter (Nelson, 2008), pH stabilizer (Tyson et al., 2004; Tyson, 2007). It is involved in the dynamics of nitrogen (Sugita et al., 2005) to be more environmentally friendly. Therefore, to overcome the old ponds that have low carrying capacity in Indonesia, multitrophic cultivation can be the right solution.

The role of aquaculture will be increasingly important, so it is important to be managed sustainably. In 2015, fish accounted for around 17% of animal protein or 7% of the world protein consumption. Besides, fish are a protein source for around 3.2 billion people, with the animal protein intake was 20% per capita. Average fish consumption per capita varies significantly in each country and region due to cultural, economic, and geographical factors. Annual fish consumption per capita varies from less than one kg to more than a hundred kg per capita in all countries. Fish consumption per capita per year grows stably in developing regions, from six kg in 1961 to 19.3 kg in 2015; and in low-income food-deficit countries (LIFDC), low income from 3.4 to 7.7 kg at the same period. In developed countries, the value is higher, at 24.9 kg in 2015. Although fish consumption level is relatively low in developing countries, developing countries have a higher portion of fish protein in their diets than in developed countries. In 2015, fish accounted for around 26% of animal protein intake in less developed countries, 19% in other developing countries, and around 16% in LIFDC (Anonymous, 2018a).

Growth in fish consumption in Asian countries, especially in the east (except Japan) and Southeast Asia, has been driven by a combination of large populations, urban growth, a dramatic expansion of fish production increased the income and international fish trade. Tiongkok, by far the largest fish-consuming country in the world, consumed 38% of the global total in 2015, with per capita consumption reaching around 41 kg. This increase was driven by improvement of domestic income and wealth. More diverse types of fish have become available to consumers in Tiongkok due to the diversion of fishery exports to the domestic market and an increase in fishery imports. If Tiongkok is excluded, the annual per capita fish consumption worldwide is around 15.5 kg in 2015, rising from 10.3 kg in 1961 (Anonymous, 2018a). These fish consumption indicate the importance of sustainability in the management of fisheries resources, especially aquaculture in a sustainable manner. Multitrophic cultivation can reduce the impact of aquaculture and mitigate the environment by absorbing CO_2 and increasing the concentration of dissolved oxygen so that the anoxic hazard can be reduced.

2.4.2 LIVELIHOOD

In 2016 there were 59.6 million people engaged in fisheries (40.3 million) and aquaculture (19.3 million). The employment in fisheries decreased from 83% to 68%, while in aquaculture increased from 17% to 32% in 26 years

(1990–2016). Fisheries governance has evolved from a focus on resource conservation and the environment, namely the biological conception of sustainability, to recognize the social institutions, welfare, and livelihoods of people working in this sector. Therefore, fisheries are not only seen as a resource; fisheries are also seen as livelihoods sources, expression sites of cultural values, and a buffer against shocks for the poor. Besides, a human rights-based approach (HRBA) was promoted to achieve this goal. HRBA in this context refers to ensuring the participation of fishers and fish workers who are non-discriminatory and effective in transparent and accountable decision-making processes and addressing the root causes of poverty such as discrimination, marginalization, exploitation, and abuse. HRBA has increasingly been recognized as a programming principle in the United Nations (UN) system, but experience with its application in small-scale fisheries is limited. The food and agriculture organization (FAO) has been involved with partners in several places to overcome this gap; Indonesia has adopted a legislative framework on protecting human rights in the fisheries, with the technical assistance of FAO (Anonymous, 2018a).

2.4.3 EQUITY AND GENDER

Capture fisheries and aquaculture are still dominated by men, in all countries. Aquaculture in Asia involves about 15% women, 76% men, and 9% unspecified. Sex-disaggregated for employment in the fishery and aquaculture sectors varies significantly among countries and regions (Anonymous, 2018a).

The 2030 Agenda calls for gender equality and women empowerment (SDG 5), particularly relevant to the fisheries sector. Participation in fisher organizations offers women an essential pathway for engaging in management. FAO supports gender mainstreaming to improve gender equality through the participation of women in fisher organizations. Case studies on fisher organizations in Barbados, Belize, Costa Rica, Indonesia, and the United Republic of Tanzania revealed that women participated as members and leaders in fisher organizations, but much less than men (Siar and Kalikoski, 2016).

The women's participation and leadership barriers in fisher organizations include lack of recognition of women's work and contribution in fisheries, particularly on male fishers, and the notion that women do not fish (Alonso-Población and Siar, 2018). Other personal obstacles are lack of time to participate, lack of confidence, and lack of formal education. Widespread bias where women are seen primarily as mothers and wives while men are seen as breadwinners and leaders. Significant gender inequality harms the performance of women and their livelihoods. In Tunisia, in 2016, women clam collectors, who typically spend six to eight hours per day in the seawater, were earning four times less than intermediaries and only 70% of the legal minimum salary in the agriculture sector. Based on the value chain, they earned only 20% of the final sale price. Strategies identified to address these issues include strengthening technical, organizational, and business management capacities of participating women; product differentiation; and fostering networking, investment in infrastructure, and access to financial services and markets, especially the rewarding international channels and institutional outlets. Priority interventions identified in Tunisia led to significant results. Women were endowed with more substantial bargaining power; advocacy at the policy level triggered more transparent marketing transactions; and a fair-trade agreement was established among an association of women clam collectors, a clam depuration and export establishment, and an international importer. In November 2017, women collectors were receiving 47% of the sale price thanks to the fair-trade agreement, from which they pay 8% of the sale price to the intermediary transporter (Anonymous, 2018a). Multitrophic cultivation can provide significant opportunities because many aspects are only done by women, such as binding and drying seaweed.

2.4.4 POVERTY

The growth of the world population and the slowdown in economic growth lead to increased poverty, especially in coastal areas. Coastal areas are a significant contributor to poverty in Indonesia. The main problem of poverty eradication is the fisheries-dependent households are ignored and marginalized, politically, and otherwise, because they do not usually appear under a given poverty line. This invisibility, in many cases, excludes them from inclusive pro-poor development interventions. Poverty alleviation can benefit from collective action, such as forming small-scale fisheries stakeholders and community organizations. This community organization is essential because small fishers and fishing communities often struggle under the dominance of powerful actors inside and outside the fisheries sector that determine fisheries governance politics (Siar and Kalikoski, 2016).

Global poverty is 736 million people (8.6%) live in extreme poverty (US$1.90 or less a day). These poverties are calculated using income and personal ability to meet basic needs. The root causes of poverty are lack of access to necessities of life such as water, food, shelter, education, and healthcare. In fragile states areas, children, and communities face higher poverty rates due to political upheaval, past or present conflict, corrupt leaders, and poor infrastructure that limit access to education, clean water, healthcare, and other necessities (Anonymous, 2020a). Global poverty will be compounded by population and economic growth, the world population's growth, and the slowdown in economic growth leading to increased poverty, especially in rural areas.

Poverty is an important issue in Indonesia. Poverty can be classified as rural and urban poverty. As an archipelago state, Indonesian poverty characteristics are widely different from an island to another island and social culture to another social culture. Indonesia has serious poverty issues due to socio-economic and physical isolations. In February 1999, 47.97 million people were classified as poor. By July 2005, the poor had been reduced to 35.10 million. Latest, the Asian Development Bank reported that Indonesia's population in 2015 was 255.46 million, 47.2% of whom are poor (Anonymous, 2020b). Indonesia's national poverty line set a consumption of Rp. 302,735 (US$25) monthly per person (82 cents daily). If the poverty line of US$2 a day is used as a reference, 40% of the population is classified as poor. Some 68 million Indonesians are classified as near-poor or vulnerable. The poor come in and out of poverty during their lifetime. In 2014, 55% of households were classified as poor, indicating a high risk of becoming poor again (Priasto, 2015). In 2014, 23.8% of the rural and 16.2% of the urban population were classified as poor. This poverty was caused by the low-productivity jobs in agriculture and low-end service sectors.

Poverty eradication efforts through small-scale fisheries governance need to empower fishing communities and get them more control over the basic conditions that determine their well-being. Collective action can take the form of organizations that help empower small-scale fishers. Once these organizations exist, collective action becomes coordinated, directed, routinized, and more robust to actively contribute to the governance process. Multitrophic cultivation can help farmers because multitrophic cultivation allows farmers to harvest various commodities; every 40 days can harvest seaweed to meet their daily needs. Every three to four months can be harvested fish and shrimp to meet more significant needs, such as school fees, clothing, and other needs. Every year can be harvested sea cucumbers, bivalves, and other cultivated organisms meet long-term needs such as houses and other needs.

2.5 ALTERNATIVE SOLUTION

2.5.1 DOMESTICATION OF POTENTIAL WILD SEAWEED

Seaweed cultivation has developed almost exclusively in Indonesia, the Philippines, Tiongkok, and Japan (Van Der Meer, 1983; Mulyani et al., 2018; Syamsuddin et al., 2019). There are six species most important, i.e., *E. cottonii, E. spinosum, K. alvarezii, Gracilaria* sp., *Porphyra* sp., and *L. japonica. E. cottonii, E. spinosum, K. alvarezii*, and *Gracilaria* sp. are four species that are cultivated a lot in Indonesia and the Philippines (Doty, 1978; Aqmal et al., 2016; Awaluddin et al., 2016; Irawati et al., 2016; Mala et al., 2016; Rima et al., 2016; Akmal et al., 2017; Mulyani et al., 2021). Cultivation of *E. cottonii* and *E. spinosum* has been initiated in the Philippines (Doty, 1978). These efforts have reached commercial production using cultivation methods that rely on cheap labor and consist, essentially, of attaching vegetative fragments of *Eucheuma* to ropes that are placed in suitable locations in the ocean. The production in 1974 reached 10,000 tons dry weight (DW), making this a locally significant but globally small mariculture effort, which is unlikely to expand unless the demand for *Eucheuma* increases dramatically. *Eucheuma* is a permanent competitor to *Chondrus crispus* and *Eucheuma* can be used as a substitute for carrageenan from *C. crispus* for many applications (Van Der Meer, 1983).

Porphyra sp. is a widely cultivated species in Japan; more than 70,000 hectares (Ha) are devoted to *Porphyra* sp. cultivation (Saito, 1976). Japanese aquaculture practices to grow *Porphyra* is the most advanced in the world. There are two alternating phases in the *Porphyra* life cycle, the filamentous shell phase and the foliose phase, the harvested *Porphyra* thallus. Much of the *Porphyra* industry's success can be attributed to the skills with which the filament phase is manipulated. This success is preserved entirely in cultivation, and by carefully regulating environmental parameters such as light, temperature, and salinity, farmers can, at will, encourage it to form spores. The net is seeded with spores that are released and placed in the sea to plant the foliose phase. Maintaining nets in the ocean and harvesting is still hard work, but various tools have been developed to help farmers. The high market value of *Porphyra* has been the key to success for farmers who now earn more than US$ 500 million for their products every year (Van Der Meer, 1983).

L. japonica is a species that is widely cultivated in Tiongkok. The Chinese have used *L. japonica* for about 1500 years as food and medical remedies. *L. japonica* has become an important dietary source of iodine and raw material for the production of alginates, iodine, and mannitol. The rise of Laminaria cultivation in Tiongkok is a remarkable story (Cheng, 1969). About 50,000 tons of dried seaweed is imported every year by Tiongkok, mainly from Japan, because *L. japonica* does not grow along the coast of Tiongkok. Ironically, Japan lost this export market after its troops accidentally introduced Laminaria into Chinese waters during the invasion and occupation of Tiongkok in the first half of the century. Efforts to grow Laminaria in Tiongkok began soon after, but had little success until the early 1950s. In 1952, the harvest reported 52 MT, in 1959 24,000 MT, and 1979 as an extraordinary 280,000 MT of DW (Tseng, 1981). Unlike Porphyra's highly mechanical industry in Japan, the Laminaria industry in Tiongkok is very labor-intensive. From spore collection to harvest, there is enormous input from manual labor, which is justified because most of the production is used as human food, but that would make the Chinese method unprofitable in developed countries with high labor costs, where algae are partly large will be used as biomass for the industry. Nevertheless, the Chinese have made a big step in understanding the practical aspects of Laminaria's growth, and it can be expected that the procedures and mechanization of labor savings will follow (Van Der Meer, 1983).

Efforts to domestication are still being pursued, and there are still many challenges that must be overcome. In North America, marine farming has not yet been pursued on a commercial scale. The most ambitious experiment involved efforts to expand the Macrocystis layer's reach by attaching plants to a floating support system in deep waters off the coast of California (North, 1977). Plant growth in this experimental structure varies, and many problems appear to be unresolved, including bad weather. There have also been several attempts to grow red algae *Iridaea cordata* using nylon nets hanging in the ocean (Mumford, 1978), but it appears that these efforts have been abandoned because researchers were focusing on efforts on the cultivation of Porphyra. Seaweed farming faces two problems that are similar to terrestrial systems. First is the cost of chemical fertilizers. Much of Laminaria's cultivation in Tiongkok depends on adding nitrogen fertilizer, which significantly increases production costs. Second, when large monocultures of selected algae are formed, there is also a widespread danger of disease.

Fortunately, the disease has not been a severe problem, but this can change if the genetic basis of the cultivated plant is reduced through inbreeding and mating, making strains uniformly susceptible to deadly pathogens (Van Der Meer, 1983).

The opportunities for the domestication of wild species from macroalgae are enormous. There are nine algae divisions, namely six divisions for Microalgae and three divisions for macroalgae or seaweed. The seaweed can be classified according to pigment content that is green algae (Chlorophyta), brown algae (Ochrophyta or Phaeophyta), and red algae (Rhodophyta). The seaweed can also be grouped according to its natural products, which are Agarophyte (producing gelatin), Alginophyte (producing alginate), Caraginophyte (producing carrageenan).

Edible seaweeds from the three divisions of macroalgae are very potential to be domesticated. Common edible seaweeds are found in all three divisions of macroalgae or seaweed (Dawson, 1966). The edible seaweeds of red algae are Carola (*Callophyllis* spp.), Carrageenan moss (*Mastocarpus stellatus*), Dulse (*Palmaria palmata*), Eucheuma (*E. spinosum* and *E. cottonii*), *Gelidiella* (*Gelidiella acerosa*), Ogonori or Gracilaria (*G. edulis* and *G. corticata*), Grapestone (*M. papillatus*), Hypnea, Irish moss (*C. crispus*), Laverbread (*Porphyra laciniata/P. umbilicalis*), Gim (*Pyropia, Porphyra*), and Nori (*Porphyra*). The edible seaweeds of green algae are Gutweed (*Ulva intestinalis*), Seagrapes or green caviar (*Caulerpa lentillifera*), and Sea lettuce (*Ulva* spp.). The edible seaweeds of brown algae are divided into three, i.e., Kelp (*Laminariales*), Fucales and Ectocarpales. The Kelp (Laminariales) are Arame (*Eisenia bicyclis*), Badderlocks (*Alaria esculenta*), Cochayuyo (*Durvillaea antarctica*), *Ecklonia cava*, Kombu (*Saccharina japonica*), Oarweed (*L. digitata*), Sea palm (*Postelsia palmaeformis*), Sea whip (*Nereocystis luetkeana*), Sugar kelp (*S. latissima*), Wakame (*U. pinnatifida*), Hiromi (*U. undarioides*), Grapestone (*M. papillatus*). The Fucales are Bladderwrack (*Fucus vesiculosus*), Channeled wrack (*Pelvetia canaliculata*), Hijiki or Hiziki (*Sargassum fusiforme*), Limu Kala (*S. echinocarpum*), Sargassum (*S. cinetum, S. vulgare, S. swartzii, S. myriocysum*), Spiral wrack (*F. spiralis*), Thongweed (*Himanthalia elongata*). The Ectocarpales is Mozuku (*Cladosiphon okamuranus*).

Successful commercial cultivation of benthic algae in enclosures is in Indonesia and Taiwan, where *Gracilaria* sp. is cultivated in land-based ponds, often associated with fish crustaceans. The *Gracilaria* is used primarily in the agar industry. The *Gracilaria* is then harvested periodically by raking it into piles and removing it from the pond for drying. There is no reason the similar approach could not be successful in other areas with similar climate, or perhaps even with different climates if appropriate organisms are selected (Van Der Meer, 1983).

Various domestication efforts are still ongoing in several countries. In North America, seaweed cultivation in a closed system is still experimental, still at the stage of a small trial designed to test physiological parameters or design principles. In Nova Scotia, Canada, two companies, with support from the Canadian government, experiment with a mass planting of *C. crispus* clones selected for carrageenan production. Although an excellent growth rate has been obtained, the process is not profitable enough to justify commercial development. On the west coast of Canada, growth experiments have been done in a new floating cultivation system that surrounds plants without bonds in large polyethylene pockets hanging in the ocean (Saunders and Lindsay, 1978). In the eastern United States, it has experimented extensively on algal cultivation, especially *Gracilaria*. It has been examined the use of algae for nutrient reclamation from wastes (DeBoer and Ryther, 1977), for phycocolloid production (DeBoer, 1978; Lapointe and Ryther, 1979), for polysaccharides (Melanie et al., 2020), for pigment (Ma et al., 2019), for bioaccumulation, bioremediation, and mitigation (Armita et al., 2017; Arbit et al., 2018; Badraeni et al., 2020), and biomass as energy crops (Hanisak, 1981; Sari et al., 2017). They have suggested that all three functions can be achieved in a well-integrated system (Ryther et al., 1978). Gracilaria yields in this experiment have been reported to be comparable to the best land plants. In the northwest of the United States, there have been several trials on the cultivation of *Iridaea* and *Gigartina* (Waaland, 1976; Waaland, 1978).

2.5.2 DEVELOPMENT OF MULTITROPHIC CULTIVATION

Current crop production is characterized by its dependence on freshwater, fertilizers, pesticides, herbicides, and land management. At the samc time, the use of very large feeds characterizes protein production from land animals. These two characteristics were not in line with the principle of sustainable development. Dependence on freshwater will cause competition between humans, livestock, and plants. An increase in food crop production will reduce the share of water for livestock and humans, while an increase in the human population will cause an increase in water demand which will reduce the share of water for livestock and humans. On the other hand, an increase in food crop production will lead to an increase in waste that will worsen environmental eutrophication due to increased use of fertilizers, increase the risk of poisoning to the environment and human and livestock food ingredients due to the use of pesticides and other active ingredients to control pests and diseases, increase the erosion rate due to tillage which will damage water quality, increasing competition between humans and livestock in the use of biomass in the form of carbohydrates and proteins.

For overcoming biomass problem, future food production requirements must be free or reduce dependence on freshwater, fertilizers, pesticides, herbicides, and land management. Multitrophic cultivation between seaweed as primary producer and fish as heterotrophs is an alternative fulfillment of future needs that meet the four requirements mentioned above. Multitrophic cultivation of seaweed and aquatic animals as a source of carbohydrates and protein practically does not require freshwater supplies. Seaweed can reduce the eutrophication of waters by absorbing fertilizer waste from land agriculture. Seaweed and aquatic animals do not need pesticides or herbicides like on land agriculture. Multitrophic cultivation of seaweed and aquatic animals also does not require land management in agriculture on land. The superiority of this multitrophic cultivation of seaweed and aquatic animal makes it a new hope in sustainably meeting human food needs. Reduced water quality is a potential outcome of intensive aquaculture. Integrated multitrophic aquaculture (IMTA) can mitigate the harmful effects of nutrients from fish farms by harvesting species that extract nutrients at adjacent sites (Hadley et al., 2016).

In addition to the comparative advantages mentioned above, multitrophic cultivation of seaweed and aquatic animals still has several other advantages than food production on land. Multitrophic cultivation of seaweed and aquatic animals are very efficient because the metabolic wastes released by aquatic animals that are cultivated together with seaweed, such as fish, shrimp, sea cucumbers, bivalves, and other aquatic animals (Tuwo and Tresnati, 2015; Hadijah et al., 2013; Tuwo et al., 2019b) can be absorbed by seaweed as nutrients so that the potential for eutrophication can be prevented. The production capacity of seaweed cultivation will increase. Also, seaweed cultivation is safer when using a closed or semi-closed system because invasive organisms can be controlled not to need pesticides or herbicides to eliminate pests. Multitrophic cultivation of seaweed and aquatic animals can also improve the welfare of farmers and fishers because pond farmers can harvest several products at different times, namely: (1) income from harvesting seaweed every 45 days can be used to meet daily needs such as food needs; (2) income from harvesting fish or shrimp or every three or four months can be used to meet needs that require a relatively high cost, such as clothing needs, school fees, and health care costs; and (3) income from harvesting sea cucumbers, bivalves, mollusks, and other animals every year can be used to meet the needs that require high costs, such as housing (Tresnati and Tuwo, 2017).

Besides, there are many more advantages of multitrophic cultivation of seaweed and aquatic animals in ponds. Ponds can function as net carbon sinks by stimulating primary productivity (Boyd et al., 2010). Other studies exclude these sinks due to uncertainties over the sequestration rates and permanence of the carbon storage; much of the sequestered carbon could be oxidized, depending on how the sludge. Also, stimulating the growth requires relatively large amounts of nitrogen and phosphorus to the water, leading to eutrophication problems. There is also a concern about the fish welfare, the nutrient input can significantly change the water quality, which may not suit some fish species (MacLeod et al., 2019).

Multitrophic cultivation of seaweed and aquatic animals are the best choice in overcoming future food insecurity because they have several advantages, namely: (i) increasing the success and performance of pond cultivation; (ii) increasing the carrying capacity of shrimp from ponds; (iii) can be applied to pond farmers with small business capital; (iv) can be applied by farmers after following a short training and mentoring; (v) facilities and production capacity, as well as the need for electrical energy, can be adjusted to the ability of the farmer's business; (vi) Business scale can be increased gradually in accordance with the skills and capabilities of the farmer's business capital; (vii) can reduce the risk of business failure in shrimp ponds; (viii) fish farmers can harvest more than one product (shrimp, seaweed, fish, sea cucumbers, etc.); and (ix) can absorb CO_2 and mitigate the environment, including heating (Tuwo, 2018).

2.5.3 IMPROVEMENT OF SUPPORT SYSTEM TECHNOLOGY

In Indonesia's case, brackish water ponds that are decades old, even some that have been more than a hundred years old, cause the carrying capacity to decrease. The high organic matter content causes a decrease in carrying capacity, because more oxygen is needed to degrade organic matter. This result in shrimp being cultivated in brackish water ponds that are only able to survive one to two months; after reaching a specific size, shrimp begin to die due to oxygen deficit.

The aquaculture business is a business that can be compared to the oxygen business, meaning that products can be harvested depending on the amount of oxygen input that can be supplied to the farm. In old aquaculture ponds, such as those in Indonesia, the supply of oxygen is necessary. Therefore, the support system technology needed for multitrophic cultivation of seaweed and aquatic animals is an oxygen generator or aerator.

So far, the oxygen supply is provided by using a pedal. Pedal wheels are not an obstacle for intensive and supra intensive ponds because investors in intensive and supra intensive ponds generally have significant capital. Pedal wheels are an obstacle to ponds with standard and semi-intensive size because these business actors are local people who have small capital. One pedal wheel is needed with a dynamo of one horsepower, or the equivalent of 750 Watts per hour to produce 500 kg of shrimp. To get 750 Watts, the farmer must upgrade his electric power, minimum to 900 Watts, because the underprivileged standard of electric power is only 450 Watts. Therefore, traditional pond farmers need innovative aerator technology that can be operated with 450 Watts of electricity.

One of the causes of traditional pond farmers' common welfare is the low selling price of dried seaweed. One of the seventeen causes of seaweed's low selling value is the low quality of dried seaweed (Tuwo, 2016). Postharvest processing is done by drying in the sun with a period of non-standard drying. Therefore, traditional pond farmers need innovation in standardized drying technology using solar power. Solar-powered dryers are more appropriate energy resources because generally, the location of the cultivation is far from electricity installation.

2.5.4 IMPROVEMENT OF SOCIO-TECHNO-ECONOMIC ASPECTS

A general description of the economic aspects of fishers and seaweed farmers in Indonesia can be seen in a case study in one of the centers of fishing and seaweed cultivation in Indonesia, precisely in the Tanakeke Islands (Tuwo, 2011). The case study results show that the majority (56%) of the population in the Tanakeke Islands only had primary school education, of which 24% did not complete primary school, only 17% had junior secondary education, and 3% had senior secondary education. The low level of education in the Tanakeke Islands is due to the limited educational facilities and infrastructure. The residents who want to continue their education to secondary level must go to the district capital to be able to follow a higher level of education.

The majority of the people's livelihoods are fishermen and seaweed cultivation, with 47% fishers, 32% seaweed farmers, 18% entrepreneurs, and 3% government employees. About 59% have been in business for three to five years; 13% pursue a business between six to ten years; 18% pursue a business between eleven to twenty years; 10% have been in business for more than twenty years. Those who claim to have been in business or work for a long time (over five years) are generally fishermen. In contrast, those who work for less than six years are generally seaweed cultivators. In managing businesses, seaweed farming, and managing trading businesses, they generally manage their businesses (57%); 28% joined the family business, and 15% claim to manage other people's businesses. The source of community business capital in the Tanakeke Islands, both for investment and operational costs, generally comes from loans (62%). Only 38% use their capital. Most people in the Tanakeke Islands are accustomed to making loans, especially to traders (50%). As many as 42% claimed to borrow from cooperatives. The rest (8%) claimed to borrow from groups or friends. None of the respondents claimed to borrow from a bank.

The income per household varies depending on the type of work or business and season. Most (86%) of the population in the Tanakeke Islands earn between Rp. 300,000 and Rp.750,000 per month. As many as 10% earn between Rp.751,000 and Rp.1,000,000 per month. Only 4% earn more than Rp.1,000,000 per month. People with an income above Rp. One million generally receive additional income from seaweed cultivation and trade. Most of the Tanakeke Islands population only rely on income from fishing in the sea by using fishing gear and fishing fleets that are still relatively traditional namely motorized boats. Also, fishing activities are influenced by the season, making it difficult for fishers to increase their income. Low-income populations generally assume that their catch is sufficient for daily living expenses, such as buying rice, kerosene, granulated sugar, and coffee. They generally do not have the motivation to develop businesses to increase income and family welfare.

Generally, the wife only takes care of the household. However, there are some of them who not only take care of the household but also help their husbands earn a living. From the female respondents, information was obtained that 35% of the women claimed only to take care of the household alone; 25% claimed to help the husband's work; 26% work for extra income by working for others, especially in seaweed farming; and 14% claimed to manage their own businesses such as trading or selling, processing fish, and others.

Around 83.3% of the population in the Tanakeke Islands occupy stilt houses as a residence. They do not have many assets. Most (63.3%) of them only have television and radio as assets. Another asset is the genet engine (100%). The case example in the Tanakeke Islands shows the need for improved socio-economic aspects to support increased production and business success.

2.6 CONCEPT IMPLEMENTATION

In terms of realizing the expected conditions and achieving outputs (sustainable development) and outcomes (national resilience increase), an implementation concept involving subjects, objects, and methods is needed. The expected condition is a high contribution of seaweed cultivation to community welfare. The subject is the regulation makers and regulations (suprastructure management), the government as the executor of the regulations (infrastructure management), and researchers, community leaders, and others (substructure management). The management objects are political institutions, government institutions, educational institutions, research institutions, community institutions, business institutions, financial/donor institutions, and industrialists. Management methods are legislation, education, research, and development, demonstration plot, product diversification, socialization or promotion, and empowerment.

Implementation concepts for domestication of potential wild seaweed for food, feeds, pharmaceuticals, or cosmetics to support diversification of seaweed cultivation are: (i) potential wild seaweed identification; where educational institutions together with research institutions performing research and development in order to identify potential wild seaweed for food, feeds, pharmaceuticals, or cosmetics to support diversification of seaweed cultivation in collaboration with financial/donors institutions; (ii) domestication techniques development; where educational institutions together with research institutions performing research and development in the context of domestication in collaboration with financial/donors institutions; (iii) manual development for domestication technique; where educational institutions together with research institutions develop manual for domestication technique of each potential seaweed species for food, feeds, pharmaceuticals, or cosmetics to support diversification of seaweed cultivation in collaboration with financial/donors institutions; (iv) dissemination of domestication technique; where educational and research institutions together with government institutions disseminate domestication techniques of each potential seaweed species for food, feed, pharmaceutical, and cosmetics to support diversification of seaweed cultivation to the local community institutions, business institutions, and industries; (v) development of laboratory scale breeding; where educational institutions together with research institutions performing research and development in the context of laboratory scale breeding for each potential seaweed species for food, feed, pharmaceutical, and cosmetics to support the diversification of seaweed cultivation in collaboration with financial/donors institutions; (vi) manual development for laboratory scale breeding technique; where educational institutions together with research institutions develop manual for laboratory scale manual breeding techniques of each potential seaweed species for food, feeds, pharmaceuticals, or cosmetics to support diversification of seaweed cultivation in collaboration with financial/donors institutions; (vii) dissemination of laboratory scale breeding technique of each potential seaweed species for food, feeds, pharmaceuticals, or cosmetics to support diversification of seaweed cultivation; where educational and research institutions together with government institutions disseminate laboratory scale breeding technique of each potential seaweed species for food, feeds, pharmaceuticals, or cosmetics to support diversification of seaweed cultivation to community and business institutions, and industries; (viii) manual development for demonstration plot for mass breeding of each potential seaweed species; where educational institutions together with research institutions develop manual for demonstration plot for mass breeding of each potential seaweed species for food, feeds, pharmaceuticals, or cosmetics to support diversification of seaweed cultivation in collaboration with financial/donors institutions; (ix) mass breeding technique dissemination; where educational and research institutions together with government institutions disseminate mass breeding technique of potential seaweed species for food, feeds, pharmaceuticals, or cosmetics to support diversification of seaweed cultivation to community institutions, business institutions, and industries; and (x) demonstration plot development for mass breeding; where educational institutions together with research institutions develop demonstration plot for mass breeding in collaboration with financial/donors institutions.

Implementation concepts for multitrophic cultivation development are: (i) development of multitrophic cultivation model; where educational and research institutions develop multitrophic cultivation models for potential wild seaweed for food, feeds, pharmaceuticals, or cosmetics to support diversification of seaweed cultivation in collaboration with financial/

donors institutions; (ii) manual development for multitrophic cultivation technique; where educational institutions together with research institutions develop manual for multitrophic cultivation technique in collaboration with financial/donors institutions; (iii) dissemination of multitrophic cultivation technique; where educational, research, government institutions disseminate multitrophic cultivation techniques to community institutions, business institutions, and industries; and (iv) development of demonstration plots; where educational institutions together with research institutions develop demonstration plots in collaboration with financial/donors institutions.

Implementation concepts for improvement support system technology are: (i) development of low cost green aerator models to support multitrophic cultivation development; where educational and research institutions develop low cost green aerator models to support multitrophic cultivation development in collaboration with industries and financial/donors institutions; (ii) mass production of low cost and environmentally friendly aerator to support multitrophic cultivation development; where educational institutions together with research institutions develop mass production of low cost green aerator to support multitrophic cultivation development in collaboration with industries and financial/donors institutions; (iii) development of low cost solar and wind power plants; where educational institutions together with research institutions develop low cost solar and wind power plants in collaboration with industries and financial/donors institutions; (iv) mass production of low cost solar and wind power plants; where educational institutions together with research institutions develop low cost solar and wind power plants in collaboration with industries and financial/donors institutions; (v) development of low cost green seaweed dryers using solar and wind power plants; where educational, and research institutions develop low cost green seaweed dryers using solar and wind power plants in collaboration with industries and financial/donors institutions; and (vi) mass production of low cost green seaweed dryers using solar and wind power plants; where educational institutions together with research institutions develop mass production of low cost green seaweed dryers using solar and wind power plants in collaboration with industries and financial/donors institutions.

Implementation concepts for improvement of socio-techno-economic aspects are: (i) skill improvement in technical seaweed cultivation to support seaweed cultivation; where educational research, and government institutions improve technical seaweed cultivation skill to support seaweed cultivation in collaboration with community institutions in collaboration with business institutions, and industries; (ii) skill improvement in post-harvest processing to support seaweed cultivation; where educational, research, and government institutions improve post-harvest processing skill to support seaweed cultivation in collaboration with community business institutions, and industries; (iii) skill improvement in application of support system technology to support seaweed cultivation; where educational and research institutions together with government institutions improve skills in application of support system technology to support seaweed cultivation in collaboration with community institutions in collaboration with business institutions, and industries; (iv) skill improvement in value-added process to support seaweed cultivation educational and research institutions together with government institutions improve skills in value-added process to support seaweed cultivation in collaboration with community institutions, business institutions, and industries; (v) skill improvement in managing seaweed business; where educational and research institutions together with government institutions improve skills in seaweed business management for the for community institutions in collaboration with business institutions, and industries; (vi) skill improvement in accessing sources of funding or seaweed business capital; where educational and research institutions together with government institutions improve skills in accessing sources of funding or seaweed business capital in collaboration with community institutions in collaboration with business institutions, and industries; (vii) regulations and incentive development that support the socio-techno-economic improvement; where government institutions together with political institutions develop regulations and incentive that support the socio-techno-economic improvement in collaboration with educational institutions, research institutions, and community institutions; and (viii) infrastructure development that supports the socio-techno-economic improvement; where government institutions together with political institutions develop basic infrastructure that supports the socio-techno-economic improvement in collaboration with financial/donors institutions, business institutions, and industries.

2.7 CONCLUSION

The holistic-integral approach and S-O-M analysis produced 28 implementation concepts to make seaweeds cultivation for community welfare in coastal areas and small islands. That is, potential wild seaweed identification, domestication techniques development, manual development for domestication technique, domestication technique dissemination, laboratory-scale breeding development, manual development for laboratory-scale breeding technique, dissemination of laboratory-scale breeding technique, manual development for demonstration plot for mass breeding; mass breeding technique dissemination; demonstration plot development for mass breeding; multitrophic cultivation model development; manual for multitrophic cultivation technique; multitrophic cultivation technique dissemination; demonstration plots development for multitrophic cultivation; development of low cost green aerator models; mass production of low cost green aerator; development of low cost solar and wind power plants; mass production of low cost solar and wind power plants; development of low cost green seaweed dryers; mass production of low cost green seaweed dryers; skill improvement in seaweed cultivation; skill improvement in post-harvest processing; skill improvement in application of support system technology; skill improvement in value-added process; skill improvement in managing seaweed business; skill improvement in accessing sources of funding or seaweed business capital; regulations and incentive development; and infrastructure development.

ACKNOWLEDGMENTS

We would like to thank the Ministry of Research and Technology/ National Research and Innovation Agency of the Republic of Indonesia for providing research and publication funding (contract number 1516/UN4.22/ PT.01.03/2020 dated May 18th, 2020 and 752/UN4.22/PT.01.03/2021 dated March 26th, 2021.

KEYWORDS

- **Food and Agriculture Organization**
- **greenhouse gas**
- **holistic-integral approach**
- **millennium development goals**
- **multitrophic cultivation**
- **seaweed's cultivation**
- **support system technology**
- **sustainable development**

REFERENCES

Akmal, A., Syamsuddin, R., Trijuno, D. D., & Tuwo, A., (2017). Morfologi, kandungan klorofil a, pertumbuhan, produksi, dan kandungan karaginan rumput laut *Kappaphycus alvarezii* yang dibudidayakan pada kedalaman berbeda. *Jurnal Rumput Laut Indonesia, 2*(2), 39–50.

Amri, K., (2003). *Budi Daya Udang Windu Secara Intensif* (p. 102). AgroMedia, Tangeran, Indonesia.

Anonymous, (1995). *Code of Conduct for Responsible Fisheries* (p. 41). FAO, Rome.

Anonymous, (2018a). *The State of World Fisheries and Aquaculture 2018 – Meeting the Sustainable Development Goals* (p. 227). FAO, Rome.

Anonymous, (2018b). *Produksi Tambak Udang Masih Jauh Dari Harapan: Gunakan Sistem Lama, Perlu Memanfaatkan Teknologi.* https://rakyatkaltara.prokal.co/read/news/14287-produksi-tambak-udang-masih-jauh-dari-harapan (accessed on 13 October 2021).

Anonymous, (2018c). *Meningkatkan Kadar Oksigen Terlarut di Dalam Tambak.* https://tambakudang.com/meningkatkan-kadar-oksigen-terlarut-di-dalam-tambak/ (accessed on 13 October 2021).

Anonymous, (2019). *Statistik Perikanan Indonesia* (p. 7). Biro Pusat Statistik Indonesia, Jakarta.

Anonymous, (2020a). *Poverty Data: Indonesia – Indonesia and Asian Development Bank.* https://www.adb.org/countries/indonesia/poverty (accessed on 13 October 2021).

Anonymous, (2020b). *Global Poverty: Facts, FAQs, and How to Help.* https://www.worldvision.org/sponsorship-news-stories/global-poverty-facts (accessed on 13 October 2021).

Aqmal, A., Tuwo, A., & Haryati, H., (2016). Analisis hubungan antara keberadaan alga filamen kompetitor terhadap pertumbuhan dan kandungan karaginan rumput laut *Kappaphycus* sp. di Provinsi Sulawesi Selatan. *Jurnal Rumput Laut Indonesia, 1*(2), 94–102.

Arbit, N. I. S., Omar, S. B. A., Soekendarsi, E., Yasir, I., Tresnati, J., & Tuwo, A., (2019). Morphological and genetic analysis of *Gracilaria* sp. cultured in ponds and coastal waters. *IOP Conf. Ser.: Earth Environ. Sci., UK, 564*, 012059.

Arbit, N. I. S., Omar, S. B., Tuwo, A., & Soekendarsi, E., (2018). Effect of global warming scenarios on carotenoid pigments *Gracilaria changii*. *International Journal of Environment, Agriculture and Biotechnology, 3*(6), 268–287.

Armita, D., Ali, S. A., Yanuarita, D., & Tuwo, A., (2017). Kualitas air di daerah dengan dan tanpa budidaya rumput laut di Dusun Malelaya, Kecamatan Mangarabombang, Kabupaten Takalar. *Jurnal Rumput Laut Indonesia, 2*(2), 74–80.

Awaluddin, A., Badraeni, B., Azis, H. Y., & Tuwo, A., (2016). Perbedaan kandungan karaginan dan produksi rumput laut *Kappaphycus alvarezii* antara bibit alam dan bibit hasil pengayaan. *Jurnal Rumput Laut Indonesia, 1*(1), 65–70.

Badraeni, B., Azis, H. Y., Tresnati, J., & Tuwo, A., (2020). Seaweed *Gracilaria changii* as a bioremediator agent for ammonia, nitrite and nitrate in controlled tanks of white leg shrimp *Litopenaeus vannamei*. *IOP Conf. Ser.: Earth Environ. Sci., UK, 564,* 012059.

Barange, M., Merino, G., Blanchard, J. L., Harle, J. S., Allison, E. H., Allen, J. I., Holt, J., & Jennings, S., (2014). Impacts of climate change on marine ecosystem production in societies dependent on fisheries. *Nature Climate Change, 4*, 211–216.

Boyd, C. E., Wood, C. W., Chaney, P. L., & Queiroz, J. F., (2010). Role of aquaculture pond sediments in sequestration of annual global carbon emissions. *Environmental Pollution, 158*(8), 2537–2540.

Cheng, T. H., (1969). Production of kelp – a major aspect of China's exploitation of the sea. *Economic Botany, 23*(3), 215–236.

Cheung, W. W. L., Lam, V. W. Y., Sarmiento, J. L., Kearney, K., Watson, R., Zeller, D., & Pauly, D., (2010). Large-scale redistribution of maximum fisheries catch potential in the global ocean under climate change. *Global Change Biology, 16*, 24–35.

Dawson, E. Y., (1966). New records of marine algae from the Gulf of California. *Journal of the Arizona Academy of Science, 4*(2), 55–66.

DeBoer, J. A., & Ryther, J. H., (1977). *The Marine Plant Biomass of the Pacific Northwest Coast* (pp. 231–239). Oregon State University Press, Corvallis.

DeBoer, J. A., (1979). Effects of nitrogen enrichment on growth rate and phycocolloid content in *Gracilaria foliifera* and *Neoagardhiella baileyi* (Florideophyceae). *Proc. Int. Seaweed Symp., UK,* 263–271.

Doty, M. S., (1978). Status of marine agronomy with special reference to the tropics. *Proc. Int. Seaweed Symp., UK,* 35–58.

Gerber, P. J., Steinfeld, H., Henderson, B., Mottet, A., Opio, C., Dijkman, J., Falcucci, A., & Tempio, G., (2013). *Tackling Climate Change Through Livestock – A Global Assessment of Emissions and Mitigation Opportunities* (p. 115). FAO, Rome.

Hadijah, H., Tuwo, A., Litaay, M., & Indrawati, E., (2013). The reproductive aspect of tropical abalone (*Haliotis asinina* L.) in the waters of Tanakeke Islands at South Sulawesi. *Aquatic Science and Technology, 1*(2), 30–44.

Hadley, S., Wild-Allen, K., Johnson, C., & Macleod, C., (2016). Quantification of the impacts of finfish aquaculture and bioremediation capacity of integrated multitrophic aquaculture using a 3D estuary model. *Journal of Applied Phycology, 28*(3), 1875–1889.

Hanisak, M. D., (1981). Methane production from the Red Seaweed *Gracilaria tikvahiae. Proc. Int. Seaweed Symp. UK,* 681–686.

Hendartyo, M., (2020). *Genjot Ekspor, KKP Targetkan Produksi Rumput Laut 10 Juta Ton.* https://bisnis.tempo.co/read/1296932/genjot-ekspor-kkp-targetkan-produksi-rumput-laut-10-juta-ton#:~:text=KKP%20mencatat%20angka%20sementara%20tahun,US%24%2010% 20miliar%20per%20tahun (accessed on 13 October 2021).

Irawati, I., Badraeni, B., Abustang, A., & Tuwo, A., (2016). Pengaruh perbedaan bobot tallus terhadap pertumbuhan rumput laut *Kappaphycus alvarezii* strain coklat yang dikayakan. *Jurnal Rumput Laut Indonesia, 1*(2), 82–87.

Kordi, K. M. G. H., (2010). *Budidaya Udang Laut* (p. 314). Lily Publisher, Yogyakarta.

Kwiatkowski, L., Bopp, L., Aumont, O., Ciais, P., Cox, P. M., Laufkötter, C., Li, Y., & Séférian, R., (2017). Emergent constraints on projections of declining primary production in the tropical oceans. *Nature Climate Change, 7*, 355–358.

Lapointe, B. E., & Ryther, J. H., (1979). Effects of nitrogen and seawater flow rate on the growth and biochemical composition of *Gracilaria foliifera* var. *angustissima* in mass outdoor cultures. *Bot. Mar. 22*, 529–538.

Larson, G., Piperno, D. R., Allaby, R. G., Purugganan, M. D., Andersson, L., Arroyo-Kalin, M., Barton, L., et al., (2014). Current perspectives and the future of domestication studies. *Proc. Natl. Acad. Sci. USA, 111*(17), 6139–6146.

Ma, Z., Khalid, N., Shu, G., Zhao, Y., Kobayashi, I., Neves, M. A., Tuwo, A., & Nakajima, M., (2019). Fucoxanthin-loaded oil-in-water emulsion-based delivery systems: Effects of

natural emulsifiers on the formulation, stability, and bioaccessibility. *ACS Omega, 4*(6), 10502–10509.
MacLeod, M., Hasan, M. R., Robb, D. H. F., & Mamun-Ur-Rashid, M., (2019). *Quantifying and Mitigating Greenhouse Gas Emissions from Global Aquaculture* (p. 36). FAO, Rome.
Mala, L., Latama, G., Abustang, A., & Tuwo, A., (2016). Analisis perbandingan pertumbuhan rumput laut *Kappaphycus alvarezii* varietas coklat yang terkena epifit di Perairan Libukang, Kabupaten Jeneponto. *Jurnal Rumput Laut Indonesia, 1*(1), 52–56.
Melanie, H., Taarji, N., Zhao, Y., Khalid, N., Neves, M. A., Kobayashi, I., Tuwo, A., & Nakajima, M., (2020). Formulation and characterization of O/W emulsions stabilized with modified seaweed polysaccharides. *International Journal of Food Science and Technology, 55*(1), 211–221.
Mouritsen, O. G., (2013). *The Science of Seaweeds*. https://www.americanscientist.org/article/the-science-of-seaweeds#:~:text=Macroalgae%20are%20 classified%20into%20three,referred%20to%20simply%20as%20kelp (accessed on 13 October 2021).
Mulyani, S., Tuwo, A., Syamsuddin, R., & Jompa, J., (2018). Effect of seaweed *Kappaphycus alvarezii* aquaculture on growth and survival of coral *Acropora muricata*. *AACL Bioflux, 11*(6), 1792–1798.
Mulyani, S., Tuwo, A., Syamsuddin, R., Jompa, J., & Cahyono, I., (2021). Relationship of the viscosity of carrageenan extracted from *Kappaphycus alvarezii* with seawater physical and chemical properties at different planting distances and depth. *AACL Bioflux, 14*(1), 328–336.
Mumford, T. F. Jr., (1978). Field and laboratory experiments with *Iridaea cordata* (Florideophyceae) grown on nylon netting. *Proc. Int. Seaweed Symp., UK,* 515–523.
Nelson, R. L., (2008). *Aquaponic Equipment: The Biofilter*. http://www.aquaponicsjournal.com/docs/Aquaponic-Equipment-The-BioFilter.pdf (accessed on 13 October 2021).
North, W. J., (1977). Possibilities of biomass from the ocean. In: Mitsui, A., Miyachi, S., San, P. A., & Tamara, S., (eds.), *Biological Solar Energy Conversion* (pp. 347–361). The Marine Farm Project, New York.
Pillay, T. V. R., (2004). *Aquaculture and the Environment* (p. 206). Blackwell Publishing, Oxford.
Priasto, A., (2015). *Summary of Indonesia's Poverty Analysis*. https://www.adb.org/publications/summary-indonesias-poverty-analysis (accessed on 13 October 2021).
Rima, R., Yunus, B., Umar, M. T., & Tuwo, A., (2016). Performa rumput laut *Kappaphycus alvarezii* pada habitat berbeda di perairan kecamatan arungkeke, kabupaten jeneponto. *Jurnal Rumput Laut Indonesia, 1*(1), 17–26.
Ryther, J. H., LaPointe, B. E., Stenberg, R. W., & Williams, L. D., (1978). Cultivation of seaweeds as a biomass source for energy. In: *Fuels from Biomass Symposium* (pp. 83–97). Illinois, Urbana, USA, University of Illinois.
Saito, Y., (1976). *Seaweed Aquaculture in the Northwest Pacific* (p. 93). FAO, Rome.
Sari, A. P., Ahmad, A., Usman, H., & Tuwo, A., (2017). Produksi bioetanol dari selulosa alga merah dengan sistem fermentasi simultan menggunakan bakteri *Clostridium acetobutylicum*. *Jurnal Rumput Laut Indonesia, 2*(2), 58–62.
Saunders, R. G., & Lindsay, J. G., (1978). Growth and enhancement of the agarophyte *Gracilaria* (Florideophyceae). *Proc. Int. Seaweed Symp. UK,* 249–256.
Seggel, A., De Young, C., & Soto, D., (2016). *Climate Change Implications for Fisheries and Aquaculture: Summary of the Findings of the Intergovernmental Panel on Climate Change Fifth Assessment Report* (p. 65). FAO, Rome.

Siar, S. V., & Kalikoski, D. C., (2016). Strengthening organizations and collective action in fisheries: Towards the formulation of a capacity development program. *Workshop Report and Case Studies* (p. 419). FAO, Barbados.

Sugita, H., Nakamura, H., & Shimada, T., (2005). Microbial communities associated with filter materials in recirculating aquaculture systems of freshwater fish. *Aquaculture, 243,* 403–409.

Syahid, M., Subhan, A., & Armando, R., (2006). *Budidaya Udang Organik Secara Polikultur* (p. 76). Penebar Swadaya, Jakarta.

Syamsuddin, R., Tuwo, A., & Aswar, N. R., (2019). Weight gain and carrageenan content of *Kappaphycus alvarezii* (Rhodophyta, Solierisceae) polyculture with *Sargassum polycystum* (Phaeophyta, Sargassaceae). *IOP Conf. Ser.: Earth Environ. Sci. UK, 253,* 012005.

Tacon, A. G. J., (1987). *The Nutrition and Feeding of Farmed Fish and Shrimp* (p. 177). FAO, Rome.

Tanno, K., & Willcox, G., (2006). How fast was wild wheat domesticated? *Science, 311*(5769), 1886–1886.

Tresnati, J., & Tuwo, A., (2017). *Multitrophic Aquaculture* (p. 20). Universitas Hasanuddin, Makassar.

Tseng, C. K., (1981). Marine phycoculture in China. *Proceedings – International Seaweed Symposium* (pp. 123–152). Walter de Gruyter & Co., Berlin.

Tuwo, A., & Tresnati, J., (2015). Sea cucumber farming in Southeast Asia (Malaysia, Philippines, Indonesia, Vietnam). In Brown, N. P., & Eddy, S. D., (eds.), *Echinoderm Aquaculture* (pp. 331–352). Wiley & Son, New Jersey.

Tuwo, A., (2011). *Pengelolaan Ekowisata Pesisir Dan Laut: Pendekatan Ekologi, Sosial-ekonomi, Kelembagaan, Dan Sarana Wilayah* (p. 412). Brilian Internasional, Surabaya.

Tuwo, A., (2016). *Fluktuasi Harga Rumput Laut di Tingkat Pembudidaya* (p. 39). Universitas Hasanuddin, Makassar.

Tuwo, A., (2018). *Marine Farming Urgency to Improve Food Security and Strengthen National Resilience* (p. 34). Asian Agriculture and Food Forum, Jakarta.

Tuwo, A., Auliana, R., Samawi, M. F., Aprianto, R., & Tresnati, J., (2020). Feasibility study of seaweed farming *Kappaphycus alvarezii* in Sub-District North Pulau Laut and Sub-District East Pulau Laut Kota baru Regency, South Borneo, Indonesia. *IOP Conf. Ser.: Earth Environ. Sci., UK, 564,* 012026.

Tuwo, A., Tresnati, J., & Yasir, I., (2018). *Kajian Model Budidaya Tambak Multitropik Yang Produktif Dan Ramah Lingkungan* (p. 66). Universitas Hasanuddin, Makassar.

Tuwo, A., Yasir, I., Tresnati, J., Aprianto, R., Yanti, A., Bestari, A. D., & Nakajima, M., (2019a). A microcosm multitrophic aquaculture system. *IOP Conf. Ser.: Earth Environ. Sci., UK, 370,* 012012.

Tuwo, A., Yasir, I., Tresnati, J., Aprianto, R., Yanti, A., Bestari, A. D., & Nakajima, M., (2019b). Evisceration rate of sandfish *Holothuria scabra* during transportation. *IOP Conf. Ser.: Earth Environ. Sci., UK, 370,* 012039.

Tyson, R. V., (2007). *Reconciling pH for Ammonia Biofiltration in a Cucumber/Tilapia Aquaponics System Using a Perlite Medium* (p. 120). University of Florida, Florida.

Tyson, R. V., Simonne, E. H., White, J. M., & Lamb, E. M., (2004). Refereed paper reconciling water quality parameters impacting nitrification in aquaponics: The pH levels. *Proc. Fla. State Hort. Soc., USA, 117,* 79–83.

Van, D. M. J. P., (1983). The domestication of seaweeds. *Bioscience, 33*(3), 172–176.

Waaland, J. R., (1976). Growth of the red alga *Iridaea cordata* (Turner) Bory in semi-closed culture. *J. Exp. Mar. Biol. Ecol., 23,* 45–53.

Waaland, J. R., (1978). Growth and strain selection in *Gigartina exasperata* (Florideophyceae). *Proc. Intl. Seaweed Symp., UK, 9,* 241–248.

Waite, R., Beveridge, M., Brummett, R., Castine, S., Chaiyawannakarn, N., Kaushik, S., Mungkung, R., et al., (2014). Improving productivity and environmental performance of aquaculture. *Working Paper, Instalment 5 of Creating a Sustainable Food Future* (p. 60). World Resources Institute, Washington.

CHAPTER 3

BIOACTIVE COMPOUNDS FROM EGYPTIAN SEAWEEDS

MADELYN NABIL MOAWAD and SOHA SHABAKA

National Institute of Oceanography and Fisheries (NIOF), Alexandria, Egypt

ABSTRACT

Out of hundreds of seaweed species present along the Mediterranean and the Red Sea coasts of Egypt, only 43 species were used to extract and/or screen their potential bioactive compounds. The high diversity of seaweeds in the Red Sea has attracted Egyptian scientists to study their bioactive compounds, especially for their antimicrobial activity, where a total of 38 seaweed species were studied. Due to the high diversity of brown algae and their interesting bioactive compounds, they have been the focus of many researchers. On the other hand, only 16 seaweed species have been studied along the Mediterranean Sea coast. Despite the extraction of novel and promising bioactive compounds from Egyptian seaweeds, there is a need for discovering more applications that serve more industrial, medicinal, and pharmaceutical purposes. Moreover, the uncertainties of the taxonomy of most species with the lack of voucher specimens in most cases will greatly affect the applicability of the extracted compounds. Accordingly, using a combination of morphological and molecular data for seaweeds used to extract bioactive compounds should be mandatory.

3.1 INTRODUCTION

Seaweeds or marine macroalgae are plant-like organisms that live in coastal areas attached to hard substrates. They belong to three groups, red algae (Class *Rhodophyceae*), brown algae (class *Phaeophyceae*), and green algae (phylum Chlorophyta, classes *Bryopsidophyceae*, *Chlorophyceae*,

Dasycladophyceae, Prasinophyceae, and *Ulvophyceae*). Green and red algae belong to the kingdom Plantae, while brown algae belong to Chromista. Seaweeds range from small and simple sheet-like or filamentous turfs to large leathery seaweeds. Some species are gelatinous others are calcified. The high genetic diversity of seaweeds is reflected in the high diversity of their morphological, ultrastructural, ecological, biochemical, and physiological traits (Rindi et al., 2012). The increased demand for new drugs and healthy food alternatives based on natural products derived from seaweed has placed the seaweed under the focus of numerous researches.

Macroalgae are subjected to several biotic and abiotic stressors resulting from the interaction with other marine organisms and the impact of environmental factors. Accordingly, seaweeds produce bioactive compounds as a defensive mechanism against these stressors. The synergetic effect of UV radiation and temperature stress could lead to the formation of free radicals and reactive oxygen species (ROS), yet, healthy seaweeds lack oxidative damage in their structural components, indicating the presence of protective antioxidant defense systems in their cells (Aguilera et al., 2002; Van De Poll et al., 2003). Accordingly, seaweeds produce compounds that are suitable as natural antioxidants for the human body. Many studies have shown that bioactive compounds from seaweeds have strong antioxidant properties (Cornish and Garbary, 2010). According to Rindi (2008), bioactive compounds have been documented for approximately 150 genera of seaweeds. The accurate taxonomic classification of the seaweed used in the extraction of biological compounds is of utmost importance, as these compounds are closely related to the genetic diversity of the algae (Rindi et al., 2012).

3.2 SEAWEEDS OF THE MEDITERRANEAN AND RED SEA COAST OF EGYPT

3.2.1 THE MEDITERRANEAN COAST OF EGYPT

The Mediterranean Coast of Egypt extends to about 970 km from Rafah in the east to Salloum in the west. The Eastern Mediterranean Coast of Egypt is a very interesting zone for studying the biodiversity of seaweeds where the passage of biota from the Red Sea via the Suez Canal is prominent and results in the settlement and spreading of alien species. Shabaka (2018) compiled historical work on seaweeds diversity and distribution along the Mediterranean Coast of Egypt and provided a checklist that included 187 accepted species (107 red, 45 green, and 35 brown algae) belonging to 80, 25, and 25 genera of red, green, and brown algae, respectively (Figures 3.1–3.3).

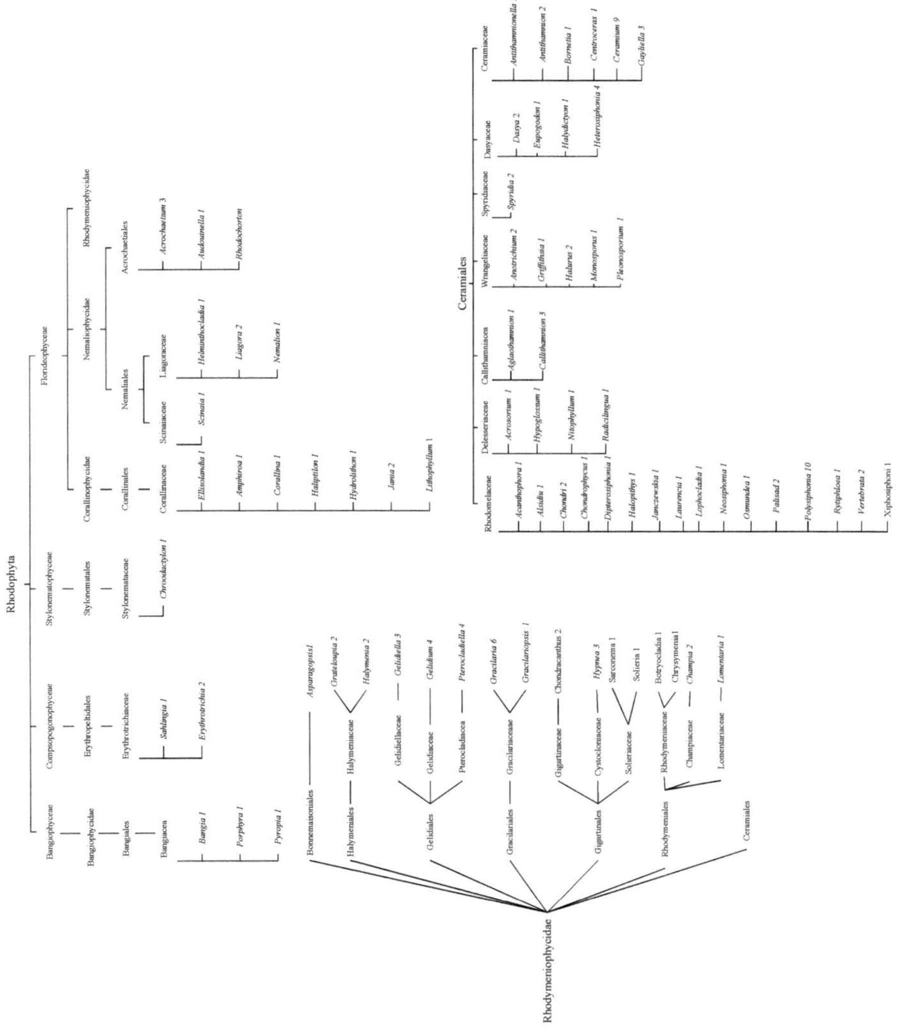

FIGURE 3.1 Classification tree of the red algae (Rhodophyta) along the Mediterranean Coast of Egypt showing the number of species. Data was based on Shabaka et al. (2018) and the classification was updated and checked from Guiry and Guiry (2021).

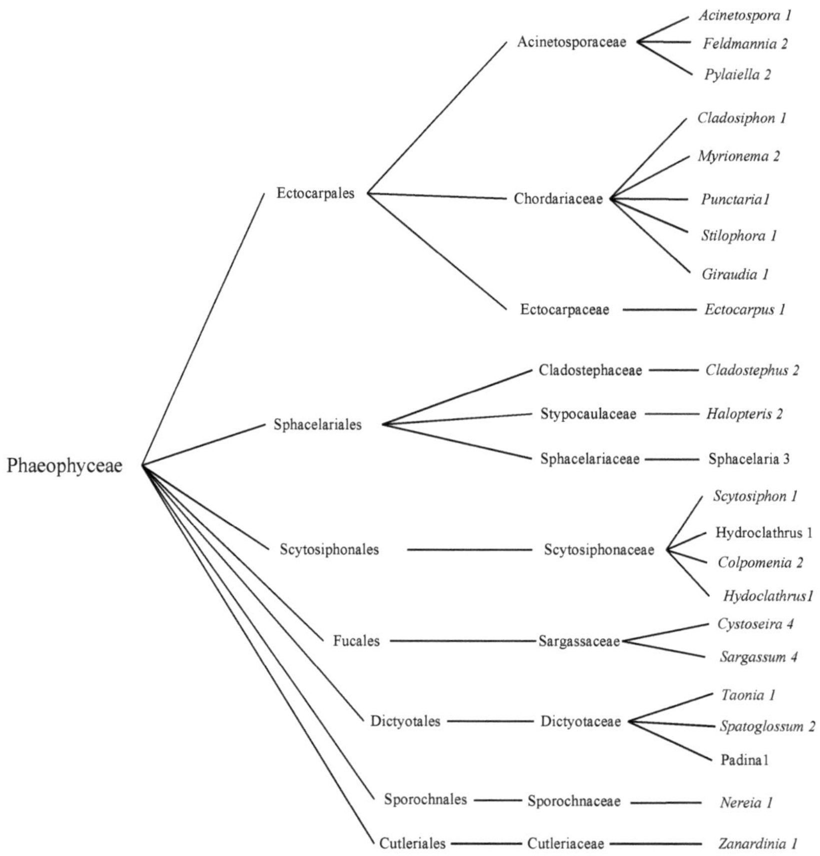

FIGURE 3.2 Classification tree of the brown algae (Phaeophyta) along the Mediterranean Coast of Egypt. Data was based on Shabaka et al. (2018) and the classification was updated and checked from Guiry and Guiry (2021).

3.2.2 THE RED SEA COAST OF EGYPT

Studying the distribution and biodiversity of macroalgae along the 1,100 km Red Sea Coast of Egypt is interesting, due to the habitat complexity (coral reefs, mangroves, reef flats, lagoons), the special topography (islands, bays, headlines), and the tropical to semi-tropical climate that is affected by the highly diverse Indo-pacific region. Aleem (1978a,b) studied the distribution of macroalgae in the vicinity of Hurghada and recorded 37 species of green algae, 25 species of brown algae, and 47 species of red algae. In Suez Canal, Aleem (1980) recorded 34 species of green algae, 20 species of brown algae, and 56 species of red algae. Negm (1988) surveyed the algal flora inhabiting the upper intertidal zone of the Egyptian coast of the Red Sea at the vicinity of Hurghada during 1979–1980 and identified a total of 57 species (18 green algae; 18 brown algae, and 21 Red algae). The dominant species were *Sargassum latifolium*, *Sargassum asperifolium*, *Sargassum crispum*, *Jania rubens*, *Sirophysalis trinodis* (as *Cystoseira trinodis*), *Ulva lactuca*, *Dictyota* sp. and *Halimeda tuna*. In the Red Sea, the mangroves ecosystem supports more than 36 species of macroalgae (Fouda, 2014). Mangroves ecosystem in Hurghada and Safaga supports 23 species of macroalgae (9 brown algae, 6 red algae, and 8 species of green algae) (El-Sharouny et al., 2001). At Nabq protected area, the reef flat is covered by macroalgae belonging to the genera *Padina*, *Caulerpa*, *Laurencia*, *Dictyota*, and *Gracilaria*, and the shallow lagoons provide habitat for *Turbinaria* sp. and *Sargassum* sp. (Galal, 1991). Mohamed et al. (2006) studied seaweeds distribution at *Hurghada, Safaga*, and *Quseir* in the Red Sea, where they reported a total of 35 species comprised of 7 species of green algae, 7 species of red algae, and 21 species of brown algae.

FIGURE 3.3 Classification tree of the green algae (Chlorophyta, Ulvophyceae). Data was based on Shabaka et al. (2018) and the classification was updated and checked from Guiry and Guiry (2021).

3.3 AN OVERVIEW OF THE EGYPTIAN SEAWEEDS USED AS SOURCES OF BIOACTIVE COMPOUNDS

The updates of species names applied here were based mainly on AlgaeBase (Guiry and Guiry, 2021). Species numbers were extracted from the published literature. A total of 43 seaweed species were used to extract bioactive compounds along the Mediterranean and Red Sea shores (Table 3.1 and 3.2). The high diversity of seaweeds in the Red Sea has attracted Egyptian scientists to study their bioactive compounds especially for their antimicrobial activity, where a total of 38 seaweed species were studied (Table 3.1). Due to the high diversity of the brown algae and their interesting bioactive compounds, they have been the focus of many Egyptian researchers, where about 20 species have been studied in comparison to red (12 species) and green algae (6 species) (Table 3.1). On the other hand, only 16 species (8 red, 3 green, and 5 brown algae) have been studied along the Mediterranean Sea Coast (Table 3.2). Eleven species were common with the Red Sea (Table 3.1). This indicates a lack of studies along both Mediterranean and Red Sea coasts; in addition, the uncertainties of the taxonomy of most species with the lack of voucher specimens, in most cases, will greatly affect the applicability of the extracted compounds. In some cases, the identification of complex species such as *Dictyota*, *Ceramium*, and *Caulerpa* is not correct. For instance, a few details are usually given when the species *Caulerpa racemosa* is used (Table 3.1 and 3.2); this species includes a great number of forms and varieties, including invasive strains. Accordingly, using a combination of morphological and molecular data for seaweeds used to extract bioactive compounds should be mandatory.

TABLE 3.1 List of Seaweed Species in the Red Sea Coast of Egypt with Reported Bioactivities (Shades: Red Algae, Green Algae, Brown Algae).

Location	Species	Effect	References
Faied and Ein Al-Sokhna 29°39’56.61”N 32°21’45.95”E	*Palisada perforata* (Bory) K. W. Nam = *Laurencia papillosa* (C. Agardh) Greville	Antiviral, anti-tumor, and antimicrobial activity	El Baz et al. (2014)
	Tricleocarpa cylindrica (J.Ellis & Solander) Huisman & Borowitzka=*Galaxaura cylindrica* (J.Ellis & Solander) J.V.Lamouroux		
Suez Canal 31° 3’21.34”N 32°18’32.37”E	*Codium tomentosum* Stackhouse	Anti-tumor, anti-bacterial, anti-inflammatory activity	Hamza et al. (2015)
	Ulva lactuca Linnaeus		
	Hypnea musciformis (Wulfen) J.V.Lamouroux		

TABLE 3.1 *(Continued)*

Location	Species	Effect	References
Hurghada 27°15'31.41"N 33°50'58.92"E	*Padina pavonica* (Linnaeus) Thivy	Anti-tumor activity	Awad et al. (2008)
	Cladostephus spongiosus (Hudson) C.Agardh	Anticandidal activity against the fungus *Candida krusei*	El Zawawy et al. (2020)
	P. perforata (Bory) K.W.Nam = *Laurencia papillosa* (C.Agardh) Greville		
	Codium arabicum Kützing		
	Jania rubens (Linnaeus) J.V.Lamouroux	Anti-tumor and anthelmintic activity	Awad (2004)
	P. pavonica (Linnaeus) Thivy	Antimicrobial activity	Awad et al. (2009b)
	Hydroclathrus clathratus (C.Agardh) M.Howe		
	Dictyopteris polypodioides (A.P.De Candolle) J.V.Lamouroux=*Dictyopteris membranacea* Batters	Anti-tumor and hypolipidemic, antiviral activity	Matloub et al. (2015)
	P. pavonica (Linnaeus) Thivy		
	Colpomenia sinuosa (Mertens ex Roth) Derbès & Solier		
	Ulva intestinalis Linnaeus		
	Corallina officinalis Linnaeus		
	Pterocladiella capillacea (S.G. Gmelin) Santelices & Hommersand		
	J. rubens (Linnaeus) J.V.Lamouroux		
	Hormophysa cuneiformis (J.F.Gmelin) P.C.Silva	Antifungal	Mohamed and Saber (2019)
	Sargassum fusiforme (Harvey) Setchell	Antibacterial	El Shafay et al. (2016)
	Sargassum vulgare C.Agardh, *nom. illeg.*		
	Ceramium rubrum		
	P. pavonica (Linnaeus) Thivy		
	Sargassum asperifolium Hering & G.Martens ex J.Agardh	Insecticidal activity	Matloub et al. (2012)
	Sargassum dentifolium (Turner) C.Agardh		
	Sargassum acinarium (Linnaeus) Setchell =*S. linifolium* C.Agardh		

TABLE 3.1 *(Continued)*

Location	Species	Effect	References
	S. dentifolium (Turner) C.Agardh	Growth stimulant activity for plant	Shaddad et al. (2014)
	Padina gymnospora (Kützing) Sonder		
	P. pavonica (Linnaeus) Thivy	Anti-tumor activity	Awad et al. (2009a)
	Hydroclathrus clathratus (C.Agardh) M.Howe		
	Polycladia myrica (S.G.Gmelin) Draima, Ballesteros, F.Rousseau & T.Thibaut=*Cystoseira myrica* (S.G.Gmelin) C.Agardh	Antiviral activity	Zaid et al. (2016)
	U. lactuca (Linnaeus)		
Safaga 26°44'33.66"N 33°56'57.17"E	*Treptacantha barbata* (Stackhouse) Orellana & Sansón in Orellana, Hernández & Sansón=*Cystoseira barbata* (Stackhouse) C.Agardh	Antimicrobial activity	Abdel-Raouf et al. (2017)
Quseir 26° 5'57.36"N 34°17'1.83"E	*Acanthophora spicifera* (M.Vahl) Børgesen	Antimicrobial and anti-inflammatory activity	Ibraheem et al. (2017)
	Digenea simplex (Wulfen) C.Agardh		
Quseir 26° 5'57.36"N 34°17'1.83"E Marsa Alam 25° 4'0.30"N 34°54'9.33"E	*Sargassum platycarpum* Montagne	Anti-tumor and antifungal activity	Abdelrheem et al. (2020a)
	Halimeda tuna (J.Ellis & Solander) J.V.Lamouroux	Anti-tumor and antimicrobial activity	Abdelrheem et al. (2020b)
	P. gymnospora (Kützing) Sonder		
	Phacelocarpus tristichus J.Agardh		
	U. lactuca (Linnaeus)	Antimicrobial activity	Ibraheem et al. (2012)
	Caulerpa racemosa		
	A. spicifera (M.Vahl) Børgesen		
	Galaxaura elongata J.Agardh		
	Ganonema farinosum (J.V.Lamouroux) K.C.Fan & Yung C.Wang =*Liagora farinosa* J.V.Lamouroux		

TABLE 3.1 *(Continued)*

Location	Species	Effect	References
	Cystoseira compressa (Esper) Gerloff & Nizamuddin		
	P. myrica (S.G.Gmelin) Draima, Ballesteros, F.Rousseau & T.Thibaut= *Cystoseira myrica* (S.G.Gmelin) C.Agardh		
	H. clathratus (C.Agardh) M.Howe		
	Turbinaria ornata (Turner) J.Agardh		
	P. pavonica (Linnaeus) Thivy		
	T. ornata (Turner) J.Agardh	*In vivo* protective effect against liver injury induced by azoxymethane (AOM)	Abdella et al. (2016)
	P. pavonica (Linnaeus) Thivy		
	T. ornata (Turner) J.Agardh	Antiproliferative and protective efficacy against AOM-induced colon carcinogenesis	Mahmoud et al. (2015)
	P. pavonica (Linnaeus) Thivy		
	Sirophysalis trinodis (Forsskal) Kützing=*Cystoseira trinodis* (Forsskål) C.Agardh	Anti-tumor and antimicrobial activity	Ahmed et al. (2020)
	U. lactuca (Linnaeus)	Antimicrobial and antimalarial activity as well as binding affinity for human opioid receptors	Elsayed et al. (2012)
	G. farinosum (J.V. Lamouroux) K.C.Fan & Yung C.Wang =*Liagora farinosa* J.V.Lamouroux		
	Padina pavonica		
	Turbinaria ornata Turn		
Wadi El Gemal 24°40'18.08"N 35° 5'46.03"E	*T. ornata* (Turner) J.Agardh	Antimicrobial activity	Madkour et al. (2019)
	P. myrica (S.G.Gmelin) Draima, Ballesteros, F.Rousseau & T.Thibaut= *Cystoseira myrica* (S.G.Gmelin) C.Agardh		
	P. pavonica (Linnaeus) Thivy		

TABLE 3.1 *(Continued)*

Location	Species	Effect	References
Red Sea	*Sarcodiotheca furcata* Setchell & N.L.Gardner	Antimicrobial activity	El-Sheekh et al. (2014)
	P. myrica (S.G.Gmelin) Draima, Ballesteros, F.Rousseau & T.Thibaut= *Cystoseira myrica* (S.G.Gmelin) C.Agardh		
	Sargassum ramifolium Kützing		
	Dictyota sp.	-	-

TABLE 3.2 List of Seaweed Species in the Mediterranean Coast of Egypt with Reported Bioactivities (Shades: Red Algae, Green Algae, Brown Algae)

Location	Species	Effect	References
Alexandria 31°12'18.90"N 29°53'7.35"E	*Gratelopia gibbessi*	Antioxidant activity	Shabaka and Moawad (2021)
Alexandria 31°18'17.83"N 30° 4'30.97"E	*U. lactuca* (Linnaeus)	Anti-tumor and antimicrobial activity	Saeed et al. (2020)
	U. lactuca (Linnaeus)=*Ulva fasciata* Delile		
	U. intestinalis (Linnaeus)		
	J. rubens (Linnaeus) J.V.Lamouroux		
	Gelidium spinosum (S.G.Gmelin) P.C.Silva		
	C. sinuosa (Mertens ex Roth) Derbès & Solier	Anti α-glucosidase activity	Moawad (2017)
	S. acinarium (Linnaeus) Setchell =*S. linifolium* C.Agardh		
	P. pavonica (Linnaeus) Thivy		
	P. capillacea (S.G.Gmelin) Santelices & Hommersand		

TABLE 3.2 *(Continued)*

Location	Species	Effect	References
	Osmundea pinnatifida (Hudson) Stackhouse=*Laurencia pinnatifida* (Hudson) J.V.Lamouroux		
	Caulerpa racemosa		
	P. capillacea (S.G. Gmelin) Santelices & Hommersand	Anti-tumor and anti α-glucosidase activity	
	P. pavonica (Linnaeus) Thivy	-	-
	U. lactuca (Linnaeus)=*Ulva fasciata* Delile	Antifouling activity	Ibrahim et al. (2019)
	U. lactuca (Linnaeus)	Antimicrobial and anti-inflammatory activity	Awad (2000)
	U. lactuca (Linnaeus)	Molluscicidal and anthelminthic activity	Saad et al. (2019)
	U. intestinalis (Linnaeus)		
	P. capillacea (S.G. Gmelin) Santelices & Hommersand		
	J. rubens (Linnaeus) J.V.Lamouroux		
	J. rubens (Linnaeus) J.V.Lamouroux	Antimicrobial activity	Mohy El-Din and El-Ahwany (2016)
	Ellisolandia elongata (J.Ellis & Solander)=*Corallina mediterranea* Areschoug		
	P. capillacea (S.G. Gmelin) Santelices & Hommersand		

TABLE 3.2 *(Continued)*

Location	Species	Effect	References
	Dictyota fasciola (Roth) J.V.Lamouroux =*Dilophus fasciola* (Roth) M.Howe	Anti-tumor, antiviral, and antimicrobial activity	El Baz et al. (2014)
	Taonia atomaria (Woodward) J.Agardh		
	U. lactuca (Linnaeus)=*Ulva fasciata* Delile		
	U. lactuca (L.)	Antifouling activity	Tadros et al. (2006, 2009)
	C. officinalis Linnaeus	Antimicrobial and hypolipidemic activity	Awad et al. (2003)
Marsa Matrouh 31°22'2.28"N 27°14'25.43"E	*Asparagopsis taxiformis* (Delile) Trevisan	Anti-tumor, antiviral, antialgal, antibacterial, and antifungal activity as well as biofertilizer	El-Baroty et al. (2007)

3.3.1 CYTOTOXIC ACTIVITY (ANTICANCER AND INSECTICIDAL ACTIVITY)

Awad (2004) has isolated and identified seven brominated diterpenes of the parguerene and isoparguerene series, from the red alga *J. rubens*; isoparguerol (1), isoparguerol-16-acetate (2), isoparguerol-7,16-diacetate (3), parguerol-16-acetate (4), parguerol-7,16-diacetate (5), deoxyparguerol (6), and deoxyparguerol-7-acetate (7) (Awad, 2004). Isoparguerol-7,16-diacetate 3 had the highest antitumor activity at concentrations of 50 and 100 μg/ml against Ehrlich ascites carcinoma cell lines with 100% inhibition of cell viability. Generally, isoparguerol derivatives were slightly more effective than parguerol derivatives (Awad, 2004). Also, potent anthelmintic activity on earthworms (*Allolobophora caliginosa*) was recorded for compound 3 followed by compounds 5, 2, 1, and 4 with death time ranged from 1 to 5 hr. Compounds 6 and 7 showed weak and normal responses with death times 20 and 18 hr, respectively (Awad, 2004).

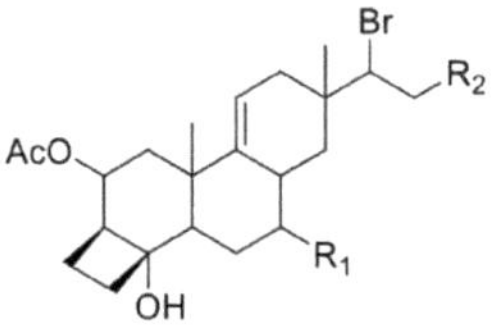

	R_1	R_2
1	OH	OH
2	OH	OAc
3	OAc	OAc

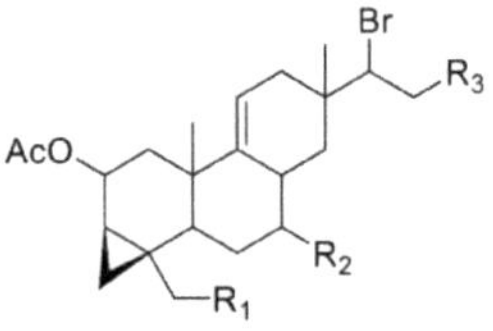

	R_1	R_2	R_3
4	OH	OH	OAc
5	OH	OAc	OAc
6	H	OH	OH
7	H	OAc	OH

Awad et al. (2008) reported the isolation of two diterpenes; 18,19-epoxyxenic-19-methoxy-18-hydroxy-4-acetoxy-6,9,13-triene (8) and 18,19-epoxyxenic-18,19-dimethoxy-4-hydroxy-6,9,13-triene (9) from hexanolic extract of a brown alga, *Padina pavonica*, collected from the Red Sea. The compounds 8 and 9 had antitumor activity against lung carcinoma (H460) with effective dose (ED_{50}) values of 13.2 and 18.4 μg/ml, respectively, compared to control Cisplatin (ED_{50}= 4.77 μg/ml) (Awad et al., 2008).

	R_1	R_2	R_3
8	OH	OCH_3	OAc
9	OCH_3	OCH_3	OH

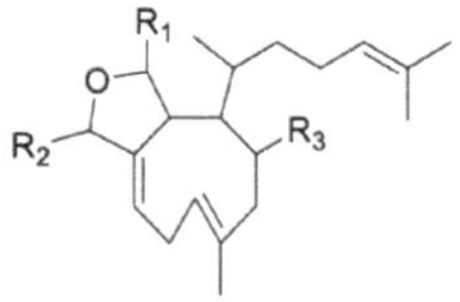

Bis-indole alkaloid, caulerpin (10), was isolated for the first time from the brown alga *Sargassum platycarpum* collected from the Red Sea (Abdelrheem et al., 2020a). The anticancer activity was evaluated *in vitro* using cell viability assay against new human hepatocellular carcinoma cell line (HepG-2) with a low half-maximal inhibitory concentration (IC_{50}) value of 24.6 μg/ml. In addition, the studied mechanism of caulerpin binding with two nucleobases (T/U) could be the first step for future experimental studies to understand and develop the performance of caulerpin as a drug candidate (Abdelrheem et al., 2020a). Caulerpin was also isolated by Moawad (2017) from the red alga *Pterocladiella capillacea*, for the first time, as it was found to be effective against melanoma cancer cell lines (SK-MEL-28) with IC_{50} of 3.98 μg/ml (Moawad, 2017).

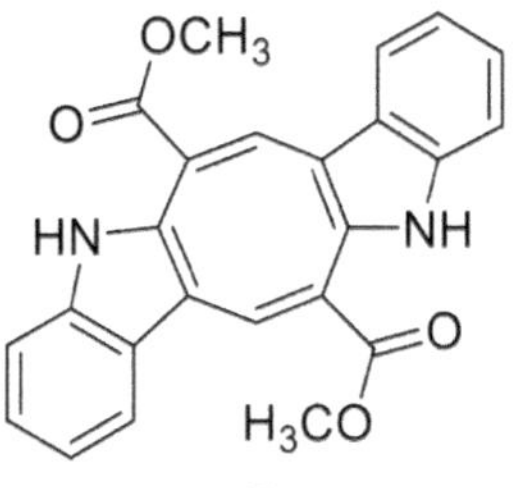

10

The fractionation of extracts from the brown alga, *Sirophysalis trinodis* (as *Cystosira trinodis*), from the Red Sea, provided nine compounds; octacosanoic acid (11), glyceryl trilinoleate (12), oleic acid (13), the epimeric mixture of saringosterols (14,15), *β*-sitosterol (16), glycoglycerolipid (17), and a mixture of kjellmanianone and loliolide (18,19). The marked activity was observed for the isolates mixture (18 and 19) against HepG-2 cell lines with an IC_{50} value of 7.27 μg/ml. Molecular docking confirmed that compounds 18 and 19 strongly interact with tyrosine-specific protein kinase and vascular endothelial growth factor receptor as significant inhibitors which supported the high cytotoxicity against human hepatocellular carcinoma (Ahmed et al., 2020).

Antiproliferative activities of extracts from the brown algae *Turbinaria ornata* and *P. pavonica* were detected against colon cancer cell line (HCT-116) with IC_{50} values of 5.1 and 47.6 μg/ml, respectively (Mahmoud et al., 2015).

A study on water extracts of *P. pavonica* and *Hydroclathrus clathratus* resulted in the isolation of mannitol, other carbohydrates of low-molecular-weight, fucose-containing polysaccharides, laminarin, and alginic acid (Awad et al., 2009a). The hot water polysaccharide extract of *P. pavonica* exhibited *in vitro* cytotoxicity against HepG-2 with a ED_{50} of 0.84 μg/ml compared to reference Doxorubicin (ED_{50}= 0.80 μg/ml). In addition, fractions II and III obtained from hot aqueous polysaccharide extract of *P. pavonica* had potent cytotoxic activity against HepG-2 with ED_{50} of 0.9 μg/ml while Fractions I and V of *H. clathratus* were more potent with ED_{50} = 0.5 μg/ml in comparison with the reference Doxorubicin (ED_{50}= 0.8 μg/ml) (Awad et al., 2009a).

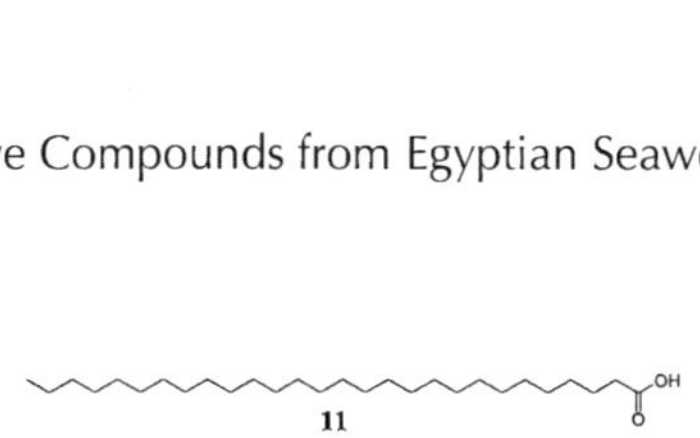
11

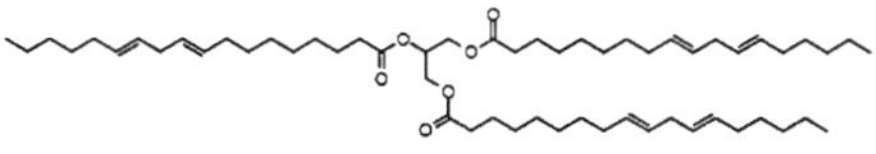
12

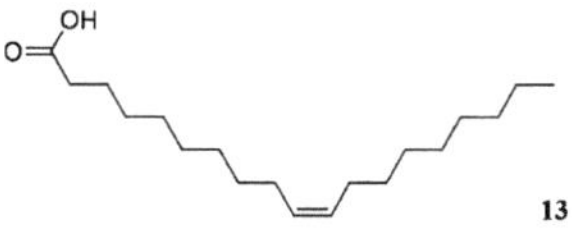
13

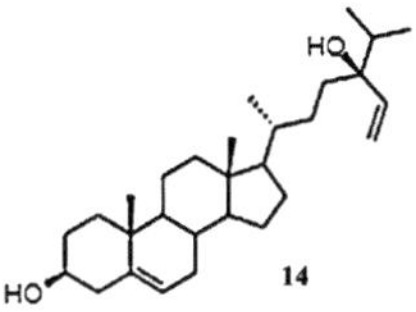
14

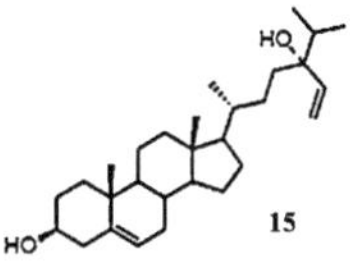
15

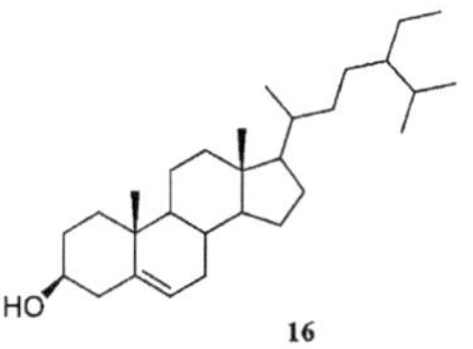
16

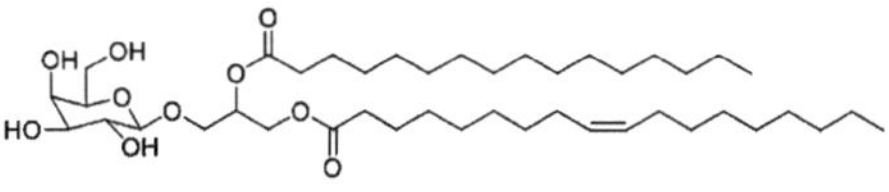
17

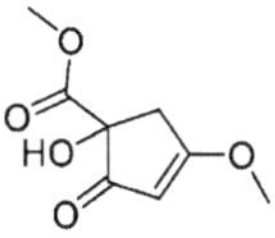
18

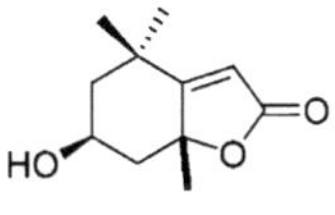
19

Another study on antihepatocarcinoma effects of polysaccharides extracts were carried out for brown algae, *Dictyopteris polypodioides* (as *D. membranacea*), *P. pavonica*, *Colpomenia sinuosa*, the green alga *Ulva intestinalis* (as *Enteromorpha intestinalis*), the red algae *Corallina officinalis*, *P. capillacea*, and *J. rubens*. Cold and hot water extracts of *J. rubens* (IC_{50}= 10.73 and 38.47 µg/ml) and hot water extract of *C. sinuosa* (IC_{50}= 38.75 µg/ml) showed cytotoxic activity against HepG-2 *in vitro* compared with reference drug Doxorubicin (IC_{50}= 29.58 µg/ml) (Matloub et al., 2015).

Four seaweed species *U. lactuca*, *U. intestinalis*, *P. capillacea*, and *J. rubens* exhibited variable molluscicidal and anthelminthic potentials against *Lymnaea natalensis* snails (intermediate host of *Fasciola gigantica*). The most potent extracts were from *P. capillacea* (LC_{50}= 111.3 µg/ml) followed by *U. lactuca* (LC_{50}= 121 μg/ml) and *J. rubens* (LC_{50}= 127 μg/ml) (Saad et al., 2019).

The cytotoxic activity of crude extracts of three macroalgae species (*Codium tomentosum*, *U. lactuca*, and *Hypnea musciformis*) was investigated on five human cell lines HepG-2, MCF-7 (breast carcinoma cell line), prostate cancer cell line (PC3), HCT-116, and HELA (cervix cancer cell line) (Hamza et al., 2015). A promising cytotoxic activity (inhibition % (I%): 70–80%) against HepG-2 and PC3 cancer cell lines were recorded for all crude extracts (Hamza et al., 2015). Furthermore, three fractions (hexane, ethyl acetate, and methanol) were obtained by successive fractionation of crude extracts. The ethyl acetate fractions of *U. lactuca, C. tomentosum,* and *H. musciformis* had IC_{50} values of 12.8, 24.5, and 17.8 μg/ml against HepG-2, respectively, compared to the positive control Doxorubicin (IC_{50}= 3.83 μg/ml) (Hamza et al., 2015). Methylene chloride/methanol fraction of the extracts from the introduced red alga *Asparagopsis taxiformis* collected from Marsa Matrouh (Mediterranean coast of Egypt), showed 100% inhibition of Erlich ascites carcinoma cells (El-Baroty et al., 2007).

Cytotoxic activities were tested for methanolic extracts of three macroalgae *H. tuna*, *Padina gymnospora*, and *Phacelocarpus tristichus* collected from the Red Sea against HepG-2 cells and MCF-7 cells (Abdelrheem et al., 2020b). *Padina gymnospora* showed higher cytotoxicity against HepG-2 and MCF-7 cell lines (IC_{50}= 9.80 and 11 μg/ml) than *P. tristichus* (IC_{50}= 21 and 29.3 μg/ml) and *H. tuna* (IC_{50}= 22.4 and 37.3 μg/ml), respectively. The significant cytotoxicity of *P. gymnospora* extract was attributed to the synergetic effects of bioactive metabolites with predominant fatty acids and sterols (Abdelrheem et al., 2020b).

The algal lipids of five marine macroalgae; *Palisada perforata* (as *Laurencia papillosa*), *Tricleocarpa cylindrica* (as *Galaxaura cylindrica*), *U. lactuca* (as *U. fasciata*), *Dictyota fasciola* (as *Dilophus fasciola*), and *Taonia atomaria*, exhibited potent inhibitory effects on both breast and human liver cancer cell lines with IC_{50} values ranged from 0.34 to 7.18 μg/ml (El Baz et al., 2014). The lipids of *T. atomaria* and *D. fasciola* showed significantly higher antitumor activity against MCF-7 (0.34 μg/ml) and HepG-2 (1.68 μg/ml) compared with that of Novantron antitumor drug (1.40 and 4.00 μg/ml), respectively (El Baz et al., 2014).

Five seaweed species collected from Abu-Qir bay were evaluated for cytotoxic assay against four tumor cell lines HepG-2, MCF-7, HELA, and PC3 cell lines. Seaweeds extracts that showed the highest antioxidant activity were selected for cytotoxic assay (Saeed et al., 2020). *Ulva lactuca* extract had strong activity against MCF-7 and HELA cell lines (IC_{50}= 10.83 and 12.43 μg/ml) compared with reference drug Doxorubicin (IC_{50}= 4.17 and 5.57 μg/ml), respectively, while *U. lactuca* (as *U. fasciata*) had strong activity against PC3 and HepG-2 cell lines (IC_{50}= 12.99 and 16.75 μg/ml) compared with reference drug Doxorubicin (IC_{50}= 8.87 and 4.50 μg/ml), respectively (Saeed et al., 2020).

Matloub et al. (2012) studied the insecticidal activity of extracts from the brown algae; *Sargassum asperifolium*, *S. dentifolium*, and *S. acinarium* (as *S. linifolium*) from the Red Sea. The hydroalcoholic extract of *S. dentifolium* had potent insecticidal activity against the moths *Spodopetra littoralis* and *Spodopetra frugiperda*, which are considered to be the most destructive agricultural lepidopterous pests, alone or in combination with baculovirus nuclear polyhedrosis virus (Matloub et al., 2012).

3.3.2 ANTIMICROBIAL ACTIVITY

Awad et al. (2003) studied the antimicrobial activity of 2 hydrocarbon fractions (F-I and F-II) obtained from *C. officinalis* (100 μg/disc) against eight microorganisms. The results revealed that the hydrocarbon fractions showed no antimicrobial activity against gram-negative bacteria (*Pseudomonas putida* and *Escherichia coli*) while the gram-positive bacteria (inhibition zone (IZ): 11–14 mm) and fungal strains (IZ: 8–17 mm) were sensitive by different levels to hydrocarbon fractions when compared to positive controls Ampicillin at concentration of 200 μg/disc (IZ: 12–25 mm) and Canestin at a concentration of 100 μg/disc (IZ: 14–25 mm), respectively. F-I showed relatively higher activity than F-II (Awad et al., 2003).

The crude extracts of the four seaweed species *U. lactuca*, *Ganonema farinosum* (as *Liagora farinosa*), *P. pavonica*, and *T. ornata* were investigated for antimicrobial and antimalarial capacity (Elsayed et al., 2012). Moderate antimicrobial activity of *P. pavonica* methanolic extract compared with other algal extracts was recorded against *Candida krusei, Cryptococcus neoformans*, *Staphylococcus aureus*, and methicillin-resistant *S. aureus*, with IC_{50} values of 22.3, 59.9, 18.9 and 16.6 μg/ml, respectively, whereas methanolic extracts of *U. lactuca* and *P. pavonica* displayed mild antimalarial activity against *Plasmodium falciparum* with IC_{50} values of 60.0 and 54.0 μg/ml for the D6 clone, and 66.0 and 56.0 μg/ml for the W2 clones, respectively (Elsayed et al., 2012).

El Zawawy et al. (2020) have investigated the suppressive effect of various seaweeds extracts; *Cladostephus spongiosus*, *P. perforata* (as *L. papillosa*), and *Codium arabicum*, collected from the Red Sea, on the virulence factors of isolated *Candida* including biofilm formation and hyphal growth in the blood that leads to candidemia infections. Acetone extract of *C. spongiosus* (AECS) showed a stronger anticandidal activity against *C. krusei* with minimum inhibitory concentration (MIC) and minimum fungicidal concentration of 80 and 320 μg/ml, respectively, compared to positive control Fuconazole which gave a fungicidal effect at very high concentration of 2000 μg/ml. Moreover, the reduction in the biofilm formation, as well as candidal Exo-polysaccharide layer, were detected and confirmed by down-regulating the expression of hyphal specific genes such as hyphal wall protein 1, Agglutinin-like protein 1, and fourth secreted aspartyl proteinase. Based on gas chromatography-mass spectroscopic (GC-MS), the three major compounds in AECS; 4-hydroxy-4-methyl-2-pentanone, n-hexadecanoic acid, and phenol,2-methoxy-4-(2-propenyl), might be responsible for the anticandidal activity.

The aqueous methanolic extracts of two red algae species, *Acanthophora spicifera*, and *Digenea simplex* were tested for antimicrobial activity against gram-positive, gram-negative bacteria, and fungal strains. Potent antimicrobial activity for both algal extracted at concentration of 10 mg/well was detected against *Streptococcus agalactiae* (IZ: 22.5 and 23.1 mm), *Streptococcus sanguis* (IZ: 20.6 and 21.2 mm), *Streptococcus pyogenes* (IZ: 21.4 and 21.6 mm), and *Corynebacterium diphtheria* (19.3 and 20.1 mm) if compared to positive control Ampicillin with IZ 22.3, 21.7, 22.7, and 20.0 mm, respectively. On the other hand, gram-negative bacteria, *Serratia marcescens* (IZ: 22.3 and 24.1 mm), and *Pseudomonas aeruginosa* (IZ: 22.3 and 22.9 mm) were sensitive to both algae extracts if compared to Gentamycin (IZ: 20.4 and 20.6 mm), respectively. Eventually, the effect of both *A. spicifera* and *D. simplex* extracts at a concentration of 1 mg/well showed the highest zone of inhibition (IZ: 23.8 and 22.4 mm) against fungal strain *Geotricum candidum* compared to Amphotericin B as a positive control (IZ: 20.3 mm) (Ibraheem et al., 2017).

Zaid et al. (2016) investigated the bioactive constituents responsible for the antiviral activity of two seaweeds; *Polycladia myrica* (as *Cystoseira myrica*) and *U. lactuca* against hepatitis A virus (HAV-H10), Coxsackie B4 virus, herpes simplex virus types-1 (HSV-1) and type-2 (HSV-1 and 2). The water extracts fractionation resulted in 13 and 15 fractions for *P. myrica* and *U. lactuca* with potent antiviral activity detected in fractions

7 and 10, respectively. The identified bioactive compounds were chloromethyl 2-(dodecahydro-4,6,7,9-tetrahydroxy-9a-methyl-1H-cyclopenta[α] naphtalen-3-yl)acetate (20) from *P. myrica* aqueous extracts and 2-((1E,3E)-4-chloro-2-meracaptobuta-1,3-dienyl)-1,2,3,4,4b,5,6,8a,9,10-decahydrophenanthrene3,5,10-triol (21) from *U. lactuca* (Zaid et al., 2016).

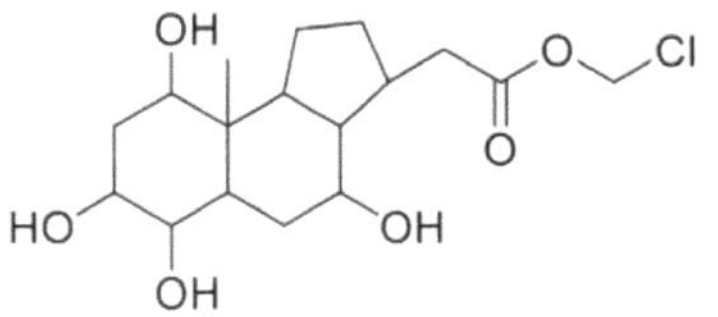

20

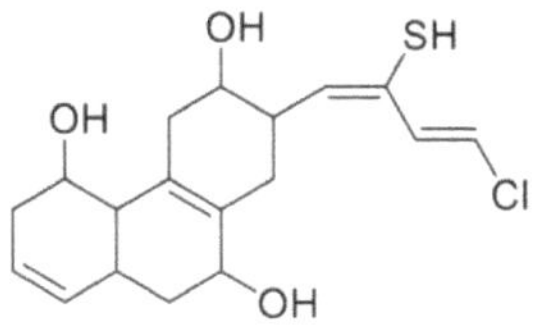

21

Hydrodistillation technique was used to extract volatile constituents from brown seaweeds *P. pavonica* and *H. clathratus* collected from the Red Sea (Awad et al., 2009b). The volatile constituents of both algae were composed of hydrocarbons, sesquiterpenes, alcohols, aldehydes, ketones, methoxy, acids, esters, phenol derivatives, and miscellaneous compounds (Awad et al., 2009b). The main component was Bis (2-ethylhexyl) phthalate (19.75 and 40.22%) for *P. pavonica* and *H. clathratus*, respectively. The volatile fraction of *P. pavonica* exhibited significant antimicrobial activity against *Bacillus cereus* (IZ: 14 mm), *P. aeruginosa* (IZ: 9 mm), and *Aspergillus niger* (IZ: 8 mm) compared to Amoxycillin as a reference drug (IZ: 10, 9, and 9 mm, respectively) (Awad et al., 2009b). On the other hand, *H. clathratus* showed pronounced antimicrobial activity against *Saccharomyces cerevisiae* (IZ: 12 mm), and *P. aeruginosa* (IZ: 8 mm) compared with positive control Canestin (IZ: 11 and 9 mm, respectively) (Awad et al., 2009b).

Caulerpin (10) isolated from *Sargassum platycarpum* exhibited moderate antifungal activity at 1 mg/well against *Cryptococcus neoformans* (IZ: 12 mm) and *Candida albicans* (IZ: 7 mm) when compared to positive control Ketoconazole (IZ: 25 and 21 mm, respectively) (Abdelrheem et al., 2020a).

3-O-*β*-D Glucopyranosyl-stigmasta-5,25-dien (22) was isolated from *U. lactuca* for the first time (Awad, 2000). The antimicrobial activity of the isolated compound was tested against 10 microorganisms. The dose of 200 μg/disc gave significant inhibition activities against *Streptococcus lactus* (IZ: 25 mm), *Bacillus subtilis* (IZ: 15 mm), and *Pseudomonas putida* (IZ: 12 mm) when compared to 200 μg/disc of Ampicillin (IZ: 25, 12, and 12 mm, respectively). Furthermore, compound 22 inhibited the growth of *Fusarium oxysporum* (IZ: 14 and 17 mm) and *S. cerevisiae* (IZ: 16 and 17 mm) at doses of 100 and 200 μg/disc which are nearly equivalent to 100 μg/disc of Canestin positive control (IZ: 14 and 15 mm), respectively (Awad, 2000).

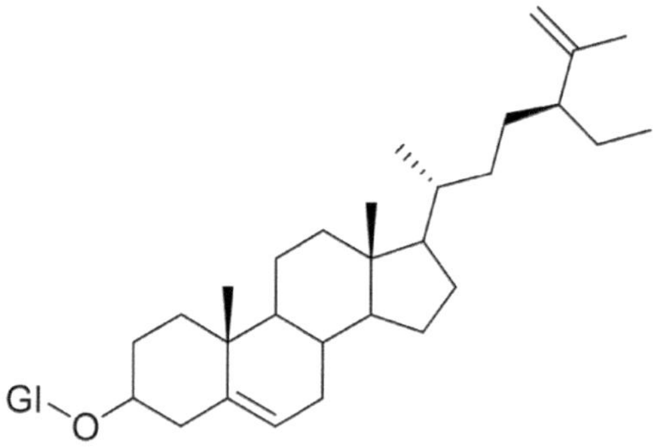

22

The antimicrobial activity of the red algae, *Treptacantha barbata* (as *Cystoseira barbata*) collected from the Red Sea, was evaluated against 10 microorganisms (Abdel-Raouf et al., 2017). The ethanolic extracts showed medium activity against *B. subtilis* (IZ: 16.00 mm), *S. aureus* (IZ: 13.30 mm), *S. marcescens* (IZ: 12.16 mm), and the fungus *C. albicans* (IZ: 10.33 mm) when compared to the positive control Norfloxcin (IZ: 26.00, 30.60, 26.00, and 20.00 mm) at a concentration of 10 mg/disc, respectively. High-performance liquid chromatography (HPLC) of the *T. barbata* ethanolic extract revealed the presence of 20 flavonoids, 20 phenolic compounds, and 10 saccharides (Abdel-Raouf et al., 2017). The crude extract and fractions of *Sirophysalis trinodis* (as *Cystoseira trinodis*) collected from the Red Sea shores between Quseir and Marsa-Alam, showed moderate inhibition activity against some of the tested microorganisms, especially the butanol fraction with the maximum IZ of 16.00 mm against *Salmonella typhimurium* at a concentration of 5 mg/ml (Ahmed et al., 2020).

The antibacterial activity of three different extraction solvents namely acetone, isopropyl alcohol, and ethanol of different seaweeds (*T. ornata*, *P. myrica*, and *P. pavonica*) was tested against three pathogens; gram-positive bacteria (*S. aureus*) and gram-negative bacteria (*P. aeruginosa* and *E. coli*) (Madkour et al., 2019). *Turbinaria ornata* had a high antibacterial activity with isopropyl alcohol extract against *S. aureus* (IZ: 15.00 mm) and *E. coli* (IZ: 14.75 mm), with acetone extract against *P. aeruginosa* (IZ: 14.50 mm), and with ethyl alcohol extract against *S.*

aureus (IZ: 15.00 mm), while, isopropyl alcohol extract of *P. myrica* showed good antimicrobial activity against *P. aeruginosa* (IZ: 15.50 mm) (Madkour et al., 2019).

Mohamed and Saber (2019) studied the in-vitro antifungal activities of different crude extracts (methanol, ethyl acetate, chloroform, and petroleum ether) of the brown seaweed *Hormophysa cuneiformis*, collected from the Red Sea coast, against 8 pathogenic fungi, including human and plant pathogens. The chloroform extract exhibited a potential antifungal activity against all tested fungal isolates (IZ: 12.00–21.67 mm) while the other extracts had relatively no remarkable effects compared to antifungal drug Amphotericin B (14.00–24.67 mm). The MIC of chloroform extract exhibited potent and significant activity against 8 fungal strains ranged between 0.78 and 6.25 μg/ml compared to standard antifungal drug Amphotericin B (MIC: 0.63–5.00 μg/ml) (Mohamed and Saber, 2019).

Four seaweed species, *Ceramium rubrum*, *Sargassum vulgare*, *Sargassum fusiforme*, and *P. pavonica* were collected from the Red Sea to investigate the antibacterial activity of diethyl ether, methanol, ethanol, and chloroform extracts against ten multidrug- resistant clinical isolates of gram-positive and gram-negative bacteria (El Shafay et al., 2016). Among the four seaweeds, the 100 μl diethyl ether extract of brown alga *S. fusiforme* and 50 μl ethanol extract of *S. vulgare* showed more inhibitory activity against *S. aureus* isolate 2 and *Klebsiella pneumoniae*, respectively. The transmission electron microscopy showed the effect of *S. fusiforme* and *S. vulgare* extracts on the morphology of *S. aureus* isolate 2 and *K. pneumoniae*, including rupture and perforation of the cell wall, leakage of cytoplasmic contents, distortion of the outer cell shape, and decreased cell size (El Shafay et al., 2016).

The antimicrobial activity of different extraction solvents was evaluated for ten marine macroalgae (*U. lactuca*, *C. racemosa*, *A. spicifera*, *Galaxaura elongata*, *G. farinosum*, *Cystoseira compressa*, *P. myrica*, *H. clathratus*, *T. ornata,* and *P. pavonica*), collected from the Red Sea, against 3 gram-positive, 3 gram-negative bacteria, 3 fungal strains, and Newcastle sense virus (NDV) (Ibraheem et al., 2012). The most effective solvent for extraction was diethyl ether followed by acetone. Broad-spectrum antibacterial activity was observed for *U. lactuca* and *C. racemosa* followed by two red algal species *G. elongata* and *A. spicifera* (Ibraheem et al., 2012). Regarding antifungal activity, *C. albicans* was significantly inhibited by acetone extract of *U. lactuca*, *G. elongata*, and *G. farinosum* (Ibraheem et al., 2012). Three acetone extracts of *G. farinosum*, *G. elongata*, and

T. ornata showed strong inhibitory activity against *F. oxysporum* (Ibraheem et al., 2012). In addition, the tested algae exhibited antiviral activity against NDV, which may be attributed to their content of polysaccharides that block viral adsorption point in the cell membrane of the host cell (Ibraheem et al., 2012).

Four macroalgae (*Dictyota* sp., *Sarcodiotheca furcata*, *P. myrica*, and *Sargassum ramifolium*) from the Red Sea coast were tested for their antimicrobial activities (El-Sheekh et al., 2014). Algal extracts were more effective against gram-negative bacteria (IZ: 0–33 mm) than gram-positive bacteria (IZ: 0–17 mm), and fungus (IZ: 0–15 mm) (El-Sheekh et al., 2014). *Sarcodiotheca furcata* showed the highest antibacterial activity against the tested gram-negative bacteria at 15 mg/well with IZ: 0–33 mm, where the extract was more effective in the growth inhibition of *Shigella flexneri* (IZ: 33 mm) followed by *E. coli* (IZ: 21 mm), *K. pneumonia*e (IZ: 15 mm), and *S. typhimurium* (IZ: 14 mm) (El-Sheekh et al., 2014). Concerning gram-positive bacteria, *S. aureus* isolates 1 was more susceptible to algae extracts of *Dictyota* sp. and *S. furcata* with an IZ of 17 mm at 15 mg/well. The antifungal activity was only observed against *C. albicans* (IZ: 9–15 mm). Gram-positive bacteria were affected by MICs (from 200–300 mg/ml) of the algal extracts, in addition, effective MIC values of 50–200 mg/ml were recorded against the tested gram-negative bacteria (El-Sheekh et al., 2014).

The antibacterial activity of *J. rubens*, *Ellisolandia elongata* (as *Corallina mediterranea*), and *P. capillacea* from Alexandria were analyzed against human pathogenic bacteria *Vibrio fluvialis* (Mohy El-Din and El-Ahwany, 2016). The extracts from *P. capillacea* were found to be the most effective against bacterial growth, where the highest zone of inhibition (3 mm) was recorded in the methanol extract at a concentration of 50 μl (Mohy El-Din and El-Ahwany, 2016).

Hamza et al. (2015) studied the antimicrobial activity of three algal extracts (*C. tomentosum*, *U. lactuca*, and *H. musciformis*) against 8 bacteria and 2 fungal strains. Both extracts of *H. musciformis* and *U. lactuca* showed a wide spectrum of antibacterial activity and the highest activity was against *K. pneumoniae* (IZ: 12.0 mm), and *E. coli* (IZ: 10.5 mm) at a concentration of 10 μl of algae extract.

Extracts of the introduced alga *A. taxiformis*, collected from the Mediterranean Coast of Egypt, showed the highest activity against influenza A virus subtype H5N1 cells (I%: 100%), respectively, (El-Baroty et al., 2007). Water fraction showed antialgal, antibacterial, and antifungal activities, while other fractions (hexane, petroleum ether, and methylene chloride: methanol) had only antibacterial and antifungal activities (El-Baroty et al., 2007).

Three gram-positive, 3 gram-negative bacteria, and 3 fungal strains were used to test the antimicrobial activity of three macroalgae; *H. tuna, P. gymnospora*, and *P. tristichus*, collected from Egyptian Red Sea shores (Quseir and Marsa Alam). The methanol extract of *P. gymnospora* exhibited both the highest values and the broadest spectrum of antimicrobial activity against tested fungal and bacterial pathogens (Abdelrheem et al., 2020b).

The antimicrobial potential of the crude lipids from *P. perforata, Tricleocarpa cylindrica* (as *Galaxaura cylindrica*), *U. lactuca, D. fasciola*, and *T. atomaria* was investigated (El Baz et al., 2014). Lipid extract of *P. perforata* had antifungal effect against *A. niger* with IZ of 12.0 mm, followed by *T. cylindrica* (IZ: 10.0 mm) at lipids concentration of 100 μg/well. Lipid extracts of *P. perforata* and *T. atomaria* induced an anti-yeast effect against *C. albicans* with IZ values of 16.0 and 15.0 mm, respectively. However, lipid extracts of *U. lactuca* and *D. fasciola* had moderate activity against both microbes. *Dictyota fasciola* lipids had the maximum antibacterial activity against *E. coli* (IZ: 13.0 mm) followed by *T. atomaria* (IZ: 11.0 mm). The maximum inhibition of HSV-1 was observed for the lipid extract of *T. atomaria* (I%: 74.40%) followed by *D. fasciola* (I%: 72.00%), *T. cylindrica* (I%: 72.00%), and *U. lactuca* (I%: 53.12%) at a concentration of 10 μg/ml (El Baz et al., 2014).

Saeed et al. (2020) studied the antimicrobial effect of five macroalgae. Chloroform extract of *U. lactuca* exhibited the highest inhibition zones (IZ: 15–24 mm) against the tested pathogenic bacteria (*K. pneumoniae* and *Proteus mirabilis*) and fungi (*Aspergillus flavus, A. fumigatus, and A. niger*) (Saeed et al., 2020). The antimicrobial activity of the tested seaweeds may be due to the damaging effect of the plasma membrane of the microbial cell (Smit, 2004; Saeed et al., 2020).

3.3.3 HYPOLIPIDEMIC ACTIVITY

The hypolipidemic activity of different extracts (alcohol, hexane, and fatty acids) of *C. officinalis* was evaluated *in vivo* on male albino rats in a dose of 400 mg extract/kg rat body weight (lower than the LD_{50} by 1/10) (Awad et al., 2003). A reduction in serum cholesterol, triglyceride, and low-density lipoprotein (LDL) in test organisms was observed from the normal control group by 15–41%, 42–58%, and 43–58% after extracts administration, respectively (Awad et al., 2003). In addition, the ratio of high-density lipoprotein (HDL)-cholesterol to LDL-cholesterol was increased by five-, four-, and two times after subcutaneous administration of alcohol, hexane, and fatty acids extracts (400 mg/kg) with respect to the normal control group, respectively. This increase in the HDL-cholesterol to LDL-cholesterol ratio is of great importance in the prevention of atherosclerosis. Additionally, these ratios are reliable risk assessments (Awad et al., 2003). A study on *C. sinuosa*, *D. polypodioides*, and *U. intestinalis* polysaccharides extract displayed an *in-vitro* hypolipidemic effect (Matloub et al., 2015).

The methanolic extracts of six macroalgal species (*C. sinuosa*, *S. acinarium*, *P. pavonica*, *P. capillacea*, *Osmundea pinnatifida* (as *Laurencia pinnatifida*), and *C. racemosa*) collected from Abu-Qir Bay (Mediterranean Sea) were tested for the anti α-glucosidase activity (Moawad, 2017). According to Moawad (2017), the most potent anti α-glucosidase activity was observed for *P. pavonica* with IC_{50} of 17.12 μg/ml, *S. acinarium* (IC_{50}= 71.75 μg/ml), *C. racemosa* (IC_{50}= 84.73 μg/ml), *P. capillacea* (IC_{50}= 92.16 μg/ml), *C. sinuosa* (IC_{50}= 112.44 μg/ml), and *O. pinnatifida* (IC_{50}= 115.11 μg/ml). Further purification steps resulted in the isolation of caulerpin (10) and 7 (*E*)-9-keto-octadec-7-enoic acid (23) from *P. capillacea* and 7 (*E*)-9-keto-hexadec-7-enoic acid (24) from *P. pavonica* for the first time. Similar anti α-glucosidase activity was observed for compounds 24 (IC_{50}= 7.1 μg/ml) and 10 (IC_{50}= 7.2 μg/ml), followed by 23 (IC_{50}= 14.3 μg/ml) (Moawad, 2017).

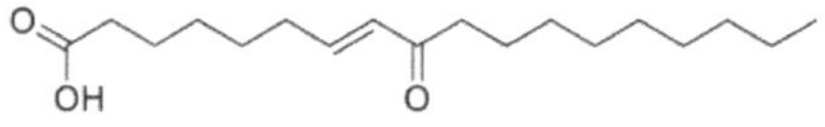

23

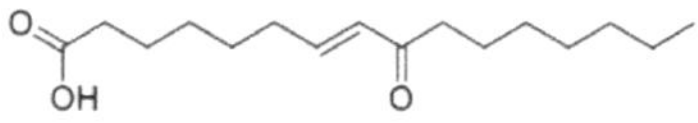

24

3.3.4 ANTINOCICEPTION ACTIVITY

The crude extracts of four seaweed species, *U. lactuca*, *G. farinosum*, *P. pavonica*, and *T. ornata* collected from Red Sea shores were investigated for binding affinity for human opioid receptors (Elsayed et al.,

2012). *Ulva lactuca* and *T. ornata* exhibited 64.7% and 52.9% binding affinity to δ-opioid receptors, respectively, while *P. pavonica* and *G. farinosum* extracts displayed 69.3% and 52.5% binding affinity to μ-opioid receptors. Opioid receptors are members of the superfamily of seven transmembrane-spanning G-protein coupled receptors and are divided into three types, μ (MOP), δ (DOP), and κ (KOP). Each opioid receptor type plays a role in antinociception, as well as other biological responses (Prisinzano, 2009).

Phytochemical analysis of these four seaweed species led to the isolation and identification of 10 compounds: fucosterol (25), stearic acid (26), palmitic acid (27), palmitoleic acid (28), oleic acid (13), myristic acid (29), *p*-hydroxybenzoic acid (30), β-sitosterol (16), glycerol-1-oleoyl-3-palmitoyl-2-galactoside (31), and glycerol-1,3-dioleyl (32) using a combination of spectroscopic methods, as well as by comparing with literature values (Elsayed et al., 2012). The compounds 31 and 32 had an affinity to bind human δ-opioid receptors with a capacity of 91.9 and 80.7%, respectively, while compound 32 showed 60.9 and 72.7% binding affinity to κ- and μ-receptors, respectively. The antimicrobial and antimalarial activities of the isolated compounds were tested and displayed no activities (Elsayed et al., 2012).

3.3.5 ANTI-INFLAMMATORY AND ANTIOXIDANT ACTIVITY

Two red seaweeds species, *A. spicifera* and *D. simplex* were collected from the west coastal area of the Red Sea (Ibraheem et al., 2017). The methanolic extracts of these two algal species exhibited similarly potent anti-inflammatory activity on inflamed liver rats treated with a dose of 500 mg/kg. Anti-inflammatory activity was indicative by retrieval for the levels of superoxide dismutase (SOD), catalase (CAT), glutathione (GSH), malondialdehyde (MDA), interleukin 6, and tumor necrosis factor α to normal levels in the control rats group (Ibraheem et al., 2017).

The topical anti-inflammatory activity of 3-O-β-D glucopyranosyl-stigmasta-5,25-dien (22) from the green alga *U. lactuca* was tested using the mouse ear edema assay as an experimental model of topical inflammation (Awad, 2000). A significant inhibition of 62.25 and 72.18% was observed when compared with 94.03% inhibition of the reference drug Indomethacin at a concentration of 1000 μg/ear (Awad, 2000).

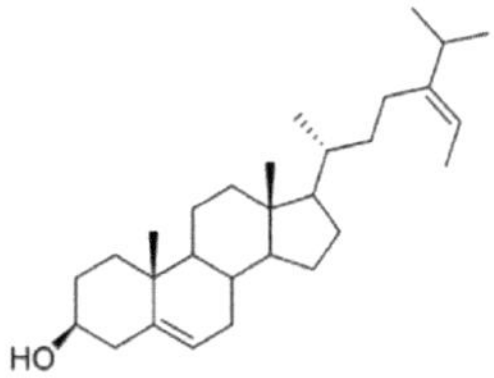

25

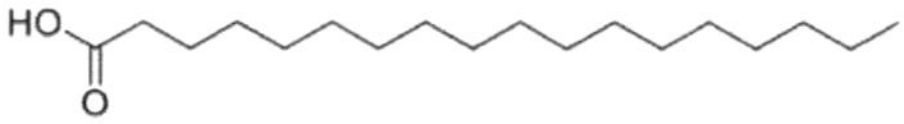

26

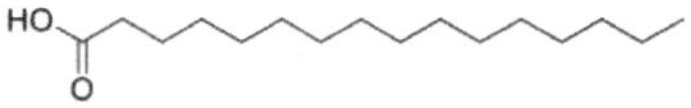

27

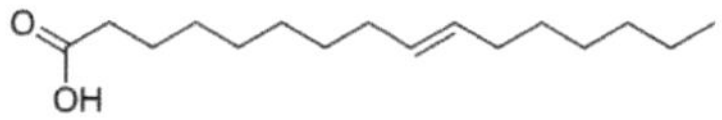

28

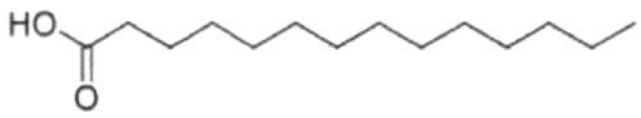

29

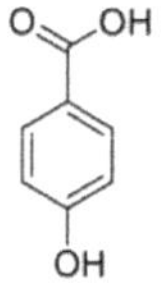

30

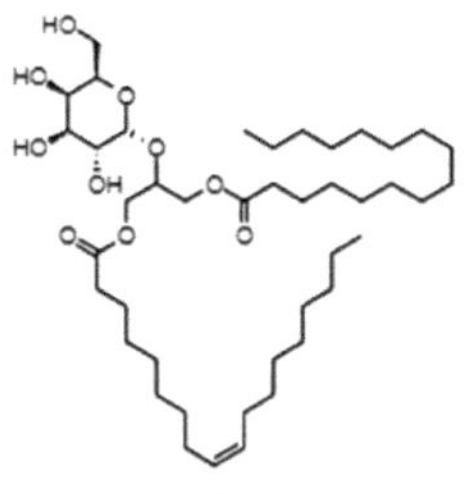

31

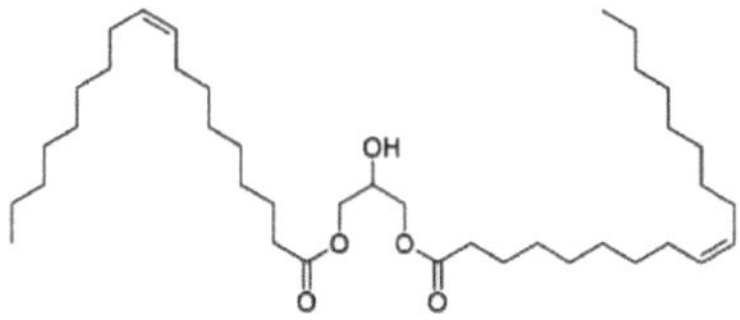

32

The protective effect of the brown seaweeds ethanolic extracts of *T. ornata* and *P. pavonica* collected from the Red Sea shores was studied against liver injury induced by the chemical carcinogen azoxymethane (AOM) on mice for thirteen weeks at 100 mg/kg of daily intake (Abdella et al., 2016). The potent antioxidant and anti-inflammatory effects of algae extracts were evidenced by a marked reduction in the elevated levels of aminotransferases activity, serum bilirubin concentration, MDA, and nitric oxide (NO), as well as a marked improvement in liver architecture of AOM induced liver injury group. One of the interesting observations is the increase in the level of serum albumin after seaweed administration which considered a rich source of protein, essential, and non-essential amino acids. Also, the increases in glutathione peroxidase (GPx) activity and peroxisome proliferator-activated receptor gamma (PPARγ) expression and the decrease in transcription factor nuclear factor kappa B (NF-κB) expression were observed after treating the AOM induced liver injury mice with seaweed extracts (Abdella et al., 2016).

Another study was conducted to determine the antiproliferative and protective efficacy of the same seaweeds (*T. ornata* and *P. pavonica*) on AOM-induced colon carcinogenesis in mice (Mahmoud et al., 2015). Administration of either algal extracts *T. ornata* or *P. pavonica* markedly alleviated the recorded alterations in the histological structure of the colon, lipid peroxidation (MDA) and NO, the declined GSH, and the reduced activity of SOD and GPx in AOM-induced mice. In addition, the protective effect of seaweed extracts was explicated by the upregulation of PPARγ and p53 mRNA expression, with concomitant downregulation of NF-κB (Mahmoud et al., 2015).

Polysaccharides from *J. rubens*, *C. sinuosa*, *C. officinalis*, and *P. pavonica* showed higher scavenging abilities against 2,2-diphenyl-1-picrylhydrazyl radical (DPPH) (9.10–58.97%) and NO (28.57–92.86%) than that of ascorbic acid and sodium nitrite standards at the same concentrations, respectively (Matloub et al., 2015).

The antioxidant and anti-inflammatory activities of three algae species (*C. tomentosum*, *U. lactuca*, and *H. musciformis*) were determined in alloxan-induced diabetic rats (Hamza et al., 2015). Results showed that marked antioxidant effect was observed for all methanol/methylene chloride extracts, especially *H. musciformis* (60.18%) compared to the reference drug vitamin E. Moreover, it showed the maximum anti-inflammatory activity in carrageenan-induced rat paw edema at 100 mg/kg of algae extract (Hamza et al., 2015).

The lipids from *P. perforata*, *T. cylindrica*, *U. lactuca*, *D. fasciola*, and *T. atomaria* exhibited moderate scavenging activity toward DPPH radical. The highest activity was found in lipids fraction of *D. fasciola* (El Baz et al., 2014). 9The antioxidant activity may be due to the presence of lipophilic antioxidants (total carotenoids and α-tocopherol, phenolic, terpenoid compounds, and polyunsaturated fatty acids) (El Baz et al., 2014).

Shabaka and Moawad (2021) studied the antioxidant activity of different extracts from the newly introduced red alga *Grateloupia gibbesii* in the Mediterranean Sea. Maximum antioxidant activity was observed for chloroform extract (199.31 μM ascorbic acid equivalent/g), followed by ethyl acetate extract (73.33 μM/g), and aqueous extract (30.85 μM/g).

3.3.6 SOIL FERTILIZER

The addition of the powder of the red alga *A. taxiformis* to the soil, as a biofertilizer, gave a significant increase in the growth of *Vicia faba* (plant height, number of internodes, main stem diameter, number of lateral branches, number of leaves per plant, number of pods per plant, seed yield, and number of seeds per plant) (El-Baroty et al., 2007).

The effect of water extracts of *S. dentifolium* and *P. gymnospora* on faba bean and wheat grown in salinized conditions was studied (Shaddad et al., 2014). Positive effects of seeds soaking in 1% seaweed water extracts was indicated by a significant increase in growth, photosynthetic pigments, and the activity of antioxidant enzymes. Higher activity was observed for *S. dentifolium* water extract. Marine algal extracts may act to reduce the salinity stress in faba bean and wheat plants by decreasing permeability of plasma membranes and membrane lipid peroxidation, and by maintaining membrane functionality (Shaddad et al., 2014).

3.3.7 ANTIFOULING ACTIVITY

Seven Seaweed species collected from Abu Qir Bay (Alexandria) were tested for their antifouling activity (Ibrahim et al., 2019). The most effective antifouling activity with the highest suppressive effect on microbial growth of 99.8 and 100% was observed for *U. lactuca* methanolic extract after immersion in seawater for several weeks. The tested algal extracts were arranged according to their antifouling activity as *U. lactuca*, *E. elongata*, *C. tomentosum*, *P. capillacea*, *C. sinuosa*, and *J. rubens* (Ibrahim et al., 2019).

Five percent of *U. lactuca* free-lipid coated-glass panel showed the most suppressive effect on *S. aureus* with an inhibition of 86.2 and 82.6% after 9 and 20 days of seawater immersion, respectively when compared to the binder-coated glass panels (Tadros et al., 2009).

The glass-coated panels with *U. lactuca* and algae free lipid containing binder together with the blank (binder coated glass) were immersed in natural seawater. The inspection of the coated surfaces after 6 months of immersion, showed good performance of the surfaces coated with *U. lactuca* and algae-free-lipids without attached epiphytes (Tadros et al., 2006).

3.4 CONCLUSION

About 43 seaweed species along the Mediterranean and the Red Sea coasts of Egypt were studied for their cytotoxic, antimicrobial, hypolipidemic, antinociception, anti-inflammatory, antioxidant, and antifouling activities. Most studies focused mainly on the antibacterial activity as well as crude extract rather than purified isolated compound. Moreover, providing DNA sequencing and depositing voucher samples in registered herbaria for the used seaweed species should be mandatory in order to provide applicable and useful resources.

KEYWORDS

- **anti-inflammatory activity**
- **azoxymethane**
- **brown seaweeds**
- **green seaweeds**
- **hepatitis A virus**
- **hypolipidemic activity**
- **Mediterranean Sea coast of Egypt**
- **Newcastle sense virus**
- **nuclear factor kappa B**
- **red seaweeds**

REFERENCES

Abdella, E. M., Mahmoud, A. M., & El-Derby, A. M., (2016). Brown seaweeds protect against azoxymethane-induced hepatic repercussions through up-regulation of peroxisome proliferator-activated receptor gamma and attenuation of oxidative stress. *Pharmaceutical Biology, 54*(11), 2496–2504.

Abdel-Raouf, N., Mohamed, H. M., Mostafa, S. S., & Ibraheem, I. B. M., (2017). Controlling of microbial growth by using *Cystoseira barbata* extract. *Egyptian Journal of Botany, 57*(3), 469–477.

Abdelrheem, D. A., El-Mageed, H. R. A., Mohamed, H. S., Rahman, A. A., Elsayed, K. N. M., & Ahmed, S. A., (2020a). Bis-indole alkaloid caulerpin from a new source *Sargassum platycarpum*: Isolation, characterization, *in vitro* anticancer activity, binding with

nucleobases by DFT calculations and MD simulation. *Journal of Biomolecular Structure and Dynamics, 39*(14):5137-5147. doi: 10.1080/07391102.2020.1784285.
Abdelrheem, D. A., Rahman, A. A., Elsayed, K. N. M., & Ahmed, S. A., (2020b). GC/MS spectroscopic approach, antimicrobial activity and cytotoxicity of some marine macroalgae from Qusier and Marsa Alam Seashore (Red Sea), Egypt. *Egyptian Journal of Aquatic Biology and Fisheries, 24*(3), 125–144.
Aguilera, J., Dummermuth, A., Karsten, U., Schriek, R., & Wiencke, C., (2002). Enzymatic defenses against photooxidative stress induced by ultraviolet radiation in arctic marine macroalgae. *Polar Biology, 25*(6), 432–441.
Ahmed, S. A., Rahman, A. A., Elsayed, K. N. M., El-Mageed, H. R. A., Mohamed, H. S., & Ahmed, S. A., (2020). Cytotoxic activity, molecular docking, pharmacokinetic properties and quantum mechanics calculations of the brown macroalga *Cystoseira trinodis* compounds. *Journal of Biomolecular Structure and Dynamics.* doi: 10.1080/07391102.2020.1774418.
Aleem, A. A., (1978a). Contributions to the study of the marine algae of the Red Sea. I. The algae in the neighborhood of Al-Ghardaqa, Egypt (Cyanophyceae, Chlorophyta, Phaeophyta). *Bulletin of Faculty of Science, K.A.U. Jeddah, 2,* 73–88.
Aleem, A. A., (1978b). Contributions to the study of the marine algae of the Red Sea. II. The algae in the neighborhood of Al-Ghardaqa, Egypt (Rhodophyta). *Bulletin of Faculty of Science, K.A.U. Jeddah, 2*, 89–98.
Aleem, A. A., (1980). Contributions to the study of the marine algae of the Red Sea. IV. The algae and seagrasses inhabiting the Suez Canal (Systematic part). *Bulletin of Faculty of Science, K.A.U. Jeddah, 4*, 31–89.
Awad, N. E., (2000). Biologically active steroid from the green alga *Ulva lactuca. Phytother. Res., 14*, 641–643.
Awad, N. E., (2004). Bioactive brominated diterpenes from the marine red alga *Jania rubens* (L.) Lamx. *Phytother. Res., 18*, 275–279.
Awad, N. E., Motawe, H. M., Selim, M. A., & Matloub, A. A., (2009a). Antitumourigenic polysaccharides isolated from the brown algae, *Padina pavonica* (L.) Gaill. and *Hydroclathrus clathratus* (C. Agardh) Howe. *Medicinal and Aromatic Plant Science and Biotechnology, 3*(1), 6–11.
Awad, N. E., Motawe, H. M., Selim, M. A., & Matloub, A. A., (2009b). Volatile constituents of the brown algae *Padina pavonia* (L.) Gaill. and *Hydroclathrus clathratus* (C. Agardh) Howe and their antimicrobial activity. *Medicinal and Aromatic Plant Science and Biotechnology, 3*(1), 12–15.
Awad, N. E., Selim, M. A., Metawe, H. M., & Matloub, A. A., (2008). Cytotoxic xenicane diterpenes from the brown alga *Padina pavonica* (L.) Gaill. *Phytother. Res., 22*, 1610–1613.
Awad, N. E., Selim, M. A., Saleh, M. M., & Matloub, A. A., (2003). Seasonal variation of the lipoidal matters and hypolipidaemic activity of the red algae *Corallina officinalis* L. *Phytother. Res., 17*, 19–25.
Cornish, M. L., & Garbary, D. J., (2010). Antioxidants from macroalgae: Potential applications in human health and nutrition. *Algae, 25*, 1–17.
El Baz, F. K., El-Baroty, G. S., Ibrahim, A. E., & Abd El, B. H. H., (2014). Cytotoxicity, antioxidants and antimicrobial activities of lipids extracted from some marine algae. *Journal of Aquaculture Research and Development, 5*, 284. doi: 10.4172/2155-9546.1000284.
El Shafay, S. M., Ali, S. S., & El-Sheekh, M. M., (2016). Antimicrobial activity of some seaweeds species from Red Sea, against multidrug resistant bacteria. *The Egyptian Journal of Aquatic Research, 42*(1), 65–74.

El Zawawy, N. A., El-Shenody, R. A., Ali, S. S., & El-Shetehy, M., (2020). A novel study on the inhibitory effect of marine macroalgal extracts on hyphal growth and biofilm formation of candidemia isolates. *Scientific Reports, 10*, 9339. doi: 10.1038/s41598-020-66000-1.

El-Baroty, G. S., Moussa, M. Y., Shallan, M. A., Ali, M. A., Sabh, A. Z., & Shalaby, E. A., (2007). Contribution to the aroma, biological activities, minerals, protein, pigments and lipid contents of the red alga: *Asparagopsis taxiformis* (Delile) Trevisan. *Journal of Applied Sciences Research, 3*(12), 1825–1834.

El-Din, S. M. M., & El-Ahwany, A. M. D., (2016). Bioactivity and phytochemical constituents of marine Red Seaweeds (*Jania rubens, Corallina mediterranea* and *Pterocladia capillacea*). *Journal of Taibah University for Science, 10*(4), 471–484.

Elsayed, K. N. M., Radwan, M. M., Hassan, S. H. M., & Abdelhameed, M. S., (2012). Phytochemical and biological studies on some Egyptian seaweeds. *Natural Product Communications, 7*(9), 1209–1210.

El-Sharouny, H. M., El-Tayeb, M. A., & Ismail, M. S., (2001). Macroalgae associated with mangrove at Hurghada and Safaga of the Egyptian Red Sea coast. *Journal of King Abdulaziz University: Marine Science, 12*(1), 241–251.

El-Sheekh, M. M., Gharieb, M. M., El-Sabbagh, S. M., & Hamza, W. T., (2014). Antimicrobial efficacy of some marine macroalgae of Red Sea. *International Journal of Microbiology and Immunology Research, 3*(3), 21–28.

Fouda, M. M., (2014). *Fifth National Report to the Convention of Biological Diversity* (p. 203). Egyptian Environmental Affairs Agency (EEAA), Nature Conservation Sector (NCS).

Galal, N., (1991). *Studies on the Coastal Ecology and Management of the Nabq Protected Area, South Sinai, Egypt.* PhD Thesis, University of York, England, p. 261.

Guiry, M. D., & Guiry, G. M., (2021). *Algae Base*. National University of Ireland, Galway. https://www.algaebase.org (accessed on 13 October 2021).

Hamza, E. H., Temraz, T. A., & Ahmed, S. A., (2015). Bioactivity of some Egyptian seaweeds extract. *Catrina, 11*(1), 17–25.

Ibraheem, I. B. M., Abd, E. B. E., Moawad, A., Hassan, H. M., Mohamed, W. A., & Abdel-Raouf, N., (2017). Antimicrobial and anti-inflammatory effects of two different marine red algae species collected from Quseir, the Red Sea, Egypt. *Asian Journal of Biology, 2*(2), 1–10.

Ibraheem, I. B. M., Abdel-Raouf, N., Abdel-Hameed, M. S., & El-Yamany, K., (2012). Antimicrobial and antiviral activities against Newcastle disease virus (NDV) from marine algae isolated from Qusier and Marsa-Alam Seashore (Red Sea), Egypt. *African Journal of Biotechnology, 11*(33), 8332–8340.

Ibrahim, A. M. M., Moawad, M. N., Attia, A. M., El-Naggar, M. M. A., & Tadros, H. R. Z., (2019). Production of antifouling paints using environmentally safe algal extracts on laboratory scale. *Egyptian Journal of Aquatic Biology and Fisheries, 23*(3), 171–184.

Madkour, F. F., El-Shoubaky, G. A., & Ebada, M. A., (2019). Antibacterial activity of some seaweeds from the Red Sea coast of Egypt. *Egyptian Journal of Aquatic Biology and Fisheries, 23*(2), 265–274.

Mahmoud, A. M., Abdella, E. M., El-Derby, A. M., & Abdella, E. M., (2015). Protective effects of *Turbinaria ornata* and *Padina pavonica* against azoxymethane-induced colon carcinogenesis through modulation of PPAR gamma, NF-κB and oxidative stress. *Phytother. Res., 29*(5), 737–748.

Matloub, A. A., Awad, N. E., & Khamiss, O. A., (2012). Chemical composition of some *Sargassum* spp. and their insecticidal evaluation on nucleopolyhedrovirus replication *in vitro* and *in vivo*. *Egyptian Pharmaceutical Journal, 11*(1), 53–58.

Matloub, A. A., Elsouda, S. S. M., El-Senousy, W. M., Hamed, M., Aly, H., Ali, S. A., Mohammed, R. S., et al., (2015). *In vitro* antiviral, cytotoxic, antioxidant and hypolipidemic activities of polysaccharide isolated from marine algae. *International Journal of Pharmacognosy and Phytochemical Research, 7*(5), 1099–1111.

Moawad, M. N., (2017). *Biochemical Activities of Some Natural Products Extracted from Marine Macroalgae* (p. 194). PhD Thesis, Faculty of Science, Alexandria University, Egypt.

Mohamed, N. A., Ibraheem, B. I., & Mohamed, S. H., (2006). Seasonal occurrence and variations in macro-algal populations in Hurghada, Safaga and Qusier coasts of Red Sea, Egypt. *Proc.4th Int. Con. Biol. Sci. (Botany),* 189–197.

Mohamed, S. S., & Saber, A. A., (2019). Antifungal potential of the bioactive constituents in extracts of the mostly untapped brown seaweed *Hormophysa cuneiformis* from the Egyptian coastal waters. *Egyptian Journal of Botany, 59*(3), 695–708.

Negm, S. N., (1988). *Ecological and Phytochemical Studies on Some Marine Algae from the Red Sea* (p. 254). PhD Thesis, Cairo University, Egypt.

Prisinzano, T. E., (2009). Natural products as tools for neuroscience: Discovery and development of novel agents to treat drug abuse. *J. Nat. Prod., 72*(3), 581–587.

Rindi, F., (2008). *A Review of the Recent Literature on Medical and Pharmaceutical Uses of Seaweeds* (p. 98). Unpublished technical report, Marine Institute of Ireland.

Rindi, F., Soler-Vila, A., & Guiry, M. D., (2012). Taxonomy of marine macroalgae used as sources of bioactive compounds. In: Hayes, M., (ed.), *Marine Bioactive Compounds: Sources, Characterization and Applications* (pp. 1–53). Springer: Boston, MA, USA.

Saad, A. A., Hassanain, M. A., Darwish, E. M., Abdel-Rahman, E. H., Sleem, S. H., & Shaapan, R. M., (2019). Molluscicidal effect of some red and green marine algae on *Lymnaea natalensis*. *Egyptian Journal of Aquatic Biology and Fisheries, 23*(4), 483–490.

Saeed, A. M., Abotaleb, S. I., Alam, N. G., El Mehalawy, A. A., & Gheda, S. F., (2020). *In vitro* assessment of antimicrobial, antioxidant and anticancer activities of some marine macroalgae. *Egyptian Journal of Botany, 60*(1), 81–96.

Shabaka, S. H., (2018). Checklist of seaweeds and seagrasses of Egypt (Mediterranean Sea): A review. *The Egyptian Journal of Aquatic Research, 44*(3), 203–212.

Shabaka, S., & Moawad, M., (2021). Ecology and biochemical composition of a newly reported non-indigenous red alga, *Grateloupia gibbesii*, in the Mediterranean Sea, with reference to edible Red Seaweeds. *Regional Studies in Marine Science, 44*, 101767. https://doi.org/10.1016/j.rsma.2021.101767.

Shaddad, M. A. K., Farghl, A. A. M., Galal, H. R., & Hassan, E. A., (2014). Influence of seaweed extracts (*Sargassum dentifolium* or *Padina gymnospora*) on the growth and physiological activities of faba bean and wheat plants under salt stress. *Egyptian Journal of Botany, 54*(1), 121–135.

Smit, A. J., (2004). Medicinal and pharmaceutical uses of seaweed natural products: A review. *Journal of Applied Phycology, 16*(4), 245–262.

Tadros, A. B., Ibrahim, H. A., Zaki, H. R., El-Naggar, M. M., & Abbas, A. E., (2009). Suppressive effect of coated surfaces contain *Ulva lactuca* free lipid on *Staphylococcus aureus* ATCC 6538. *The Egyptian Journal of Aquatic Research, 35*(4), 405–412.

Tadros, A. B., Zaghoul, F. A., Zaki, H. R., Abbas, A. E., & Kandeel, M. M., (2006). Impact of coatings containing algae on some seawater parameters and epiphytes plankton. *The Egyptian Journal of Aquatic Research, 32*, 24–37.

Van De, P. W. H., Bischof, K., Buma, A. G., & Breeman, A. M., (2003). Habitat related variation in UV tolerance of tropical marine red macrophytes is not temperature dependent. *Physiol. Plant, 118*(1), 74–83. doi: 10.1034/j.1399-3054.2003.00090.x.

Zaid, S. A. A., Hamed, N. N. E., Abdel-Wahab, K. S. E., El-Magd, E. K. A., & El-Din, R. A. S., (2016). Antiviral activities and phytochemical constituents of Egyptian marine seaweeds (*Cystoseira myrica* (S.G. Gmelin) C. Agardh and *Ulva lactuca* Linnaeus) aqueous extract. *The Egyptian Journal of Hospital Medicine, 64*(1), 422–429.

CHAPTER 4

RECENT ADVANCES AND RESEARCH CHALLENGES IN BIOPROSPECTING OF BROWN SEAWEEDS

SREEKALA KANNIKULATHEL GOPIDAS and
NAGARAJ SUBRAMANI

Centre for Advanced Studies in Botany, University of Madras, Guindy Campus, Chennai – 600025, Tamil Nadu, India

ABSTRACT

The brown algae (Phaeophyta) are estimated to be comprised of 1,500 to 2,000 species which is almost entirely marine, as compared to the green algae (Chlorophyta), comprising of 6,000 to 8,000 species out of which 90% are freshwater rather than marine, and the red algae (Rhodophyta) comprising of approximately 4,000 to 5,000 species of which 90% are marine. While the red algae are used commonly as sources of industrial gelling agents and as feed in aquaculture, and the green algae for consumption, the brown algae are harnessed for extraction of many valuable chemicals of pharmaceutical, nutraceutical, and cosmeceutical significance. There are numerous reports on the functions and utilities of brown seaweeds from countries worldwide, and seaweeds have become one of the most celebrated entities in the modern world. Although intense research is going on in various aspects of seaweeds, the release of finished products or drugs is limited to a few, due to some key challenges that need to be addressed while processing these seaweeds. This chapter discusses the bioactive compounds obtained from brown seaweeds, their processing, economic significance, recent research advances leading to the effective utilization of seaweeds, and the key challenges involved in the successful bioprospecting of these seaweeds.

4.1 INTRODUCTION

Seaweeds are marine, benthic macroalgae that are multicellular and form complex macrothalli, visible as attached on rocks on the shore or as the giant kelps in the bottom of the sea. They are lower cryptograms, categorized into three phyla based on their pigmentation as, the Phaeophyceae or brown seaweeds, Rhodophyceae or red algae, and Chlorophyceae or green algae. They constitute approximately 10,000 species which are diverse in forms and sizes. There are about 1500 brown, 4000 red, and 900 green seaweed species worldwide. Red and green seaweeds are commonly found in the tropical and subtropical waters while the brown mostly occupies the cool, temperate waters (Rengasamy et al., 2014). Altogether, about 7.5 to 8 billion tons of wet seaweeds are produced in the coastal regions around the world annually. They are a significant renewable resource in the marine ecosystem and have been a part of human civilization since ancient times. Although the first reports on seaweed utilization date back to 2500 years ago in Chinese literature, it was not until the Second World War that its use became popularized (Rao and Mantri, 2006).

Seaweeds are exposed to very harsh conditions like extreme salinity, temperature, light, radiation, nutrients, oxygen deficit, *etc*., in the marine environment. In response to such factors and to withstand extremities, the seaweeds develop a chemical defense, which results in the production of unique secondary metabolites (Garcia-Vaquero et al., 2017). These include pigments, polysaccharides, polyphenols, peptides, alkaloids, steroids, polyunsaturated fatty acids (PUFAs) like omega-3 fatty acids, minerals, and mycosporine-like amino acids (MAAs), some of which are exclusive to seaweeds. Seaweeds are the only natural resource for chemicals like agar, carrageenans, and alginates. These secondary metabolites possess diverse biological functions, as explored by the recent advances in the field of algal biotechnology. The growing interest in the commercial applications of seaweeds comes from the fact that they can be easily grown without the necessity for land, freshwater, or fertilizers. Due to increasing demands as raw material for food, pharmaceutical, fertilizer, cosmetics, and hydrocolloids industries, they have been under cultivation in about fifty countries in the world. They are either consumed as such or used as a starting material for the extraction of economically important chemicals and products (Charoensiddhi et al., 2017).

4.2 BIOACTIVE COMPOUNDS FROM BROWN ALGAE

Figure 4.1 demonstrates the various types of biochemical constituents present in brown macroalgae. The polysaccharides such as cellulose, alginic acid, fucoidan, and sargassans form the cell wall structural materials, and laminaran form the storage material in brown algae. Their physicochemical and biofunctional properties have invoked huge industrial and research interests. Alginate is an unbranched homo- or hetero- polymer made of mannuronic or glucuronic acid units, with a molecular weight ranging from 32000 to 400000 g mol^{-1}. Fucoidans are $\alpha(1\rightarrow2)$, $\alpha(1\rightarrow3)$, and/or $\alpha(1\rightarrow4)$ linked sulfated polysaccharides (PLS) with L-fucose as the main monosaccharide unit along with glucose, galactose, mannose, and xylose and sulfate groups at C_2 or C_4 position (Fernando et al., 2018). It is found in the extracellular matrix and performs functions like tissue hydration, cell wall integrity, osmotic adjustment, cell-to-cell communication, and defense mechanisms. Laminarin or Laminaran is a polysaccharide with β-(1–3)-glucan main chain and β-(1–6)-glucose branches of 20–25 units. It may be water-soluble or insoluble based on the level of branching (Kadam et al., 2015). Sargassan is made of glucuronic acid, galactose, and mannose residues, with neutral and partially sulfated residues of galactose, fucose, and xylose side chains, along with protein moieties. They have α-L-glycosidic and/or β-D-glycosidic linkages (Abdel-Fattah et al., 1974).

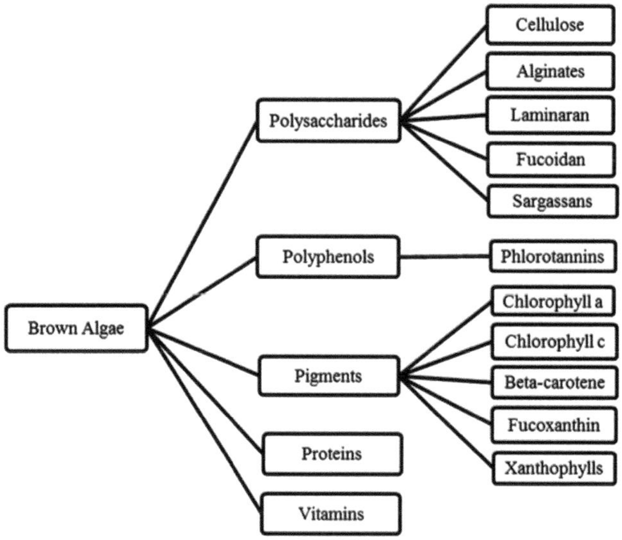

FIGURE 4.1 The biochemical constituents of brown seaweeds.

The main polyphenol present in brown algae is the phlorotannins, which are polymers of phloroglucinol or 1, 3, 5-trihydroxy benzene and make up to 5–12% of the DW of some algae. They are of three types as fucols, with aryl-aryl linkage; phlorethols and fuhalols, with aryl-ether linkage; and eckols and carmalols, with dibenzodioxin linkage. Around 150 types of phlorotannins have been isolated so far, and their molecular weight ranges from 126 to 6500 Daltons (Jégou et al., 2015).

The pigments of brown algae include chlorophyll a and c, β-carotene, xanthophylls, and fucoxanthin. Fucoxanthin is the major carotenoid produced by brown seaweeds, which imparts the peculiar brown color to the Phaeophyceae. It has both photosynthetic as well as protective roles (Gutiérrez-Rodríguez et al., 2018). Besides this, the brown seaweeds possess certain proteins, essential, and non-essential amino acids, vitamins such as vitamin A, B1, B2, B3, B6, B12, C, D, E, folic acid and pantothenic acid, and minerals (e.g., magnesium, calcium, potassium, sodium, iodine, phosphorus, chromium, nickel, selenium, manganese, zinc, iron, copper, cadmium, lead, mercury, and arsenic) that make them nutritional as well (Roohinejad et al., 2017).

4.3 PROCESSING OF BROWN SEAWEEDS

The fresh algal biomass collected from the marine sites is usually washed, dried, and milled before processing. Further, the milled biomass is sequentially treated with selected non-polar or less polar solvents such as hexane, chloroform, diethyl ether, methanol, ethanol, *etc.* to extract out pigments, lipids, and other non-polar compounds, and then with carefully optimized aqueous solutions or multi-solvent systems to extract out polyphenols and polysaccharides. The structural composition and the chemical characteristics of these components allow selected precipitation or extraction of the components. The recently emerged extraction techniques include enzyme-assisted extraction (EAE), ultrasound-assisted extraction (UAE) and microwave-assisted extraction (MAE). The post extracted crude compounds are further purified by ion-exchange chromatography, affinity chromatography, size exclusion chromatography, gel filtration, or membrane filtration techniques. The active fractions from chromatography are dialyzed against highly purified water to remove salts, proteins, and other smaller contaminants that may get along with the compound of interest during the extraction process, and then the lyophilized, purified compounds have to be characterized using spectroscopic techniques such as Fourier transform infra-red (FTIR) spectroscopy, ^{1}H and ^{13}C nuclear magnetic resonance (NMR) spectroscopy, and advanced chromatographic techniques such as high-performance liquid chromatography (HPLC), gas chromatography (GC), and ultra performance liquid chromatography quadruple time-of-flight (UPLC Q-TOF) after which the compounds are subjected to in-depth research following applications in various fields (Figure 4.2).

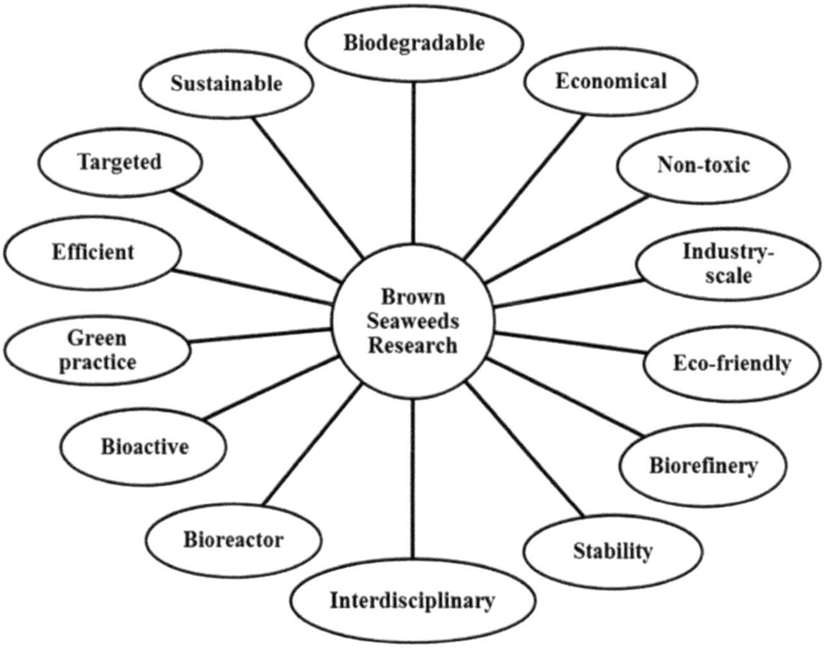

FIGURE 4.2 The core values to be adopted in brown seaweed research to achieve the best benefits.

4.4 ECONOMIC SIGNIFICANCE

The nutritional value of brown algae is higher than other algae due to the higher amounts of fucosterol, α-tocopherol, unsaturated fatty acids (UFAs), and polyphenol. Hence they are considered as one of the most favorable functional foods for improving human health. The Phaeophyta also exhibit a higher radical scavenging property and there exists a correlation between this activity and the amount of α-tocopherol and polyphenol in them (Ito et al., 2018).

The polysaccharides obtained from brown macroalgae are the source of hydrocolloids and find utility in food, pharmaceutical, and cosmeceutical industries. Alginates possess good gelling properties in the presence of calcium ions. While fucoidan possesses a broad range of bioactivities including antioxidant, anticancer, anti-inflammatory, immunomodulatory properties (Fernando et al., 2018). Among these, the anticancer property of fucoidan has achieved global attention due to its *in vitro* and *in vivo* activity against a broad range of cancer cell lines. For e.g., the fucoidan isolated from *Undaria pinnatifida* effectively curbed T47, SK-MEL-28, RPMI-7951, T47D, A549, HeLa, and PC-3 cancer cell lines (Gutiérrez-Rodríguez et al., 2018).

The phlorotannins have been reported for various biological activities *viz*., antioxidant, antimicrobial, anti-inflammatory, anticancer, antifungal, antiprotozoal, antimalarial, anti-arthritic, and antidiabetic, and so they are used as ingredients in drugs, diet supplements, nutraceuticals, cosmetics, fertilizers, and animal feeds (Wang et al., 2017). Phlorotannins from *Eisenia bicyclis*, *Ecklonia kurome*, *Ecklonia cava,* and *Hizikia fusiformis* exhibited potent antioxidant effects. The antioxidant property of phlorotannins is 2–10 folds superior to the commercial antioxidants like ascorbic acid, α-tocopherol and catechin (Suleria et al., 2016). The phlorotannins can also effectively inhibit cancer cells. Phlorotannins from *Ecklonia cava* inhibited MCF-7 and MDA-MB-231 cells while phlorotannin-rich extracts from *Ascophyllum nodosum* inhibited HT-29 colon cancer cells (Gutiérrez-Rodríguez et al., 2018).

The fucoxanthin from *U. pinnatifida* has a significant effect on prostate cancer cells, while its deacetylated derivative fucoxanthinol inhibits Jurkat cells. On the other hand, the fucoxanthin isolated from *Laminaria japonica* inhibited human colon adenocarcinoma WiDr cells while fucoxanthinol inhibited proliferation of drug-resistant HT-29-derived cells (Gutiérrez-Rodríguez et al., 2018). Apart from these conventional applications, the bioactives from brown algae have also found utility in several other novel products and arenas. The essential characteristics such as high availability, biocompatibility, better liquid absorbency, biodegradability, high strength, non-toxicity, alterability, and multiple bioactivities make these compounds ideal candidates for employment in specific areas of different biotechnological and biomedical applications. The divergent realms of applications of the Phaeophyceae and examples have been summed up in Table 4.1.

TABLE 4.1 The Fields of Application and Examples of Brown Algae

Field of application	Bioactives used	Source seaweeds	References
Pharmaceuticals, drugs			
Antioxidant	Fucoidan	*Ascophyllum nodosum*	Mohan et al., 2018
	Alginate	*Cystoseira barbata*	Mohan et al., 2018
Antimicrobial	Laminarin	*Laminaria hyperborea*	Mohan et al., 2018
Anticancer	Fucoidan	*Saccharina cichorioides*	Mohan et al., 2018
Anti-inflammatory	Fucoidan	*Cystoseira compressa*	Mohan et al., 2018
Immunomodulatory	Laminarin	*Laminaria digitata*	Cesário et al., 2018
Anticoagulant	Laminarin	*Laminaria digitata*	Centella et al., 2017
Antiviral	Fucoidan Laminaran	*Macrocystis pyrifera, Ascophyllum nodosum*	Deniaud-Bouët et al., 2017
Tissue engineering	Alginates	*Ecklonia maxima, Lessonia nigrescens*	Chandika et al., 2015
Foods, nutraceuticals			
Thickeners/ stabilizers	Alginates	*Laminaria hyperborean, Ascophyllum nodosum*	Centella et al., 2017
Preservatives	Phlorotannins	*Fucus vesiculosus*	Roohinejad et al., 2017
Functional foods	Alga	*Undaria pinnatifida*	Roohinejad et al., 2017
Packaging	Alginates	*Laminaria* spp.	Fernando et al., 2018
Dietary fiber	Laminarin	*Laminaria* spp.	Cesário et al., 2018
Diet supplements	Minerals and Vitamins	*Undaria pinnatifida, Fucus vesiculosus*	Cesário et al., 2018
Cosmetics			
Antioxidant and anti-inflammatory	Oligosaccharide	*Ecklonia cava, Ishige sinicola*	Anyanwu et al., 2018
Animal feed			
Antioxidant and antimicrobial	Fucoidan Laminarin	*Laminaria* spp.	Rajauria et al., 2016
Aquaculture	Alginates	*Laminaria digitata*	Mohan et al., 2018
Diet supplements	Fucoidan	*Cladosiphon okamuranus*	Mohan et al., 2018

TABLE 4.1 *(Continued)*

Field of application	Bioactives used	Source seaweeds	References
Agriculture			
Fertilizers and soil conditioners	Extracts	*Ascophyllum nodosum*, *Ecklonia cava*	Ditchburn and Carballeira, 2019
Bioremediation	Alga	*Saccharina latissima*	Anyanwu et al., 2018
Bio-indication	Alga	*Padina gymnospora*, *Dictyota bartayresiana*	Ditchburn and Carballeira, 2019
Bioengineering, biofuel			
Biomethane	Oils	*Ascophyllum nodosum*, *Macrocystis pyrifera*, *Saccharina latissima*	Anyanwu et al., 2018
Bioethanol	Oils	*Padina tetrastromatica*, *Laminaria hyperborean*, *Saccharina japonica*	Anyanwu et al., 2018
Biohydrogen	Alga	*Hizikia fusiforme*, *Laminaria japonica*, *Fucus serratus*	Chia et al., 2018
Bioplastics	Alginates	*Laminaria hyperborean*, *Ascophyllum nodosum*	Ditchburn and Carballeira, 2019
Bionanotechnology			
Antibacterial	Iron-NPs	*Dictyota dichotoma*	Ditchburn and Carballeira, 2019
Hemolytic	Platinum-NPs	*Padina gymnospora*	Ditchburn and Carballeira, 2019
Drug delivery	Alginates Nanocomposites	*Laminaria* spp.	Fernando et al., 2018
Bio-flocculent	Alginates	*Laminaria* spp., *Macrocystis pyrifera*	Salehizadeh et al., 2018

4.5 RECENT RESEARCH ADVANCES IN ITS UTILIZATION

In quest of desirably designed target compounds, the natural macroalgal biopolymers such as alginates and fucoidan have been recently modified structurally by partial degradation, co-polymerization, oxidation, cross-linking, and functional group substitutions, to improve its functionalities as drug delivery systems, in wound healing, tissue engineering, and as fillers and gelling agents to control viscosity. Such modifications also impart enhanced physicochemical properties which in turn ameliorates the biofunctions of these compounds as antioxidant, immunomodulatory, anticoagulant, antilipidemic agents and so on. This kind of activity is a prelude to biomaterials research aiming to meet the special needs of the medical field (Fernando et al., 2018). The direct application of such a modified biopolymer was reported wherein the fucoidan from *Laminaria japonica* modified by carboxymethylation applied on strawberries was found to reduce water loss and fruit decay during cold storage through improved antioxidative and preservative effect (Duan et al., 2019). Edible biofilms were developed combining alginate molecules and alginate-fucoidan blends extracted from *Sargassum latifolium* with the help of calcium cross-linking. The films so obtained were less soluble in water due to the incorporation of fucoidan and calcium, while they showed good antioxidant properties, water vapor and oxygen permeability (Gomaa et al., 2018).

In course of time, the relationship between the structure as well as the molecular weight of the biopolymers and their bioactivity was established based on which, the fucoidan obtained from hydrothermal extraction was fractionated using membranes to get low molecular weight fucoidan (LMWF) of 5–30kDa with improved molecular properties like binding, biocompatibility, diffusibility, mobility, and biodegradability. This technique also imparted higher bioactivity (Álvarez-Viñas et al., 2019). Abraham et al. (2019) developed a two-step biorefinery system involving an acidic extraction step before the conventional alkaline extraction step that could extract fucoidan, laminarin, and alginate from the giant bull kelp *Durvillaea potatorum*. Subcritical water extraction (SCWE) is yet another modern technique employed in the extraction of value-added compounds from brown seaweeds. Using this technique, fucoidan was successfully extracted from *Nizamuddinia zanardinii* and it showed good antioxidant and anticancer properties. The SCWE developed thus offered a competent method over the conventional method in the fact that it provides higher yield and quality within a shorter period, with lower energy consumption. Besides, it is a green process as it utilizes water as the solvent (Alboofetileh et al., 2019). Other recent techniques evolved include MAE, UAE and enzyme-assisted extraction (EAE). These techniques involve breaking of brown algal cell walls using physical shear or enzyme thereby rendering higher yields of target compounds within a shorter time. As they proceed with no use of solvents or chemicals, they are considered green processes – both human and environment-friendly. Fucoidan from *Ascophyllum nodosum* was extracted by both conventional and these methods and it was found that the MAE gave better physicochemical properties, but the bioactivity remained the same (Okolie et al., 2019).

Another interesting area of research that has emerged anew is the development of chemically modified electrodes (CMEs) incorporating polysaccharides like alginates obtained from brown seaweeds. These polysaccharides are used due to their high mechanical strength, low cost, easy availability, and non-toxic environment-friendly nature, thus following a sustainable and green chemistry principle. Such electrodes with polysaccharide coating find application in electrochemical sensors with high sensitivity and selectivity towards the detection of pollutants and contaminants in the environment (Maciel et al., 2019).

Marine-derived polysaccharides (MDPs) such as alginate, fucoidan, and laminarin are administered as a dietary supplement in growing aquatic animals such as fish, crab, shrimp, lobster, *etc.* in aquaculture which proved to be an excellent replacement for chemotherapeutics and antibiotics in common use. These components act as a prebiotic and enhance disease resistance, immune response, gut microbiota and survival in these animals, thus ensuring antibiotic-free aquatic products (Mohan et al., 2018).

The most recent topic that churned out attention worldwide was the bioengineering aspect of macroalgae. It was shown that bioengineering is not only limited to microalgae when it comes to biofuels, biopolymers, or even bionanotechnology. Brown algae such as *Macrocystis pyrifera* and *Saccharina latissima* are thrice methanogenic than terrestrial plants, and hence a transesterification of their oils yield biodiesel. Also, bio-crude and bio-chars are produced from macroalgal biomass through novel techniques like pyrolysis, hydrothermal liquefaction, and catalytic hydrothermal gasification, which generates higher energy output than conventional fuels. Direct combustion of algal biomass to obtain gases is also a common method. Bioplastics were developed by combining alginates with paper or calcium fibers to obtain biodegradable natural fiber composites for making furniture or cloth material, respectively. Also, metal, and metal oxide nanoparticles (NPs) with improved activities were synthesized using extracts of brown algae (Fe-*Dictyota dichotoma* and Pt-*Padina gymnospora*). The metal-NPs exhibited efficient antimicrobial and antioxidant properties while the metal oxides – seaweed NPs found utility in electronics, catalysis, solar energy conversion, and magnetic storage media (Ditchburn and Carballeira, 2019).

4.6 KEY CHALLENGES IN SUCCESSFUL BIOPROSPECTING

1. Brown seaweeds possess a highly complex structure encompassing a blend of biochemical constituents like polysaccharides, phenols, proteins, minerals, and vitamins, which makes the separation of these components a strenuous task. The polysaccharides especially form the part of the thick cell wall and are less accessible and soluble. The existing extraction and purification techniques are inefficient in the maximum exploitation of these components. Hence a more suitable biorefinery process for the sequential extraction and purification of multiple components needs to be developed and optimized (Abraham et al., 2019).
2. The molecular weight of polysaccharides such as fucoidan has been reported as one of the main characteristics that influence its biological activities, but a single universal relationship between these factors could not be established so far (Álvarez-Viñas et al., 2019).
3. The brown algal bio-constituents are numerous and only a few of them have been brought to the limelight as bioactives with multiple applications. While coming to the development of such components to the industrial level, only a still fewer number has been attended to. For instance, many of the polymers like laminaran and fucoidan has been shown to effectively inhibit varied lines of cancer cells yet the release of a complete drug from them is still far from the reality due to the comprehensive assessments and clinical studies required to further tailor them (Wali et al., 2019).
4. Although the structural modifications of biopolymers like fucoidan and alginates enhance their properties and functions, their stability is still to be ensured. In some cases like over sulfated fucoidan, the excess sulfates lead to undesired reactions within the systems hindering the target outcomes. Hence it demands further diligent designing of these components (Fernando et al., 2018).
5. The conventional extraction methods of bioactives from Phaeophyceae involve the use of water or solvents or a combination of both along with high temperatures and pH. It results in low efficiency, lesser yields, high expenses, and longer extraction periods, which leads to the loss of structure and function in them. Therefore, the development of a novel, optimized method that surpasses all these shortcomings is a need of the hour, for the maximum exploitation of marine resources (Alboofetileh et al., 2019).

6. The modern methods such as MAE, EAE, and UAE provide good biochemical yield within a short duration, the high amounts of heat produced by microwave or sound waves may affect the structural stability of these compounds. Besides, the installation of a microwave or ultrasound facility may be too expensive at industrial levels. Hence further research to explore the effects of such treatments on the products of interest might provide a better understanding to fine-tune these techniques (Okolie et al., 2019).
7. Several intrinsic and extrinsic factors influence the quality and quantity of biochemicals extracted from brown algae. Although the intrinsic factors such as the extraction and purification conditions can be designed according to the demands, the extrinsic factors such as the location and the type of species harvested, seasonal, and climatic variations, environmental pollutions, *etc*., are uncontrollable and hence open up a probability for altered structures, quantity, quality, and bioactivity of the components isolated. Hence a sustainable, eco-friendly, and cost-effective bioreactor culturing system has to be developed for optimized growth, biochemical production, and extraction from macroalgal systems (Murray et al., 2013).
8. In the bioengineering of brown seaweeds, the biofuel production strategies needs to be strengthened with more of economic funding as well as qualitative research, so that the macroalgae-derived biofuels may reach the consumers as early as possible especially in this scenario where conventional fuels are running short, and sticking to microalgal sources or plant sources alone is not an ideal option to meet the demands of the rising population. In the case of seaweed NPs, there is a requirement to elucidate their biosynthesis mechanisms and effects (Ditchburn and Carballeira, 2019).
9. In the case of food products, the incorporation of the algae as such or its biochemical constituents has been undertaken to improve its texture, organoleptic, and sensory characteristics, nutritional quality as well as storage span. But in all the cases reported so far, the development of standardized specific formulations meeting the food standards or the special consumer demands has not yet been achieved (Ito et al., 2018). Such works confined to the lab scale should be developed to the industrial level so that a wide population can be targeted. Also, the less utilized brown seaweeds need to be explored for their edibility.

10. Many brown algae-derived extracts and compounds have found utility in the pharmaceutical and food industry and a new product is coming up every day. But still, there are no Food and Drug Administration (FDA) approved anticancer drugs or products from these compounds, which indicate the crucial necessity to undertake practical, in-depth, and multidisciplinary approaches in this regard, rather than the surface level research (Gutiérrez-Rodríguez et al., 2018).
11. When coming to the extraction of fucoidan, there is an inherent heterogeneity in its monosaccharide composition, sulfate pattern, glycosidic linkages, and molecular weights depending on the type of brown algal species. The type of extraction method chosen may also cause variation in these features. Besides, the high molecular weight and complex structure will also make fucoidan therapeutically inapplicable. These are the reasons why fucoidan failed to raise from the level of a drug candidate to the level of a drug, and this needs to be tackled first so that we can develop fucoidan into a promising anticancer drug. The elucidation of metabolic pathways of fucoidan biosynthesis also needs to be carried out so that the switches for regulating its chemical diversity can be determined (Deniaud-Bouët et al., 2017).
12. A lack of understanding of the life cycle, growth parameters, symbiotic relationships, and exact nutritional requirements of the brown seaweeds is also a challenge in the path to their complete exploration.

4.7 CONCLUSION

The bioactive compounds derived from brown seaweeds are highly valuable products that have left no field untouched with their applications. Table 4.1 gives an overall view of the diverse applications of these compounds, for instance. With constant research and explorations worldwide, we were able to understand the biology of brown seaweeds and it enabled us to put them and their products/extracts into various areas of day-to-day life. Yet a very detailed expedition into the brown seaweed research so far, projects out certain shortcomings, gaps or challenges that need to be cleverly addressed for the absolute utilization of their benefits for mankind and other organisms. Each bioactive component must be efficiently extracted, precisely characterized, and then desirably designed to meet the demands of any kind. Thus, in a nutshell, a sustainable, cost-effective, environment-friendly, non-toxic, targeted, interdisciplinary, and green approach (Figure 4.2) should be undertaken for a standardized and universal utilization of brown algae.

KEYWORDS

- **bioprospecting**
- **Food and Drug Administration**
- **Fourier transform infra-red**
- **fucoidan**
- **gas chromatography**
- **high-performance liquid chromatography**
- **microwave-assisted extraction**
- **Phaeophyceae**
- **phlorotannins**
- **ultrasound-assisted extraction**

REFERENCES

Abdel-Fattah, A. F., Hussein, M. M. D., & Salem, H. M., (1974). Studies of the purification and some properties of sargassan, a sulfated heteropolysaccharide from *Sargassum linifolium*. *Carbohydrate Research, 33*(1), 9–17.

Abraham, R. E., Su, P., Puri, M., Raston, C. L., & Zhang, W., (2019). Optimization of biorefinery production of alginate, fucoidan and laminarin from brown seaweed *Durvillaea potatorum*. *Algal Research, 38*, 101389.

Alboofetileh, M., Rezaei, M., Tabarsa, M., Rittà, M., Donalisio, M., Mariatti, F., You, S., et al., (2019). Effect of different non-conventional extraction methods on the antibacterial and antiviral activity of fucoidans extracted from *Nizamuddinia zanardinii*. *International Journal of Biological Macromolecules, 124*, 131–137.

Álvarez-Viñas, M., Flórez-Fernández, N., González-Muñoz, M. J., & Domínguez, H., (2019). Influence of molecular weight on the properties of *Sargassum muticum* fucoidan. *Algal Research, 38*, 101393.

Anyanwu, R., Rodriguez, C., Durrant, A., & Olabi, A. G., (2018). In: Hashmi, S., (ed.), *Reference Module in Materials Science and Materials Engineering*. Elsevier B.V. https://doi.org/10.1016/B978-0-12-803581-8.09259-6.

Centella, M. H., Arevalo-Gallegos, A., Parra-Saldivar, R., & Iqbal, H. M. N., (2017). Marine-derived bioactive compounds for value-added applications in bio- and non-bio sectors. *Journal of Cleaner Production, 168*, 1559–1565.

Cesário, M. T., Da Fonseca, M. M. R., Marques, M. M., & De Almeida, M. C. M., (2018). Marine algal carbohydrates as carbon sources for the production of biochemicals and biomaterials. *Biotechnology Advances, 36*(3), 798–817.

Chandika, P., Ko, S. C., & Jung, W. K., (2015). Marine-derived biological macromolecule-based biomaterials for wound healing and skin tissue regeneration. *International Journal of Biological Macromolecules, 77*, 24–35.

Charoensiddhi, S., Conlon, M. A., Franco, C. M., & Zhang, W., (2017). The development of seaweed-derived bioactive compounds for use as prebiotics and nutraceuticals using enzyme technologies. *Trends in Food Science and Technology, 70*, 20–33.

Chia, S. R., Ong, H. C., Chew, K. W., Show, P. L., Phang, S. M., Ling, T. C., Nagarajan, D., Lee, D. J., & Chang, J. S., (2018). Sustainable approaches for algae utilization in bioenergy production. *Renewable Energy, 129*, 838–852.

Deniaud-Bouët, E., Hardouin, K., Potin, P., Kloareg, B., & Hervé, C., (2017). A review about brown algal cell walls and fucose-containing sulfated polysaccharides: Cell wall context, biomedical properties and key research challenges. *Carbohydrate Polymers, 175*, 395–408.

Ditchburn, J. L., & Carballeira, C. B., (2019). Versatility of the humble seaweed in biomanufacturing. *Procedia Manufacturing, 32*, 87–94.

Duan, Z., Duan, W., Li, F., Li, Y., Luo, P., & Liu, H., (2019). Effect of carboxymethylation on properties of fucoidan from *Laminaria japonica*: Antioxidant activity and preservative effect on strawberry during cold storage. *Postharvest Biology and Technology, 151*, 127–133.

Fernando, I. P. S., Kim, D., Nah, J. W., & Jeon, Y. J., (2018). Advances in functionalizing fucoidans and alginates (bio) polymers by structural modifications: A review. *Chemical Engineering Journal, 355*, 33–48.

Garcia-Vaquero, M., Rajauria, G., O'Doherty, J. V., & Sweeney, T., (2017). Polysaccharides from macroalgae: Recent advances, innovative technologies and challenges in extraction and purification. *Food Research International, 99*, 1011–1020.

Gomaa, M., Fawzy, M. A., Hifney, A. F., & Abdel-Gawad, K. M., (2018). Use of the brown seaweed *Sargassum latifolium* in the design of alginate-fucoidan based films with natural antioxidant properties and kinetic modeling of moisture sorption and polyphenolic release. *Food Hydrocolloids, 82*, 64–72.

Gutiérrez-Rodríguez, A. G., Juarez-Portilla, C., Olivares-Banuelos, T., & Zepeda, R. C., (2018). Anticancer activity of seaweeds. *Drug Discovery Today, 23*(2), 434–447.

Ito, M., Koba, K., Hikihara, R., Ishimaru, M., Shibata, T., Hatate, H., & Tanaka, R., (2018). Analysis of functional components and radical scavenging activity of 21 algae species collected from the Japanese coast. *Food Chemistry, 255*, 147–156.

Jégou, C., Kervarec, N., Cérantola, S., Bihannic, I., & Stiger-Pouvreau, V., (2015). NMR use to quantify phlorotannins: The case of *Cystoseira tamariscifolia*, a phloroglucinol-producing brown macroalga in Brittany (France). *Talanta, 135*, 1–6.

Kadam, S. U., Álvarez, C., Tiwari, B. K., & O'Donnell, C. P., (2015). Extraction of biomolecules from seaweeds. In: Tiwari, B. K., & Troy, D. J., (eds.), *Seaweed Sustainability* (pp. 243–269). Academic Press, Cambridge.

Maciel, J. V., Durigon, A. M. M., Souza, M. M., Quadrado, R. F. N., Fajardo, A. R., & Dias, D., (2019). Polysaccharides derived from natural sources applied to the development of chemically modified electrodes for environmental applications: A review. *Trends in Environmental Analytical Chemistry*, e00062. https://doi.org/10.1016/j.teac.2019.e00062.

Mohan, K., Ravichandran, S., Muralisankar, T., Uthayakumar, V., Chandirasekar, R., Seedevi, P., Abirami, R. G., & Rajan, D. K., (2018). Application of marine-derived polysaccharides as immunostimulants in aquaculture: A review of current knowledge and further perspectives. *Fish and Shellfish Immunology, 86*, 1177–1193. https://doi.org/10.1016/j.fsi.2018.12.072.

Murray, P. M., Moane, S., Collins, C., Beletskaya, T., Thomas, O. P., Duarte, A. W., Nobre, F. S., et al., (2013). Sustainable production of biologically active molecules of marine based origin. *New Biotechnology, 30*(6), 839–850.

Okolie, C. L., Mason, B., Mohan, A., Pitts, N., & Udenigwe, C. C., (2019). The comparative influence of novel extraction technologies on *in vitro* prebiotic-inducing chemical properties of fucoidan extracts from *Ascophyllum nodosum. Food Hydrocolloids, 90*, 462–471.

Rajauria, G., Draper, J., McDonnell, M., & O'Doherty, J. V., (2016). Effect of dietary seaweed extracts, galactooligosaccharide and vitamin E supplementation on meat quality parameters in finisher pigs. *Innovative Food Science and Emerging Technologies, 37*, 269–275. doi: 10.1016/j.ifset.2016.09.007.

Rao, P. V. S., & Mantri, V. A., (2006). Indian seaweed resources and sustainable utilization: Scenario at the dawn of a new century. *Current Science,* 164–174.

Rengasamy, K. R. R., Kulkarni, M. G., Stirk, W. A., & Van, S. J., (2014). Advances in algal drug research with emphasis on enzyme inhibitors. *Biotechnology Advances, 32*(8), 1364–1381.

Roohinejad, S., Koubaa, M., Barba, F. J., Saljoughian, S., Amid, M., & Greiner, R., (2017). Application of seaweeds to develop new food products with enhanced shelf-life, quality and health-related beneficial properties. *Food Research International, 99*, 1066–1083.

Salehizadeh, H., Yan, N., & Farnood, R., (2018). Recent advances in polysaccharide bio-based flocculants. *Biotechnology Advances, 36*(1), 92–119.

Suleria, H. A. R., Gobe, G., Masci, P., & Osborne, S. A., (2016). Marine bioactive compounds and health promoting perspectives; innovation pathways for drug discovery. *Trends in Food Science and Technology, 50*, 44–55.

Wali, A. F., Majid, S., Rasool, S., Shehada, S. B., Abdulkareem, S. K., Firdous, A., Beigh, S., et al., (2019). Natural products against cancer: Review on phytochemicals from marine sources in preventing cancer. *Saudi Pharmaceutical Journal, 27*(6), 767–777. http://doi.org/10.1016/j.jsps.2019.04.013.

Wang, H. M. D., Li, X. C., Lee, D. J., & Chang, J. S., (2017). Potential biomedical applications of marine algae. *Bioresource Technology, 244*(Pt 2), 1407–1415.

CHAPTER 5

BROWN SEAWEED POLYSACCHARIDES: A ROADMAP AS BIOMOLECULES

DIANA PACHECO,[1] JOÃO COTAS,[1] ADRIANA LEANDRO,[1] SARA GARCÍA-POZA,[1] ANA MARTA MENDES GONÇALVES,[1,2] and LEONEL PEREIRA[1]

[1]*Department of Life Sciences, University of Coimbra, MARE-Marine, and Environmental Sciences Centre, Coimbra – 3001 456, Portugal*

[2]*Department of Biology and CESAM, University of Aveiro, Aveiro – 3810–193, Portugal*

ABSTRACT

Marine macroalgae are the source of a wide range of compounds that have demonstrated a myriad of potential uses. Sulfated and non-sulfated polysaccharides are some of the major compounds found in these organisms, and they could be applied in many industries due to their diverse potentials. Seaweeds are a valuable source of sulfated polysaccharides, which are distinctive characteristics between taxonomic groups. For instance, brown seaweeds (Phaeophyceae) are known to contain alginate, fucoidan, laminarin, and ascophyllan. Sulfated carbohydrates are raising interest among the scientific community due to their beneficial properties. This chapter discusses these brown seaweed polysaccharides from their discovery, chemical structure, methodologies of extraction to the most important areas in the research, and also possible applications in several industries.

5.1 INTRODUCTION

The ocean and coastal ecosystems hold rich biodiversity under different environmental conditions, thus being considered a big source of different biomolecules (Fitton et al., 2019). Marine homo- and heteropolysaccharides are extracted from marine organisms, which represent a large part of global biodiversity; among these are included algal polysaccharides (Hamed et al., 2015). Marine sulfated polysaccharides (PLS) derived from seaweed are a widespread source of bioactive compounds (Rinaudo 2008; Pereira et al., 2009). Seaweed polymers are complex, due to the diversity of chemical structures, which differ according to the species, season, extraction, and isolation methods selected (Costa et al., 2010).

Sulfated polysaccharides are common in seaweed and they are characteristic according to their taxonomic group. For instance, alginate, fucoidan, and laminarin are the main polysaccharides found in brown algae (Phaeophyceae); carrageenan and agar are sulfated galactans characteristic from red algae (Rhodophyta), while ulvan occur in green macroalgae (Chlorophyta) (Jiao et al., 2011; Rodrigues et al., 2015; Freile-Pelegrín and Tasdemir, 2019).

Polysaccharides are the major component of seaweed's biomass, with a structural and biological role (Lahaye and Robic, 2007; Hidari et al., 2008; Venkatesan et al., 2016). Seaweed polysaccharides act as dietary fibers and in the brown algae are common and diverse (e.g., alginates, fucoidans, and laminarins) forming the main components (over 75%) of the dried thalli weight (De Jesus Raposo et al., 2015; Gobet et al., 2018; Maneein et al., 2018).

The majority of algal polysaccharides are non-starch fiber, which contributes to balance blood glucose levels, and also helps to the reference nutrient intake of 30 g fiber day^{-1} (BNF, 2016; Shannon and Abu-Ghannam, 2019). Among the most studied compounds, therapeutic applications of laminarin and fucoidan have been linked with bioactivities, such as antioxidant, antitumoral, anticoagulant, anti-inflammatory, and antiviral (Zhou et al., 2004; Jiao et al., 2011; Ngo and Kim, 2013; Cunha and Grenha, 2016).

Laminarin, alginate, and fucoidan are bioactive polysaccharides derived from brown algae that are already employed in functional foods, cosmeceutical, and pharmaceutical products in the last years. The extraction methods of these polysaccharides have many complex and time-requiring steps and the right adjustment of extraction parameters (e.g., time, temperature, power, pressure, solvent, and sample to solvent ratio) highly affects the yield, their biological activities and their chemical, biochemical, and physical features (Dobrinčić et al., 2020).

Biological activities of PLS are directly affected by chemical structure, molecular weight and chain conformations (Ye et al., 2009). Polysaccharides play many roles in living cells such as structural functions, whereas cellulose and chitin are the main components of the different cell wall matrix, energy storage, hydration, and signaling functions (Ghanem et al., 2010; Ball et al.,

2011; Helle et al., 2018; Amos and Mohnen, 2019; Lampugnani et al., 2019; Shukla et al., 2019).

The attention of the researchers has been directed to polysaccharides derived from marine flora (i.e., marine algae) due to their abundance, low extraction cost, singular physicochemical and biological properties, such as wound healing, anticancer, antioxidant, antibacterial activities, anti-inflammatory or immunostimulatory (Lee et al., 2013; Ayoub et al., 2015; Lee et al., 2017). This is why these marine polysaccharides, which include fucoidan, laminarin, and alginate, have been widely employed in pharmaceutical and biomedical industries (Ye et al., 2009; Aravamudhan et al., 2014; Cardoso et al., 2016; Ruocco et al., 2016; Elbi et al., 2017; Sanjeewa et al., 2017). For instance, ascophyllan is one of the least studied seaweeds' polymers; however, it presents high potential, such as antitumoral, anti-inflammatory, and immunomodulatory bioactivities (Nakayasu et al., 2009; Mohsin et al., 2013). Also sargassan is a polymer that remains least studied, even though some research, with positive results, has been conducted in order to evaluate this carbohydrate's anticoagulant bioactivity (Vavilala and D'Souza, 2015).

Brown seaweeds have abundant amounts of bioactive compounds such as polysaccharides, peptides, omega-3 fatty acids, carotenoids, phenolics, vitamins, and minerals (Kadam et al., 2015). In this chapter, brown seaweed polysaccharides are explored, since the extraction until the final application in biomedical, food, and cosmetic field, which are the most important areas in the research of these polymers.

5.2 ALGINATE

5.2.1 GENERAL OVERVIEW

Alginate is the major anionic polymer synthesized in brown algae (Phaeophyceae), which is present mainly on the cell walls of these organisms, being one of the most important structural elements in the key skeletal brown algae compounds, comparable with the cellulose purpose in terrestrial plants, with the polymer placed in the cell walls and in the intercellular matrix convening the mechanical strength and flexibility required to endure the energy of the weather and seawater agitation where the seaweed grows (Kannan and Sivaperumal, 2017). In addition, this alginate's role is mirrored in the compositional modification and dissimilarity of the alginates from different seaweed, even from the same species due to environmental changes.

The alginate content can vary from 20% to 60% of the brown seaweed dry matter, although the brown seaweed has, on average, 40%. The alginate in brown seaweed is gel formed by a matrix of alginic acid with various ions, such as sodium (Na), calcium (Ca), strontium (Sr), magnesium (Mg), and barium (Ba) ions (Haug and Smidsrod, 1967). Alginate is not a specific polymer of brown algae due to various species of bacteria that synthesize alginate identical to the brown algae, but, nowadays, all commercial alginate comes from the algal biomass (Draget et al., 2005).

Alginate industrial usage is consequent from several interesting properties that alginate and alginate-based solutions have, such as gelation, viscosity, and stabilizer properties, that perform in the final solutions and products in which it is applied. Typically, the alginate applied in industry is a mixture of alginic acid with specific cations, for example, calcium, sodium, ammonia, potassium, or magnesium. These ions provide bigger stability to the alginic acid molecule; the divalent cations provide alginate a very rigid conformation and a stable structure unlike the alginate with monovalent cations.

The innovations in the biotechnological usage of alginate are based on specific characteristics of the alginate-based molecules and their chemical variations depending on the conformation of the alginate-based matrix with the ions or other compounds (e.g., calcium, magnesium or polyethylene glycol). These modifications allow a great number of applications in various industries with different objectives, so the alginates are easy to be modified in the physical-chemical, and so, they are one of the natural polymers most investigated for various purposes, such as agriculture, food, textile, and biomedical areas.

Alginates and their variations have been highly investigated and are trending for specialized knowledge areas as pharmaceutical or biomedical ingredients, or as a component for advanced and high-tech biotechnology. These studies are turnoff to have a more comprehensive learning about the properties and structures of alginates, leading to scientific innovation that, associated with empirical knowledge, will increase the value of the traditional techniques of alginate exploitation.

5.2.2 EXTRACTION AND ISOLATION METHODS

The alginic acid extraction method is based on the transformation of the insoluble blend of alginic acid salts of the brown algae cell wall into a soluble salt (alginate). This can be extracted from several genera of brown algae (e.g., mixed Fucales and Laminariales), that are commonly exploited as raw resources by alginate industrial producers. Species that produce alginate are from the genus *Macrocystis*, *Laminaria*, *Lessonia*, *Ascophyllum*, *Alaria*, *Ecklonia*, *Eisenia*, *Nereocystis*, *Sargassum*, *Cystoseira*, and *Fucus*, where the *Macrocystis pyrifera* and *Ascophyllum nodosum* are the main resources of the alginate industry. The intercellular mucilage in brown algae has been considered as the main location of alginic acid, even though alginate can also be found in the cell walls and intracellular (Fischer and Dörfel, 1955; Arunkumar, 2017).

For the alginate industry, a wide range of brown algae are harvested and even cultivated in offshore aquaculture (for example, *Laminaria*, and *Alaria*), between southern and northern hemispheres, being one of the major utilization of seaweeds. Countries such as Argentina, Austria, Chile, Canada, China, Mexico, Japan, Norway, Scotland, the United States, South Africa, and the United Kingdom are the main alginate producers. With the growth of the alginate market, some countries, like China and some other northern hemisphere countries, are cultivating brown algae, which is an expensive approach for alginate extraction. One of the examples is *Laminaria*, whose cultivation is mostly used as food, but if additional volumes are cultivated, then it will be used for alginate production.

Brown algae that grow in cold water (up to 20°C) are more convenient for alginate extraction due to higher alginic acid content because in more turbulent waters, seaweeds will synthesize more alginic acid than the same species in calmer waters. The seaweeds from hotter waters have lower alginic acid content and high volatile content when compared with cold habitat brown algae.

To extract alginate, it is required to transform the alginic acid into its soluble salt formula, for example as sodium or potassium alginate (Clare, 1993). The alginate is extracted through alternating insoluble and soluble alginic acid in a solvent variation, conducing to ion-exchange reactions to isolate the alginate from the other compounds of the brown algae. As huge molecules must be drawn-out from the algae tissues, the algae should be, if possible, ground into small particles as a necessary preliminary step. After that, the alginate extraction first step is to wash (dry, if necessary) and grind the seaweed. Alginate extraction and isolation is basically an ion-exchange procedure, and alginate is conveyed into an aqueous solution as sodium alginate by using a strong alkali to solubilize the alginic acid into a soluble alginate, this occurs after a pre-treatment with hydrochloric acid before the alginate extraction with sodium carbonate (McHugh, 1987; Sime, 1990). There are other several methods to separate the alginate from other soluble substances from the crude alginate extract solution. One of them, the addition of an alcohol (Haug and Smidsrod, 1967), would precipitate out sodium alginate; other main method is the addition of a solution of calcium chloride to the alginate extraction solution, with good stirring would precipitate out calcium alginate, whereas adding hydrochloric acid would precipitate out alginic acid.

The pre-treatment to the extraction of alginate is through a mild acid solution treatment that eliminates unwanted composites (normally hydrochloric acid) and changes the cell wall alginate into alginic acid to obtain high yield extraction (Pereira et al., 2013). The alginic acid is recuperated as a soluble sodium formula by neutralizing the residues of the pre-treatment with sodium carbonate or sodium hydroxide. After the reaction of the alginic acid with sodium-based alkali solution, the insoluble residue is discharged by filtration, flotation, or centrifugation and the solubilized alginate is precipitated by transforming into alginic acid or calcium/sodium alginate, depending on the alginate future application. The alginic acid can be converted into the necessary counter ion by neutralizing the alginic acid solution with appropriate ions hydroxides or chlorites. The modification in the alginate recovery process depends on the source and structure constituents of alginate (McHugh, 1987).

The trouble of the alginate extraction is mainly in physical separations between stages. For example, the separation of the muddy residues after the extraction with sodium carbonate viscous solution can be a hard and slow filtration process, due to the high viscosity. It can also happen that gelatinous-like precipitates retain a high quantity of fluid in their basic structure, resisting, and hardening the filtration and centrifugation overall process.

5.2.3 CHEMICAL STRUCTURE

Alginic acid was firstly identified by the British Chemical Scientist E.C.C. Stanford, conducting to alginic acid patent. This work permitted the discovery of the alginic acid chemical structure (Draget et al., 2005). The Stanford patent explained the primary alginate extraction method, which has a pre-treatment by soaking the brown algae with water or an aqueous solution with diluted acid, and then extraction was done with sodium carbonate, following precipitation of the alginate solubilized in the extraction solution by the addition of an acid (Hoagland and Lieb, 1915).

During the first decade of the twentieth century, research groups were studying separately the alginate, establishing that uronic acid was the chemical base of the constituents of alginic acid. These researches steered to the discovery of D-mannuronic acid in hydrolyzed alginate samples. The L-guluronic acid in the alginic acid was only reported in the 1950s by the chromatographic study of uronic acids of Fischer and Dörfel (1955), through a chromatographic study of uronic acids. This acid had a substantial quantity in the alginic acid samples analyzed, and so a quantitative method was developed to determine the two acids present in alginate: mannuronic and guluronic acids. After this study, the alginic acid was determined to be a binary copolymer composed of residues of guluronic and mannuronic acids; however, the alginic acid composition was chemically homogeneous and had equal chemical structure, independent of the raw material from which it was extracted.

Therefore, the alginate structure analysis was done by fractionation techniques, mainly by Haug et al. (1974), one of the researchers that helped to characterize alginate from various species as a similar block copolymer and established a correlation between the alginate block structure and the physicochemical properties of an alginate with a specific composition.

The alginates of seaweeds comprise the monomers of 1,4-β-D mannuronic acid (M) and 1,4-α-L-guluronic acid (G), with a homogeneous (poly-G, poly-M) or heterogeneous (MG) block composition (Figure 5.1). So, the alginate synthesized by different brown algae species may show different alginate copolymers alignments and differences in the ratio of mannuronic and guluronic acid blocks, varying the composition and the sequence (Szekalska et al., 2016). The quantities of the two acids show a discrepancy from species to species and even from diverse locations of the same seaweed (Haug et al., 1974).

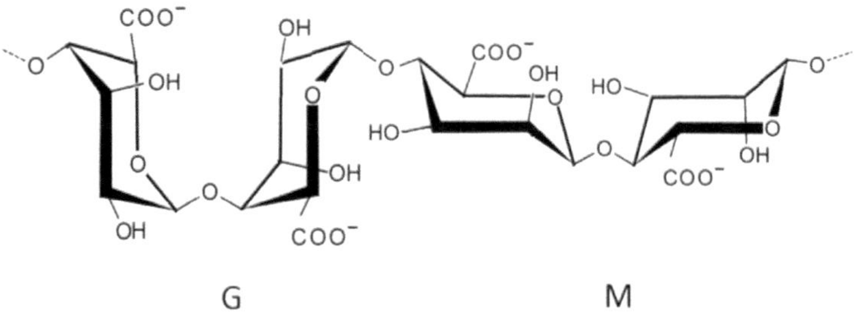

FIGURE 5.1 Idealized structure of the chemical units of alginic acid.

It was already proven that the alginates do not have consistent repeating units and that the distribution of the monomers lengthwise the polymer chain could not be described henceforth the knowledge of the monomeric composition was not enough to determine the complete chemical structure of alginates from diverse species, only the monomers ratio.

Alginic acid is found in seaweeds bonded to salts of diverse metals, mainly sodium and calcium, and the alginic acid biological properties in the brown algae are different by the structural and ion exchange type. Alginic acid high in polymannuronic acid monomers is found in young cell wall tissue and/or intercellular regions, whereas polyguluronic-rich alginic acid is found in the cell wall having a high affinity for Ca^{2+}, which is mainly responsible for gel strength (Arunkumar, 2017). Alginic acid polymer is synthesized in the cytoplasm and then transported to the cell surface (Abe et al., 1973). The major distinction from the algal to bacterial alginates is at the molecular level, due to the occurrence of O-acetyl groups at C2 and/or C3 in the bacterial alginates (Skjåk-Bræk et al., 1986).

5.2.4 APPLICATIONS

The various types of alginates are being applied in pharmaceutical, biomedical, food, cosmetic, and other industries, representing the most important and used seaweed colloid in terms of volume. There is a wide range of alginates and purities that are available for diverse applications, resulting in diverse prices depending on the alginate characteristics (Gillespie and Critchley, 1999). Alginates are mostly applied as thickeners and stabilizers in the pharmaceutical, food, and cosmetic areas, due to alginates being easy to use, having low-cost production even with a high purity, and being well tolerated in the human organism and also, they can be straightforwardly adjusted for specific objectives and in the diverse areas of application. The request for alginates in the food area will be increasing because of a growing intake of foods such as ice-creams, beer, and yogurt, which is projected to enhance the alginate market growth.

5.2.4.1 BIOMEDICAL

The alginates in the biomedical field are used as thickening, gel-forming, and stabilizing agents, but the main alginate application nowadays is to form hydrogel, as wound healing (Pielesz et al., 2011), drug delivery systems and tissue engineering applications. The alginate hydrogels are composed of hydrophilic polymers with high water-retaining capacity. These alginate-based hydrogels are biocompatible and can be inserted into the human body without wide/great invasive techniques of administration (Sakiyama-Elbert and Hubbell, 2001). Alginate's biocompatibility has been extensively studied, because impurities compromise the alginate-inherent biocompatibility, which affects the biomaterial to insert in the human body. Thus, it is necessary, in the biomedical industry, to use the highest purity of alginate to reduce the hypothesis of secondary effects (Lee and Mooney, 2012). The overall ranking of alginate's characteristics to decide its usage in the biomedical industry are: the lower cost, the biocompatibility and the availability. So, the study that has been developed permits to understand that modifications in physiochemical in alginate and derivatives are easy. Alginate is already largely applied in the biomedical area, with an excellent track record in clinical trials applications as wound dressing material and medicament ingredient. More applications are identified and described in the following points.

5.2.4.1.1 Biomaterials

The alginate-based biomaterials are being explored and they are the subject of an extensive research, because they are easy to obtain and cheap and have inherent biocompatibility, low toxicity, and mild gelation by the adding of divalent cations such as Ca^{2+} (Gombotz and Wee, 1998). Furthermore, the natural polymers are easy to work for mimicking the extracellular cellular matrix, where the biocompatibility is very important. Still, alginate hydrogels have a low mechanical stiffness and to surpass this, there is a need to define the alginate-based hydrogels to the final objective. In this context, the comprehension of how ions (e.g., Ca, Na) links to alginic acid in order to obtain hydrogels with higher mechanical resistance. Another alternative to achieve this goal is mixing alginate with other polymers (Lee and Mooney, 2012). This occurs not only in biomaterials inserted in the human body, but also in wound dressing.

Alginate wound dressings preserve the physiological humid microenvironment, decreasing the occurrence of microbial infections at the wound location, and assist wound healing. Drug molecules from various classes were tested with success in the alginate-based wound dressings, furthermore there was slow time drug-release in a controlled manner, and the velocity was dependent on the chemical structure of the wound dressing (Queen et al., 2004).

Alginate-based dressings are normally made recurring to calcium alginate ionic solution to form a stable gel and adding a fibrous non-woven dressing; this type of wound dressing is normally dried applied thus re-absorbing the wound fluids to re-jellification. Thus, the gel maintains the wound location humidity, maintaining the microenvironment and reducing the hypothesis of infection in the wound location. There is also the hypothesis to add drugs in the alginate-based wound dressings to help with tissue growth, rapid epithelialization, and healing. Various alginate-based wound dressings are already in the market (Pielesz et al., 2011; Lee and Mooney, 2012; Aderibigbe and Buyana, 2018).

Other potential of alginate as a biomaterial is in tissue engineering. This is to provide man-made tissue and organ replacements *in vitro* or *in vivo* and then transplant them to patients that have acute problems in the tissue or organs (Langer and Vacanti, 1993). The alginate-based hydrogels are applied to deliver cells to the desired location for the tissue growth and with control of the structure and normal functioning of this new tissue (Lee and Mooney, 2001). The flaw of alginate-based materials in this area is derived from the fast exchange within divalent and monovalent ions, lowering the matrix stability, where the exception is the calcium alginate, where is reported the promotion of homeostasis and the alginate matrix is used to facilitate the aggregation of platelets and erythrocytes (Suzuki et al., 1998).

Various combinations were chemically assayed to understand and increase the quality of the alginate-based hydrogels to the tissue engineering and reduce the risk of the matrix deformation (Zhao et al., 2010). Currently, the alginate hydrogels are gaining interest in the bone regeneration, by the easiness of transforming into the ideal matrix with the system of drug delivery, of bone stem cells or osteoinductive factor. So, alginate-based have potential because of their inherent characteristics and, furthermore, their ability to fill irregular spaces. However, the biodegradability and the lower mechanical stiffness are the problems that need to be solved, consequently, there are ongoing studies to mixture various natural polymers to tackle these mechanical problems (Kolambkar et al., 2011; Im et al., 2017).

Other biomedical area where the alginate biomaterials are used is the cell culture. The alginate hydrogels are mainly used as model system for cell culture. These hydrogels can easily adapt to 2-D or 3-D cell culture systems; the alginate inherent properties turn the hydrogels into blank slate with a versatile system to incorporate proteins or cells adhesions factors. These findings from *in vitro* can be easily transposed to *in vivo* studies and applications, due to the biocompatibility and easy insertion of alginate into the body (Lee and Mooney, 2012).

Furthermore, the alginate-based matrix is reversible, and so can be repeated unlimited times, where the cell-cell interactions are dominant. The matrix, when broken, can recover within minutes. This can be important during the insertion cell culture in the body, where injecting the alginate hydrogel can be in gel-form and solidify once it is placed in the targeted area in the body (Drury et al., 2005).

5.2.4.1.2 Drug Delivery Systems

Alginate is a non-degradable compound in mammals, but the ionically cross-linked alginate hydrogels can be dissolved by the releasing of the divalent ions in the surrounding media to the chemical reactions, involving the ions such as sodium or potassium or calcium. But even if this occurs, it is proven that alginate is not totally removed from the human body. Thus, there are new techniques involving the oxidation of alginate chemical structure so the alginate hydrogels can be dissolved in aqueous media and removed from the human body. Also, this technique has been proved to have potential as a vehicle for drug delivery for various compounds and applications (Al-Shamkhani and Duncan, 1995; Bouhadir et al., 2001).

These alginate-based drug delivery systems can support multiple drugs and perform sequential or simultaneous delivery, being important the study of chemical structure and incorporation system in the hydrogel to see the release kinetics to diminish the release time errors and faulty drug delivery systems (Bouhadir et al., 2001). For this case, alginates have been explored in drug delivery systems, in combination with complementary polymers, like chitosan (Sandford and Steinnes, 1991).

5.2.4.1.3 Other Biomedical Applications

Alginate is a natural agent against heavy metals poisoning by the natural chelating activity, and it is demonstrated to be a coadjuvant operational in food poisoning medical cases (Eliaz et al., 2007). In the medical area, there is a steady growth of awareness to use alginate as pharmaceutical compound to give response against some diseases or health problems, such as diabetes, cholesterol, obesity and digestive tract problems, because the dietary fiber compartment of alginate can regulate the intestinal tract (Mariya and Ravindran, 2013; Brown et al., 2014).

5.2.4.2 FOOD

The application of alginate in the food industry is approved by the food regulatory agencies in all of the world, including the Food and Drug Administration (FDA) (USA), and the European Food Safety Authority (EFSA) (EU), which drives a high demand for food manufacturing industry as stabilizer, emulsifier, and thickening agents (OMRI, 2015). Alginates are usually applied in food industry as natural additives, so alginates have been coded from European Union (EU) as food additives to be identified in the commercially available food, and alginate codes diverge within the ion type linked to the alginic acid (Pereira, 2012; Pereira et al., 2013). So, the European codes for alginates approved for the food industry are the alginic acid-E400, sodium alginate-E401, potassium alginate-E402, ammonium alginate-E403, calcium alginate-E404 and propylene glycol alginate (PGA)-E405.

Alginates are extensively applied in manufacturing food processing as food additives, in foods like jams, jellies, ice creams, dairy products, etc., to increase the overall quality and stability of the food. Also, alginates can be used as thickeners and homogenizers in water-in-oil emulsions, such as mayonnaise and salad dressings. However, the alginate needs to be wisely chosen, due to, for example, the sodium alginate not being suitable when the emulsion is acidic or due to the insoluble alginic acid that rapidly forms in the acidic solutions. So, for these emulsions, the PGA is widely applied, due to the firm form and low reaction in mildly acidic conditions (Pereira, 2012).

The advantage of alginate application in the food industry is that the human system does not have enzymes to breakdown the molecule and, consequently, alginate acts as a dietary fiber in the human digestive system, augmenting the satiety symptom and decreasing the food consumption in humans, dropping the energy ingestion by humans and, consequently, inhibiting the obesity growth (Brown et al., 2014).

5.2.4.3 COSMETIC

Alginate is considered an omnipresent ingredient in the cosmetics industry. Alginates are doing a trustworthy employment in quietness and far away from the all-too-fleeting active agents and, due to this, alginate is not been recognized by the major part of the humans. Alginates are normally applied as emulsifier, stabilizer, and thickening ingredients in cosmetic products, making moisture-retaining surface films. They also have an active effect based on their water-retain inherent capacity, which is important to maintain the skin hydrated (Pereira, 2018b).

Moreover, alginates are water-insoluble molecules, nevertheless, alginate can absorb water efficiently (water-retaining properties), property identical to the hyaluronic acid application. Therefore, alginate has quite a lot of cosmetic purposes at the same time.

The sodium and potassium alginates are water-soluble forms and are applied in the cosmetic area as a primary emulsifier. Enhancing the steadiness of alginate, by mixing diverse types of alginates or also, adding other polymers, can help to give stability to the oil phase in the cosmetic products, devoid of subjoining synthetic ingredients to the cosmetic product. Generally, the human skin does not absorb alginates. PGA (E 405) is a mixture of an ester of propylene glycol with alginic acid and normally is applied identically to the alginic acid, being a usual ingredient to replace synthetic ingredients, such as Polyethylene glycol (PEG).

The alginates are normally ingredients applied in hand jellies and lotions, ointment bases, pomades, and other equivalent products, greaseless creams, dentifrices, and other products (Malinowska, 2011; Fabrowska et al., 2015). Alginate has moreover been reported with a bioactivity, the antioxidant, and so they are being used to avoid skin aging and cutaneous disorders. Furthermore, anti-oxidative agents can support the maintenance of the organoleptic possessions of the cosmetic formulae by deterring the lipid oxidation, as a consequence evading deviations in look, odor, and flavor of the cosmetic product (Wang et al., 2015).

5.3 FUCOIDAN

5.3.1 GENERAL OVERVIEW

Since the discovery and isolation of fucoidan for the first time by Kylin (1913), there have been many studies highlighting how to extract and isolate this compound, in order to elucidate the chemical structure and analyze the bioactivities (Morya et al., 2012; Fletcher et al., 2017). Nowadays, with the evolving purification methodologies, molecular characterization and bioactivities screening equipment's (Morya et al., 2012), several patents were developed regarding this field of research (Kearl and Mina, 2007; Johnson, 2008; Shakleej et al., 2008; Kitano and Hamaguchi, 2010; Ki, 2012; Oh, 2012).

Even though the main source of fucoidan are brown seaweeds (Morya et al., 2012), this biomolecule is also present in other marine invertebrates, such as sea urchins, sea cucumbers and seagrass (Fitton et al., 2015; Chollet et al., 2016). Fucoidan is part of the seaweed cell wall constitution and it is characterized as a complex sulfated polysaccharide (Ale et al., 2011). Within brown seaweeds as a fucoidans' source, the most highlighted by researchers have been the members of the order Fucales and Laminariales (Ale et al., 2011; Rodriguez-Jasso et al., 2011; Yang et al., 2013; Zhang et al., 2020).

Thus, several different seaweed species have already proven the presence of fucoidan with biotechnological applications, namely *Cladosiphon* sp., *Macrocystis pyrifera, Ecklonia radiata, Seirococcus axillaris, Saccharina latissima, Fucus vesiculosus, Fucus serratus, Fucus evanescens, Fucus distichus, Fucus spiralis, Laminaria digitata, Dictyosiphon foeniculaceus, Dictyopteris delicatula, Undaria pinnatifida,* (*Saccharina*) *Laminaria japonica, Ascophyllum nodosum, Durvillaea potatorum, Sargassum* sp., *Padina boryana* (formerly *Padina commersonii*) (Cumashi et al., 2007; Fitton et al., 2015; Schneider et al., 2015; Sanjeewa et al., 2019).

However, the availability and chemical constitution of this molecule in seaweeds is susceptible to biotic or abiotic parameters and not only differs seasonally as well as according to the species and geolocation (Rioux et al., 2007; Foley et al., 2011; Morya et al., 2012). Recently, the biological properties of this polysaccharide have been tested and demonstrated a high potential at therapeutical and nutraceutical levels, revealing its relevance among the polysaccharides synthesized by brown seaweeds (Fitton et al., 2019).

Fucoidan's chemical composition and structure can be affected by the previously mentioned biotic and abiotic parameters, but also by further downstream processes (Qu et al., 2014). For instance, Ponce et al. (2003) revealed that with different extraction methods, it was possible to extract two types of fucoidans. Therefore, it is important to take into account that changes in fucoidan's chemical structure can lead to different or the absence of bioactivities (Qiu et al., 2006).

5.3.2 EXTRACTION AND ISOLATION METHODS

Commonly, seaweeds undergo a pre-treatment where they are washed with tap or distilled water. This step will allow the removal of epiphytes, salt, and other impurities, to guarantee the purity of the polysaccharide extracted. Thereafter, the processes of drying or freezing will enable feedstock conservation. Another process that could support higher yields of extraction is the milling process. By reducing the particle size, the increase of the contact area with the solvent can enhance the yield (Dobrinčić et al., 2020). In some cases, to enhance fucoidan purity, the biomass is submitted to an ethanolic pre-extraction process, whereas pigments and proteins are removed (Rioux et al., 2007; Foley et al., 2011).

Nowadays, the selection of the extraction and isolation methods are key processes in order to obtain bio-compounds with biotechnological applications. According to the selected extraction method, fucoidan structure could be affected and so could its bioactivity (Ale and Meyer, 2013). Traditionally, polysaccharide extraction is performed with heated aqueous or acidic solutions for several hours (Hifney et al., 2016). However, these methods have evolved in order to be faster, while increasing yields and being eco-friendly.

For instance, microwave-assisted extraction (MAE) causes cell disruption through heated microwave radiation that leads to the increase of mass transfer of the bioactive compounds to the cold solvent (Eskilsson and Björklund, 2000; Srogi, 2006). Nevertheless, in this method, some parameters can differ, such as extraction time, pressure, and ratio between macroalgae/solvent (Zayed and Ulber, 2019). Previous research showed that fucoidan extracted from *Fucus vesiculosus* through MAE under 1 minute of extraction, 120 psi and 1 g macroalgae/25 ml water was possible to achieve a yield of extraction of 18% (Rodriguez-Jasso et al., 2011). Thus, with this technique, it is possible to reduce the time of extraction and consequently reduce the energy requirements (Khalil et al., 2018).

Another strategy to cause cell disruption efficiently is using ultrasound-assisted extraction. This method consists in reducing the size of particle using sound waves to promote the polysaccharides released to the solvent. However, there are different parameters that could affect fucoidans' yield, quality, and further industrial use, such as temperature, pressure, extraction time and the ultrasonic power (Thao et al., 2020). New techniques are being developed, such as subcritical water extraction, supercritical fluid extraction and pressurized fluid extraction. These methods are based in high temperature and pressure (Wu, 2017) and the yield could be different according to the variation of technical and operational parameters like temperature, pressure or the chosen solvent (Kadam et al., 2013).

In supercritical fluid extraction, the solvent used is normally carbon dioxide, which is considered an eco-friendly solvent, because it is recoverable from the extract and it decreases the waste (Rustad and Hayes, 2012; Kadam et al., 2013a). This technique is efficient and feasible to upscale. However, the implementation of this method involves a high initial investment and one of the drawbacks of the method is that it is not suitable to polar compound extraction (Crampon et al., 2011). Alternatively, pressurized fluid extraction requires less solvent consumption, because it involves high temperature and pressure (Kadam et al., 2013a). Temperature increase leads to the increase of mass transfer in a short period of time (Shang et al., 2011). This technique was already applied successfully to obtain a fucoidan rich-extract from *Himanthalia elongata* (Santoyo et al., 2011), although it is not suitable at an industrial scale due to high energy demands (Michalak and Chojnacka, 2014).

Recently, fucoidan extracts can also be obtained through enzymatic processes with high selectivity (Lakmal et al., 2015). The enzymes capable of performing this extraction are called fucoidanases, such as alpha-L-fucosidase that can be found in fungi or in marine mollusks (Berteau et al., 2002;

Shvetsova et al., 2017). Although it is a method with a high selectivity, it is very expensive and so this technique is not industrially feasible (Michalak and Chojnacka, 2014). Afterwards, to isolate this bioactive polysaccharide with high chemical stability usually is used column chromatography. This separation could be achieved through anion or cation exchange chromatography and through gel filtration chromatography, depending on fucoidans' molecular weight (Cong et al., 2016; Hu et al., 2016; Li et al., 2018).

5.3.3 CHEMICAL STRUCTURE

Fucoidans' first chemical structure characterization was around 100 years ago (Kylin, 1913). This biomolecule's molecular weight can range between 7 kDa and 2300 kDa (Zvyagintseva et al., 1999; Rioux et al., 2007). Their chemical complexity and diversity, due to the branching structure, is still an issue under research (Morya et al., 2012; Fernando et al., 2019). Sulfate groups and other minor elements in fucoidan structure are pivotal in order to assess this fucan bioactivities (Hidari et al., 2008; Pomin, 2012; Thinh et al., 2013).

Further research was already done to study fucoidan chemical structure through methylation analysis and NMR (Li et al., 2008b; Anastyuk et al., 2017), but, in a general way, it is constituted by a fucose skeleton and other carbohydrates such as mannose, glucose or xylose (Li et al., 2008; Ale et al., 2011; Fitton, 2011). Research showed that the main general structure was characterized by a bone of (1→2) linked 4-O-sulfated fucopyranose residues. It was also determined that 3- linked fucose with 4- sulfated groups are around fucose residues (Figure 5.2). Besides that, fucoidan molecule has a branch every 2–3 fucose residues (Chevolot et al., 2001; Pereira et al., 2013).

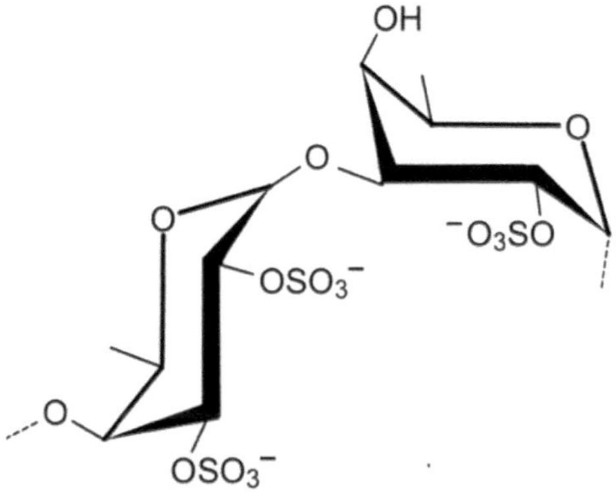

FIGURE 5.2 Idealized structure of the chemical units of fucoidan.

Still, fucoidan's chemical structure differs according to the species selected. Some examples are the macroalgae *F. evanescens* and *F. serratus* (order Fucales), composed by a backbone of α (1→3)- and α (1→4)-linked L-fucopyranose, which could be replaced by sulfate or acetate groups (Bilan et al., 2002; Cumashi et al., 2007). While in other seaweeds belonging to the same order (*A. nodosum* and *F. vesiculosus*) it has a repeating structure of α(1→3)-linked L-fucopyranose residues with sulfate at the C-2 position, with the α(1→3)-linked L-fucopyranose residues linked to α(1→4)-L-fucopyranose residues with disulfate at C-2 and C-3 (Chevolot et al., 1999). It is necessary to consider that sulfated polysaccharide extraction methods also affect their chemical structure. However, there are many parameters that could influence this fucan structure, such as phosphorylation, alkylation or the sulphation degree (Morya et al., 2012). For this reason, there is a need to tackle the gap between fucoidan structure and their bioactivities.

5.3.4 APPLICATIONS

Seaweeds' fucans have gained attention from the scientific community. Among them, fucoidan has received attention for its bioactive properties (Dobrinčić et al., 2020). This polysaccharide presents a wide range of bioactive properties, according to its biological and physicochemical characteristics, such as molecular weight, chemical structure, sugar composition and sulfate groups (Qu et al., 2014). As it was revised by Ale et al. (2011b), the relation between fucoidan structure and bioactivity is strictly related and it depends on several factors. As it was already mentioned, species selection, harvesting season and extraction method selection are critical drivers to obtain this bioactive compound (Li et al., 2008b; Lim and Wan Aida, 2017). Several studies showed evidence of fucoidan bioactivities, such as biomedical, nutraceutical, and cosmeceutical that will be further thorough presented (Fitton and Dragar, 2006; Holtkamp et al., 2009; Morya et al., 2012; Schneider et al., 2015; Zhao et al., 2018; Delma et al., 2019).

5.3.4.1 BIOMEDICAL

5.3.4.1.1 Antioxidant

Aerobic metabolic pathways bring the increase of reactive oxygen species (ROS), whose concentration is controlled through the antioxidant system. Radical scavenging capacity is a property that enables oxidation protection, thus preventing many diseases due to oxidative stress. There are several synthetic antioxidants available in the market, but consumer awareness and preference of natural products lead to the development of novel natural products. Thus, when studies pointed that marine polymers hold that capability, brown seaweed polysaccharides were further investigated. Researchers found evidences that high purity fucoidan rich extracts from *Sargassum tenerrimum* exhibited antioxidant activity (Ashayerizadeh et al., 2020). Researchers studied another seaweed of the same genus, *Sargassum fusiforme*, and evaluated its antioxidant activity, promoting stress resistance and cellular protection (Zhang et al., 2019). Fucoidan acts as a promoter of enzymes that belong to the antioxidant mechanism of action, such as superoxide dismutase (SOD) and glutathione peroxidase (GPx) (Chen et al., 2018).

Antioxidant activity of 6 brown seaweeds' fucoidan was evaluated through DPPH test. The selected macroalgae were *Fucus vesiculosus*, *F. serratus*, *F. evanescens*, *Dictyosiphon foeniculaceus*, *Laminaria digitata* and *Saccharina latissima*. Still, *F. serratus* was the one that exhibited the highest radical scavenging capacity (24.6%) (Bittkau et al., 2020). Although fucan's bioactivities are directly related with sulphation degree and molecular weight, the antioxidant activity could also be related to the coextraction of phenolic compounds.

5.3.4.1.2 Antitumor

Despite fucoidan's chemical diversity, further research is justified for its therapeutical applications. For instance, this complex polysaccharide enhances the immune system and triggers antitumor mechanisms in order to induce carcinogenic cells apoptosis (Freile-Pelegrín and Tasdemir, 2019). However, this mechanism is not yet fully elucidated. In 2018, researchers selected (*Saccharina*) *Laminaria japonica* as fucoidan source and therefore, after polysaccharide purification, antitumor potential was effectively demonstrated against HeLa cells (Zhao et al., 2018). However, through the chemical analysis of several fractions of the extracts, researchers concluded that antitumor ability was related to sulfate concentration, the ratio between fucose and sulfate, as also varying according to the monosaccharide present (Foley et al., 2011). Fucoidans' dose of 1–1000 μg/mL extracted from *S. latissima* also showed anticancer activity against Raji cells (Burkitt lymphoma; B–lineage). However, in this study fucoidan purification process lead to the loss of bioactivities (Schneider et al., 2015). An experiment in rats also demonstrated that commercially available fucoidan purified from *F. vesiculosus* was effective to inhibit lung cancer through its ingestion (Huang et al., 2015).

5.3.4.1.3 Immunomodulatory

Compounds that present this bioactivity could be used to reduce some drugs side effects. Immunomodulation is a complex mechanism that regulates the immune system (Prendergast and Jaffee, 2007). Researchers isolated a low and high molecular weight fucoidan from *U. pinnatifida*, in order to establish a correlation between the physicochemical properties of this polysaccharide and its immunomodulatory activity. In the mentioned study, high molecular weight fucoidan had better results with NK-92MI cells, a human natural killer (NK), exhibiting immunostimulant and immunosuppressive activity in mice (Yoo et al., 2019).

Murine macrophage (RAW264.7) are in the first line of immune system protection (Klimp et al., 2002). Investigators provided evidences that in a concentration between 1 and 50 μg/mL, low molecular weight fucoidan (LMWF) extracted from *U. pinnatifida* was effective to enhance this murine macrophage activation (Bi et al., 2018). In this context, tests were performed with mouse and researchers concluded that fucoidan is capable of stimulating dendritic cells and T cells activation, promoting protection against tumors and virus (Zhang et al., 2015). These cells will be responsible for the increase of pro-inflammatory cytokine synthesis (Hayashi et al., 2008).

5.3.4.1.4 Anti-Inflammatory

Tissue inflammation is a natural process that acts as a defense response; however, if this mechanism is not controlled, it can lead to serious cell damage (Taniguchi and Karin, 2018). Despite the mechanism not being fully understood, seaweeds' polysaccharides can act as anti-inflammatory agents (Cumashi et al., 2007). A study of Sanjeewa et al. (2019) demonstrated evidences supporting that fucoidan (1–200 μg/mL) extracted from *P. boryana* showed anti-inflammatory activity in RAW 264.7 cell lines. Fucoidan extracts from the macroalgae *S. fusiforme* (previously known as *Hizikia fusiformis*) revealed to be a potential target to develop biomedical products, by showing its anti-inflammatory bioactivities. This was evidenced through the reduction of prostaglandin E_2 and pro-inflammatory cytokines synthesis, as well as the suppression of NO synthase and cyclooxygenase-2 (COX-2) in cell lines (RAW 264.7) (Wang et al., 2020). The previous investigation also highlights enzymatic extraction as a feasible extraction method of bioactive crude fucoidan extracts from *E. cava* with therapeutic applications. In this study, those extracts exhibited anti-inflammatory activity through the inhibition of NO synthase (Lee et al., 2012).

5.3.4.1.5 Antiviral

Seaweed polysaccharides are able to modify cell superficial properties in order to prevent enveloped viruses infection (Chen et al., 1997). For instance, fucoidan presents antiviral activity against dengue virus type-2, as it was reported by Hidari et al. (2008), which has demonstrated that *Cladosiphon okamuranus* fucoidan interacts directly with virus envelope proteins. *In vivo* tests were performed in mice (5 mg/day every 8 h, three times per day) with fucoidan provided by *U. pinnatifida* that revealed their antiviral bioactivity against herpes simplex virus type 1 (HSV-1). Thus, fucoidan oral ingestion was effective in infection protection (Hayashi et al., 2008). *Sargassum tenerrimum* is also a potential candidate for the development of natural viral treatment approach. This species' extracted fucoidan has shown antiviral activity against Herpes simplex viruses type-1 with concentrations up to 1000 μg/ml with no toxicity for the cells (Sinha et al., 2010). Hepatitis-B virus is an infection with a global dimension; thus, research has been focusing in a natural approach for this infection treatment. *In vivo* assays were performed with mice, in order to evaluate fucoidan potential to be a natural treatment through an intraperitoneal injection (Li et al., 2017). In this context, researchers concluded that fucoidan was effective in inhibiting viral replication.

5.3.4.1.6 Neuroprotection

In Asia, where macroalgae are part of the daily diet, traditionally they are also used to prevent neurological diseases (Olasehinde et al., 2019). For instance, Alzheimer's disease is characterized by the high concentration of beta-amyloid plaques stress caused by a cascade reaction that leads to neurodegeneration (Paris et al., 2011). Seaweed synthesizes secondary metabolites with antioxidant and neuroprotective activity, specially fucoidan (Alghazwi et al., 2019; Olasehinde et al., 2019). For example, it was experimented in rats, *F. vesiculosus* commercially available fucoidan at a 1 μM protects neurons of beta-amyloid in high concentration (Jhamandas et al., 2005). By the inhibition of ROS, fucoidan has anti-amyloid activity (Alghazwi et al., 2019). Another study points that low molecular fucoidan (from *E. radiata*) could have potential in brain injuries' neuroprotection due to its antioxidant and anti-inflammatory properties (Alghazwi et al., 2020). Other evidences of neuroprotection in rat models from fucoidan isolated from *S. japonica* are highlighted through dopamine system and mitochondrial function preservation, thus preventing Parkinson's disease (Zhang et al., 2018).

5.3.4.1.7 Anticoagulant and Antithrombotic

In vitro and *in vivo* research on fucoidan anticoagulant properties have been focused mainly in three species, namely *U. pinnatifida*, *S. japonica* and *F. vesiculosus* (Morya et al., 2012). This bioactivity could be not only correlated with sulfate groups and its position (Nishino et al., 1989), but also with their molecular weight (Chandía and Matsuhiro, 2008). Fucoidan has the capability of promoting endothelialization (Yao et al., 2020). Besides that, fucoidan application as anticoagulant or antithrombotic properties has the advantage of not causing hemorrhage (Min et al., 2012). Another biomedical application is that fucoidan inhibits arterial smooth muscle multiplication, which is a key process in vascular remodeling, being relevant for instance in organ transplants (Akyürek, 1998; Religa et al., 2000).

5.3.4.2 FOOD

Fucoidans' food application and therapeutic potential is highlighted by several authors (Jiao et al., 2011; Pereira, 2018; Freile-Pelegrín and Tasdemir, 2019). Within these bioactivities, antibacterial is one of fucoidans' properties with a wide range of applications, from therapeutic to food safety (Li et al., 2008). The high concentration of sulfuric acid and glucuronic acid present in fucoidan depolymerization is one of the drivers that leads to bacteria apoptosis induction (Zhao et al., 2018). Antioxidant fucoidan bioactivity is also important in the food industry to avoid food products' oxidation, thus increasing their shelf-life time (Balboa et al., 2013). European food safety competent authorities considered fucoidan from the seaweeds' species *U. pinnatifida* and *F. vesiculosus* as a novel food product (Regulation (EU) 2018/1023 2020). Also, in America, fucoidan is generally recognized as safe for human consumption (Adams, 2016). Besides fucoidan nutraceutical effects, this polysaccharide holds emulsifying properties, thus promoting quality in food products (Hifney et al., 2016).

5.3.4.3 COSMETIC

Seaweeds are a rich source of bioactive compounds with cosmetic application (Kim et al., 2018). For instance, *Chnoospora minima* and *Sargassum polycystum* fucoidan was isolated and chemically characterized, showing a high sulfate content and low protein or phenolic compounds. Those extracts exhibited a wide range of bioactivities for cosmetic applications, such as skin protection through a UV barrier and anti-wrinkling, as others beneficial to the consumer, like anti-inflammatory and antioxidant (Fernando et al., 2020). Other studies performed with fucoidan isolated from several seaweed species from *Sargassum* genus highlight antioxidant, anti-wrinkle, and the ability to absorb and preserve moisture (Jesumani et al., 2019; Shannon and Abu-Ghannam, 2019).

5.4 LAMINARIN

5.4.1 GENERAL OVERVIEW

Laminarin is a polysaccharide present in brown macroalgae. Laminarin is a low-molecular-weight reserve polysaccharide and bioactive compound from brown algae, extracted like a product of photosynthesis and placed in vacuoles of cells (Kadam et al., 2015a), giving flexibility to the algae and allowing them to adjust to diverse water movements (Rinaudo, 2007). Laminarin was for the first time isolated in 1885 by Oswald Schmiedeberg from a brown algae from the family *Laminariaceae* of the phylum Phaeophyta (Bull, 1966). Laminarin can be principally discovered in *Laminaria, Saccharina* or *Eisenia* genus located in Asian and some European countries (Kadam et al., 2015a; Kraan, 2012).

Dry weight (DW) mass of laminarin extracted from a species relies on the origin and harvesting season (Zia et al., 2017). Laminarin is located in the fronds of *Laminaria/Saccharina* and to a lesser extent in *Ascophyllum* and *Fucus* genus, as well as *Undaria* (Kraan, 2012). The content of laminarin in seaweeds changes like other polysaccharides present in algae. Normally, laminarin form about 10% of algal dry mass but, dependent on the season, its content can be of over 32% (Holdt and Kraan, 2011). Algae produce their biomass reserve after the rapid growth phase in spring for surviving the winter where photosynthesis barely occurs (Painter, 1983; Jiang et al., 2011). So, as storage polysaccharides, their concentration gets its maximum in the autumn and goes down to its minimum in the spring (Holdt and Kraan, 2011). As a result, in spring laminarin is absent while in fall and winter can represent more than 35% of the dried weight of the fronds (Rinaudo, 2007). As shown in Table 5.1, the content of laminarin changes in relation to species, habitat, harvesting season, and method of extraction (Kadam et al., 2015). These changes that depend of the season have to be taken into account to select the best harvesting period for the extraction of laminarin for commercial or industrial application (Ledesma-Hernandez et al., 2013). Anticoagulant activity was first reported in brown algae in 1941, when *Laminaria* presented anticoagulant properties of its active compound (Shanmugam and Mody, 2000). Laminarin simply exhibits anticoagulant activity once it suffers structural modifications as sulphation, oxidation or reduction and the anticoagulant activity is higher chemically by growing the degree of sulphation (Shanmugam and Mody, 2000; Kraan, 2012).

TABLE 5.1 Algal Yield of Laminarin Sources.

Seaweed	Content	Reference
Saccharina latissima	33% of dry weight	Jensen (1956)
Laminaria hyperboria	32% of dry weight	Holdt and Kraan (2011)
Laminaria digitata	36% of dry weight based on the season	Devillé et al. (2004)
Fucus vesiculosus	84% of total sugars	Rioux et al. (2007)
Undaria pinnatifida	3% of dry weight	Je et al. (2009)
Ascophyllum nodosum	4.5% of dry weight	Macartain et al. (2007)
Ecklonia maxima	9.57% of dry weight	Rocher et al. (2021)
Ascoseira mirabilis	1-15% of dry weight	Gómez and Wiencke (1998)
Fucus distichus	10-20% of dry weight	Dethier and Williams (2009)
Sargassum muticum	0.3% of dry weight	Gorham and Lewey (1984)
Pelvetia canaliculata	0.03% of dry weight	Graiff et al. (2016)
Fucus serratus	0.38% of dry weight	Graiff et al. (2016)
Sargassum mcclurei	8.49% of dry weight	Xuan Cuong (2020)
Ecklonia radiata	7.9–27.9% of dry weight	Charoensiddhi et al. (2016)
Cystoseira barbata	7.27% of dry weight	Sellimi et al. (2018)

5.4.2 CHEMICAL STRUCTURE

Laminarin chemical structure is made of branched glucose sections connected by β-1,3-glycosidic bonds with several β-1,6-glycosidic bonds (Edgar, 2008; Choi et al., 2012) (Figure 5.3). The common ratio of β-1→3 to β-1→6 bonds in laminarin is 3:1 (Kraan, 2012).

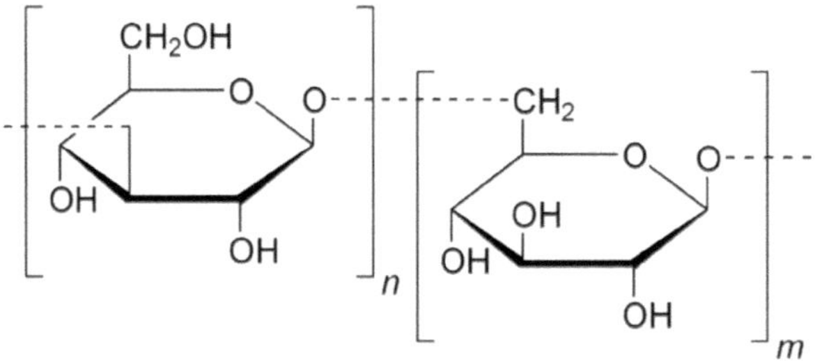

FIGURE 5.3 Chemical structure of β-1,3-1,6-glucan (Laminarin, Laminaran) (Edgar, 2008).

Laminarin is an uncharged polysaccharide at neutral pH that can form a triple-helical conformation (Young et al., 2000). The normal structure of laminarin is plain but it can exhibit some heterogeneity on the reducing end: the extent of the principal chain modifies and M-type chain will terminate with 1-O-substituted D-Mannitol group, while G-type chain will end with D-glucose units (Read et al., 1996; Kadam et al., 2015b). The majority of laminarins are low molecular weight compounds with 25 to 50 glucose units, but the weight of the molecule is related to algal source, degree of polymerization (Nelson and Lewis, 1974; Alderkamp et al., 2007), and extraction conditions (Rioux et al., 2010). Besides to molecular weight and ratio of types of laminarin chains, environmental factors, like frond age and nutrients, water temperature, salinity, oceanic currents, waves, and depth, alter the structure and composition of laminarin too (Chizhov et al., 1999). In addition, the biological activity of laminarin is also influenced by factors like the method of extraction (Rinaudo, 2007).

The branching level conditions laminarin solubility, being branched laminarians more soluble than unbranched ones (Rupérez et al., 2002). Laminarin differs from other macroalgae polysaccharides (such as agar, alginate, and carrageenan) as it does not have thickening and gelling properties (Miao et al., 1995) nor does it form viscous solutions (Ledesma-Hernandez et al., 2013). The principal utility of the compound lies in the food industry and biomedicine because of its various biological properties (Kadam et al., 2015b). Modification of laminarin backbone using techniques such as irradiation, sulphation, oxidation, and reduction can lead to an enhanced bioactivity of the modified molecule (Zargarzadeh et al., 2020). In spite of more studies on the structure-activity relationship being needed, the researchers suggested the hypothesis that the introduction of sulfate groups modifies the spatial conformation of the chain, and also leads to repulsion between anionic groups that can lengthen the sugar chain and contribute to form hydrogen bonds. All this results in a helical structure and adopts an active conformation (Zargarzadeh et al., 2020). For instance, Ji et al. (2013) exhibited that sulfated laminarin showed higher antitumor activity than plain laminarin in human colorectal adenocarcinoma cells using the same conditions. An enhanced antioxidant activity is also observed when this laminarin is irradiated with gamma rays (Choi et al., 2011). In addition, Shanmugam and Mody (2000) reported that anticoagulant activity could be chemically enhanced through tuning the sulphation degree of laminarin.

5.4.3 EXTRACTION METHODS

The extraction of macroalgal polysaccharides (Table 5.2) is normally carried out using pre-treated dried macroalgae or defatted biomass with several solvents at different temperatures ranging from room temperature to 120 °C, for various hours (Garcia-Vaquero et al., 2017). In comparison to other polysaccharides, laminarin has an average low molecular weight (around 4–5 kDa) which makes it available to process and solubilize in aqueous medium (ca. 4–40°C) or organic solvents, whereas its degree of branching can affect the solubility in cold or hot water (Miao et al., 1995; Custódio et al., 2016). The solubility of laminarin depends on the degree of branching in its structure (Miao et al., 1995). Highly branched laminarin is soluble in cold water and hot water (Kadam et al., 2015a) whereas laminarin with low level of branching is soluble only in hot water (Miao et al., 1995; Rupérez et al., 2002). Laminarin exists either as a soluble polysaccharide extractable with cold water as in *Laminaria digitata* and *Fucus serratus* or in insoluble form as in *Laminaria cloustoni* and *Laminaria hyperborea*, which is extractable with hot water (Kim, 2015).

TABLE 5.2 Different Reported Methods for Extraction of Laminarin.

Seaweed	Method	Reference
Laminaria saccharina	After two hours of extraction with 0.09 M HCl at 4 °C, the supernatant was precipitated by adding 100% ethanol.	Black et al. (1953)
Laminaria saccharina	After two hours of extraction with 0.09 M HCl at 4 °C, the supernatant was neutralised with 1 M NaOH.	Black et al. (1953)
Sargassum linifolium	Extraction using HCl or oxalic acid solutions, followed by neutralisation with saturated sodium carbonate solution.	Abdel-Fattah and Hussein (1973)
Laminaria saccharina, *Laminaria digitata*	Extracted with 0.3 % H_2SO_4 at 50 °C for 1 hour, followed by 2.5 hours of extraction with 0.09 M H_2SO_4 (1:14 w/v) at 70 °C and precipitated with absolute ethanol.	Voronova et al. (1991)

TABLE 5.2 *(Continued)*

Seaweed	Method	Reference
Laminaria saccharina	After two hours of extraction with 0.09 M HCl at 4 °C, the supernatant was precipitated by adding 100% ethanol.	Yvin and Vasseur (1999)
Laminaria saccharina	Extracted for 2.5 hours at 70 °C with 0.09 M HCl (1:14 w/v) and precipitated with 100% ethanol.	Devillé et al. (2004)
Saccharina longicruris	Extracted with 1% $CaCl_2$ at 85°C for 4 hours, filtrate was combined with 2% NaCl and 95% ethanol in a 1:2 ratio, and dialysis was performed for 48 hours using a 15 kDa cut-off membrane.	Rioux et al. (2010)
Laminaria sp.	Pretreatment with ethanol, extraction with water at 50 °C for 2 hours, and precipitation with ethanol.	Cheng et al. (2011)
Laminaria japonica	Extracted at 60 °C for 1 hour using a 1:50 dried seaweed-to-water ratio.	Zha et al. (2012)
Eisenia bicyclis	Defatted algal fronds were extracted twice with 0.1 M HCl at 60 °C for 2 hours and precipitated with four volumes of 96 % ethanol.	Ermakova et al. (2013)
Sargassum fusiforme	Extracted twice with 0.1 N HCl at room temperature for 2 hours.	Jin et al. (2014)
Ecklonia maxima	Extracted with 0.09 M HCl at 70 °C for 2.5 h and treated with an equal volume absolute ethanol to precipitate the laminarin.	Rocher et al. (2021)
Pelvetia canaliculata	Extracted in ultrapure water or THF (8 mL) on a shaker (250 rpm) for 5 h.	Graiff et al. (2016)
Fucus serratus	Extracted in ultrapure water or THF (8 mL) on a shaker (250 rpm) for 5 h.	Graiff et al. (2016)
Fucus distichus	Extraction by algal tissue was soaked in 0.1N NaOH overnight and homogenized in a tissue grinder, after which 2N HCl was added to precipitate alginates.	Dethier and Williams (2009)
Ecklonia radiata	The dried seaweed powder was added to solvent ratio of 1:100 (w/v). Then incubated for 10 min under continuous shaking, and the pH adjusted to pH at 6.0. The enzyme preparation was then added, at 50 °C for 24 h. The extract was centrifuged at 8000g for 20 min at 4 °C.	Charoensiddhi et al. (2016)
Cystoseira barbata	Seaweed was treated twice with 0.1 M HCl for 2 h at 60 °C under constant stirring (250 rpm). Then centrifugated (5000 rpm, 10 min) and precipitated by addition of two volumes of absolute ethanol at 4 °C for 24 h.	Sellimi et al. (2018)

Normal extraction methods include hot and mildly acidic conditions followed by precipitation in ethanol, though increasingly more efficient and "green" processes as microwaves, ultrasounds or super-critical fluids are being investigated to improve the extraction rate and yield (Kadam et al., 2013a, 2015b). Many methods use hot and mildly acidic conditions to extract laminarin followed by precipitation with ethanol. Hydrochloric acid and sulfuric acid at a concentration range of 0.09–0.1 M are usually employed (Kadam et al., 2015). The molecular weight of the laminarin extracted may be affected by the conditions employed (e.g., solvent and time) in these processes (Rioux et al., 2010; Benito-Román et al., 2013). Through one of the two extraction methods developed by Black et al. (1953), laminarin can be extracted from macroalgae. The first technique uses precipitation of laminarin from its aqueous solutions using ethanol or other organic solvents. This product is called cold-water-soluble laminarin due to its solubility in cold water. The second technique uses precipitation of laminarin from acidic solutions and hydrochloric and sulfuric acids with pH up to 2.4 are the most typically acids used. This product is called cold water-insoluble laminarin which is insoluble in cold water but easily dissolves in warm water (Jard et al., 2013; Zia et al., 2017).

Deville et al. (2004) studied different yields in four extraction conditions resulting that yield decreased in the following order: hot HCl extract > cold HCl extract > hot H_2SO_4 extract > cold HCl (Devillé et al., 2004). Je et al. (2009) employed carbohydrate and protein hydrolyzing enzymes for the preparation of extracts from *Undaria pinnatifida* yielding a laminarin content of 3.2 ± 0.9 g per 100 g of DW. The result showed that enzymatic digestion of algae generated high bioactive yield and reported better biological activity compared to solvent extraction method (Je et al., 2009).

5.4.4 APPLICATIONS

5.4.4.1 BIOMEDICAL APPLICATIONS

Laminarin does not form a gel or gels any viscous solution, and the commercial use of laminaran on an industrial scale has so far been limited; medical and pharmaceutical uses are the principal potential applications (Kraan, 2012).

5.4.4.1.1 Antioxidant and Immunostimulatory

Laminarin shows high therapeutic potential due to its intrinsically increased antioxidant properties (Choi et al., 2011; Choi et al., 2012; Giese et al., 2015; Liu et al., 2017). Combating oxidative damage caused by ROS and free radicals is the main antioxidant laminarin activity presented until nowadays (Zargarzadeh et al., 2020). A study conducted in rats by Cheng et al. (2011) showed the effect of laminarin as a pulmonary oxidation and lipid peroxidation protective agent. Another work reported the mitigation of ROS generation by the administration of laminarin to the *in vitro* culture of porcine early-stage embryos (Jiang et al., 2018).

β-glucans present a low ability to collect free radicals and stimulate macrophages (Tsiapali et al., 2001), and an increase in the number of transcription factors involved in immune reactions was also shown, which points toward the immunostimulatory properties of laminarin (Lee et al., 2012). Alteration of the molecule may improve these effects. Choi et al. (2012) showed the effect of gamma irradiation on the antioxidant ability of laminarins and proved that a lower molecular weight (degraded) product enlarges the antioxidant potential. Laminarin-rich macroalgae was used in porcine and fish feed, showing a reduction in lipid oxidation of muscles in laminarin-fed pigs and increasing the shelf life of pork meat, keeping it fresh for a longer time. In addition, incorporation of a laminarin-rich diet in *Epinephelus* species (typical farmed fish in China and South Asian countries) increases the immune responses and stimulates the growth of fish (Moroney et al., 2012; Yin et al., 2014).

5.4.4.1.2 Anti-Inflammatory

Laminarin presents anti-inflammatory activity which has been related with the activation of immune responses (Kim et al., 2006; Neyrinck et al., 2007; Fuentes et al., 2011; Lee et al., 2012; O'Shea et al., 2016; Smith et al., 2018). Laminarin stimulates the immune system and shows immunomodulatory properties that help prevent inflammation (Zia et al., 2017). Laminarin can be considered as a potential therapeutic agent in case of systemic inflammatory syndrome and associated with hepatic damage (Neyrinck et al., 2007). In addition, Laminarin helps wound healing by promoting the adhesion of human skin fibroblast cells and favors bone formation by inducing proliferation of human osteoblast cells (Rioux et al., 2010).

5.4.4.1.3 Anticancer

Laminarin also reported antitumorigenic and antiproliferative properties that can lead to its potential application as an anticancer drug. In the last years, many studies showed potential anticancer effects of laminarin involving improvement apoptotic cellular death, colony formation inhibition and angiogenic potential inhibition (Zargarzadeh et al., 2020). Laminarin's role in preventing colon cancer may be due to its functional properties (as dietary fiber); however, the compound also exhibits antitumor activity via its capacity to promote apoptosis in cancer cells. Laminarin fermentation in the intestine is considered to inhibit compounds such as indole, ρ-cresol, and sulfur, which are indicators of colon cancer risk, besides that in experimental tests it showed a capacity to cause apoptosis in cancerous cells (Kuda et al., 2005).

Ji et al. (2012) showed laminarin can be considered as an effective antitumor agent due to its capacity to inhibit human colon adenocarcinoma (LoVo) cell proliferation and induce apoptosis through mitochondrial or death receptor pathways (Ji et al., 2012; Ji and Ji, 2014). The same group also showed the effect of sulfation on the antitumor activity of laminarin. The structural characterization revealed clear differences in the surface conformation of the sulfated laminarin, which proved an enhanced antitumor activity (Ji et al., 2013).

Ermakova et al. (2011) tested the bioactivity of laminarin and fucoidan extracted from *Alaria marginata* and *Alaria angusta* seaweeds in various cancer lines. Laminarin at the concentration of 100 μg/mL from *A. angusta* had a stronger effect against colony formation of HT-29 cells, whereas laminarin from *A. marginata* was more effective on T-47D cells (Usoltseva et al., 2016). Remya et al. (2018) created laminarin-based silver nanoparticles (AgNPs) (ranging from 21 to 34 nm) to study their effect on the apoptosis of retinoblastoma Y79 cells. The particles showed a moderate DPPH radical scavenging activity, but they had the ability to increase the cellular death in a dose-dependent manner (IC50 ~16.46 μg/mL) (Remya et al., 2018).

Malyarenko et al. (2017) proved the laminarin angiogenic potential inhibition of cancer cell lines. Laminarins obtained from *Sacharina cichorioides*, *Saccharina japonica* and *Fucus evanescens* were employed to investigate the *in vitro* capacity to inhibit colony formation, proliferation, and migration of human colorectal adenocarcinoma (HCT 116), malignant melanoma (SK-MEL-5) and breast adenocarcinoma (MDA-MB-231) cells. Sulfated *F. evanescens* laminarin showed the biggest inhibitory effect on colony formation of HCT 116 and MDA-MB-231 cell lines (> 50%) (Malyarenko et al., 2017). Other experimental test by Park et al. (2012) also proved its anticancer effects in HT-29 human colon cells, where laminarin induced apoptosis in a dose-dependent manner by its effect on the IGF-IR pathway. Besides colon cancer prevention, it is proposed that a diet based on *Laminaria* spp. (rich source of laminarin) may be responsible for a lower rate of breast cancer in the Japanese population (Teas, 1982).

5.4.4.1.4 Antibacterial and Antiviral

Kuda et al. (2015) tested the effect of laminarin and sodium alginate on the infections of *Salmonella typhimurium*, *Listeria monocytogenes*, and *Vibrio parahaemolyticus* in the human enterocyte-like HT-29-Luc cells. The conclusion was that laminarin prevents infection by preventing the adhesion of *Salmonella typhimurium*, *Listeria monocytogenes* and *Vibrio parahaemolyticus* bacteria on the surface of human enterocyte-like HT-29-Luc cells (Kuda et al., 2015). Another study also reported significant antimicrobial activity against *Staphylococcus aureus*, *Escherichia coli*, *Salmonella typhimurium*, and *Listeria monocytogenes* through purified laminarin extracts from the Irish brown seaweeds *Laminaria hyperborea* and *Ascophyllum nodosum* (Kadam et al., 2015). In addition, laminarin was reported to inhibit HIV adsorption on the surface of lymphocytes, and inhibit the activity of HIV reverse transcriptase, which proves its potential to be developed as a new antiviral medicine highly efficient and with lower toxicity (Wang et al., 2012).

Laminaran from *Fucus*, *Laminaria*, *Ascophyllum nodosum* and *Undaria pinnatifida* showed to have an effect on pathogenic bacteria (Hoffman et al., 1995). Due to their antibacterial properties, Laminarin, and fucoidan, can be efficiently applied in piglets at weaning (stage in which it occurs damage to the intestinal villi, which are a protective barrier against *Escherichia coli*) (Kim, 2015). O'Doherty et al. (2010) reported that it is possible to achieve a better performance in weaning piglets by supplementation with lactose extract from seaweed (*Laminaria* spp.) containing laminarin and fucoidan. The results proved that the use of lactose (150 g kg^{-1}), in combination with an extract of *Laminaria* spp., reduced *E. coli* levels by 8% and slightly increased the number of beneficial species of *Lactobacillus* (bacteria) in stool samples. In addition, this seaweed extract increased the coefficient of the total tract's apparent digestibility of feed components in comparison with a diet not containing an extract derived from *Laminaria* spp. (O'Doherty et al., 2010). The same results were shown by Dillon et al. (2010), who confirmed that the addition of an extract from *Laminaria* spp. (rich in laminarin and fucoidan) helps to achieve a significant reduction ($P > 0.05$) of *E. coli* populations in the gut and growth of the *Lactobacillus* population in comparison with a diet without the addition of seaweed extract. The utilization of laminarin as substratum for prebiotic bacteria seems to be a good commercial application (Devillé et al., 2004).

5.4.4.2 COSMETIC APPLICATIONS

The use of laminarin in cosmetics does not depend on their physical properties but on their broad range of bioactive features (Fernando et al., 2019).

Laminarin oligosaccharides obtained by the hydrolysis of laminarin from *Laminaria digitata* with laminarinase (α,β-(1–3)-glucanase) have stimulating, regenerating, energizing, and conditioning effect on human dermis fibroblasts and human epidermis keratinocytes (Yvin et al., 1999). Structural modification of laminarin can improve its functional features (Fernando et al., 2019). Photo-cross-linkable laminarin derived by chemical modification with glycidyl methacrylate has hydro gelling features and the capacity to encapsulate human-adipose-derived stem cells despite laminarin not being able to create hydrogels in its native state (Custódio et al., 2016). Turgeon et al. (2012) synthesized an antisenescence preparation from *S. longicruris* suggested to be employed in scar treatments like acne scars, surgical scars, burn scars, and stretch mark wounds. The procedure of laminarin stimulating wound repair involves the glucan specific receptors of human fibroblasts, which can regulate cellular function and subsequent interaction of glucan with the ligand (Browder, 2002). Laminaran sulfate inhibits heparanase activity, which is commonly expressed in skin and in the wound granulation tissue (Kim, 2015).

Yvin et al. (2006) proved that laminarin has anti-inflammatory action due to the inhibition of neutrophil polynuclears chemotaxy. They have patented a laminarin sulfate formulation having a degree of sulfation not less than 1.9 and between 2 and 2.5 that has a degree of polymerization similar to that of the natural molecule (20–30) for the treatment of non-specific inflammations (Yvin et al., 2006). Yvin et al. (1999) patented a laminarin formulation that can be employed for cosmetic and pharmaceutical uses as a healing unguent. Laminarin composition has been effective in the treatment of hydrofluoric acid corneal burns tested in rabbits (Hatipoglu et al., 2008). It has been shown to be a safe surgical dusting powder (Miao et al., 1999). Nevertheless, commercial application of laminarin is restricted and it is mainly used as a dietary fiber and an immunomodulant for oral application (Kim, 2015). Laminarin obtained from *S. longicruris* was employed in combination with an ascorbate as a substrate to foment matrix deposition of human skin fibroblasts. The thickness of the dermis increased because of the following growth in the deposition of collagen, proposing a potential use of this compound in cosmetic or therapeutic skin preparations (Ayoub et al., 2015).

5.5 ASCOPHYLLAN

Ascophyllan is a fucan, firstly extracted from the species *A. nodosum* (L.) by Le Jol, and was described by Larsen et al. (1966). Beyond *A. nodosum*, ascophyllan also occurs in *Padina tetrastromatica* (Larsen et al., 1966; Mohsin et al., 2013), the extraction methodology of this polymer, described by Larsen et al. (1966), is a mild acid-hydrolyze.

The seaweed *A. nodusum* must be milled and pre-treated with cold and diluted hydrochloric acid to remove inorganic cations. Then, acidic polysaccharides neutralization is performed by adding sodium hydroxide to the water-suspended residue. The resultant viscous mixture should be filtered and then de-alginated by: (a) digestion with alginase, followed by dialyze; (b) acidification adjusting the pH to 2 with hydrochloric acid, followed by centrifugation; and (c) addition of excess of calcium chloride and removal of the precipitated calcium alginate by centrifugation. The resulting precipitates (the alginates) could be discarded, and the supernatant, containing ascophyllan, should be further dialyzed against distilled water and consequently suspended and precipitated the indiffusible with 0.167 M sodium chloride, in one volume of ethanol (1% w/v) to obtain the final ascophyllan extract (Larsen et al., 1966).

Nakayasu et al. (2009) described the current applied methodology, presenting some adaptations of the previous one. In this case, after the pre-treatment with diluted hydrochloric acid (inorganic cations removal), the solution is submitted to filtration. The resultant residue is diluted in distilled water and then incubated for 20 h, at 20°C, with aqueous sodium hydroxide. After the solution is filtered again, filtered is collected for further steps and the residue is kept. This residue is resuspended in distilled water, incubated (2h, 100°C) and filtered. The filtrates are combined, and the pH is adjusted to 1.3 with hydrochloric acid. It is centrifuged and the alginate fraction (precipitate) is removed. The supernatant is filtrated again, and the liquid phase is neutralized with sodium hydroxide. The neutralized solution is concentrated by reduced pressure-evaporation, then dialyzed extensively against distilled water to eliminate generated salt by neutralization. Subsequently, alginate lyase is added to the solution to degrade residual alginate. After incubation at 30°C for 20h, the solution is dialyzed extensively against distilled water. Five parts of this solution are mixed with one part of a sodium chloride solution (in one molar concentration) and six parts of ethanol. This process of dissolution and precipitation with ethanol is repeated twice. Ascophyllan fraction (precipitate) is separated by centrifugation. The pellet is then dissolved in water, dialyzed extensively against water, and lyophilized (Nakayasu et al., 2009).

Ascophyllan (Figure 5.4) is a sulfated heteropolysaccharide (xylofucoglycuronan) composed by a backbone of hexuronic acids (glucuronic and mannuronic acid), fucose, and xylose (Larsen et al., 1966; Medcalf and Larsen, 1977). Ascophyllan is one of the least studied seaweeds' polymers; however, it presents high potential. So far, most of the investigations demonstrated antitumoral, anti-inflammatory, and immunomodulatory bioactivities (Nakayasu et al., 2009; Mohsin et al., 2013). Most likely, first study published on ascophyllan bioactivities was performed by Nakayasu et al. (2009). That investigation revealed cytotoxicity against a human myeloid leukemia (U937) cell line, in a dose-dependent manner. Ascophyllan-treated U937 cells, concretely with 62.5μg/ml, showed increased DNA fragmentation and consequent apoptosis of 50% of cells, after incubation of 72 hours (Nakayasu et al., 2009).

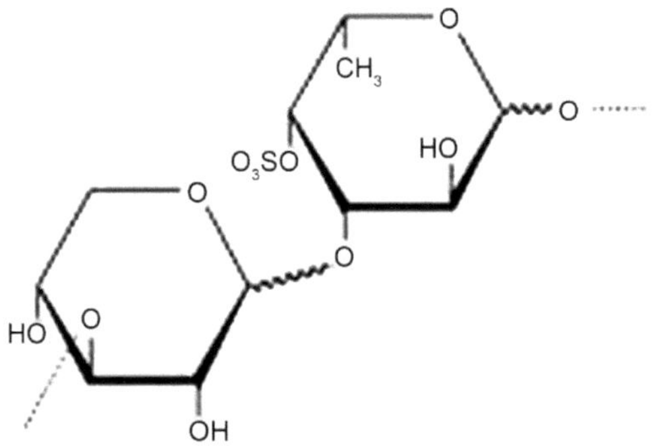

FIGURE 5.4 Chemical structure of 3-*O*-D-xylosyl-L-fucose-4-sulfate (ascophyllan).

Ascophyllan, still in a concentration-dependent approach, inhibited the migration and adhesion of B16 melanoma cells by reducing the expression of N-cadherin and enhancing the expression of E-cadherin in mice (Abu et al., 2015). Furthermore, the number of lung surface metastatic nodules in ascophyllan-treated mice was significantly reduced and enhanced NK cell activity, suggesting that ascophyllan may exhibit *in vivo* anti-metastatic activity on B16 melanoma cells through activation of the host immune system in addition to a direct effect on cancer cells (Abu et al., 2015).

In vitro and *in vivo* mice essays demonstrated enhanced NK cell activation and proliferation. These ascophyllan-activated NK cells produced IFN-γ and exhibited cytotoxic activity against cancer cells (Zhang et al., 2019). This compound features other immunomodulatory actions, improving the secretion of tumor necrosis factor-alpha (TNF-α) granulocyte colony-stimulating factor from mouse macrophage RAW264.7 cell line (Jiang et al., 2011, 2013). In the same cell line, ascophyllan induced ROS production, through the activation of NADPH oxidase. As nitric oxide, as well as iNOS, inducible-nitric oxide synthase, mRNA, and protein levels were significantly increased in ascophyllan-treated cells (Wang et al., 2013).

Ascophyllan promoted the maturation of mouse bone marrow dendritic cells, increased the expression of co-stimulatory molecules CD40, CD80, CD86 and MHC class II, and consequently enhanced Th1 and Tc1 responses (Zhang et al., 2014). Ascophyllan extracted from *P. tetrastromatica* inhibited carrageenan-induced inflammation, through TNF-α and interleukin-6 (IL-6) production, preventing cytokine stimulation. This anti-inflammatory of the natural extract was higher, when compared to the reference drug, diclofenac (Mohsin et al., 2013). These effects may be due to the high content of sulfate present in the algal polysaccharide. Therefore, ascophyllan could be used as an anti-inflammatory and immunostimulatory molecule interesting to the pharmaceutical sector, for the treatment of cancer and infectious diseases. However, the composition of the polysaccharides isolated from marine sources varies with tidal waves, and consequently, the biological activity will also vary (Renaud and Luong-Van 2006).

5.6 SARGASSAN

The polymer sargassan was discovered in 1973, by Abel-Fattah et al. (1973) from *Sargassum acinarium* (formerly, *Sargassum linifolium*) (Figure 5.5). The information about this polysaccharide is scarce. Abdel-Fattah et al. (1973) described the extraction method which consists in incubating the seaweed *S. linifolium* in water with the pH adjusted to 1, with hydrochloric acid (or 0.5N oxalic acid) (1 g:10 ml), at 80°C and for 2–3 h. Subsequently, the solution is neutralized with sodium carbonate and dialyzed against distilled water for 48 hours. The supernatant, which corresponds to the water-soluble polysaccharides fraction, is then percolated through a column of Lewatit S 100 (H^+) resin and the effluent dialyzed against distilled water, for 48 hours. Then, the dialyzed solution must be concentrated, under reduced pressure, at 40 °C, and treated with 4 volumes of absolute ethanol. Then, the extract needs to be centrifuged and the supernatant dried under low pressure and at room temperature. The precipitate isolated by centrifugation is then dissolved in water, and trichloroacetic acid solution is added to give a final concentration of 10%. The precipitated proteins are centrifuged out and removal of excess trichloroacetic acid from the supernatant is achieved by extraction (3×) with equal volume of ethyl ether. The aqueous layer is then separated and dialyzed for 2 days against distilled water. Thereafter, the dialyzed solution is concentrated, under reduced pressure at 40 °C, to half its volume and treated with 4 volumes of ethanol. The purified polysaccharide is isolated by centrifugation, washed with ethanol, ethyl ether and finally dried under vacuum at 40 °C (Abdel-Fattah et al., 1973, 1974c). At this point, it has been partially purified, acid-extractable, water-soluble polysaccharide fraction to further isolate sargassan. Abdel-Fattah et al. (1974c) also explained the procedures for the next step, the fraction of the foregoing polysaccharide. That could be performed in four alternative ways: (a) with lead acetate and barium hydroxide; (b) with cetylpyridinium chloride; (c) with copper acetate and ethanol; or (d) by column chromatography on DEAE-cellulose and DEAE-Sephadex (A-50, Cl-). By one of these procedures, it is obtained a partially purified fraction of polysaccharide material. Then, polysaccharide fraction, in a 0.3% aqueous sodium sulfate, is treated with 2.5% aqueous cetylpyridinium chloride, and the solution is centrifuged, and the precipitate is isolated. The resultant supernatant is dialyzed against distilled water and freeze-dried. Polysaccharide material is recovered from each of the precipitated fractions by several extractions of the Celite with 4M potassium chloride. Excess cetylpyridinium chloride is removed by precipitation with 1M potassium thiocyanate, followed by dialysis of the supernatant against distilled water. The dialyzed solution is concentrated to 25 ml and treated with 4 vol. of ethanol. Finally, the pure sargassan precipitate is isolated and dried (Abdel-Fattah, 1974c). Sargassum is a laminarin-like glucan and water-soluble sulfated heteropolysaccharide composed of glucuronic acid, mannose, galactose, xylose, and fucose residues, and a protein moiety (Abdel-Fattah et al., 1973, 1974 a,b).

FIGURE 5.5 *Sargassum acinarium* (formerly *S. linifolium*).

The researchers that discovered this polymer found that it has an anticoagulant effect. They concluded that sargassum has higher anticoagulating activity than the reference drug, heparin. Under similar conditions, plasma coagulated after 1 h in the presence of heparin, while, with sargassum, the coagulation did not occur until 72 h after (Abdel-Fattah et al., 1974c). This indicates that sargassum could be useful as a natural drug, an anticoagulant and, as this molecule has a similar structure as other fucans/polymers, it is presumable that has other bioactivities. There is more to research to conduct on sargassans' potential effects and applications.

5.7 CONCLUSION

Traditionally, humans have explored seaweed since ancient times due to their beneficial properties, such as medicinal, nutritional for humans and animal or fertilizers. For instance, alginate has a long history in the food area as a food additive, emulsifying, gelling, and stabilizing agent. Alginate exclusive characteristics and properties have opened new fields and industries where it can be harnessed and used with success.

Through the analysis of the literature, fucoidan is a polysaccharide present in several seaweed species, but only a few of them are explored commercially. Thus, the complexity and diversity of chemical structure elucidated leads to different bioactivities, which is also a future challenge. Within this context, fucoidan applications go from pharmaceutical to food industries, being considered for human consumption by American and European food security agencies. Taking into consideration fucoidan complexity, there is yet a lot to discover about this biomolecule and its applications. Brown seaweeds polymers are low-cost natural polymers and because of these particularities, alginate, and fucoidan are the most researched polymer among all seaweed polysaccharides, but laminarin have gaining also interest.

This chapter demonstrated the fully potential of the brown seaweeds polymers in several areas, namely biomedical, food, and cosmetic industry, although the roadmap of these carbohydrates as biomolecules is not ended and more research is needed, further the research already conducted in referred areas. Moreover, there are new approaches and areas that are being investigated to understand the potential of these polymers to tackle some challenges, mainly to substitute synthetic polymers.

ACKNOWLEDGMENTS

This work is financed by national funds through FCT – Foundation for Science and Technology, I.P., within the scope of the projects UIDB/04292/2020 – MARE – Marine and Environmental Sciences Centre and UIDP/50017/2020+UIDB/50017/2020 (by FCT/MTCES) granted to CESAM – Centre for Environmental and Marine Studies. Adriana Leandro thanks FCT for the financial support provided through the doctoral grant SFRH/BD/143649/2019 funded by National Funds and Community Funds through FSE. Sara García-Poza thanks to the project MENU – Marine Macroalgae: Alternative recipes for a daily nutritional diet (FA_05_2017_011) which co-financed this research, funded by the Blue Fund under Public Notice No. 5 – Blue Biotechnology. João Cotas thanks MARE, UC, and Universidade Nova de Lisboa, through its Chemistry Department of the Faculty of Sciences and Technology within the scope of the research project "Algae in Gastronomy – Development of innovative techniques for conservation and use – Alga4Food" (MAR-01.03.01-FEAMP -0016). Diana Pacheco thanks to PTDC/BIA-CBI/31144/2017-POCI-01 project-0145-FEDER-031144-MARINE INVADERS, co-financed by the ERDF through POCI (Operational Program Competitiveness and Internationalization) and by the Foundation for Science and Technology (FCT, IP). Ana M.M. Gonçalves acknowledges the University of Coimbra for the contract IT057–18–7253.

KEYWORDS

- **brown seaweeds**
- **European Food Safety Authority**
- **Food and Drug Administration**
- **fucoidan**
- **laminarin**
- **microwave-assisted extraction**
- **reactive oxygen species**
- **silver nanoparticles**
- **sulfated polysaccharides**

REFERENCES

Abdel-Fattah, A. F., & Hussein, M. M., (1973). Isolation of water insoluble laminaran-like polysaccharide from *Sargassum linifolium*. *Qualitas Plantarum et Materiae Vegetabiles, 22*(2), 181–187. doi: 10.1007/BF01100686.

Abdel-Fattah, F. A., Hussein, M. D., & Salem, H. M., (1973). Sargassan: A sulfated heteropolysaccharide from *Sargassum linifolium*. *Phytochemistry, 12*(8), 1995–1998. doi: 10.1016/S0031-9422(00)91522-X.

Abdel-Fattah, F. A., Hussein, M. M., & Mohamed, S. H. M., (1974a). Constitution of sargassan, a sulfated heteropolysaccharide from *Sargassum linifolium*. *Carbohydrate Research, 33*, 209–215.

Abdel-Fattah, F. A., Hussein, M. M., & Salem, H. M., (1974b). Some structural features of sargassan, a sulfated heteropolysaccharide from *Sargassum linifolium*. *Carbohydrate Research, 33*(1), 19–24. doi: 10.1016/S0008-6215(00)82936-5.

Abdel-Fattah, F. A., Hussein, M. M., & Salem, H. M., (1974c). Studies of the purification and some properties of sargassan, a sulfated heteropolysaccharide from *Sargassum linifolium*. *Carbohydrate Research, 33*(1), 9–17. doi: 10.1016/S0008-6215(00)82935-3.

Abdul, K. H. P. S., Lai, T. K., Tye, Y. Y., Rizal, S., Chong, E. W. N., Yap, S. W., Hamzah, A. A., et al., (2018). A review of extractions of seaweed hydrocolloids: Properties and applications. *Express Polymer Letters, 12*(4), 296–317. doi: 10.3144/expresspolymlett.2018.27.

Abe, K., Sakamoto, T., Sasaki, S. F., & Nisizawa, K., (1973). *In vivo* studies of the synthesis of alginic acid in *Ishige okamurai*. *Botanica Marina, 16*(4), 229–234. doi: 10.1515/botm.1973.16.4.229.

Abu, R., Jiang, Z., Ueno, M., Isaka, S., Nakazono, S., Okimura, T., Cho, K., et al., (2015). Anti-metastatic effects of the sulfated polysaccharide ascophyllan isolated from *Ascophyllum nodosum* on B16 melanoma. *Biochemical and Biophysical Research Communications, 458*(4), 727–732. doi: 10.1016/j.bbrc.2015.01.061.

Adams, M., (2016). *GRAS Notice No. GRN 000661*. https://www.cfsanappsexternal.fda.gov/scripts/fdcc/index.cfm?set=GRASNotices&id=661&sort=GRN_No&order=DESC&startrow=1&type=basic&search=661 (accessed on 13 October 2021).

Aderibigbe, B. A., & Buyana, B., (2018). Alginate in wound dressings. *Pharmaceutics, 10*(2), 42. doi: 10.3390/pharmaceutics10020042.

Akyürek, M. L., (1998). Cellular and molecular mechanisms in transplant arteriosclerosis. *International Journal of Molecular Medicine, 1*(1), 279–288. doi: 10.3892/ijmm.1.1.279.

Alderkamp, A. C., Van, R. M., & Bolhuis, H., (2007). Characterization of marine bacteria and the activity of their enzyme systems involved in degradation of the algal storage glucan laminarin. *FEMS Microbiology Ecology, 59*(1), 108–117. doi: 10.1111/j.1574-6941.2006.00219.x.

Ale, M. T., & Meyer, A. S., (2013). Fucoidans from brown seaweeds: An update on structures, extraction techniques and use of enzymes as tools for structural elucidation. *RSC Advances, 3*, 8131–8141. doi: 10.1039/c3ra23373a.

Ale, M. T., Maruyama, H., Tamauchi, H., Mikkelsen, J. D., & Meyer, A. S., (2011). Fucoidan from *Sargassum* sp. and *Fucus vesiculosus* reduces cell viability of lung carcinoma and melanoma cells *in vitro* and activates natural killer cells in mice *in vivo*. *International Journal of Biological Macromolecules, 49*(3), 331–336. doi: 10.1016/j.ijbiomac.2011.05.009.

Ale, M. T., Mikkelsen, J. D., & Meyer, A. S., (2011a). Important determinants for fucoidan bioactivity: A critical review of structure-function relations and extraction methods for fucose-containing sulfated polysaccharides from brown seaweeds. *Marine Drugs, 9*(10), 2106–2130. doi: 10.3390/md9102106.

Alghazwi, M., Charoensiddhi, S., Smid, S., & Zhang, W., (2020). Impact of *Ecklonia radiata* extracts on the neuroprotective activities against amyloid beta ($A\beta_{1-42}$) toxicity and aggregation. *Journal of Functional Foods, 68*(11), 103893. doi: 10.1016/j.jff.2020.103893.

Alghazwi, M., Smid, S., Karpiniec, S., & Zhang, W., (2019). Comparative study on neuroprotective activities of fucoidans from *Fucus vesiculosus* and *Undaria pinnatifida*. *International Journal of Biological Macromolecules, 122*, 255–264. doi: 10.1016/j.ijbiomac.2018.10.168.

Al-Shamkhani, A., & Duncan, R., (1995). Radioiodination of alginate via covalently-bound tyrosinamide allows monitoring of its fate *in vivo*. *Journal of Bioactive and Compatible Polymers, 10*(1), 4–13. doi: 10.1177/088391159501000102.

Amos, R. A., & Mohnen, D., (2019). Critical review of plant cell wall matrix polysaccharide glycosyltransferase activities verified by heterologous protein expression. *Frontiers in Plant Science, 10*, 915. doi: 10.3389/fpls.2019.00915.

Anastyuk, S. D., Shevchenko, N. M., Usoltseva, R. V. (Menshova), Silchenko, A. S., Zadorozhny, P. A., Dmitrenok, P. S., & Ermakova, S. P., (2017). Structural features and anticancer activity *in vitro* of fucoidan derivatives from brown alga *Saccharina cichorioides*. *Carbohydrate Polymers, 157*(10), 1503–1510. doi: 10.1016/j.carbpol.2016.11.031.

Aravamudhan, A., Ramos, D. M., Nada, A. A., & Kumbar, S. G., (2014). Natural polymers: Polysaccharides and their derivatives for biomedical applications. natural and synthetic biomedical polymers. In: Kumbar, S. G., Laurencin, C. T., & Deng, M., (eds.), *Natural and Synthetic Biomedical Polymers* (pp. 67–89). Elsevier Inc. doi: 10.1016/B978-0-12-396983-5.00004-1.

Arunkumar, K., (2017). Extraction, isolation, and characterization of alginate. In: *Industrial Applications of Marine Biopolymers* (pp. 19–35). Boca Raton, FL: CRC Press.

Asanka, S. K. K., Jayawardena, T. U., Kim, H. S., Kim, S. Y., Shanura, F. I. P., Wang, L., Abetunga, D. T. U., et al., (2019). Fucoidan isolated from *Padina commersonii* inhibit LPS-induced inflammation in macrophages blocking TLR/NF-KB signal pathway. *Carbohydrate Polymers, 224*, 115195. doi: 10.1016/j.carbpol.2019.115195.

Ashayerizadeh, O., Dastar, B., & Pourashouri, P., (2020). Study of antioxidant and antibacterial activities of depolymerized fucoidans extracted from *Sargassum tenerrimum*. *International Journal of Biological Macromolecules, 151*, 1259–1266. doi: 10.1016/j.ijbiomac.2019.10.172.

Ayoub, A., Pereira, J. M., Rioux, L. E., Turgeon, S. L., Beaulieu, M., & Moulin, V. J., (2015). Role of seaweed laminaran from *Saccharina longicruris* on matrix deposition during dermal tissue-engineered production. *International Journal of Biological Macromolecules, 75*, 13–20. doi: 10.1016/j.ijbiomac.2015.01.017.

Balboa, E. M., Conde, E., Moure, A., Falqué, E., & Domínguez, H., (2013). *In vitro* antioxidant properties of crude extracts and compounds from brown algae. *Food Chemistry, 138*(2, 3), 1764–1785. doi: 10.1016/j.foodchem.2012.11.026.

Ball, S., Colleoni, C., Cenci, U., Raj, J. N., & Tirtiaux, C., (2011). The evolution of glycogen and starch metabolism in eukaryotes gives molecular clues to understand the establishment of plastid endosymbiosis. *Journal of Experimental Botany, 62*(6), 1775–1801. doi: 10.1093/jxb/erq411.

Benito-Román, Ó., Alonso, E., Gairola, K., & Cocero, M. J., (2013). Fixed-bed extraction of β-glucan from cereals by means of pressurized hot water. *Journal of Supercritical Fluids, 82*, 122–128. doi: 10.1016/j.supflu.2013.07.003.

Berteau, O., Mccort, I., Goasdoué, N., Tissot, B., & Daniel, R., (2002). Characterization of a new α-L-fucosidase isolated from the marine mollusk *Pecten maximus* that catalyzes the hydrolysis of α-L-fucose from algal fucoidan (*Ascophyllum nodosum*). *Glycobiology, 12.*

Bi, D., Yu, B., Han, Q., Lu, J., White, W. L., Lai, Q., Cai, N., et al., (2018). Immune activation of RAW264.7 macrophages by low molecular weight fucoidan extracted from New Zealand *Undaria pinnatifida. Journal of Agricultural and Food Chemistry, 66*(41), 10721–10728. doi: 10.1021/acs.jafc.8b03698.

Bilan, M. I., Grachev, A. A., Ustuzhanina, N. E., Shashkov, A. S., Nifantiev, N. E., & Usov, A. I., (2002). Structure of a fucoidan from the brown seaweed *Fucus evanescens* C. Ag. *Carbohydrate Research, 337*(8), 719–730. doi: 10.1016/S0008-6215(02)00053-8.

Bittkau, K. S., Neupane, S., & Alban, S., (2020). Initial evaluation of six different brown algae species as source for crude bioactive fucoidans. *Algal Research, 45*, 101759. doi: 10.1016/j.algal.2019.101759.

Black, W. A. P., Cornhill, W. J., Dewar, E. T., & Woodward, F. N., (1953). Manufacture of algal chemicals. VI. – laboratory-scale isolation of L-fucose from brown marine algae. *Journal of the Science of Food and Agriculture, 4*(2), 85–91. doi: 10.1002/jsfa.2740040205.

Bouhadir, K. H., Alsberg, E., & Mooney, D. J., (2001). Hydrogels for combination delivery of antineoplastic agents. *Biomaterials, 22*(19), 2625–2633.

Bouhadir, K. H., Lee, K. Y., Alsberg, E., Damm, K. L., Anderson, K. W., & Mooney, D. J., (2001). Degradation of partially oxidized alginate and its potential application for tissue engineering. *Biotechnology Progress, 17*(5), 945–950. doi: 10.1021/bp010070p.

British Nutrition Foundation (BNF), (2016). *Nutrition Requirements: Reference Nutrient Intakes for Minerals*. https://www.nutrition.org.uk/attachments/article/261/Nutrition%20 Requirements_Revised%20August%202019.pdf (accessed on 13 October 2021).

Brown, E. M., Allsopp, P. J., Magee, P. J., Gill, G. I., Nitecki, S., Strain, C. R., & Mcsorley, E. M., (2014). Seaweed and human health. *Nutrition Reviews, 72*(3), 205–216. doi: 10.1111/ nure.12091.

Bull, A. T., & Chesters, C. G. C., (1966). The biochemistry of laminarin and the nature of laminarase. In: Nord, F. F., (ed.), *Advances in Enzymology – and Related Areas of Molecular Biology*. https://doi.org/10.1002/9780470122730.ch5.

Cardoso, M., Costa, R., & Mano, J., (2016). Marine origin polysaccharides in drug delivery systems. *Marine Drugs, 14*(2), 34. doi: 10.3390/md14020034.

Chandía, N. P., & Matsuhiro, B., (2008). Characterization of a fucoidan from *Lessonia vadosa* (Phaeophyta) and its anticoagulant and elicitor properties. *International Journal of Biological Macromolecules, 42*(3), 235–240. doi: 10.1016/j.ijbiomac.2007.10.023.

Charoensiddhi, S., Lorbeer, A. J., Lahnstein, J., et al., (2016). Enzyme-assisted extraction of carbohydrates from the brown alga *Ecklonia radiata:* Effect of enzyme type, pH and buffer on sugar yield and molecular weight profiles. *Process Biochem., 51*, 1503–1510. https:// doi.org/10.1016/j.procbio.2016.07.014.

Chen, L., Chen, P., Liu, J., Hu, C., Yang, S., He, D.,Yu, P., Wu, M., & Zhang, X., (2018). *Sargassum fusiforme* polysaccharide SFP-F2 activates the NF-KB signaling pathway via CD14/IKK and P38 axes in RAW264.7 cells. *Marine Drugs, 16*(8), 264. doi: 10.3390/ md16080264.

Chen, Y., Maguire, T., Hileman, R. E., Fromm, J. R., Esko, J. D., Linhardt, R. J., & Marks, R. M., (1997). Dengue virus infectivity depends on envelope protein binding to target cell heparan sulfate. *Nature Medicine, 3*(8), 866–871. doi: 10.1038/nm0897-866.

Cheng, D., Bin, L. B., Mingxin, L. M., & Jin, M., (2011). Influence of laminarin polysaccahrides on oxidative damage. *International Journal of Biological Macromolecules, 48*(1), 63–66. doi: 10.1016/j.ijbiomac.2010.09.011.

Chevolot, L., Alain, F. A., Frederic, C. F., Nelly, K. N., Corinne, S. C., Anne, M. F. A. M., & Boisson-Vidal, C., (1999). Further data on the structure of brown seaweed fucans: Relationships with anticoagulant activity. *Carbohydrate Research, 319*(1–4), 154–165. doi: 10.1016/S0008-6215(99)00127-5.

Chevolot, L., Barbara, M. B., Jacqueline, R. J., Alain, F. A., & Colliec-Jouault, S., (2001). A disaccharide repeat unit is the major structure in fucoidans from two species of brown algae. *Carbohydrate Research, 330*(4), 529–535. doi: 10.1016/S0008-6215(00)00314-1.

Chizhov, A. O., Anne, D. A., Morris, H. R., Haslam, S. M., McDowell, R. A., Shashkov, A. S., Nifantev, N. E., et al., (1999). A study of fucoidan from the brown seaweed *Chorda filum*. *Carbohydrate Research, 320*(1, 2), 108–119. doi: 10.1016/S0008-6215(99)00148-2.

Choi, J. I., Hyun, J. K. H. J., & Lee, J. W., (2011). Structural feature and antioxidant activity of low molecular weight laminarin degraded by gamma irradiation. *Food Chemistry, 129*(2), 520–523. doi: 10.1016/j.foodchem.2011.03.078.

Choi, J. I., Hyun, J. K. H. J., Jae, H. K. J. H., & Lee, J. W., (2012). Enhanced biological activities of laminarin degraded by gamma-ray irradiation. *Journal of Food Biochemistry, 36*(4), 465–469. doi: 10.1111/j.1745-4514.2011.00552.x.

Chollet, L., Pierre, S. P., Cédric, C. C., Jean-Noël, V. J. N., Didier, L. D., & Chaubet, F., (2016). Fucoidans in nanomedicine. *Marine Drugs, 14*(8), 145. doi: 10.3390/md14080145.

Clare, K., (1993). Algin. In: Whistler, R. L., & Bemiller, J. N., (eds.), *Industrial Gums* (pp. 105–143). Elsevier. doi: 10.1016/B978-0-08-092654-4.50010-3.

Commission Implementing Regulation (EU) 2018/1023 of 23 July 2018 Correcting implementing regulation (EU) 2017/2470 establishing the union list of novel Foods. (2020). https://ec.europa.eu/food/sites/food/files/safety/docs/novel-food_authorisation_2015_auth-letter_krill-oil_en.pdf (accessed on 13 October 2021).

Cong, Q., Huanjun, C. H., Wenfeng, L. W., Fei, X. F., Peipei, W. P., Yi, Q. Y., Dong, Q., & Ding, K., (2016). Structural characterization and effect on anti-angiogenic activity of a fucoidan from *Sargassum fusiforme*. *Carbohydrate Polymers, 136*, 899–907. doi: 10.1016/j.carbpol.2015.09.087.

Costa, L. S., Fidelis, G. P., Cordeiro, S. L., Oliveira, R. M., Sabry, D. A., Câmara, R. B. G., Nobre, L. T. D. B., et al., (2010). Biological activities of sulfated polysaccharides from tropical seaweeds. *Biomedicine and Pharmacotherapy, 64*(1), 21–28. doi: 10.1016/j.biopha.2009.03.005.

Crampon, C., Olivier, B. O., & Badens, E., (2011). Supercritical carbon dioxide extraction of molecules of interest from microalgae and seaweeds. *Industrial and Engineering Chemistry Research, 50*(15), 8941–8953. doi: 10.1021/ie102297d.

Cumashi, A., Ushakova, N. A., Preobrazhenskaya, M. E., D'Incecco, A., Piccoli, A., Totani, L., Tinari, N., et al., (2007). A comparative study of the anti-inflammatory, anticoagulant, antiangiogenic, and antiadhesive activities of nine different fucoidans from brown seaweeds. *Glycobiology, 17*(5), 541–552. doi: 10.1093/glycob/cwm014.

Cunha, L., & Grenha, A., (2016). Sulfated seaweed polysaccharides as multifunctional materials in drug delivery applications. *Marine Drugs, 14*(3), 42. doi: 10.3390/md14030042.

Custódio, C. A., Reis, R. L., & Mano, J. F., (2016). Photocrosslinked laminarin based hydrogels for biomedical applications. *Biomacromolecules, 17*(5), 1602–1609. doi: 10.1021/acs.biomac.5b01736.

Delma, C. R., Thirugnanasambandan, S., Srinivasan, G. P., Raviprakash, N., Manna, S. K., Natarajan, M. N., & Aravindan, N., (2019). Fucoidan from marine brown algae attenuates pancreatic cancer progression by regulating P53–NFκB crosstalk. *Phytochemistry, 167*, 112078. doi: 10.1016/j.phytochem.2019.112078.

Dethier, M. N., & Williams, S. L. (2009). Seasonal stresses shift optimal intertidal algal habitats. *Mar Biol., 156,* 555–567. https://doi.org/10.1007/s00227-008-1107-8.

Devillé, C., Damas, J., Forget, P., Dandrifosse, G., & Peulen, O., (2004). Laminarin in the dietary fibre concept. *Journal of the Science of Food and Agriculture, 84*(9), 1030–1038. doi: 10.1002/jsfa.1754.

Dillon, S., Sweeney, T., Figat, S., Callan, J. J., & O'Doherty, J. V., (2010). The effects of lactose inclusion and seaweed extract on performance, nutrient digestibility and microbial populations in newly weaned piglets. *Livestock Science, 134*(1–3), 205–207. doi: 10.1016/j.livsci.2010.06.142.

Dobrinčić, A., Balbino, S., Zorić, Z., Pedisić, S., Kovačević, D. B., Garofulić, I. E., & Dragović-Uzelac, V., (2020). Advanced technologies for the extraction of marine brown algal polysaccharides. *Marine Drugs, 18*(3), 168. doi: 10.3390/md18030168.

Draget, K. I., Bræk, S. G., & Smidsrød, O., (1994). Alginic acid gels: The effect of alginate chemical composition and molecular weight. *Carbohydrate Polymers, 25*(1), 31–38. doi: 10.1016/0144-8617(94)90159-7.

Draget, K. I., Smidsrød, O., & Skjåk-Bræk, G., (2005). Alginates from algae. In: Vandamme, E. J., De Baets, S., & Steinbüchel, A., (eds.), *Biopolymers* (pp. 1–30). Wiley, Germany. doi: 10.1002/3527600035.bpol6008.

Drury, J. L., Boontheekul, T., & Mooney, D. J., (2005). Cellular cross-linking of peptide modified hydrogels. *Journal of Biomechanical Engineering, 127*(2), 220. doi: 10.1115/1.1865194.

Edgar, G., (2008). *Laminaran/Laminarin – Chemical Structure of β-1,3-1,6-Glucan.* https://commons.wikimedia.org/w/index.php?curid=5274129 (accessed on 13 October 2021).

Edmond, G. M., Han, R. M., Classen, B., Quetin-Leclerq, J., Mahy, G., Ruan, G. J., Qin, P., et al., (2010). Mucilage and polysaccharides in the halophyte plant species *Kosteletzkya virginica*: Localization and composition in relation to salt stress. *Journal of Plant Physiology, 167*(5), 382–392. doi: 10.1016/j.jplph.2009.10.012.

Elbi, S., Nimal, T. R., Rajan, V. K., Baranwal, G., Biswas, R., Jayakumar, R., & Sathianarayanan, S., (2017). Fucoidan coated ciprofloxacin loaded chitosan nanoparticles for the treatment of intracellular and biofilm infections of *Salmonella. Colloids and Surfaces B: Biointerfaces, 160*, 40–47. doi: 10.1016/j.colsurfb.2017.09.003.

Eliaz, I., Weil, E., & Wilk, B., (2007). Integrative medicine and the role of modified citrus pectin/alginates in heavy metal chelation and detoxification – five case reports. *Forschende Komplementarmedizin, 14*(6), 358–364. doi: 10.1159/000109829.

Ermakova, S., Menshova, R., Vishchuk, O., Kim, S. M., Um, B. H., Isakov, V., & Zvyagintseva, T., (2013). Water-soluble polysaccharides from the brown alga *Eisenia bicyclis*: Structural characteristics and antitumor activity. *Algal Research, 2*(1), 51–58. doi: 10.1016/j.algal.2012.10.002.

Fabrowska, J., Łęska, B., Schroeder, G., Messyasz, B., & Pikosz, M., (2015). Biomass and extracts of algae as material for cosmetics. In: Kim, S. K., & Chojnacka, K., (eds.), *Marine*

Algae Extracts: Processes, Products, and Applications (pp. 681–706). Wiley, Germany. doi: 10.1002/9783527679577.ch38.

Fernando, I. P. F., Kim, K. N., Kim, D., & Jeon, Y. J., (2019). Algal polysaccharides: Potential bioactive substances for cosmeceutical applications. *Critical Reviews in Biotechnology, 39*(1), 99–113. doi: 10.1080/07388551.2018.1503995.

Fernando, I. P. S., Kim, D., Nah, J. W., & Jeon, Y. J., (2019). Advances in functionalizing fucoidans and alginates (bio)polymers by structural modifications: A review. *Chemical Engineering Journal, 355*, 33–48. doi: 10.1016/j.cej.2018.08.115.

Fischer, F. G., & Dörfel, H., (1955). Die polyuronsäuren der braunalgen (kohlenhydrate der algen I). *Hoppe-Seyler's Zeitschrift Für Physiologische Chemie, 302*, 186–203. doi: 10.1515/bchm2.1955.302.1–2.186.

Fitton, J. H., (2011). Therapies from fucoidan; multifunctional marine polymers. *Marine Drugs, 9*(10), 1731–1760. doi: 10.3390/md9101731.

Fitton, J. H., Stringer, D. N., Park, A. Y., & Karpiniec, S. S., (2019). Therapies from fucoidan: New developments. *Marine Drugs, 17*(10), 571. doi: 10.3390/md17100571.

Fitton, J., & Dragar, C., (2006). *Method and Composition for the Treatment of a Viral Infection*. https://patents.google.com/patent/US7056520B2/en (accessed on 13 October 2021).

Fitton, J., Stringer, D., & Karpiniec, S., (2015). Therapies from fucoidan: An update. *Marine Drugs, 13*(9), 5920–5946. doi: 10.3390/md13095920.

Fletcher, H. R., Biller, P., Ross, A. B., & Adams, J. M. M., (2017). The seasonal variation of fucoidan within three species of brown macroalgae. *Algal Research, 22*, 79–86. doi: 10.1016/j.algal.2016.10.015.

Foley, S. A., Mulloy, B., & Tuohy, M. G., (2011). An unfractionated fucoidan from *Ascophyllum nodosum*: Extraction, characterization, and apoptotic effects *in vitro*. *Journal of Natural Products, 74*(9), 1851–1861. doi: 10.1021/np200124m.

Freile-Pelegrín, Y., & Tasdemir, D., (2019). Seaweeds to the rescue of forgotten diseases: A review. *Botanica Marina, 62*(3), 211–226. doi: 10.1515/bot-2018-0071.

Fuentes, A. L., Millis, L., & Sigola, L. B., (2011). Laminarin, a soluble beta-glucan, inhibits macrophage phagocytosis of zymosan but has no effect on lipopolysaccharide mediated augmentation of phagocytosis. *International Immunopharmacology, 11*(11), 1939–1945. doi: 10.1016/j.intimp.2011.08.005.

Garcia-Vaquero, M., Rajauria, G., O'Doherty, J. V., & Sweeney, T., (2017). Polysaccharides from macroalgae: Recent advances, innovative technologies and challenges in extraction and purification. *Food Research International, 99*, 1011–1020. doi: 10.1016/j.foodres.2016.11.016.

Giese, E. C., Jacob, G. J., Gianluca, A. G., Barbosa, A. M., Da Cunha, M. A. A., & Dekker, R. F. H., (2015). Free-radical scavenging properties and antioxidant activities of botryosphaeran and some other β-D-glucans. *International Journal of Biological Macromolecules, 72*, 125–130. doi: 10.1016/j.ijbiomac.2014.07.046.

Gillespie, R. D., & Critchley, A. T., (1999). Phenology of *Sargassum* spp. (Sargassaceae, Phaeophyta) from reunion rocks, KwaZulu-Natal, South Africa. *Hydrobiologia, 398*, 210. doi: 10.1023/A:1017082032748.

Gobet, A., Barbeyron, T., Matard-Mann, M., Magdelenat, G., Vallenet, V., Duchaud, E., & Michel, G., (2018). Evolutionary evidence of algal polysaccharide degradation acquisition by *Pseudoalteromonas carrageenovora* 9^T to adapt to macroalgal niches. *Frontiers in Microbiology, 9*, 2740. doi:10.3389/fmicb.2018.02740.

Gómez, I., & Wiencke, C., (1998). Seasonal changes in C, N and major organic compounds and their significance to morpho-functional processes in the endemic Antarctic brown alga *Ascoseira mirabilis*. *Polar Biol., 19*, 115–124. https://doi.org/10.1007/s003000050222.

Gombotz, W. R., & Wee, S., (1998). Protein release from alginate matrices. *Advanced Drug Delivery Reviews, 31*(3), 267–285. doi: 10.1016/S0169-409X(97)00124-5.

Gorham, J., & Lewey, S. A., (1984). Seasonal changes in the chemical composition of *Sargassum muticum*. *Mar Biol., 80*, 103–107. https://doi.org/10.1007/BF00393133.

Graiff, A., Ruth, W., Kragl, U., & Karsten, U., (2016). Chemical characterization and quantification of the brown algal storage compound laminarin - A new methodological approach. *J Appl Phycol., 28*, 533–543. https://doi.org/10.1007/s10811-015-0563-z.

Hamed, I., Özogul, F., Özogul, Y., & Regenstein, J. M., (2015). Marine bioactive compounds and their health benefits: A review. *Comprehensive Reviews in Food Science and Food Safety, 14*(4), 446–465. doi: 10.1111/1541-4337.12136.

Hatipoglu, F., Ogurtan, Z., Sezer, A. D., Uney, K., Erol, M., Ozdemir, O., Bas, A. L., & Akbuga, J., (2008). Effect of laminarin and chitosan gel formulations on the treatment of hydrofluoric acid induced corneal burns in the rabbits. *Revue de Medecine Veterinaire, 159*(4), 207–214.

Haug, A., & Smidsrød, O., (1967). Strontium-calcium selectivity of alginates. *Nature, 215*(5102), 757. doi: 10.1038/215757a0.

Haug, A., Larsen, B., & Smidsrød, O., (1974). Uronic acid sequence in alginate from different sources. *Carbohydrate Research, 32*(2), 217–225. doi: 10.1016/S0008-6215(00)82100-X.

Hayashi, K., Nakano, T., Hashimoto, M., Kanekiyo, K., & Hayashi, T., (2008). Defensive effects of a fucoidan from brown alga *Undaria pinnatifida* against herpes simplex virus infection. *International Immunopharmacology, 8*(1), 109–116. doi: 10.1016/j.intimp.2007.10.017.

Helle, S., Bray, F., Verbeke, J., Devassine, S., Courseaux, A., Facon, M., Tokarski, C., et al., (2018). Proteome analysis of potato starch reveals the presence of new starch metabolic proteins as well as multiple protease inhibitors. *Frontiers in Plant Science, 9*, 746. doi: 10.3389/fpls.2018.00746.

Hidari, K. I. P. J., Takahashi, N., Arihara, M., Nagaoka, M., Morita, K., & Suzuki, T., (2008). Structure and anti-dengue virus activity of sulfated polysaccharide from a marine alga. *Biochemical and Biophysical Research Communications, 376*(1), 91–95. doi: 10.1016/j.bbrc.2008.08.100.

Hifney, A. F., Fawzy, M. A., Abdel-Gawad, K. M., & Gomaa, M., (2016). Industrial optimization of fucoidan extraction from *Sargassum* sp. and its potential antioxidant and emulsifying activities. *Food Hydrocolloids, 54*, 77–88. doi: 10.1016/j.foodhyd.2015.09.022.

Hoagland, D. R., & Lieb, L. L., (1915). The complex carbohydrates and forms of sulphur in marine algae of the pacific coast. *Journal of Biological Chemistry, 23*(1), 287–297.

Hoffman, R., Paper, D. H., Donaldson, J., Alban, S., & Franz, G., (1995). Characterization of a laminarin sulfate which inhibits basic fibroblast growth factor binding and endothelial cell proliferation. *Journal of Cell Science, 108*(11), 3591–3598.

Holdt, S. L., & Kraan, S., (2011). Bioactive compounds in seaweed: Functional food applications and legislation. *Journal of Applied Phycology, 23*, 543–597. doi: 10.1007/s10811-010-9632-5.

Holtkamp, A. D., Kelly, S., Ulber, R., & Lang, S., (2009). Fucoidans and fucoidanases-focus on techniques for molecular structure elucidation and modification of marine polysaccharides. *Applied Microbiology and Biotechnology, 82*, 1–11. doi:10.1007/s00253-008-1790-x.

Hu, P., Li, Z., Chen, M., Sun, Z., Ling, Y., Jiang, J., & Huang, C., (2016). Structural elucidation and protective role of a polysaccharide from *Sargassum fusiforme* on ameliorating learning and memory deficiencies in mice. *Carbohydrate Polymers, 139*, 150–158. doi: 10.1016/j.carbpol.2015.12.019.

Huang, T., Chiu, Y., Chan, Y., Chiu, Y., Wang, H., Huang, K., Li, T., et al., (2015). Prophylactic administration of fucoidan represses cancer metastasis by inhibiting vascular endothelial growth factor (VEGF) and matrix metalloproteinases (MMPs) in Lewis tumor-bearing mice. *Marine Drugs, 13*(4), 1882–1900. doi: 10.3390/md13041882.

Im, J. H., Choi, C. H., Mun, F., Lee, J. H., Kim, H., Jung, W. K., Jang, C. H., & Kim, G. H., (2017). A polycaprolactone/fish collagen/alginate biocomposite supplemented with phlorotannin for hard tissue regeneration. *RSC Advances, 7*(4), 2009–2018. doi: 10.1039/c6ra25182j.

Jard, G., Marfaing, H., Carrère, H., Delgenes, J. P., Steyer, J. P., & Dumas, C., (2013). French Brittany macroalgae screening: Composition and methane potential for potential alternative sources of energy and products. *Bioresource Technology, 144*, 492–498. doi: 10.1016/j.biortech.2013.06.114.

Je, J. Y., Park, P. J., Kim, E. K., Park, J. S., Yoon, H. D., Kim, K. R., & Ahn, C. B., (2009). Antioxidant activity of enzymatic extracts from the brown seaweed *Undaria pinnatifida* by electron spin resonance spectroscopy. *LWT – Food Science and Technology 42*(4), 874–878. doi: 10.1016/j.lwt.2008.10.012.

Jensen, A., (1956). *Geographical and Seasonal Variation in the Chemical Composition of Laminaria hyperborea and Laminaria digitata from the Norwegian Coast.* Akademisk trykningssentral, Oslo.

Jesumani, V., Du, H., Pei, P., Zheng, C., Cheong, K. L., & Huang, N., (2019). Unravelling property of polysaccharides from *Sargassum* sp. as an anti-wrinkle and skin whitening property. *International Journal of Biological Macromolecules, 140*, 216–224. doi: 10.1016/j.ijbiomac.2019.08.027.

Jhamandas, J. H., Wie, M. B., Harris, K., MacTavish, D., & Kar, S., (2005). Fucoidan inhibits cellular and neurotoxic effects of β-amyloid (aβ) in rat cholinergic basal forebrain neurons. *European Journal of Neuroscience, 21*(10), 2649–2659. doi: 10.1111/j.1460-9568.2005.04111.x.

Ji, C. F., Ji, Y. B., & Meng, Y., (2013). Sulfated modification and anti-tumor activity of laminarin. *Experimental and Therapeutic Medicine, 6*(5), 1259–1264. doi: 10.3892/etm.2013.1277.

Ji, Y. B., Ji, C. F., & Zhang, H., (2012). Laminarin induces apoptosis of human colon cancer LOVO Cells through a mitochondrial pathway. *Molecules, 17*(8), 9947–9960. doi: 10.3390/molecules17089947.

Jiang, H., Liang, S., Yao, X. R., Jin, Y. X., Shen, X. H., Yuan, B., Zhang, J. B., & Kim, N. H., (2018). Laminarin improves developmental competence of porcine early stage embryos by inhibiting oxidative stress. *Theriogenology, 115*, 38–44. doi: 10.1016/j.theriogenology.2018.04.019.

Jiang, Z., Okimura, T., Yamaguchi, K., & Oda, T., (2011). The potent activity of sulfated polysaccharide, ascophyllan, isolated from *Ascophyllum nodosum* to induce nitric oxide and cytokine production from mouse macrophage RAW264.7 cells: Comparison between ascophyllan and fucoidan. *Nitric Oxide, 25*(4), 407–415. doi: 10.1016/j.niox.2011.10.001.

Jiang, Z., Ueno, M., Nishiguchi, T., Abu, R., Isaka, S., Okimura, T., Yamaguchi, K., & Tatsuya, O. T., (2013). Importance of sulfate groups for the macrophage-stimulating activities of ascophyllan isolated from the brown alga *Ascophyllum nodosum. Carbohydrate Research, 380*, 124–129. doi: 10.1016/j.carres.2013.05.018.

Jiao, G., Yu, G., Zhang, J., & Ewart, S., (2011). Chemical structures and bioactivities of sulfated polysaccharides from marine algae. *Marine Drugs, 9*(2), 196–233. doi: 10.3390/md9020196.

Jin, W., Zhang, W., Wang, J., Ren, S., Song, N., Duan, D., & Zhang, Q., (2014). Characterization of laminaran and a highly sulfated polysaccharide from *Sargassum fusiforme. Carbohydrate Research, 385,* 58–64. doi: 10.1016/j.carres.2013.12.009.

Kadam, S. U., O'Donnell, C. P., Rai, D. K., Hossain, M. B., Burgess, C. M., Walsh, D., & Tiwari, B. K., (2015). Laminarin from Irish brown seaweeds *Ascophyllum nodosum* and *Laminaria hyperborea. Marine Drugs, 13*(7), 1247–1247. doi: 10.3390/md1307427.

Kadam, S. U., Tiwari, B. K., & O'Donnell, C. P., (2013a). Application of novel extraction technologies for bioactives from marine algae. *Journal of Agricultural and Food Chemistry, 61*(20), 4667–4675. doi: 10.1021/jf400819p.

Kadam, S. U., Tiwari, B. K., & O'Donnell, C. P., (2015a). Extraction, structure and biofunctional activities of laminarin from brown algae. *International Journal of Food Science and Technology, 50*(1), 24–31. doi: 10.1111/ijfs.12692.

Kannan, K., & Sivaperumal, P., (2017). Industrial applications of marine biopolymers: Biological activities of marine biopolymers. In: Sudha, P. N., (ed.), *Industrial Applications of Marine Biopolymers*. CRC Press, Boca Raton. doi: 10.4324/9781315313535.

Kearl, T., & Sandi, M., (2007). *Fucoidan Delivery System*. US 2007.0003.669A1.

Ki, K., (2012). *Fucoidan for Thrombolysis Derived from Seaweed*. https://patents.google.com/patent/WO2012102555A3/en (accessed on 13 October 2021).

Kim, J. H., Lee, J. E., Kim, K. H., & Kang, N. J., (2018). Beneficial effects of marine algae-derived carbohydrates for skin health. *Marine Drugs, 16*(11), 1–20. doi: 10.3390/md16110459.

Kim, K. H., Kim, Y. W., Kim, H. B., Lee, B. J., & Lee, D. S., (2006). Anti-apoptotic activity of laminarin polysaccharides and their enzymatically hydrolyzed oligosaccharides from *Laminaria japonica. Biotechnology Letters, 28*(6), 439–446. doi: 10.1007/s10529-005-6177-9.

Kim, S. K., & Chojnacka, K., (2015). *Marine Algae Extracts: Processes, Products, and Applications*. Wiley, USA.

Kitano, K., Hamaguchi, S., & Aoki, H., (2010). *Fucoidan Compositions and Methods for Dietary and Nutritional Supplements*. US 2010.0330211A1. https://patentimages.storage.googleapis.com/3b/c9/82/c283c7b24afe69/US20100019677A1.pdf (accessed on 13 October 2021).

Klimp, A. H., De Vries, E. G. E., Scherphof, G. L., & Daemen, T., (2002). A potential role of macrophage activation in the treatment of cancer. *Critical Reviews in Oncology/Hematology, 44*(2), 143–161. doi: 10.1016/S1040-8428(01)00203-7.

Kolambkar, Y. M., Dupont, K. M., Boerckel, J. D., Huebsch, N., Mooney, D. J., Hutmacher, D. W., & Guldberg, R. E., (2011). An alginate-based hybrid system for growth factor delivery in the functional repair of large bone defects. *Biomaterials, 32*(1), 65–74. doi: 10.1016/j.biomaterials.2010.08.074.

Kraan, S., (2012). Algal polysaccharides, novel applications and outlook. In: Chang, C., (ed.) *Carbohydrates – Comprehensive Studies on Glycobiology and Glycotechnology*. Intechopen, United Kingdom. doi: 10.5772/51572.

Kuda, T., Kosaka, M., Hirano, S., Kawahara, M., Sato, M., Kaneshima, T., Nishizawa, M., Takahashi, H., & Kimura, B., (2015). Effect of sodium-alginate and laminaran on *Salmonella typhimurium* infection in human enterocyte-like HT-29-Luc cells and BALB/c mice. *Carbohydrate Polymers, 125*, 113–119. doi: 10.1016/j.carbpol.2015.01.078.

Kuda, T., Yano, T., Matsuda, N., & Nishizawa, M., (2005). Inhibitory effects of laminaran and low molecular alginate against the putrefactive compounds produced by intestinal microflora *in vitro* and in rats. *Food Chemistry, 91*(4), 745–749. doi: 10.1016/j.foodchem.2004.06.047.

Kylin, H., (1913). Zeitschrift fur physikalische chemie. *Biochemistry of Sea Algae, 83*, 171–197.

Lahaye, M., & Robic, A., (2007). Structure and function properties of ulvan, a polysaccharide from green seaweeds. *Biomacromolecules, 8*(6), 1765–1774. doi: 10.1021/bm061185q.

Lakmal, H. H. C., Lee, J. H., & Jeon, Y. J., (2015). Enzyme-assisted extraction of a marine algal polysaccharide, fucoidan and bioactivities. In Ramawat, K., & Mérillon, J. M., (eds.), *Polysaccharides* (pp. 1065–1077). Springer, Cham. doi: 10.1007/978-3-319-16298-0_46.

Lampugnani, E. R., Flores-Sandoval, E., Tan, Q. W., Mutwil, M., Bowman, J. L., & Persson, S., (2019). Cellulose synthesis – central components and their evolutionary relationships. *Trends in Plant Science, 24*(5), 402–412. doi: 10.1016/j.tplants.2019.02.011.

Langer, R., & Vacanti, J., (1993). Tissue engineering. *Science, 260*(5110), 920–926. doi: 10.1126/science.8493529.

Larsen, B., Haug, A., & Painter, T. J., (1966). Sulfated polysaccharides in brown algae – I. isolation and preliminary characterization of three sulfated polysaccharides from *Ascophyllumm nodosum* (L.) Le Jol. *Acta Chemica Scandinavica, 20*(1), 219–230. doi: 10.3891/acta.chem.scand.20-0219.

Ledesma-Hernández, B., & Herrero, M., (2013). *Bioactive Compounds from Marine Foods.* John Wiley & Sons, United Kingdom. doi: 10.1002/9781118412893.

Lee, J. C., Hou, M. F., Huang, H. W., Chang, F. R., Yeh, C. C., Tang, J. Y., & Chang, H. W., (2013). Marine algal natural products with anti-oxidative, anti-inflammatory, and anti-cancer properties. *Cancer Cell International, 13*(1), 1–7. doi: 10.1186/1475-2867-13-55.

Lee, J. Y., Kim, Y. J., Kim, H. J., Kim, Y. S., & Park, W., (2012). Immunostimulatory effect of laminarin on RAW 264.7 mouse macrophages. *Molecules, 17*(5), 5404–5411. doi: 10.3390/molecules17055404.

Lee, K. Y., & Mooney, D. J., (2001). Hydrogels for tissue engineering. *Chemical Reviews, 101*(7), 1869–1879.

Lee, K. Y., & Mooney, D. J., (2012). Alginate: Properties and biomedical applications. *Progress in Polymer Science (Oxford), 37*(1), 106–126. doi: 10.1016/j.progpolymsci.2011.06.003.

Lee, S. H., Ko, C. I., Ahn, G., You, S., Kim, J. S., Heu, M. S., Kim, J., et al., (2012). Molecular characteristics and anti-inflammatory activity of the fucoidan extracted from *Ecklonia cava*. *Carbohydrate Polymers, 89*(2), 599–606. doi: 10.1016/j.carbpol.2012.03.056.

Lee, Y. E., Kim, H., Seo, C., Park, T., Lee, K. B., Yoo, S. Y., Hong, S. C., et al., (2017). Marine polysaccharides: Therapeutic efficacy and biomedical applications. *Archives of Pharmacal Research, 40*(9), 1006–1020. doi: 10.1007/s12272-017-0958-2.

Li, B., Lu, F., Wei, X., & Zhao, R., (2008). Fucoidan: Structure and bioactivity. *Molecules, 13*(8), 1671–1695. doi: 10.3390/molecules13081671.

Li, H., Li, J., Tang, Y., Lin, L., Xie, Z., Zhou, J., Zhang, L., et al., (2017). Fucoidan from *Fucus vesiculosus* suppresses Hepatitis B virus replication by enhancing extracellular signal-regulated kinase activation. *Virol. J., 14*(1), 178. doi: 10.1186/s12985-017-0848-8.

Li, Y. T., Chen, B. J., Wu, W. D., Ge, K., Wei, X. Y., Kong, L. M., Xie, Y. Y., et al., (2018). Antioxidant and antimicrobial evaluation of carboxymethylated and hydroxamated degraded polysaccharides from *Sargassum fusiforme*. *International Journal of Biological Macromolecules, 118*, 1550–1557. doi: 10.1016/j.ijbiomac.2018.06.196.

Lim, S. J., & Aida, W. M. W., (2017). Extraction of sulfated polysaccharides (Fucoidan) from brown seaweed. In: Venkatesan, J., Anil, S., & Kim, S., (eds.), *Seaweed Polysaccharides: Isolation, Biological and Biomedical Applications* (pp. 27–46). Elsevier. doi: 10.1016/B978-0-12-809816-5.00003-7.

Liu, X., Liu, H., Zhai, Y., Li, Y., Zhu, X., & Zhang, W., (2017). Laminarin protects against hydrogen peroxide-induced oxidative damage in MRC-5 cells possibly via regulating NRF2. *Peer J., 7*, 1–18. doi: 10.7717/peerj.3642.

Macartain, P., Gill, C. I. R., Brooks, M., Campbell, R., & Rowland, I. R., (2007). Nutritional value of edible seaweeds. *Nutrition Reviews, 65*(12), 535–543. doi: 10.1301/nr.2007.dec.535.

Malinowska, P., (2011). Algae extracts as active cosmetic ingredients. *Zeszyty Naukowe/Uniwersytet Ekonomiczny w Poznaniu, 212*, 123–129.

Malyarenko, O. S., Usoltseva, R. V., Shevchenko, N. M., Isakov, V. V., Zvyagintseva, T. N., & Ermakova, S. P., (2017). *In vitro* anticancer activity of the laminarans from far eastern brown seaweeds and their sulfated derivatives. *Journal of Applied Phycology, 29*(1), 543–553. doi: 10.1007/s10811-016-0915-3.

Maneein, S., Milledge, J. J., Nielsen, B. V., & Harvey, P. J., (2018). A review of seaweed pre-treatment methods for enhanced biofuel production by anaerobic digestion or fermentation. *Fermentation, 4*(4), 100. doi: 10.3390/fermentation4040100.

Mariya, V., & Ravindran, V. S., (2013). Biomedical and pharmacological significance of marine macro algae – review. *Indian Journal of Marine Sciences, 42*(5), 527–537.

McHugh, D. J., (1987). *Production and Utilization of Products from Commercial Seaweeds*. Food and Agriculture Organization of the United Nations, Rome.

Medcalf, D. G., & Larsen, B., (1977). Structural studies on ascophyllan and the fucose-containing complexes from the brown alga *Ascophyllum nodosum*. *Carbohydrate Research, 59*(2), 539–546. doi: 10.1016/S0008-6215(00)83191-2.

Miao, H. Q., Elkin, M., Aingorn, E., Ishai-Michaeli, R., Stein, C. A., & Vlodavsky, I., (1999). Inhibition of heparanase activity and tumor metastasis by laminarin sulfate and synthetic phosphorothioate oligodeoxynucleotides. *International Journal of Cancer, 83*(3), 424–431. doi: 10.1002/(SICI)1097-0215(19991029)83:3<424::AID-IJC20>3.0.CO;2-L.

Miao, H. Q., Ishai-Michaeli, R., Peretz, T., & Israel, V. I., (1995). Laminarin sulfate mimics the effects of heparin on smooth muscle cell proliferation and basic fibroblast growth factor-receptor binding and mitogenic activity. *Journal of Cellular Physiology, 164*(3), 482–490. doi: 10.1002/jcp.1041640306.

Michalak, I., & Chojnacka, K., (2014). Algal extracts: Technology and advances. *Engineering in Life Sciences, 14*(6), 581–591. doi: 10.1002/elsc.201400139.

Min, S. K., Kwon, O. C., Lee, S., Park, K. H., & Kim, J. K., (2012). An antithrombotic fucoidan, unlike heparin, does not prolong bleeding time in a murine arterial thrombosis model: A comparative study of *Undaria pinnatifida* sporophylls and *Fucus vesiculosus*. *Phytotherapy Research, 26*(5), 752–757. doi: 10.1002/ptr.3628.

Mohsin, S., Kurup, M., & Mahadevan, R., (2013). Effect of ascophyllan from brown algae *Padina tetrastromatica* on inflammation and oxidative stress in carrageenan-induced rats. *Inflammation, 36*(6), 1268–1278. doi: 10.1007/s10753-013-9665-4.

Moroney, N. C., O'Grady, M. N., O'Doherty, J. V., & Kerry, J. P., (2012). Addition of seaweed (*Laminaria digitata*) extracts containing laminarin and fucoidan to porcine diets: Influence on the quality and shelf-life of fresh pork. *Meat Science, 92*(4), 423–429. doi: 10.1016/j.meatsci.2012.05.005.

Morya, V. K., Kim, J., & Kim, E. K., (2012). Algal fucoidan: Structural and size-dependent bioactivities and their perspectives. *Applied Microbiology and Biotechnology, 93*(1), 71–82. doi: 10.1007/s00253-011-3666-8.

Nakayasu, S., Oegima, R., Yamaguchi, K., & Oda, T., (2009). Biological activities of fucose-containing polysaccharide ascophyllan isolated from the brown alga *Ascophyllum nodosum. Bioscience, Biotechnology, and Biochemistry, 73*(4), 961–964. doi: 10.1271/bbb.80845.

Nelson, T. E., & Lewis, B. A., (1974). Separation and characterization of the soluble and insoluble components of insoluble laminaran. *Carbohydrate Research, 33*(1), 63–74. doi: 10.1016/S0008-6215(00)82940-7.

Neyrinck, A. M., Mouson, A., & Delzenne, N. M., (2007). Dietary supplementation with laminarin, a fermentable marine β (1-3) glucan, protects against hepatotoxicity induced by LPS in rat by modulating immune response in the hepatic tissue. *International Immunopharmacology, 7*(12), 1497–1506. doi: 10.1016/j.intimp.2007.06.011.

Ngo, D. H., & Kim, S. K., (2013). Sulfated polysaccharides as bioactive agents from marine algae. *International Journal of Biological Macromolecules, 62*, 70–75. doi: 10.1016/j.ijbiomac.2013.08.036.

Nishino, T., Yokoyama, G., Dobashi, K., Fujihara, M., & Nagumo, T., (1989). Isolation, purification, and characterization of fucose-containing sulfated polysaccharides from the brown seaweed *Ecklonia kurome* and their blood-anticoagulant activities. *Carbohydrate Research, 186*(1), 119–129. doi: 10.1016/0008-6215(89)84010-8.

O'Doherty, J. V., Dillon, S., Figat, S., Callan, J. J., & Sweeney, T., (2010). The effects of lactose inclusion and seaweed extract derived from *Laminaria* spp. on performance, digestibility of diet components and microbial populations in newly weaned pigs. *Animal Feed Science and Technology, 157*(3, 4), 173–180. doi: 10.1016/j.anifeedsci.2010.03.004.

O'Shea, C. J., O'Doherty, J. V., Callanan, J. J., Doyle, D., Thornton, K., & Sweeney, T., (2016). The effect of algal polysaccharides laminarin and fucoidan on colonic pathology, cytokine gene expression and *Enterobacteriaceae* in a dextran sodium sulfate-challenged porcine model. *Journal of Nutritional Science, 5*(6), 1–9. doi: 10.1017/jns.2016.4.

Oh, K., (2012). *Degradation of Brown Alga-Derived Fucoidan.* https://patents.justia.com/patent/8202988#history (accessed on 13 October 2021).

Olasehinde, T. A., Olaniran, A. O., & Okoh, A. I., (2019). Macroalgae as a valuable source of naturally occurring bioactive compounds for the treatment of Alzheimer's disease. *Marine Drugs, 17*(11), 609. doi: 10.3390/md17110609.

OMRI, (2015). *Alginates: Handling/Processing.* Organic Materials Review Institute. https://www.ams.usda.gov/sites/default/files/media/Alginates%20TR%202015.pdf (accessed on 13 October 2021).

Painter, T. J., (1983). Algal polysaccharides. In: Aapinall, G. O., (ed.) *The Polysaccharides* (pp. 195–285). Elsevier, Netherlands.

Paris, D., Beaulieu-Abdelahad, D., Bachmeier, C., Reed, J., Ait-Ghezala, G., Bishop, A., Chao, J., et al., (2011). Anatabine lowers Alzheimer's aβ production *in vitro* and *in vivo. European Journal of Pharmacology, 670*(2, 3), 384–391. doi: 10.1016/j.ejphar.2011.09.019.

Park, H. K., Kim, I. H., Kim, J., & Nam, T. J., (2012). Induction of apoptosis by laminarin, regulating the insulin-like growth factor-IR signaling pathways in HT-29 human colon cells. *International Journal of Molecular Medicine, 30*(4), 734–738. doi: 10.3892/ijmm.2012.1084.

Pereira, L., (2011). A review of the nutrient composition of selected edible seaweeds. In: Pomin, V. H., (ed.), *Seaweed: Ecology, Nutrient Composition and Medicinal Uses*. Nova Science, New York.

Pereira, L., (2018). Biological and therapeutic properties of the seaweed polysaccharides. *International Biology Review, 2*(2), 1–50. doi: 10.18103/ibr.v2i2.1762.

Pereira, L., (2018b). Seaweeds as source of bioactive substances and skin care therapy – cosmeceuticals, algotheraphy, and thalassotherapy. *Cosmetics, 5*(4), 68. doi: 10.3390/cosmetics5040068.

Pereira, L., Amado, A. M., Critchley, A. T., Velde, F. V., & Ribeiro-Claro, P. J. A., (2009). Identification of selected seaweed polysaccharides (phycocolloids) by vibrational spectroscopy (FTIR-ATR and FT-Raman). *Food Hydrocolloids, 23*(7), 1903–1909. doi: 10.1016/j.foodhyd.2008.11.014.

Pereira, L., Gheda, S. F., & Ribeiro-Claro, P. J. A., (2013). Analysis by vibrational spectroscopy of seaweed polysaccharides with potential use in food, pharmaceutical, and cosmetic industries. *International Journal of Carbohydrate Chemistry, 2013*(6), 1–7. doi: 10.1155/2013/537202.

Pielesz, A., Machnicka, A., & Ewa, S. E., (2011). Antibacterial activity and scanning electron microscopy (SEM) examination of alginate-based films and wound dressings. *Ecological Chemistry and Engineering S – Chemia I Inzynieria Ekologiczna, S18*(2), 197–210.

Pomin, V. H., (2012). Fucanomics and galactanomics: Current status in drug discovery, mechanisms of action and role of the well-defined structures. *Biochimica et Biophysica Acta – General Subjects, 1820*(12), 1971–1979. doi: 10.1016/j.bbagen.2012.08.022.

Ponce, N. M. A., Pujol, C. A., Damonte, E. B., Flores, M. L., & Stortz, C. A., (2003). Fucoidans from the brown seaweed *Adenocystis utricularis*: Extraction methods, antiviral activity and structural studies. *Carbohydrate Research, 338*(2), 153–165. doi: 10.1016/S0008-6215(02)00403-2.

Prendergast, G. C., & Jaffee, E. M., (2007). Cancer immunologists and cancer biologists: Why we didn't talk then but need to now. *Cancer Research, 67*(8), 3500–3504. doi: 10.1158/0008-5472.CAN-06-4626.

Qiu, X. D., Amarasekara, A., & Doctor, V., (2006). Effect of over sulfation on the chemical and biological properties of fucoidan. *Carbohydrate Polymers, 63*(2), 224–228. doi: 10.1016/j.carbpol.2005.08.064.

Qu, G., Liu, X., Wang, D., Yuan, Y., & Han, L., (2014). Isolation and characterization of fucoidans from five brown algae and evaluation of their antioxidant activity. *Journal of Ocean University of China, 13*(5), 851–856. doi: 10.1007/s11802-014-2260-y.

Queen, D., Orsted, H., Sanada, H., & Sussman, G., (2004). A dressing history. *International Wound Journal, 1*(1), 59–77. doi: 10.1111/j.1742-4801.2004.0009.x.

Raposo, J., Filomena, M., De Morais, A. M. B., & De Morais, R. M. S. C., (2015). Marine polysaccharides from algae with potential biomedical applications. *Marine Drugs, 13*(5), 2967–3028. doi: 10.3390/md13052967.

Read, S. M., Currie, G., & Bacic, A., (1996). Analysis of the structural heterogeneity of laminarin by electrospray-ionization-mass spectrometry. *Carbohydrate Research, 281*(2), 187–201. doi: 10.1016/0008-6215(95)00350-9.

Religa, P., Kazi, M., Thyberg, J., Gaciong, Z., Swedenborg, J., & Hedin, U., (2000). Fucoidan inhibits smooth muscle cell proliferation and reduces mitogen-activated protein kinase activity. *European Journal of Vascular and Endovascular Surgery, 20*(5), 419–426. doi: 10.1053/ejvs.2000.1220.

Remya, R. R., Radhika, R. S. R., Suman, T. Y., Aranganathan, L., Gayathri, S., Gobalakrishnan, M., & Karthih, M. G., (2018). Laminarin based AgNPs using brown seaweed *Turbinaria ornata* and its induction of apoptosis in human retinoblastoma Y79 cancer cell lines. *Materials Research Express, 5*(3), 1–18. doi: 10.1088/2053-1591/aab2d8.

Renaud, S. M., & Luong-Van, J. T., (2006). Seasonal variation in the chemical composition of tropical Australian marine macroalgae. *Journal of Applied Phycology, 18*(3–5), 381–387. doi: 10.1007/s10811-006-9034-x.

Rinaudo, M., (2007). Seaweed polysaccharides. *Comprehensive Glycoscience: From Chemistry to Systems Biology, 2–4*, 691–735. doi: 10.1016/B978-044451967-2/00140-9.

Rinaudo, M., (2008). Main properties and current applications of some polysaccharides as biomaterials. *Polymer International, 57*(3), 397–430. doi: 10.1002/pi.2378.

Rioux, L. E., Turgeon, S. L., & Beaulieu, M., (2007). Characterization of polysaccharides extracted from brown seaweeds. *Carbohydrate Polymers, 69*(3), 530–537. doi: 10.1016/j.carbpol.2007.01.009.

Rioux, L. E., Turgeon, S. L., & Martin, B. M., (2010). Structural characterization of laminaran and galactofucan extracted from the brown seaweed *Saccharina longicruris*. *Phytochemistry, 71*(13), 1586–1595. doi: 10.1016/j.phytochem.2010.05.021.

Rocher, D. F., Cripwell, R. A., & Viljoen-Bloom, M., (2021). Engineered yeast for enzymatic hydrolysis of laminarin from brown macroalgae. Algal Res., *54*, 102233. https://doi.org/10.1016/j.algal.2021.102233.

Rodrigues, D., Freitas, A. C., Pereira, L., Rocha-Santos, T. A. P., Vasconcelos, M. W., Roriz, M., Rodríguez-Alcalá, L. M., et al., (2015). Chemical composition of red, brown and green macroalgae from Buarcos bay in Central West coast of Portugal. *Food Chemistry, 183*, 197–207. doi: 10.1016/J.FOODCHEM.2015.03.057.

Rodriguez-Jasso, R. M., Mussatto, S. I., Pastrana, L., Aguilar, C. N., & Teixeira, J. A., (2011). Microwave-assisted extraction of sulfated polysaccharides (fucoidan) from brown seaweed. *Carbohydrate Polymers, 86*(3), 1137–1144. doi: 10.1016/j.carbpol.2011.06.006.

Ruocco, N., Costantini, S., Guariniello, S., & Costantini, M., (2016). Polysaccharides from the marine environment with pharmacological, cosmeceutical and nutraceutical potential. *Molecules, 21*(5), 551. doi: 10.3390/molecules21050551.

Rupérez, P., Ahrazem, O., & Leal, J. A., (2002). Potential antioxidant capacity of sulfated polysaccharides from the edible marine brown seaweed *Fucus vesiculosus*. *Journal of Agricultural and Food Chemistry, 50*(4), 840–845. doi: 10.1021/jf010908o.

Rustad, T., & Hayes, M., (2012). Marine bioactive peptides and protein hydrolysates: Generation, isolation procedures, and biological and chemical characterizations protein and peptides from marine byproducts. In: Hayes, M., (ed.), *Marine Bioactive Compounds: Sources, Characterization and Applications* (pp. 99–113). Springer, USA. doi: 10.1007/978-1-4614-1247-2_3.

Sakiyama-Elbert, S. E., & Hubbell, J. A., (2001). Functional biomaterials: Design of novel biomaterials. *Annual Review of Materials Research, 31*(1), 183–201. doi: 10.1146/annurev.matsci.31.1.183.

Sandford, P. A., & Steinnes, A., (1991). Biomedical applications of high-purity chitosan. In: Shalaby, W. S., McCormick, C. L., & Butler, G. B., (eds.), *Water-Soluble Polymers* (pp. 430–445). American Chemical Society, USA. doi: 10.1021/bk-1991-0467.ch028.

Sanjeewa, K. K. A., Lee, J. S., Kim, W. S., & Jeon, Y. J., (2017). The potential of brown-algae polysaccharides for the development of anticancer agents: An update on anticancer effects reported for fucoidan and laminaran. *Carbohydrate Polymers, 177*, 451–459. doi: 10.1016/j.carbpol.2017.09.005.

Santoyo, S., Plaza, M., Jaime, L., Ibañez, E., Reglero, G., & Señorans, J., (2011). Pressurized liquids as an alternative green process to extract antiviral agents from the edible seaweed *Himanthalia elongata. Journal of Applied Phycology, 23*(5), 909–917. doi: 10.1007/s10811-010-9611-x.

Schneider, T., Ehrig, K., Liewert, I., & Alban, S., (2015). Interference with the CXCL12/CXCR4 axis as potential antitumor strategy: Superiority of a sulfated galactofucan from the brown alga *Saccharina latissima* and fucoidan over heparins. *Glycobiology, 25*(8), 812–824. doi: 10.1093/glycob/cwv022.

Sellimi, S., Maalej, H., Rekik, D. M., et al., (2018). Antioxidant, antibacterial and in vivo wound healing properties of laminaran purified from *Cystoseira barbata* seaweed. *Int J Biol Macromol., 119*, 633–644. https://doi.org/10.1016/j.ijbiomac.2018.07.171.

Shaklee, P. N., Bahr-Davidson, J., Prasad, S., & Kirk, J., (2008). *Process Methods for Fucoidan Purification from Seaweed Extracts*. https://patents.google.com/patent/EP2121767A1/en (accessed on 13 October 2021).

Shang, Y. F., Kim, S. M., Lee, W. J., & Um, B. H., (2011). Pressurized liquid method for fucoxanthin extraction from *Eisenia bicyclis* (Kjellman) setchell. *Journal of Bioscience and Bioengineering, 111*(2), 237–241. doi: 10.1016/j.jbiosc.2010.10.008.

Shanmugam, M., & Mody, K. H., (2000). Heparinoid-active sulfated polysaccharides from marine algae as potential blood anticoagulant agents. *Current Science, 79*(12), 1672–1683.

Shannon, E., & Abu-Ghannam, N., (2019). Seaweeds as nutraceuticals for health and nutrition. *Phycologia, 58*(5), 563–577. doi: 10.1080/00318884.2019.1640533.

Shanura, F. I. P., Asanka, S. K. K., Samarakoon, K. W., Kim, H. S., Gunasekara, U. K. D. S. S., Park, Y. J., Abeytunga, D. T. U., et al., (2020). The potential of fucoidans from *Chnoospora minima* and *Sargassum polycystum* in cosmetics: Antioxidant, anti-inflammatory, skin-whitening, and antiwrinkle activities. *Journal of Applied Phycology, 30*, 3223–3232. doi: 10.1007/s10811-018-1415-4.

Shukla, P. S., Mantin, E. G., Adil, M., Bajpai, S., Critchley, A. T., & Prithiviraj, B., (2019). *Ascophyllum nodosum*-based biostimulants: Sustainable applications in agriculture for the stimulation of plant growth, stress tolerance, and disease management. *Frontiers in Plant Science, 10*, 655. doi: 10.3389/fpls.2019.00655.

Shvetsova, S. V., Shabalin, K. A., Bobrov, K. S., Ivanen, D. R., Ustyuzhanina, N. E., Krylov, V. B., Nifantiev, N. E., et al., (2017). Characterization of a new α-l-fucosidase isolated from *Fusarium proliferatum* LE1 that is regioselective to α-(1 → 4)-l-fucosidic linkage in the hydrolysis of α-l-fucobiosides. *Biochimie, 132*, 54–65. doi: 10.1016/j.biochi.2016.10.014.

Sime, W. J., (1990). Alginates. In: Harris, P., (ed.), *Food Gels* (pp. 53–78). Springer, Dordrecht. doi: 10.1007/978-94-009-0755-3_2.

Sinha, S., Astani, A., Ghosh, T., Schnitzler, P., & Ray, B., (2010). Polysaccharides from *Sargassum tenerrimum*: Structural features, chemical modification and anti-viral activity. *Phytochemistry, 71*(2, 3), 235–242. doi: 10.1016/j.phytochem.2009.10.014.

Skjåk-Bræk, G., Grasdalen, H., & Larsen, B., (1986). Monomer sequence and acetylation pattern in some bacterial alginates. *Carbohydrate Research, 154*(1), 239–250. doi: 10.1016/S0008-6215(00)90036-3.

Smith, A. J., Graves, B., Child, R., Rice, P. J., Ma, Z., Lowman, D. W., Ensley, H. E., et al., (2018). Immunoregulatory activity of the natural product laminarin varies widely as a result of its physical properties. *The Journal of Immunology, 200*(2), 788–799. doi: 10.4049/jimmunol.1701258.

Sparr, E. C., & Björklund, E., (2000). Analytical-scale microwave-assisted extraction. *Journal of Chromatography A, 902*(1), 227–250. doi: 10.1016/S0021-9673(00)00921-3.

Srogi, K., (2006). A review: Application of microwave techniques for environmental analytical chemistry. *Analytical Letters, 39*(7), 1261–1288. doi: 10.1080/00032710600666289.

Suzuki, Y., Nishimura, Y., Tanihara, M., Suzuki, K., Nakamura, T., Shimizu, Y., Yamawaki, Y., & Kakimaru, Y., (1998). Evaluation of a novel alginate gel dressing: Cytotoxicity to fibroblasts *in vitro* and foreign-body reaction in pig skin *in vivo*. *Journal of Biomedical Materials Research, 39*(2), 317–322.

Szekalska, M., Puciłowska, A., Szymańska, E., Ciosek, P., & Winnicka, K., (2016). Alginate: Current use and future perspectives in pharmaceutical and biomedical applications. *International Journal of Polymer Science,* 1–17. doi: 10.1155/2016/7697031.

Taniguchi, K., & Karin, M., (2018). NF-B, inflammation, immunity and cancer: Coming of age. *Nature Reviews Immunology, 18*(5), 309–324. doi: 10.1038/nri.2017.142.

Teas, J., (1982). The dietary intake of *Laminaria*, a brown seaweed, and breast cancer prevention. *Nutrition and Cancer, 4*(3), 217–222. doi: 10.1080/01635588209513760.

Thao, M., Le, P., Sung, V. V., Dat, T. D., Nam, H. M., Phong, M. T., & Hieu, N. H., (2020). Ultrasound-assisted extraction of fucoidan from Vietnamese brown seaweed *Sargassum mcclurei* and testing bioactivities of the extract. *Chemistry Select, 5*(14), 4371–4380. doi: 10.1002/slct.201903818.

Thinh, P., Menshova, R., Ermakova, S., Anastyuk, S., Ly, B., & Zvyagintseva, T., (2013). Structural characteristics and anticancer activity of fucoidan from the brown alga *Sargassum mcclurei*. *Marine Drugs, 11*(5), 1456–1476. doi: 10.3390/md11051456.

Tsiapali, E., Whaley, S., Kalbfleisch, J., Ensley, H. E., Browder, I. W., & Williams, D. L., (2001). Glucans exhibit weak antioxidant activity, but stimulate macrophage free radical activity. *Free Radical Biology and Medicine, 30*(4), 393–402. doi: 10.1016/S0891-5849(00)00485-8.

Turgeon, S. L., Rioux, L. E., & Beaulieu, M., (2012). *Use of Skin Care Compositions Comprising Laminaraceae Extract for Treatment of Skin Aging Signs*. https://patentscope.wipo.int/search/en/detail.jsf?docId=WO2010145008 (accessed on 13 October 2021).

Usoltseva (Menshova), R. V., Anastyuk, S. D., Shevchenko, N. M., Zvyagintseva, T. N., & Ermakova, S. P., (2016). The comparison of structure and anticancer activity *in vitro* of polysaccharides from brown algae *Alaria marginata* and *A. angusta*. *Carbohydrate Polymers, 153*, 258–265. doi: 10.1016/j.carbpol.2016.07.103.

Vavilala, S. L., & D'Souza, J. S., (2015). Algal polysaccharides and their biological applications. In: Kim, S. K., & Chojnacka, K., (eds.), *Marine Algae Extracts: Processes, Products, and Applications* (pp. 411–452). Wiley-Verlag. doi: 10.1002/9783527679577.ch26.

Venkatesan, J., Anil, S., Kim, S. K., & Shim, M. S., (2016). Seaweed polysaccharide-based nanoparticles: Preparation and applications for drug delivery. *Polymers, 8*(2), 1–25. doi: 10.3390/polym8020030.

Voronova, Y. G., Rekhina, N. I., Nikolaeva, T. A., Tiunova, N. A., Zaikina, I. V., Ya Kobzeva, N., & Valiente, O., (1991). Extraction of carbohydrates from *Laminaria* and their utilization. *Journal of Applied Phycology, 3*(3), 243–245. doi: 10.1007/BF00003582.

Wang, H. M. D., Chen, C. C., Huynh, P., & Chang, J. S., (2015). Exploring the potential of using algae in cosmetics. *Bioresource Technology, 184*, 355–362. doi: 10.1016/j.biortech.2014.12.001.

Wang, L., Oh, J. Y., Thilina, U., Jayawardena, T. U., Jeon, Y. J., & Ryu, B. M., (2020). Anti-inflammatory and anti-melanogenesis activities of sulfated polysaccharides isolated from *Hizikia fusiforme*. *International Journal of Biological Macromolecules, 142*, 545–550. doi: 10.1016/j.ijbiomac.2019.09.128.

Wang, W., Wang, S. X., & Guan, H. S., (2012). The antiviral activities and mechanisms of marine polysaccharides: An overview. *Marine Drugs, 10*(12), 2795–2816. doi: 10.3390/md10122795.

Wang, Y., Jiang, Z., Kim, D., Ueno, M., Okimura, T., Yamaguchi, K., & Oda, T., (2013). Stimulatory effect of the sulfated polysaccharide ascophyllan on the respiratory burst in RAW264.7 macrophages. *International Journal of Biological Macromolecules, 52*, 164–169. http://hdl.handle.net/10069/31123%0A (accessed on 13 October 2021).

Wei, D., Williams, D., & Browder, W., (2002). Activation of AP-1 and SP1 correlates with wound growth factor gene expression in glucan-treated human fibroblasts. *International Immunopharmacology, 2*(8), 1163–1172. doi: 10.1016/S1567-5769(02)00082-6.

Wu, S. C., (2017). Antioxidant activity of sulfated seaweeds polysaccharides by novel assisted extraction. In: Xu, Z., (ed.), *Solubility of Polysaccharides*. IntechOpen, United Kingdom. doi: 10.5772/intechopen.69633.

Xuan Cuong, D., (2020). Laminarin (Beta-glucan) of brown algae *Sargassum mcclurei:* Extraction, antioxidant activity, lipoxygenase inhibition activity, and physicochemistry properties. *World J Food Sci Technol., 4,* 31. https://doi.org/10.11648/j.wjfst.20200401.15.

Yang, L., Wang, P., Wang, H., Li, Q., Teng, H., Liu, Z., Yang, W., Hou, L., & Zou, X., (2013). Fucoidan derived from *Undaria pinnatifida* induces apoptosis in human hepatocellular carcinoma SMMC-7721 cells via the ROS-mediated mitochondrial pathway. *Marine Drugs, 11*(6), 1961–1976. doi: 10.3390/md11061961.

Yao, Y., Zaw, A. M., Anderson, D. E. J., Hinds, M. T., & Yim, E. K. F., (2020). Fucoidan functionalization on poly(vinyl alcohol) hydrogels for improved endothelialization and hemocompatibility. *Biomaterials, 249*, 120011. doi: 10.1016/j.biomaterials.2020.120011.

Ye, H., Zhou, C., Sun, Y., Zhang, X., Liu, J., Hu, Q., & Zeng, X., (2009). Antioxidant activities *in vitro* of ethanol extract from brown seaweed *Sargassum pallidum*. *European Food Research and Technology, 230*(1), 101–109. doi: 10.1007/s00217-009-1147-4.

Yin, G., Li, W., Lin, Q., Lin, X., Lin, J., Zhu, Q., Jiang, H., & Huang, Z., (2014). Dietary administration of laminarin improves the growth performance and immune responses in *Epinephelus coioides*. *Fish and Shellfish Immunology, 41*(2), 402–406. doi: 10.1016/j.fsi.2014.09.027.

Yoo, H. J., You, D. J., & Lee, K. W., (2019). Characterization and immunomodulatory effects of high molecular weight fucoidan fraction from the sporophyll of *Undaria pinnatifida* in cyclophosphamide-induced immunosuppressed mice. *Marine Drugs, 17*(8), 447. doi: 10.3390/md17080447.

Young, S. H., Dong, W. J., & Jacobs, R. R., (2000). Observation of a partially opened triple-helix conformation in 1→3-β- glucan by fluorescence resonance energy transfer spectroscopy. *Journal of Biological Chemistry, 275*(16), 11874–11879. doi: 10.1074/jbc.275.16.11874.

Yvin, J. C., & Vasseur, F. L., (1999). *Use of Laminarin and Oligosaccharides Derived Therefrom in Cosmetics and for Preparing a Skin Treatment Drug.* https://patentimages.

storage.googleapis.com/fc/8e/bf/7300c483519f0d/US5980916.pdf (accessed on 13 October 2021).

Yvin, J. C., Alban, S., & Franz, G., (2006). *Anti-Inflammatory and Healing Medicine Based on Laminarin Sulfate*. https://patents.google.com/patent/US7008931B2/en (accessed on 13 October 2021).

Zargarzadeh, M., Amaral, A. J. R., Custódio, C. A., & Mano, J. F., (2020). Biomedical applications of laminarin. *Carbohydrate Polymers, 232*(115774). doi: 10.1016/j.carbpol.2019.115774.

Zayed, A., & Ulber, R., (2019). Fucoidan production: Approval key challenges and opportunities. *Carbohydrate Polymers, 211*, 289–297. doi: 10.1016/j.carbpol.2019.01.105.

Zha, X. Q., Xiao, J. J., Zhang, H. N., Wang, J. H., Pan, L. H., Yang, X. F., & Luo, J. P., (2012). Polysaccharides in *Laminaria japonica* (LP): Extraction, physicochemical properties and their hypolipidemic activities in diet-induced mouse model of atherosclerosis. *Food Chemistry, 134*(1), 244–252. doi: 10.1016/j.foodchem.2012.02.129.

Zhang, L., Hao, J., Zheng, Y., Su, R., Liao, Y., Gong, X., Liu, L., & Wang, X., (2018). Fucoidan protects dopaminergic neurons by enhancing the mitochondrial function in a rotenone-induced rat model of Parkinson's disease. *Aging and Disease, 9*(4), 590. doi: 10.14336/AD.2017.0831.

Zhang, S., Zhang, H., Jin, Z., Wang, S., Wang, Y., Zhu, L., Sun, W., & Yan, B., (2020). Fucoidan inhibits tooth movement by promoting restorative macrophage polarization through the STAT3 pathway. *Journal of Cellular Physiology, 235*, 5938–5950. doi: 10.1002/jcp.29519.

Zhang, W., Du, J. Y., Jiang, Z., Okimura, T., Oda, T., Yu, Q., & Jin, J. O., (2014). Ascophyllan purified from *Ascophyllum nodosum* induces Th1 and tc1 immune responses by promoting dendritic cell maturation. *Marine Drugs, 12*(7), 4148–4164. doi: 10.3390/md12074148.

Zhang, W., Minseok, K. M., Hae, B. P. H. B., Takasi, O. T., Tatsuya, O. T., Lee, P. C. W. L., & Jun, O., Jin, J. O., (2019). Activation of human dendritic cells by ascophyllan purified from *Ascophyllum nodosum*. *Marine Drugs, 17*(1), 66. doi: 10.3390/md17010066.

Zhang, W., Oda, T., Yu, Q., & Jin, J. O., (2015). Fucoidan from *Macrocystis pyrifera* has powerful immune-modulatory effects compared to three other fucoidans. *Marine Drugs, 13*(3), 1084–1104. doi: 10.3390/md13031084.

Zhang, W., Okimura, T., Oda, T., & Jin, J. O., (2019). Ascophyllan induces activation of natural killer cells in mice *in vivo* and *in vitro*. *Marine Drugs, 17*(4), 197. doi: 10.3390/md17040197.

Zhang, Y., Xu, M., Hu, C., Liu, A., Chen, J., Gu, C., Zhang, X., You, C., Tong, H., Wu, M., & Chen, P., (2019). *Sargassum fusiforme* fucoidan SP2 extends the lifespan of *Drosophila melanogaster* by upregulating the Nrf2-mediated antioxidant signaling pathway. *Oxidative Medicine and Cellular Longevity, 8918914*, 1–15. doi: 10.1155/2019/8918914.

Zhao, D., Xu, J., & Xu, X., (2018). Bioactivity of fucoidan extracted from *Laminaria japonica* using a novel procedure with high yield. *Food Chemistry, 245*(59), 911–918. doi: 10.1016/j.foodchem.2017.11.083.

Zhao, X., Huebsch, N., Mooney, D. J., & Suo, Z., (2010). Stress-relaxation behavior in gels with ionic and covalent crosslinks. *Journal of Applied Physics, 107*(6), 63509. doi: 10.1063/1.3343265.

Zhao, Y., Zheng, Y., Wang, J., Ma, S., Yu, Y., White, W. L., Yang, S., et al., (2018). Fucoidan extracted from *Undaria pinnatifida*: Source for nutraceuticals/functional foods. *Marine Drugs, 16*(9), 321. doi: 10.3390/md16090321.

Zhou, G., Sun, Y. P., Xin, H., Zhang, Y., Li, Z., & Xu, Z., (2004). *In vivo* antitumor and immunomodulation activities of different molecular weight lambda-carrageenans from *Chondrus ocellatus*. *Pharmacological Research, 50*(1), 47–53. doi: 10.1016/j.phrs.2003.12.002.

Zia, K. M., Zuber, M., & Ali, M., (2017). *Algae Based Polymers, Blends, and Composites: Chemistry, Biotechnology and Materials Science*. Elsevier, Netherlands. doi: 10.1016/C2015-0-06688-6.

Zvyagintseva, T. N., Shevchenko, N. M., Popivnich, I. B., Isakov, V. V., Scobun, A. S., Sundukova, E. V., & Elyakova, L. A., (1999). A new procedure for the separation of water-soluble polysaccharides from brown seaweeds. *Carbohydrate Research, 322*(1, 2), 32–39. doi: 10.1016/S0008-6215(99)00206-2.

CHAPTER 6

THE GLOBAL SEAWEED INDUSTRY: APPLICATIONS AND EMERGING MARKETS

ZAHIRA BELATTMANIA, ABDELTIF REANI, and BRAHIM SABOUR

Phycology Research Unit, Department of Biology, Faculty of Sciences, University Chouaib Doukkali, PO Box 20, El Jadida – 24000, Morocco

ABSTRACT

The marine environment offers extreme biodiversity of seaweeds. These marine organisms are generally rich in fiber, polysaccharides, proteins, and lipids. Marine macroalgae are commonly known as interesting bioresources used for human consumption, hydrocolloids production, animal feed, and fertilizers. They are also rich in chemically active secondary metabolites with antioxidant, antimicrobial, anti-inflammatory, and antitumor properties. Due to their immense genetic and chemical diversity and complexity, seaweeds are described among the richest and most promising marine bioresources offering a wide variety of applications and many potential opportunities for exploitation. In particular, recent trends in research on drugs and cosmetic products from natural sources indicate that algae constitute a promising source to provide new biologically active substances and original biomaterials with possibilities of medical, pharmaceutical, and environmental applications.

6.1 INTRODUCTION

The marine macroalgae were used worldwide in diverse food and nonfood applications. Their use dates back several centuries, especially in Asian regions, where seaweeds have been used for human consumption (McHugh, 2003). In the West (Europe and North America) since the 16th Century, seaweeds have been considered in coastal regions as a raw material for agricultural use. Subsequently, other applications have emerged (e.g., the extraction of iodine, the recovery of soda and potash used in soap and glass production, and more recently for technologically innovative products (Mouritsen and Johansen, 2013). The seaweed exploitation at industrial scale was changed over the years. Nowadays, the seaweed processing industry is comprised of different areas (Tiwari and Troy, 2015). Approximately 83% of the worldwide produced biomass is used for human consumption, and the rest of the biomass is mainly utilized for fertilizers, animal feed additives and increasingly, for medical and biotechnological applications (Loureiro et al., 2015).

The global seaweeds production exceeds currently 30.4 million tons, where 1.1 million tons derive from natural deposits and 29.4 million tons originated from seaweed farming (FAO, 2018). The leading producer's countries were Chile, China, and Norway for species harvested from wild, and China, Indonesia, the Republic of Korea and the Philippines for cultured species (FAO, 2018). Aquatic plant farming (mainly seaweeds) is taking place in about 50 countries (FAO, 2016). This activity recorded a growth rate expanding 8% per year over the past decade, compared to 6.2% during the previous decade (Figure 6.1).

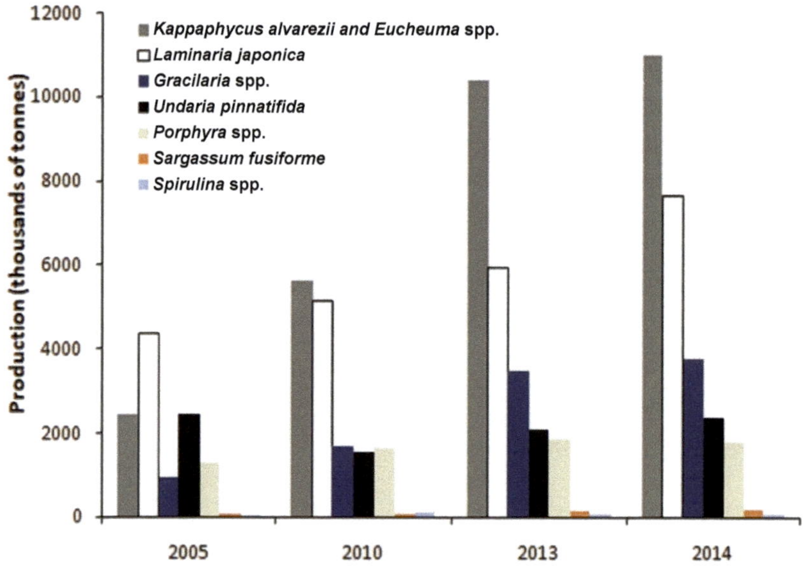

FIGURE 6.1 World cultured seaweed production (FAO, 2016).

The global seaweed industry has a market value of around US $ 6 billion per year of which some 85% comprises food products for human consumption (FAO, 2018). Therefore, macroalgae are regarded as important bioresource with interesting economic value driving to a raise of scientific research in different fields. This chapter provides an overview on the traditional uses and emerging markets for seaweed worldwide.

6.2 USES OF SEAWEEDS AND EMERGING MARKETS

6.2.1 SEAWEEDS AS HUMAN FOOD

The seaweeds have been employed as food since ancient times, however, their commercial exploitation of a global scale only began a few decades after WWII (Kilinç et al., 2013). The significant nutritional value of seaweeds makes them traditionally used as food, especially in Asian countries (Mabeau and Fleurence, 1993; Norziah and Ching, 2000; McHugh, 2003). Actually, many macroalgal species (Table 6.1) are used as food. The seaweeds components afford important nutritional value serving human nutrients, e.g., proteins, minerals, and vitamins (Marinho-Soriano et al., 2006; Marsham et al., 2007; Manivannan et al., 2008; de Oliveira et al., 2009).

TABLE 6.1 Some Edible Seaweed from the World (Mouritsen and Johansen 2013; Pereira 2016).

Taxonomic group	Species	Common name
Ochrophyta – Phaeophyceae	*Alaria esculenta*	Winged kelp
	Ascophyllum nodosum	Rockweed
	Fucus vesiculosus	Bladderwrack
	Himanthalia elongata	Hogweed
	Laminaria digitata	Sea-girdles
	Laminaria hyperborea	Tangleweed
	Macrocystis pyrifera	Giant kelp
	Nereocystis luetkeana	Bullwhip kelp
	Saccharina japonica	Japanese kombu
	Saccharina latissima	Sugar kelp
	Saccharina longicruris	Wild Atlantic kombu
	Undaria pinnatifida	Wakame

TABLE 6.1 *(Continued)*

Taxonomic group	Species	Common name
Rhodophyta	*Chondrus cripus*	Irish Moss
	Chondrus elatus	Dee-horn vegetable
	Gracilaria gracilis	Slender wart weed
	Gracilaria verrucosa	Ogo
	Palmaria palmata	Dulse
	Porphyra capensis	Cape laver
	Porphyra dioica	Black laver
	Porphyra laciniata	Laver sloke
	Porphyra purpurea	Purple laver
	Porphyra umbilicalis	Purple laver
	Pyropia tenera	Nori
	Pyropia yezoensis	Nori
Chlorophyta	*Ulva lactuca*	Sea lettuce
	Ulva intestinalis	Green string lettuce

Proteins are important and essential factors that determine the nutritional value of food. Generally, the red and green algae have relatively high protein contents with an average of 10–30% dry weight (DW) (Mabeau and Fleurence, 1993; Ramos et al., 2000; Burtin, 2003). In some cases, they can exceed 30% dw; For example, the proteins content of the red algae *Palmaria palmata* and *Porphyra tenera* are between 35 and 47% dw (Burtin, 2003; Marsham et al., 2007). However, the protein contents in brown algae are usually between 5 and 15% dw (Burtin, 2003; Dawczynski et al., 2007). The proteins quality is based on bioavailability and amino acid composition (Levesque et al., 2010). The amino acid profile has been determined for several macroalgal species (Wong and Cheung, 2000; Marrion et al., 2005; Dawczynski et al., 2007). Among the 20 natural amino acids, most are commonly found in macroalgae, including all essential amino acids (e.g., histidine, leucine, methionine) (Cerna, 2011; Holdt and Kraan, 2011). The essential amino acids can represent up to 48% of the total amino acid content (Wong and Cheung, 2000).

Typically, the lipid contents in seaweeds are relatively low (1–5% of the dry matter), most of them are polyunsaturated fatty acids (PUFA) like eicosapentaenoic acid (EPA, n-3, C 20: 5) and arachidonic acid (AA, n-6, C 20: 4) (Norziah and Ching, 2000; Ortiz et al., 2006; Maeda et al., 2008;

Belattmania et al., 2018a). For example, the PUFAs content reached 74% of total fatty acids of the red alga *Gracilaria changii* with 33.1% EPA and 12.9% docosahexaenoic acid (DHA) (n-6, C 22: 3) (Norziah and Ching, 2000). According to Ortiz et al. (2006), unsaturated fatty acids (UFAs) represent 72.5% of total fatty acids for the brown alga *Durvillaea antarctica*, with a high content of arachidonic acid (11.2%). Nonetheless, all diets containing essential fatty acids (ω6 and ω3) are not considered healthy. For example, the ω6/ω3 ratio in the Western diets was between 15/1 and 17/1 due to the excessive amounts of omega-6. As a result, Western diets are considered as deficient in ω3 fatty acids (Simopoulos, 2002). Similarly, for the American diet, the ω6/ω3 ratio was about 15–20:1, which was considered as "being deficient" in ω3 (Simopoulos, 2008). The use of macroalgae as nutrient sources in food products, commonly consumed in the western world, may provide human beings with beneficial FA in their diets (Pereira et al., 2012).

Whilst chemical composition can vary according to species and growth environment, macroalgae contain many vitamins such as L-ascorbic acid (vitamin C), thiamine (vitamin B1), riboflavin (vitamin B2), cobalamin (vitamin B12), folic acid (e.g., 5-methyltetrahydrofolic acid), tocopherols (vitamin E) and carotenoids (Misurcova, 2011). Vitamin C content ranges from 500 to 3000 mg/kg DW. *Eucheuma denticulatum* has 2000 mg of vitamin C per kg of DW (McDermi and Stuercke, 2003) and the green alga *Enteromorpha flexuosa* presents 3000 mg of vitamin C per kg of dry matter (McDermi and Stuercke, 2003). Vitamins B group, mainly riboflavin and thiamine, are recorded in both red and brown seaweeds (Hegedüs et al., 1985; MacArtain et al., 2007). The concentration of vitamin E is generally higher in brown macroalgae compared to red and green seaweeds (Ortiz et al., 2006). For example, the vitamin E content in *Durvillaea antarctica* (245.9 mg/kg of dry matter) is ten times higher than that *Ulva lactuca* (25.8 mg/kg of dry matter). Seaweeds have also significant levels of carotenoids (mainly are provitamins A). The main carotenoids in red algae are α- and β-carotene and their derivatives like lutein and zeaxanthin (Norziah and Ching, 2000), while the main carotenoids in brown algae consist of β-carotene, lutein, violaxanthin, anteraxanthin, zeaxanthin, and neoxanthin (Burtin, 2003). The red alga *Gracilaria changii* seems to be a potential source of β-carotene with a content greater than 5 mg/100 g, with comparable value found in carrots (6.8 mg/100 g) (Norziah and Ching, 2000). Seaweeds are also known as a vital source of minerals due to their potential to attract inorganic substances from their environment. The mineral fraction of macroalgae could represent up to 36% dw (Burtin, 2003). For example, there is more iron in 8 g dry matter of *Palmaria palmata* than in 100 g of beef (MacArtain et al., 2007). Seaweeds contribute significantly to the daily intake of some of the minerals: Na (5.3%), K (2.5%), Ca (2.8%), Mg (5.5%), Li (6.9%), Ba (5.1%), Fe (2.5%), Co (2.5%) and Cd (8.1%) (Shiraishi, 2005, 2006). Consequently, seaweeds could be used as a food complement to improve the recommended daily adult intake of some minerals. The interesting nutritional value of dietary seaweed has encouraged the development of several foods and food applications. López-López et al. (2009) added 5.6% edible seaweeds *Himanthalia elongata*, *Undaria pinnatifida*, and *Porphyra umbilicalis* to meat emulsions. Their analyzes have shown that the addition of these algae leads to an increase in PUFA (ω3) and a decrease in the ω6/ω3 ratio with an improvement in the contents of Ca, K, Mg, and Mn. In addition, *Porphyra* considerably enhances the levels of certain essential amino acids. In other words, seaweeds have been incorporated into dairy products. For example, cottage cheeses (home-made and industrially produced) have been made with 3%, 9% and 15% of *Undaria pinnatifida* and *Laminaria japonica*. Cheeses with 15% added algae had higher level of calcium, iron, and magnesium, while the textural quality (consistency and firmness) was better with the addition of 9% *U. pinnatifida* or *L. japonica* (Lalic and Berkovic, 2005). On a similar line, edible algae have been used as optional ingredients in pasta products. The effect of different concentrations (0–30%) of *U. pinnatifida* on the sensory, cooking, nutritional, and biofunctional quality of pasta was investigated by Prabhasankar et al. (2009). These authors have shown that pasta containing up to 20% algae exhibited adequate sensory traits and improved biofunctional characteristics.

6.2.2 PHYCOCOLLOIDS INDUSTRY

The international seaweed market dynamics are mostly due to phycocolloids products, which are traded and used widely in many countries for food and nonfood application (FAO, 2018). The phycocolloid production is the second-largest macroalgae exploitation sector in the world (Figure 6.2) make up almost 40% of the global hydrocolloid market in terms of foods. For the last few years, China has led the world seaweeds market as an importer of raw material and exporter of semi-processed products. Recently, there has been a developing focus on highlighting other countries in Asia and Latin America to match the rising demand for food-grade carrageenan and agars (FAO, 2018).

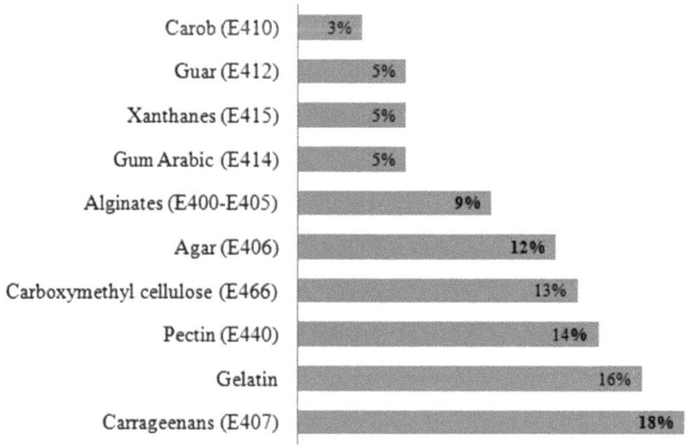

FIGURE 6.2 Importance of phycocolloids production in the global colloid market.

6.2.2.1 AGAR

Agar or agar-agar is a polysaccharide made up of two polysaccharides units: agarose and agaropectin (Figure 6.3). Agarose is a neutral homopolymer of agarobiose, consisting of a D-galactose of a 3,6-anhydro-L-galactopyranose. Agaropectin contains both pure agarobiose units and agarobiose units substituted by sulfate, methyl, pyruvate groups, etc. (Murano, 1995; Stanley, 2006). Agar is mainly extracted from *Gelidium* and *Gracilaria* (McHugh, 2003). Generally, the sulfate and methyl pyruvate contents are higher in *Gracilaria* compared to *Gelidium* species, affecting the gelling properties (Murano, 1995). Agar is used as stabilizer and gelling agents in bakery products, aqueous gels, confectionery, dairy products, and canned fish products (Imeson, 2009). It is used for other biotechnological applications, notably within culture medium in microbiology (Cardozo et al., 2007).

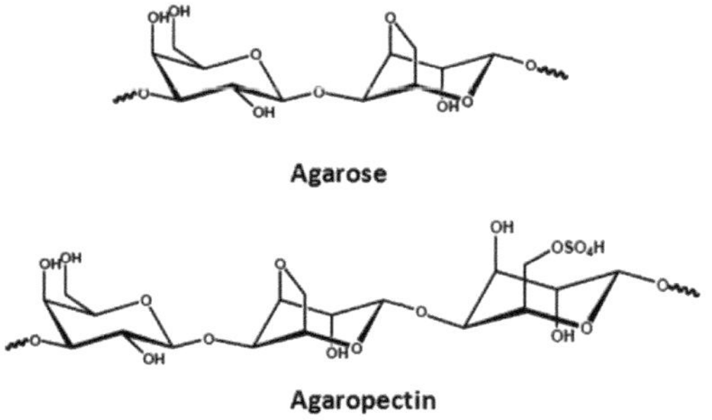

FIGURE 6.3 Structure of the two polysaccharides constituting agar: agarose and agaropectin.

6.2.2.2 CARRAGEENANS

The carrageenans (Figure 6.4) are partially sulfated β-D-galactose and 3,6-anhydro-α-D-galactose units (FAO, 2014). They differ from agars by their high content of sulfated groups OSO_3^- (24–26% against 5%; Pérez, 1997). The main structures of carrageenan are in the form of κ-carrageenans (kappa), ι-carrageenans (iota) and λ-carrageenans (lambda). In the 1960s, carrageenans were mainly extracted from the Rhodophyta *Chondrus crispus*. Afterward, this species was substituted by *Kappaphycus alvarezii* and *Eucheuma denticulatum* (McHugh, 2003). Currently, the main commercially exploitable sources are *Kappaphycus*, *Eucheuma*, *Chondrus*, *Gigartina*, *Hypnea*, *Gynmogongrus*, *Phyllophora*, *Iridaea*, *Furcellaria*, *Ahnfeltia*, *Anatheca*, and *Meristotheca* (FAO, 2014). Carrageenans are used much more in the food industry in water-based or dairy products, for example, ice creams, cakes, custards, foams, gelled yogurts and sauce. Carrageenans are also applied in various cosmetic formulations (Samaraweera et al., 2011).

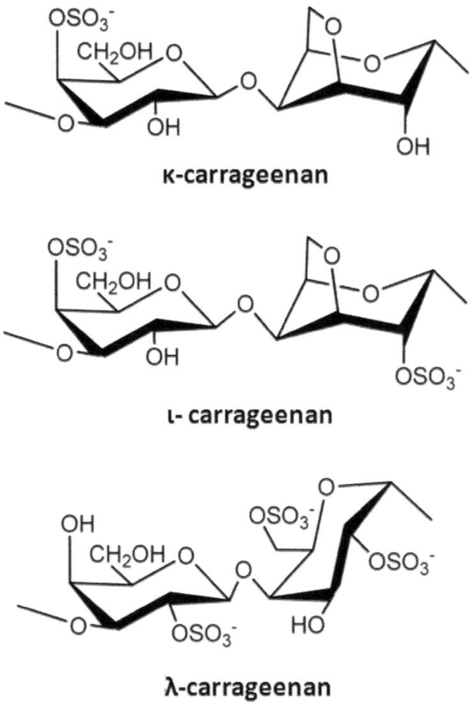

FIGURE 6.4 Chemical Structures of carrageenans Kappa (κ), Iota (ι) and Lambda (λ).

6.2.2.3 ALGINATES

Alginates are linear copolymers consisting of β-D-mannuronic acid (M) and α-L-guluronic acid (G) (Figure 6.5). Alginate's structure varies following the arrangement of the monomers on the chains. Homogeneous segments correspond to monomers from MM or GG type blocks, while MG or GM blocks are described as mixed segments (Rioux and Turgeon, 2015). Alginates are naturally present in the cell wall of brown seaweeds and it constitute up to 40% of the dry algal biomass (Rioux and Turgeon, 2015). Several species of brown algae (*Durvillaea antarctica, Durvillaea potatorum, Ascophyllum nodosum, Laminaria digitata, Laminaria hyperborea, Laminaria saccharina, Laminaria japonica, Ecklonia maxima, Macrocystis pyrifera, Lessonia nigrescens*, and *Lessonia trabeculata*) are industrially exploited to extract alginates (McHugh 2003; Helgerud et al., 2009). Industrial production of alginates on a global scale remains, at present, based on brown algae from cultivation or natural deposits (Chèze-Lange et al., 2002; Hashimoto et al., 2009). Alginates are commonly exploited for food industry as stabilizing, emulsifying, gelling, and thickening agents (Rioux and Turgeon, 2015). They are used as thickening agents in medical industries, impression material in dentistry and anticoagulant material in toothpaste (Hanzhi et al., 2011). Alginates are also employed for paint production, building materials, glue, paper, photos, and in the textile industry (Person, 2010).

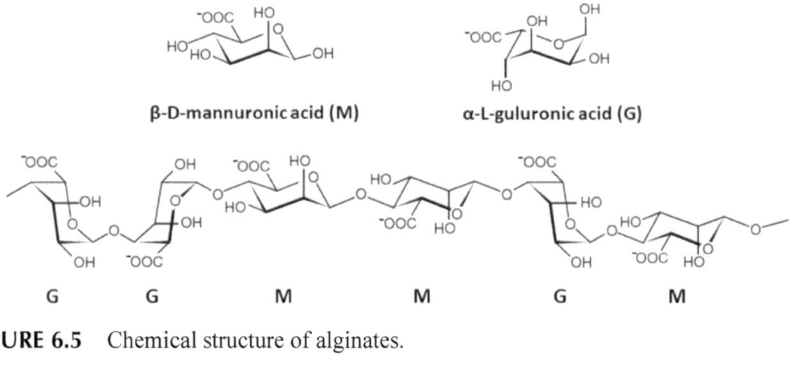

FIGURE 6.5 Chemical structure of alginates.

6.2.3 SEAWEEDS AS ANIMAL FEED

Macroalgae are traditionally used as feed supplements in animals farming (Hansen et al., 2003). Although the algal biomass is rich in minerals and vitamins, their proteins and fatty acid profiles that make them potential feed alternative. Various literature reports indicated that seaweeds and their extracts have many benefits on aquatic animals, poultry, and livestock diets (Cruz-Suarez et al., 2009; Bendary et al., 2013). The algal species commonly used in animal feed belong to the genera *Laminaria, Sargassum*, *Gracilaria*, *Ulva*, *Macrocystis*, and *Ascophyllum* (Hansen et al., 2003; Zhou et al., 2006; Cruz-Suárez et al., 2009). The addition of seaweeds to animal feed enhances the nutritional value of diets, resulting in high quality of produced food, e.g., meat, eggs, milk (Hwang et al., 2014). The incorporation of seaweed with other food can contribute a high nutritional value to feed and make a nearly "complete diet" (Rajauria et al., 2015). The macroalgae are used in aquaculture as complete feed or as food supplement (Spolaore et al., 2006). Some algae species are co-cultivated with shrimp as a natural food source, reducing the use of conventional commercial foods (Rajauria et al., 2015). Previous study carried out on ruminants shows that feeding involving algal biomasses significantly improved the profile of ω3 fatty acids in milk without disturbing feed intake (Stamey et al., 2012). Likewise, the addition of algae biomass could also adjust the fatty acid profile. For instance, EPA and DHA contents have been improved in the muscles of lambs fed on macroalgae diet (Cooper et al., 2004).

6.2.4 APPLICATIONS OF SEAWEED IN AGRICULTURE

Seaweeds have historically and traditionally been used as organic fertilizer in coastal regions (Milledge et al., 2016). Algal constituents recover the mineral composition of impoverished soils and increase the water retention capacity of soils (Person, 2010). Recent progress has been made in developing algal liquid extracts with various outcomes on the improvement of early seed germination, performance, and yield of crops, resistance to biotic and abiotic stress and post-harvest shelf life (Hong et al., 2007; Khan et al., 2009; Craigie, 2011; Mattner et al., 2013). The development of liquid seaweed biofertilizers made possible the use of these products by farmers in offshore areas (Sartal et al., 2011). The commonly commercial seaweed products are based on *Ascophyllum*, *Macrocystis*, *Ecklonia*, and *Durvillaea* (Khan et al., 2009). Due to their polysaccharides contents, seaweeds are also used to refine the soil microbiological and physical properties (Haslam and Hopkins, 1996). It has been reported that alginates combine with metallic ions present in the soil to provide complexes with high molecular weight which retain soil moisture and improve the aggregate structure, thus, inducing raise in soil aeration and capillary activity (Khan et al., 2009).

Seaweeds are also applied as biostimulants of the plant growth (Chojnacka, 2011). The beneficial effects observed in plant growth bioassays suggested that algae extracts could hold more than one group of growth-promoting substances or hormones (Mooney and van Staden, 1986; Crouch and van Staden, 1993). Seaweed biostimulants consist of nutrients (macro and microelements), amino acids, vitamins, cytokinins, auxins, abscisic acid-like growth substances (Wajahatullah et al., 2009). The zeatin (with *cis* forms) and isopentenyl adenine (iP) derivatives are the main type of isoprenoid cytokinins observed in macroalgae (Stirk et al., 2003). In addition to the improvement of germination, vegetative growth and flowering correlated with their exceptional phytohormone contents (cytokinins, gibberellins, and auxins) and the occurrence of sulfated polysaccharides (PLS) (fucoidans), laminarins, and oligoalginates (mannuronic acids). The brown algae extracts can even be used as biological protection agents against plant diseases by acting as elicitors, thus stimulating the natural defenses of plants and making it possible to decrease the frequency and the importance of curative phytosanitary treatments with fungicidal chemicals or pesticides (Person, 2010).

6.2.5 BIOMEDICAL APPLICATIONS OF SEAWEEDS

Trends in research on drugs from natural sources indicate that seaweeds are a promising source providing new biologically active substances with potential medical and pharmaceutical applications (Mayer and Hamann 2004; Singh et al., 2005; Mercurio et al., 2015).

6.2.5.1 ANTIVIRAL ACTIVITY

Some biochemical compounds originating from seaweeds have shown significant antiviral activities. The sulfated xylomannan isolated from the red alga *Nothogenia fastigiata* inhibits efficiently the replication *in vitro* of Herpes Simplex Virus 1 (HSV1), and other enveloped viruses such as HSV2, human cytomegalovirus (HCMV), respiratory syncytial virus (RSV) and the influenza A and B virus, but, it is less active for HIV-1, HIV-2 (Damonte et al., 1994). Fucoidan has also shown antiviral activity against RSV (Malhotra et al., 2003), HSV 1 and 2, and HCMV (Ponce et al., 2003). Certain algae polysaccharides have inhibitory actions against viral infection targeting the viral attachment step and generally adopting two approaches: interacting directly with virions or imitate the virus associated protein (VAP) to connect to the sustainable receptors (Shi et al., 2017). Furthermore, other algal compounds, including terpenes, reveal interesting antiviral activities. Citing, for example, (6R)-6-hydroxy dichotoma-3,14-diene-1,17-dial and (6R)-6-acetoxidichotoma-3,14-diene-1,17-dial, isolated from the brown alga *Dictyota menstrualis* and successfully tested against replication of the HIV-1 Virus (Pereira et al., 2004). The compound *N*-palmitoyl-2-amino 1,3,4,5-tetyrahydroxyoctadecane, isolated from *Ulva fasciata*, has demonstrated antiviral activity and *in vivo* protection against Semliki Forest virus (SFV) (Garg et al., 1992).

6.2.5.2 ANTIOXIDANT ACTIVITY

Antioxidants are attractive because of their ability to attenuate oxidative stress in the body and thus, disease risk reduction and health promotion (Shahidi et al., 2012). Currently, there has been a rising attention in natural antioxidants as alternate of synthetic compounds sense of increased safety concerns (Parapurath et al., 2011; Corsetto et al., 2020). Particularly, seaweeds have shown a great potential as a promising natural source of antioxidants (Cho et al., 2007; Ganesan et al., 2008; Marinho et al., 2019; Corsetto et al., 2020). Their anti-radical activity has been mainly linked to tocopherols (Miyashita and Takagi, 1987), carotenoids (Hosokawa et al., 2009), and polyphenols (Parys et al., 2007). For the latter group, brown algae were identified as the main source (Park et al., 2004; Heo et al., 2005; Siriwardhana et al., 2008) followed by red algae (Wang et al., 2010; Cian et al., 2012). Some PLS from algae have also been identified as antioxidant compounds (Rupérez et al., 2002; Zhao et al., 2008).

6.2.5.3 ANTI-INFLAMMATORY ACTIVITY

Inflammation is identified as a protective strategy that develops in high-level organisms in response to threats interfering with body homeostasis (Saraswati et al., 2019). However, extended inflammation could be potentially injurious to the body and might succeed to several complications like cardiovascular and pulmonary diseases, diabetes, arthritis, and cancer (Thomas et al., 2011). In this context, natural compounds with anti-inflammatory action could be interesting candidates to develop successful therapeutic approaches (Barbalace et al., 2019).

The PUFAs ω3 are known for their anti-inflammatory effects, which are linked to their competition as substrates of cyclooxygenase (COX) and lipoxygenase, causing a drop in the synthesis of inflammatory mediators, namely prostaglandins and leukotrienes (James et al., 2000). In fact, algae rich in PUFA ω3 could be a promising source of anti-inflammatory agents. Phlorotannins extracted from brown algae have also an important role in the mechanisms involved against inflammatory mediators (Nguyen et al., 2011); citing, for example, phlorofucofuroeckol A isolated from *Ecklonia tolonifera* which has shown a significant inhibition of the production of nitric oxide (NO) and prostaglandins (Nguyen et al., 2011). Likewise, phloroglucinol extracted from *Ecklonia cava*, reduces the levels of expression of interleukin-1β, interleukin 6 and prostaglandins E2 (Kim and Kim, 2010). PLS, alkaloids (Caulerpin), carotenoids (fucoxanthin and astaxanthin), phaeophytin A and Vidalols A and B are also confirmed as anti-inflammatory compounds, isolated from seaweeds (Jaswir and Monsur, 2011).

6.2.5.4 ANTITUMOR ACTIVITY

In traditional Chinese medicine, several extracts of brown algae have been used in the treatment of cancer (Yamamoto et al., 1981). Actually, several researches have also proven that substances isolated from algae have an anticancer power. For example, Kahalalide F, isolated from green algae *Bryopsis* sp., has been shown to have anticancer properties (Hamann et al., 1996). Their efficacy has been recorded in lung, colon, and prostate cancer (Horgen et al., 2000; Nuijen et al., 2000; Sparidams et al., 2001). Chondriamide A, a substance isolated from a red alga *Chondria atropurpurea* has been shown to be effective against human nasopharyngeal and colorectal cancer cells (Palermo et al., 1992). Likewise, phlorotannins derived from the brown alga *Laminaria japonica* have shown considerable antiproliferative activity on the human hepatocellular carcinoma cell line BEL-7402 and on the murine leukemic cell line P388 (Yang et al., 2010). Dioxinodehydroeckol (phlorotannins) isolated from the kelp *Ecklonia cava* exhibited a remarkable antiproliferative effect on MCF-7 breast cancer cells (Kong et al., 2009).

6.2.5.5 ANTIMICROBIAL ACTIVITY

Natural products derived from marine algae have widely proven to be a rich source of antimicrobial activities (Ioannou and Roussis, 2009; Osman et al., 2010; Rosaline et al., 2012). The antimicrobial compounds were recorded in green, brown, and as well as red seaweeds. Manivannan et al. (2011) screened the antibacterial effects of extracts from *Padina gymnospora*, *Sargassum tenerrimum* and *Turbinaria conoides* with promising results on a series of pathogenic bacteria (e.g., *Klebsiella*, *Escherichia coli*, *Enterococcus faecalis*, *Enterococcus* sp., *Pneumonia*, *Proteus* sp., *Staphylococcus aureus*, *Streptococcus* sp.). In another study, extracts of the red alga *Ceramium rubrum* and the brown algae *Padina pavonica*, *Sargassum vulgare* and *S. fusiforme* were highly active against 10 clinical isolates of Gram-positive and Gram-negative bacteria (El Shafay et al., 2016). Similarly, the methanolic extract of the brown alga *Dictyopteris polypodioides* has shown high activity against the pathogenic bacteria *Staphylococcus aureus*, *Bacillus cereus*, *Klebsiella pneumoniae*, *Pseudomonas aeruginosa*, *Escherichia coli*, *Enterococcus faecalis* and *Salmonella* sp. (Belattmania et al., 2016). Recently, the ethanolic extracts of the red seaweeds (*Gracilariopsis lemaneiformis* and *Gloiopeltis furcata*), the brown seaweed *Sargassum fusiforme* and the green seaweeds (*Ulva pertusa* and *U. prolifera*) inhibited *in vitro* growth of the gram-negative *Escherichia coli*, *Aeromonas hydrophila*, *Pseudomonas aeruginosa*, and *Vibrio alginolyticus*, and Gram-positive *Staphylococcus aureus* (Li et al., 2018).

6.2.6 MACROALGAE AS A SOURCE FOR COSMETIC COMPOUNDS

The integration of macroalgal bioactive molecules for cosmetics applications has been the topic of various studies (e.g., Mercurio et al., 2015; Dixit and Reddy, 2017; Pereira, 2018; Pimentel et al., 2018). The leading attractive groups of macroalgae bioactive compounds include those intended for anti-aging care (Bedoux et al., 2014; Wang et al., 2015; Ariede et al., 2017). Algal polysaccharides are widely used in cosmeceuticals as ingredients with moisturizing and antioxidant capacity and as gelling agents and stabilizers in emulsions (Pereira, 2018). Mycosporine-like amino acids (MAAs) are efficient photoprotector agents with interesting potential as natural compounds used for skin protection (Harnedy et al., 2011; Bedoux et al., 2014; Pandey et al., 2017). Some brown seaweeds derived phlorotannins have shown a significant potential as a bioactive ingredient to recover the extracellular tissue matrix (ETM) components-collagen functions (Shibata et al., 2002).

6.2.7 SEAWEEDS AS POTENTIAL SOURCE OF BIOENERGY

Over the past five decades, the world's population has almost doubled, which has led to a sharp increase in primary energy consumption, particularly the use of energy derived from fossil fuels (Jones and Mayfield, 2012). These increases will cause enormous pressure on the limited supply of fossil fuel-based energy, exacerbating global concerns about energy security, the environmental impacts of fossil fuels and the cost of energy (Dwivedi et al., 2009). The exploitation of biomass for energy production is readily and generally accessible in most countries due to its cheap upfront investment costs and straightforward technology (Marquez et al., 2015).

Several types of biomass including seaweeds are regarded as alternative resources for biofuel production. Seaweeds represent deposits with high potential, given their specific composition, their very high yield per hectare, least competition with food production (their cultivation does not require arable land and freshwater) (Marfaing et al., 2010; Marquez et al., 2015). The invention and filing of patents in this sector are exponential ranging from 5 patents (2003) to 120 patents in 2009 (Marfaing et al., 2010). Actually, there is a new wave of R&D projects on the large-scale production of macro-algae for transformation into biofuels using fermentation processes. Several potential tracks are possible, but some already seem more successful than others. Most macroalgae have a low lipid content, which excludes the production of biodiesel using diester-type processes. It is however possible to select pyrolysis process to transform the macroalgal biomass into bio-oil. In this case, species with high content of polyphenols (such as brown algae) are well suited. Furthermore, high levels of seaweed polysaccharides, making them suitable biomass for the production of bio-alcohols (ethanol, n-butanol, isobutanol) and bio-methane via fermentation process. Macroalgae contain both common polysaccharides, like glucans (starch, laminarane, cellulose) for which the processes of transformation into bioethanol are well known, and more original polysaccharides (alginates, carrageenans, ulvans, agar-agar) for which R&D works are still necessary before considering their transformation by fermentation into bio-alcohols (Adams et al., 2009; Person, 2010; Yung-Bum et al., 2010).

6.2.8 USE OF SEAWEED IN WATER DETOXIFICATION

The expansion of seaweeds is particularly important in coastal marine ecosystems subject to high nutrient loads. Previous works highlighted the ability of seaweeds to soak up nutrients and their usefulness for natural self-purification as well as for the treatment of wastewater (Goldman, 1974; Oswald, 1988). The benefit of using algae for wastewater treatment has been demonstrated since the late 1970s in mixtures of wastewater and seawater (Goldman, 1974; Prince, 1974; Guist and Humm, 1976). Also, some authors are particularly interested in the treatment of effluents from marine fish farms with macroalgae species such as *Gracilaria* and *Ulva* (Vandermeulen and Gordin, 1989; Cohen and Neori, 1991; Jiménez del Rio et al., 1994). The algal production resulting from these treatment systems provides great possibilities to exploit the produced biomass. Unlike microalgae, macroalgae allowing easier harvest suggesting good possibilities for the recovery of effluents from marine fish farming. Furthermore, seaweeds are particularly interesting for the biosorption of diverse pollutants including heavy metals. The later cause a serious ecological risk when they are released into the environment. Chemical precipitation, ion exchange, adsorption, flocculation, solvent extraction and membrane processes are the frequent techniques used to eliminate heavy metals (Cochrane et al., 2006; Satapathy et al., 2006; Llanos et al., 2010). These conventional methods generally have certain drawbacks, such as the generation of a high rate of toxic sludge or the high energy consumption (Gupta et al., 2000; Wang and Chan, 2009). In the last decades, biosorption has emerged as a potential alternative method to take out the pollutants from effluents. The biological materials are easier and less expensive to obtain, and their use as biosorbents has proven to be an economical and effective process. Among the various biosorbents reported in the literature, algal biomass is identified as a promising biosorbent, given their high absorption capacity (He and Chen, 2014). The biosorption of heavy metals using macroalgae is mainly attributed to their cell wall. Its constituents offer numerous functional groups (e.g., amine groups, hydroxyls, carboxyls, sulfates, etc.) which can be involved in the chemical bond with metal ions and could therefore play an important role in biosorption processes. Various works on biosorption have been focused on the removal of heavy metals by brown algae (Lodeiro et al., 2005; Luna et al., 2010; Kleinubing et al., 2011), green algae (Deng et al., 2009; Zakhama et al., 2011; Wan Maznah et al., 2012) and red algae (Ibrahim 2011). Brown algae, in particular, have shown promising adsorption capacities citing, for example, *Ecklonia maxima, Lessonia flavicans, Durvillaea potatorum* (Brouers, 2002), *Sargassum* spp. (e.g., Volesky et al., 2003; Cruz et al., 2004; El Atouani et al., 2016; Belattmania et al., 2017), *Halopteris scoparia* and *Dictyopteris polypodioides* (Belattmania et al., 2015a, b).

6.2.9 SEAWEEDS AND NANOTECHNOLOGY

The nanotechnology is one of the most attractive areas of research in materials science. Nanoparticles are of immense importance owing to their various applications. Actually, there is a growing requirement to develop sustainable, efficient, bio-originated, and eco-friendly methods for the synthesis of nanoparticles (NPs). The diversity of marine originated biomolecules, offer significant potential for green synthesis of NPs (Asmathunisha and Kathiresan, 2013; Ramkumar et al., 2016). Various researches highlighted the use of seaweed derivatives in the synthesis of gold (Vijayaraghavan et al., 2011; Rajathi et al., 2012), silver (Govindaraju et al., 2009; Mahdavi et al., 2013; Belattmania et al., 2018b) and iron oxide NPs (Yousefzadi et al., 2014). Noble metal NPs are exploited for many applications essentially in medical and pharmaceutical fields. Silver nanoparticles (AgNPs) are widely used as antimicrobial agents in medical products (Rai et al., 2009; Ravindra et al., 2010), such as dressings and bone cements (Chaloupka et al., 2010). It has been reported that gold NPs can be used as targeting agents for pancreatic cancer (Patra et al., 2010). Iron oxide NPs have been used to improve the diagnosis of cancer by magnetic resonance imaging (MRI) (Park et al., 2008). Besides of the medical field, silicon dioxide NPs have a potential application in the coating of lithium-ion anodes of batteries to increase battery power and reduce recharge time (Sasidharan et al., 2011). The zinc NPs are employed in food packaging as antimicrobial agents (Perez-Espitia et al., 2012) and in industrial coatings to protect certain materials (wood, plastic) from exposure to UV rays (Kathirvelu et al., 2009).

6.3 MARINE ALGAL RESOURCES AT MOROCCAN CONTEXT

The Moroccan Atlantic and Mediterranean coasts provide a significant taxonomic richness exceeding 600 species including around 160 taxa with potential economic value. However, the current Moroccan production of seaweeds, essentially intended for the phycocolloid industry. The full harvested biomass is derived from natural deposits. The latest annual export quota account mainly for the agarophyte *Gelidium* (1559 tons), the carrageenophytes *Gigartina/Chondracanthus* (300 tons) and the alginophytes kelps (50 tons). Furthermore, the authorized export quota of locally produced agar is 1247 tons annually (DGC, 2019). Morocco is the world's fifth-largest exporter of agar, following Chile, China, Indonesia, and Spain. Between 2012 and 2016, the annual export weight of agar-agar varied from 905 tons to 1066 tons (Figure 6.6), with the major markets in 2016 being Japan (240 tons) and the USA (184) tons (FAO, 2018).

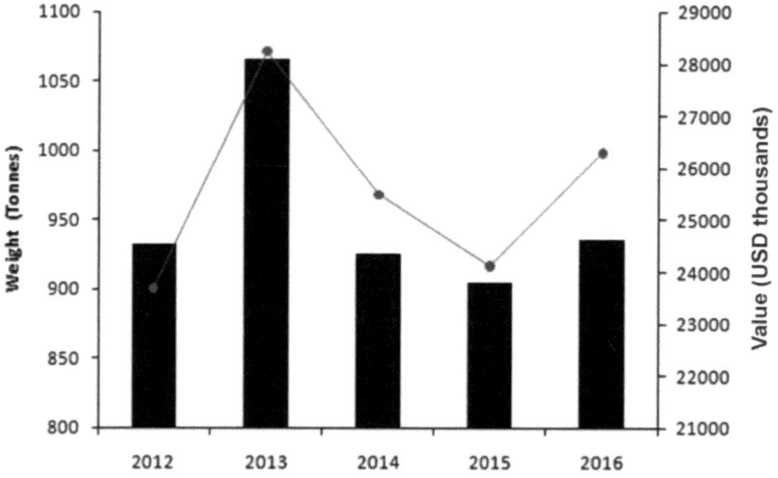

FIGURE 6.6 Annual evolution of weight (bars, primary axis) and value (line, secondary axis) of agar export in Morocco (FAO, 2018).

Given the constantly increasing world demand and the immediate government interest on aquaculture projects on a national scale, particularly in southern Morocco (Dakhla Bay – Gulf of Cintra: 878 production units, of which 106 concessions dedicated to seaweed farming), the seaweeds sector in Morocco could have potential regional development in the coming years (Sabour et al., 2016). Seaweed farming in Morocco is one of the 16 major projects of the HALIEUTIS strategy, developed in accordance with the strategic directions of His Majesty King Mohammed VI. It is therefore positioned at the level of the sustainability axis as a privileged sector and brought to constitute a lever of growth and job creation among the other fields of the fishery sector. In the context of the "Integrated Management of Coastal Zones of the Kingdom of Morocco (ICZM)," and within the framework of the "Sustainable Med" program, aquaculture activities are developed by the National Agency for the Development of Aquaculture (NADA), mainly for seaweed farming (e.g., the cultivation of *Gracilaria* sp.) in the Nador region (Marchika lagoon). Indeed, in addition to the definite economic interest, the development of farming seaweeds sector will undoubtedly contribute to reducing the pressure on natural algal resources and will ensure attractive incomes for the local populations. Notwithstanding the algal diversity and the presence of suitable sites (in terms of hydrology and climate conditions) for seaweed farming, it is crucial to point up the importance of a socio-economic advance, a reasoning approach of sustainable development as well an assessment of the expected environmental impacts of the farming systems to be used. Furthermore, the Moroccan government is collaborating in another initiative "Blue Belt" a project with several African nations working together with FAO and other partners. The Blue Belt initiative is a collaborative platform to act together and put into practice innovative solutions for adapting the fishing and aquaculture sector to climate change and for ocean and climate resilience. AT-SEA, a project developed within the framework of Blue Belt, consists in conducting a pilot project on the phyto-technical possibility of some macroalgae cultivation using new techniques.

6.4 CONCLUSION

Marine macroalgae, usually recognized as seaweeds, are considered a natural resource with a broad spectrum of applications in human life due to their interesting biochemical composition. In addition to their uses in human food as vegetables and in the food industry as texturizers, the seaweeds are getting a significant rising interest in agriculture (fertilizer and livestock feed) and pharmaceutical industry as well as in emerging fields such as cosmetology and bioenergy. The worldwide production of these marine macrophytes shows an unremitting increase exceeding currently 30 million tons, of which only 1.1 million tons derive from natural deposits and about 29 million tons from seaweeds farming. The seaweed industries constitute a diversified production in terms of species and compounds with versatile applications.

KEYWORDS

- **seaweed application**
- **phycocolloids**
- **edible seaweeds**
- **biostimulant**
- **seaweed bioactive compounds**
- **nanoparticles**
- **Moroccan algal resources**

REFERENCES

Adams, J. M., Gallagher, J. A., & Donnison, I. S., (2009). Fermentation study on *Saccharina latissima* for bioethanol production considering variable pre-treatments. *Journal of Applied Phycology, 21*(5), 569–574.

Ariede, M. B., Candido, T. M., Jacome, A. L. M., Velasco, M. V. R., De Carvalho, J. C. M., & Baby, A. R., (2017). Cosmetic attributes of algae: A review. *Algal Research, 25*, 483–487.

Asmathunisha, N., & Kathiresan, K., (2013). A review on biosynthesis of nanoparticles by marine organisms. *Colloids and Surfaces B: Biointerfaces, 103*, 283–287.

Barbalace, M. C., Malaguti, M., Giusti, L., Lucacchini, A., Silvana, H., & Angeloni, C., (2019). Anti-inflammatory activities of marine algae in neurodegenerative diseases. *International Journal of Molecular Sciences, 20*(12), 3061.

Bedoux, G., Hardouin, K., Burlot, A. S., & Bourgougnon, N., (2014). Bioactive components from seaweeds: Cosmetic applications and future development. *Advances in Botanical Research, 71*, 345–378.

Belattmania, Z., Bentiss, F., Jama, C., et al., (2018b). Biosynthesis and characterization of silver nanoparticles using sodium alginate from the invasive macroalga *Sargassum muticum*. *BioNanoScience, 8*(2), 617–623.

Belattmania, Z., Engelen, A. H., Pereira, H., et al., (2018a). Fatty acid composition and nutraceutical perspectives of brown seaweeds from the Atlantic coast of Morocco. *International Food Research Journal, 25*(4), 1520–1527.

Belattmania, Z., Reani, A., & Barakate, M., (2016). Potential antimicrobial, antioxidant and alginate properties of *Dictyopteris polypodioides* (Dictyotales, Phaeophyceae) collected from the Moroccan Atlantic coast. *Der Pharma Chemica, 8*(2), 216–226.

Belattmania, Z., Tahiri, S., Bentiss, F., et al., (2017). The residue of alginate extraction from *Sargassum muticum* (Brown seaweed) as low-cost adsorbent for hexavalent chromium removal in aqueous solutions. *Desalination and Water Treatment, 75*, 107–114.

Belattmania, Z., Tahiri, S., Zrid, R., et al., (2015b). Bioremoval of hexavalent chromium from aqueous solutions by the brown seaweed *Dictyopteris polypodioides*. *Research Journal of Environmental Toxicology, 9*(5), 218–230.

Belattmania, Z., Zrid, R., Reani, A., et al., (2015a). Phycoremediation of hexavalent chromium in simulated wastewater by *Halopteris scoparia* (Ochrophyta, Phaeophyceae). *Journal of Materials and Environmental Science, 6*(8), 2229–2238.

Bendary, M. M., Bassiouni, M. I., Ali, M. F., Gaafar, H. M., & Shamas, A. S., (2013). Effect of premix and seaweed additives on productive performance of lactating Friesian cows. *International Research Journal of Agricultural Science and Soil Science, 3*, 174–181.

Brouers, M., (2002). Les algues efficaces pour la biosorption des métaux. *Algorithme, 58*, 3–5.

Burtin, P., (2003). Nutritional value of seaweeds. *Electronic Journal of Environmental, Agricultural and Food Chemistry, 2*, 498–503.

Cardozo, K. H. M., Guaratini, T., Barros, M. P., et al., (2007). Metabolites from algae with economical impact. *Comparative Biochemistry and Physiology, Part C, 146*, 60–78.

Cerna, M., (2011). Seaweed proteins and amino acids as nutraceuticals. *Advances in Food and Nutrition Research, 64*, 297–312.

Chaloupka, K., Malam, Y., & Seifalian, A. M., (2010). Nanosilver as a new generation of nanoproduct in biomedical applications. *Trends in Biotechnology, 28*, 580–588.

Chèze-Lange, H., Beunard, D., Dhulster, P., et al., (2002). Production of microbial alginate in a membrane bioreactor. *Enzyme and Microbial Technology, 30*, 656–661.

Cho, S. H., Kang, S. E., Cho, J. Y., et al., (2007). The antioxidant properties of brown seaweed (*Sargassum siliquastrum*) extracts. *Journal of Medicinal Food, 10*, 479–485.

Chojnacka, K., (2011). Using the biomass of seaweeds in the production of components of feed and fertilizers. In: Kim, S. K., (ed.), *Handbook of Marine Macroalgae: Biotechnology and Applied Phycology* (pp. 478–490). United Kingdom: John Wiley and Sons.

Cian, R. E., Martinez-Augustin, O., & Drago, S. R., (2012). Bioactive properties of peptides obtains by enzymatic hydrolysis from protein byproducts of *Porphyra columbina*. *Food Research International, 49*, 364–372.

Cochrane, E. L., Lu, S., Gibb, S. W., & Villaescusa, I., (2006). A comparison of low-cost biosorbents and commercial sorbents for the removal of copper from aqueous media. *Journal of Hazardous Materials, 137*, 198–206.

Cohen, I., & Neori, A., (1991). *Ulva lactuca* biofilters for marine fishpond effluents. I. Ammonia uptake kinetics and nitrogen content. *Botanica Marina, 34*, 475–482.

Cooper, S. L., Sinclair, L. A., Wilkinson, R. G., Hallett, K. G., Enser, M., & Wood, J. D., (2004). Manipulation of the 3-polyunsaturated fatty acid content of muscle and adipose tissue in lambs. *Journal of Animal Science, 82*(5), 1461–1470.

Corsetto, P. A., Montorfano, G., & Zava, S., et al., (2020). Characterization of antioxidant potential of seaweed extracts for enrichment of convenience food. *Antioxidants, 9*(3), 249.

Craigie, J. S., (2011). Seaweed extract stimuli in plant science and agriculture. *Journal of Applied Phycology, 23*, 371–393.

Crouch, I. J., & Van, S. J., (1993). Evidence for the presence of plant growth regulators in commercial seaweed products. *Plant Growth Regulation, 13*, 21–29.

Cruz, C. C. V., Costa, A. C. A., Henriques, C. A., & Luna, A. S., (2004). Kinetic modeling and equilibrium studies during cadmium biosorption by dead *Sargassum* sp. biomass. *Bioresource Technology, 91*, 249–257.

Cruz-Suarez, L. E., Tapia-Salazar, M., Nieto-Lopez, M. G., Guajaro-Barbosa, C., & Ricque-Marie, D., (2009). Comparison of *Ulva clathrata* and the kelps *Macrocystis pyrifera* and *Ascophyllum nodosum* as ingredients in shrimp feeds. *Aquaculture Nutrition, 15*, 421–430.

Damonte, E., Neyts, J., Pujol, C. A., et al., (1994). Antiviral activity of a sulfated polysaccharide from the Red Seaweed *Nothogenia fastigiata. Biochemical Pharmacology, 47*, 2187–2192.

Dawczynski, C. H., Schubert, R., & Jahreis, G., (2007). Amino acids, fatty acids, and dietary fibre in edible seaweed products. *Food Chemistry, 103*, 891–899.

De Oliveira, M. N., Freitas, A. L. P., & Carvalho, A. F. U., (2009). Nutritive and non-nutritive attributes of washed-up seaweeds from the coast of Ceará Brazil. *Food Chemistry, 115*, 254–259.

Deng, L., Zhang, Y., Qin, J., Wang, X., & Zhu, X., (2009). Biosorption of Cr(VI) from aqueous solutions by nonliving green algae *Cladophora albida. Minerals Engineering, 22*(4), 372–377.

DGC, (2019). Direction générale du commerce (DGC)- Ministre de l'Industrie, du Commerce, de l'Investissement et de l'Economie Numérique. 2019. Avis no 16/19 aux exportateurs des algues brutes et d'agar agar. http://www.mcinet.gov.ma/ce/exportateurs/DocExport/avis_16-19.pdf (accessed on 13 October 2021).

Dixit, D., & Reddy, C. R. K., (2017). Non-targeted secondary metabolite profile study for deciphering the cosmeceutical potential of red marine macroalga *Jania rubens* – An LCMS-based approach. *Cosmetics, 4*, 45.

Dwivedi, P., Alavalapati, J. R. R., & Lal, P., (2009). Cellulosic ethanol production in the United States: Conversion technologies, current production status, economics, and emerging developments. *Energy for Sustainable Development, 13*, 174–182.

El Atouani, S., Tahiri, S., Reani, A., et al., (2016). Hexavalent chromium uptake from aqueous solutions using raw biomass of the invasive brown seaweed *Sargassum muticum* from the Moroccan shorelines: Kinetics and isotherms. *European Scientific Journal, 12*(30), 243–262.

El Shafay, S. M., Ali, S. S., & El-Sheekh, M. E., (2016). Antimicrobial activity of some seaweeds species from Red sea, against multidrug-resistant bacteria. *Egyptian Journal of Aquatic Research, 42*, 65–74.

Engelen, A., & Santos, R., (2009). Which demographic traits determine population growth in the invasive brown seaweed *Sargassum muticum*? *Journal of Ecology, 97*, 675–684.

FAO, (2014). *Compendium of Food Additive Specifications: Carrageenan*. Rome.

FAO, (2016). La situation mondiale des pêches et de l'aquaculture 2016. Contribuer à la sécurité alimentaire et à la nutrition de tous. Rome.

FAO, (2018). *The Global Status of Seaweed Production, Trade and Utilization Globefish Research Program* (Vol. 124). Rome.

Ganesan, P., Kumar, C. S., & Bhaskar, N., (2008). Antioxidant properties of methanol extract and its solvent fractions obtained from selected Indian Red Seaweeds. *Bioresource Technology, 99*, 2717–2723.

Garg, H. S., Sharma, M., Bhakuni, D. S., Pramanik, B. N., & Bose, A. K., (1992). An antiviral sphingosine derivative from the green alga *Ulva fasciata. Tetrahedron Letters, 33*, 1641–1644.

Goldman, J. C., Tenore, K. R., & Stanley, H. I., (1974). Inorganic nitrogen removal in a combined tertiary treatment-marine aquaculture system – II. Algal bioassays. *Water Research, 8*, 55–59.

Govindaraju, K., Kiruthiga, V., Ganesh, K. V., & Singaravelu, G., (2009). Extracellular synthesis of silver nanoparticles by a marine alga, *Sargassum wightii* Greville and their antibacterial effects. *Journal of Nanoscience and Nanotechnology, 9*, 5497–5501.

Guist, G. G., & Humm, H. J., (1976). Effect of sewage effluent on growth of *Ulva lactuca*. *Florida Science, 39*, 267–271.

Gupta, R., Ahuja, P., Khan, S., Saxena, R. K., & Mohapatra, H., (2000). Microbial biosorbents: Meeting challenges of heavy metal pollution in aqueous solutions. *Current Science, 78*, 967–973.

Hamann, M. T., Otto, C. S., Scheuer, P. J., & Dunbar, D. C., (1996). Kahalalides: Bioactive peptides from a marine mollusk *Elysia rufescens* and its algal diet *Bryopsis* sp. *The Journal of Organic Chemistry, 61*, 6594–6600.

Hansen, H. R., Hector, B. L., & Feldmann, J., (2003). A qualitative and quantitative evaluation of the seaweed diet of North Ronaldsay sheep. *Animal Feed Science and Technology, 105*, 21–28.

Hanzhi, L., Song, Q., & Peng, J., (2011). Biotechnology of seaweeds: Facing the coming decade. In: Kim, S. K., (ed.), *Handbook of Marine Macroalgae: Biotechnology and Applied Phycology* (pp. 424–430). United Kingdom: John Wiley and Sons.

Harnedy, P. A., & FitzGerald, R. J., (2011). Bioactive proteins, peptides, and amino acids from macroalgae. *Journal of Phycology, 47*, 218–232.

Hashimoto, W., Maruyama, Y., Itoh, T., Mikami, B., & Murata, K., (2009). Bacterial system for alginate uptake and degradation. In: Rehms, B. H. A., (ed.), *Alginates: Biology and Applications* (pp. 1–53). Berlin, Heidelberg: Springer-Verlag.

Haslam, S. F. I., & Hopkins, D. W., (1996). Physical and biological effects of kelp (seaweed) added to soil. *Applied Soil Ecology, 3*, 257–261.

He, J., & Chen, J. P., (2014). A comprehensive review on biosorption of heavy metals by algal biomass: Materials, performances, chemistry, and modeling simulation tools. *Bioresource Technology, 160*, 67–78.

Hegedüs, M., Pedersen, B., & Eggum, B. O., (1985). The influence of milling on the nutritive value of flour from cereal grains. vitamins and tryptophan. *Plant Foods for Human Nutrition, 35*, 175–180.

Helgerud, T., Gåserød, O., Fjæreide, T., Andersen, P. O., & Larsen, C. K., (2009). Alginates. In: Imeson, A., (ed.), *Food Stabilizers, Thickeners and Gelling Agents* (pp. 50–72). United Kingdom: Wiley-Blackwell.

Heo, S. J., Park, H. S., Lee, K. W., & Jeon, Y. J., (2005). Antioxidant activities of enzymatic extracts from brown seaweeds. *Bioresource Technology, 96*, 1613–1623.

Holdt, S. L., & Kraan, S., (2011). Bioactive compounds in seaweed; functional food applications and legislation. *Journal of Applied Phycology, 23*, 543–597.

Hong, D. D., Hien, H. M., & Son, P. N., (2007). Seaweeds from Vietnam used for functional food, medicine and biofertilizer. *Journal of Applied Phycology, 19*, 817–826.

Horgen, F. D., Delos, S. D. B., Goetz, G., Sakamoto, B., Kan, Y., Nagai, H., & Scheuer, P. J., (2000). A new depsipeptide from the sacoglossan mollusk *Elysia ornata* and the green alga *Bryopsis* species. *Journal of Natural Products, 63*, 152–154.

Hosokawa, M., Okada, T., Mikami, N., Konishi, I., & Miyashita, K., (2009). Bio-functions of marine carotenoids. *Food Science and Biotechnology, 18*(1), 1–11.

Hwang, J. A., Islam, M. M., Ahmed, S. T., et al., (2014). Seamustard (*Undaria pinnatifida*) improves growth, immunity, fatty acid profile and reduces cholesterol in Hanwoo steers. *Asian-Australasian Journal of Animal Sciences, 27*, 1114–1123.

Ibrahim, W. M., (2011). Biosorption of heavy metal ions from aqueous solution by red macroalgae. *Journal of Hazardous Materials, 192*(3), 1827–1835.

Imeson, A., (2009). Agar. In: Imeson, A., (ed.), *Food Stabilizers, Thickeners and Gelling Agents* (pp. 31–49). United Kingdom: Wiley-Blackwell.

Ioannou, E., & Roussis, V., (2009). Natural products from seaweeds. In: Osbourn, A. E., & Lanzotti, V., (eds.), *Plant Derived Natural Products* (pp. 51–81). Germany: Springer Science.

James, M. J., Gibson, R. A., & Cleland, L. G., (2000). Dietary polyunsaturated fatty acids and inflammatory mediator production. *The American Journal of Clinical Nutrition, 71*, 343–348.

Jaswir, I., & Monsur, H. A., (2011). Anti-inflammatory compounds of macroalgae origin: A review. *Journal of Medicinal Plants Research, 5*(33), 7146–7154.

Jimenez, D. R. M., Ramazanov, Z., & Garcia-Reina, G., (1994). Optimization of yield and biofiltering efficiencies of *Ulva rigida* C. Ag. cultivated with *Sparus aurata* L. wastewaters. *Scientia Marina, 58*, 329–335.

Jones, C. S., & Mayfield, S. P., (2012). Algae biofuels: Versatility for the future of bioenergy. *Current Opinion in Biotechnology, 23*, 346–351.

Kathirvelu, S., D'Souza, L., & Dhurai, B., (2009). UV protection finishing of textiles using ZnO nanoparticles. *Indian Journal of Fibre and Textile Research, 34*, 267–273.

Khan, W., Rayirath, U. P., Subramanian, S., et al., (2009). Seaweed extracts as biostimulants of plant growth and development. *Journal of Plant Growth Regulation, 28*, 386–399.

Kılınç, B., Cirik, S., Turan, G., Tekogul, H., & Koru, E., (2013). Seaweeds for food and industrial applications. In: Muzzalupo, A., (ed.), *Food Industry* (pp. 735–748). Croatia: InTech.

Kim, M. M., & Kim, S. K., (2010). Effect of phloroglucinol on oxidative stress and inflammation. *Food and Chemical Toxicology, 48*, 2925–2933.

Kleinubing, S. J., Da Silva, E. A., Da Silva, M. G. C., & Guibal, E., (2011). Equilibrium of Cu(II) and Ni(II) biosorption by marine alga *Sargassum filipendula* in a dynamic system: Competitiveness and selectivity. *Bioresource Technology, 102*(7), 4610–4617.

Kong, C. S., Kim, J. A., Yoon, N. Y., & Kim, S. K., (2009). Induction of apoptosis by phloroglucinol derivative from *Ecklonia cava* in MCF–7 human breast cancer cells. *Food and Chemical Toxicology, 47*, 1653–1658.

Lalic, L. M., & Berkovic, K., (2005). The influence of algae addition on physicochemical properties of cottage cheese. *Milchwissenschaft, 60*, 151–154.

Levesque, C. L., Moehn, S., Pencharz, P. B., & Ball, R. O., (2010). Review of advances in metabolic bioavailability of amino acids. *Livestock Science, 133*, 4–9.

Li, Y., Sun, S., Pu, X., et al., (2018). Evaluation of antimicrobial activities of seaweed resources from Zhejiang coast, China. *Sustainability, 10*, 2158.

Llanos, J., Williams, P. M., Cheng, S., et al., (2010). Characterization of a ceramic ultrafiltration membrane in different operational states after its use in a heavy-metal ion removal process. *Water Research, 44*, 3522–3530.

Lodeiro, P., Cordero, B., Barriada, J. L., Herrero, R., & Sastre De, V. M. E., (2005). Biosorption of cadmium by biomass of brown marine macroalgae. *Bioresource Technology, 96*(16), 1796–1803.

López-López, I., Bastida, S., Ruiz-Capillas, C., et al., (2009). Edible Japanese seaweed, wakame (*Undaria pinnatifida*) as an ingredient in pasta: Chemical, functional, and structural evaluation. *Food Chemistry, 115*, 501–508.

Loureiro, R., Gachon, C. M. M., & Rebours, C., (2015). Seaweed cultivation: Potential and challenges of crop domestication at an unprecedented pace. *New Phytologist, 206*(2), 489–492.

Luna, A. S., Costa, A. L. H., Da Costa, A. C. A., & Henriques, C. A., (2010). Competitive biosorption of cadmium(II) and zinc(II) ions from binary systems by *Sargassum filipendula. Bioresource Technology, 101*(14), 5104–5111.

Mabeau, S., & Fleurence, J., (1993). Seaweed in food products: Biochemical and nutritional aspects. *Trends in Food Science and Technology, 4*, 103–107.

MacArtain, P., Gill, C. I. R., Brooks, M., Campbell, R., & Rowland, I. R., (2007). Nutritional value of edible seaweed. *Nutrition Reviews, 65*, 535–543.

Maeda, H., Tsukui, T., Sashima, T., Hosokawa, M., & Miyashita, K., (2008). Seaweed carotenoid, fucoxanthin, as multi-functional nutrient. *Asia Pacific Journal of Clinical Nutrition, 17*, 196–199.

Mahdavi, M., Namvar, F., Bin, A. M., & Mohamad, R., (2013). Green biosynthesis and characterization of magnetic iron oxide (Fe_3O_4) nanoparticles using seaweed (*Sargassum muticum*) aqueous extract. *Molecules, 18*, 5954–5964.

Malhotra, R., Ward, M., Bright, H., et al., (2003). Isolation and characterization of potential respiratory syncytial virus receptor(s) on epithelial cells. *Microbes and Infection, 5*, 123–133.

Manivannan, K., Karthikai, D. G., Anantharaman, P., & Balasubramanian, T., (2011). Antimicrobial potential of selected brown seaweeds from Vedalai coastal waters, Gulf of Mannar. *Asian Pacific Journal of Tropical Biomedicine, 1*(2), 114–120.

Manivannan, K., Thirumaran, G., Devi, G. K., Hemalatha, A., & Anantharaman, P., (2008). Biochemical composition of seaweeds from Mandapam coastal regions along the southeast coast of India. *American-Eurasian Journal of Botany, 1*, 32–37.

Marfaing, H., Menguy, A., Sassi, J. F., & Lerat, Y., (2010). Les algues: Des applications traditionnelles aux marchés émergents. *Industries Alimentaires et Agricoles, 5, 6*, 29–30.

Marinho, G. S., Sørensen, A. M., Safafar, H., Pedersen, A. H., & Holdt, S. L., (2019). Antioxidant content and activity of the seaweed *Saccharina latissima*: A seasonal perspective. *Journal of Applied Phycology, 31*, 1343–1354.

Marinho-Soriano, E., Fonseca, P. C., Carneiro, M. A. A., & Moreira, W. S. C., (2006). Seasonal variation in the chemical composition of two tropical seaweeds. *Bioresource Technology, 97*, 2402–2406.

Marquez, G. P. B., Santiañez, W. J. E., Trono, J. C., De La Rama, S. R. B., Takeuchi, H., & Hasegawa, T., (2015). Seaweeds: A sustainable fuel source. In: Tiwari, B. K., & Troy, D. J., (eds.), *Seaweed Sustainability: Food and Non-Food Applications* (pp. 421–458). United Kingdom: Academic Press.

Marrion, O., Fleurence, J., Schwertz, A., et al., (2005). Evaluation of protein *in vitro* digestibility of *Palmaria palmata* and *Gracilaria verrucosa. Journal of Applied Phycology, 17*, 99–102.

Marsham, S., Scott, G. W., & Tobin, M. I., (2007). Comparison of nutritive chemistry of a range of temperate seaweeds. *Food Chemistry, 100*, 1331–1336.

Mattner, S. W., Wite, D., Riches, D. A., Porter, I. J., & Arioli, T., (2013). The effect of kelp extract on seedling establishment of broccoli on contrasting soil types in southern Victoria, Australia. *Biological Agriculture and Horticulture, 29*, 258–270.

Mayer, A. M. S., & Hamann, M. T., (2004). Marine pharmacology in 2000: Marine compounds with antibacterial, anticoagulant, antifungal, anti-inflammatory, antimalarial, antiplatelet, antituberculosis, and antiviral activities; affecting the cardiovascular, immune, and nervous system and other miscellaneous mechanisms of action. *Marine Biotechnology, 6*, 37–52.

McDermi, K. J., & Stuercke, B., (2003). Nutritional composition of edible Hawaiian seaweeds. *Journal of Applied Phycology, 15*, 513–524.

McHugh, D. J., (2003). *A Guide to the Seaweed Industry* (p.105). FAO Fisheries technical paper N° 441, Rome.

Mercurio, D. G., Wagemaker, T. A. L., Alves, V. M., et al., (2015). *In vivo* photoprotective effects of cosmetic formulations containing UV filters, vitamins, *Ginkgo biloba* and red algae extracts. *Journal of Photochemistry and Photobiology B: Biology, 153*, 121–126.

Milledge, J. J., Nielsen, B. V., & Bailey, D., (2016). High-value products from macroalgae: The potential uses of the invasive brown seaweed, *Sargassum muticum. Reviews in Environmental Science and Bio/Technology, 15*, 67–88.

Misurcova, L., (2011). Chemical composition of seaweeds. In: Kim, S. K., (ed.), *Handbook of Marine Macroalgae: Biotechnology and Applied Phycology* (pp. 173–192). United Kingdom: John Wiley and Sons.

Miyashita, K., & Takagi, T., (1987). Tocopherol content of Japanese algae and its seasonal variation. *Agricultural and Biological Chemistry, 51*, 3115–3118.

Mooney, P. A., & Van, S. J., (1986). Algae and cytokinins. *Journal of Plant Physiology, 123*, 1–21.

Moore, K. K., (2004). Using seaweed compost to grow bedding plants. *BioCycle, 45*, 43, 44.

Mouritsen, J. D., & Johansen, M., (2013). *Seaweeds: Edible, Available and Sustainable*. London: The University of Chicago Press, Ltd.

Murano, E., (1995). Chemical structure and quality of agars from *Gracilaria. Journal of Applied Phycology, 7*, 245–254.

Nguyen, P. H., Choi, W., Kim, S. K., & Jung, W. K., (2011). Immune regulatory effects of phlorotannins derived from marine brown algae (Phaeophyta). In: Kim, S. K., (ed.), *Handbook of Marine Macroalgae: Biotechnology and Applied Phycology* (pp. 340–347). United Kingdom: John Wiley and Sons.

Norziah, M. H., & Ching, C. H. Y., (2000). Nutritional composition of edible seaweed *Gracilaria changgi. Food Chemistry, 68*, 69–76.

Nuijen, B., Bouma, M., Talsma, H., et al., (2000). Development of a lyophilized parental pharmaceutical formulation of the investigational polypeptide marine anticancer agent kahalalide F. *Drug Development and Industrial Pharmacy, 27*, 767–780.

Ortiz, J., Romero, N., Robert, P., et al., (2006). Dietary fiber, amino acid, fatty acid and tocopherol contents of the edible seaweeds *Ulva lactuca* and *Durvillaea antarctica. Food Chemistry, 99*, 98–104.

Osman, M. E. H., Abushady, A. M., & Elshobary, M. E., (2010). *In vitro* screening of antimicrobial activity of extracts of some macroalgae collected from Abu-Qir bay Alexandria, Egypt. *African Journal of Biotechnology, 9*(12), 7203–7208.

Oswald, W. J., (1988). Micro-algae and waste-water treatment. In: Borowitzka, M. A., & Borowitzka, L. J., (eds.), *Micro-Algal Biotechnology* (pp. 305–328). Cambridge: Cambridge University Press.

Palermo, J. A., Flower, B. P., & Seldes, A. M., (1992). Chondriamides A and B, new indolic metabolites from the red alga *Chondria* sp. *Tetrahedron Letters, 33*, 3097–3100.

Pandey, A., Pandey, S., Pathak, J., et al., (2017). Mycosporine-like amino acids (MAAs) profile of two marine red macroalgae, *Gelidium* sp. and *Ceramium* sp. *International Journal of Applied Biotechnology and Biochemistry, 5*, 12–21.

Parapurath, S. N., Bernard, H. E., Subramaniam, D. M., & Ramamurthy, R., (2011). A dimensional investigation on seaweeds: Their biomedical and industrial applications. In: Kim, S. K., (ed.), *Handbook of Marine Macroalgae: Biotechnology and Applied Phycology* (pp. 532–540). United Kingdom: John Wiley and Sons.
Park, J. H., Maltzahn, G. V., Zhang, L., Schwartz, M. P., Ruoslahti, E., Bhatia, S. N., & Sailor, M. J., (2008). Magnetic iron oxide nanoworms for tumor targeting and imaging. *Advanced Materials, 20*, 1630–1635.
Park, P. J., Shahidi, F., & Jeon, Y. J., (2004). Antioxidant activities of enzymatic extracts from an edible seaweed *Sargassum horneri* using ESR spectrometry. *Journal of Food Lipids, 11*, 5–27.
Parys, S., Rosenbaum, A., Kehraus, S., Reher, G., Glombitza, K. W., & Konig, G. M., (2007). Evaluation of quantitative methods for the determination of polyphenols in algal extracts. *Journal of Natural Products, 70*, 1865–1870.
Patra, C. R., Bhattacharya, R., Mukhopadhyay, D., & Mukherjee, P., (2010). Fabrication of gold nanoparticles for targeted therapy in pancreatic cancer. *Advanced Drug Delivery Reviews, 62*, 346–361.
Pereira, H. S., Leao-Ferreira, L. R., Moussatché, N., et al., (2004). Antiviral activity of diterpenes isolated from the Brazilian marine alga *Dictyota menstrualis* against human immunodeficiency virus type 1 (HIV–1). *Antiviral Research, 64*, 69–76.
Pereira, H., Barreira, L., Figueiredo, F., et al., (2012). Polyunsaturated fatty acids of marine macroalgae: Potential for nutritional and pharmaceutical applications. *Marine Drugs, 10*, 1920–1935.
Pereira, L., (2016). *Edible Seaweeds of the World*. Boca Raton: CRC Press.
Pereira, L., (2018). Seaweeds as source of bioactive substances and skincare therapy – cosmeceuticals, algotherapy, and thalassotherapy. *Cosmetics, 5*(4), 68.
Pérez, R., (1997). *Ces Algues Qui Nous Entourent. Conception Actuelle, Rôle Dans la Biosphère, Utilisations, Culture*. France: Ifremer.
Perez-Espitia, P. J., Ferreira-Soares, N. F., Reis-Coimbra, J. S. D., De Andrade, N. J., Cruz, R. S., & Alves-Medeiros, E. A., (2012). Zinc oxide nanoparticles: Synthesis, antimicrobial activity and food packaging applications. *Food and Bioprocess Technology, 5*, 1447–1464.
Person, J., (2010). *Algues, Filières du Futur*. France: Adebiotech.
Pimentel, F. B., Alves, R. C., Rodrigues, F., & Oliveira, M. B. P. P., (2018). Macroalgae-derived ingredients for cosmetic industry – An update. *Cosmetics, 5*(1), 2.
Pompani, S. A., (2001). The oceans and human health: The discovery and development of marine-derived drugs. *Oceanography, 14*, 78–87.
Ponce, N. M. A., Pujol, C. A., Damonte, E. B., Flores, M. L., & Stortz, C. A., (2003). Fucoidans from the brown seaweed: *Adenocystis utricularis*: Extract methods, antiviral activity and structural studies. *Carbohydrate Research, 338*, 153–165.
Prince, J. S., (1974). Nutrient assimilation and growth of some seaweeds in mixtures of seawater and secondary sewage treatment effluents. *Aquaculture, 4*, 69–79.
Radulovich, R., Neori, A., Valderrama, D., Reddy, C. R. K., Cronin, H., & Forster, J., (2015). Farming of seaweeds. In: Tiwari, B. K., & Troy, D. J., (eds.), *Seaweed Sustainability: Food and Non-Food Applications* (pp. 27–59). United Kingdom: Academic Press, Elsevier.
Rai, M., Yadav, A., & Gade, A., (2009). Silver nanoparticles as a new generation of antimicrobials. *Biotechnology Advances, 27*, 76–83.
Rajathi, A. F. A., Parthiban, C., Ganesh-Kumar, V., & Anantharaman, P., (2012). Biosynthesis of antibacterial gold nanoparticles using brown alga, *Stoechospermum marginatum*

(Kützing). *Spectrochimica Acta Part A: Molecular and Biomolecular Spectroscopy, 99*, 66–173.

Rajauria, G., Cornish, L., Ometto, F., Msuya, F. E., & Villa, R., (2015). Identification and selection of algae for food, feed, and fuel applications. In: Tiwari, B. K., & Troy, D. J., (eds.), *Seaweed Sustainability: Food and Non-Food Applications* (pp. 315–344). United Kingdom: Academic Press.

Ramkumar, V. S., Prakash, S., Ramasubburayan, R., et al., (2016). Seaweeds: A resource for marine bionanotechnology. *Enzyme and Microbial Technology, 95*, 45–57.

Ramos, M. V., Monteiro, A. C. O., Moreira, R. A., & Carvalho, A. F. F. U., (2000). Amino acid composition of some Brazilian seaweed species. *Journal of Food Biochemistry, 24*, 33–39.

Ravindra, S., Mohan, Y. M., Reddy, N. N., & Raju, K. M., (2010). Fabrication of antibacterial cotton fibres loaded with silver nanoparticles via "green approach". *Colloids and Surfaces A: Physicochemical and Engineering Aspects, 367*, 31–40.

Rioux, L. E., & Turgeon, S. L., (2015). Seaweed carbohydrates. In: Tiwari, B. K., & Troy, D. J., (eds.), *Seaweed Sustainability: Food and Non-Food Applications* (pp. 141–192). United Kingdom: Academic Press.

Rosaline, X. D., Sakthivelkumar, S., Rajendran, K., & Janarthanan, S., (2012). Screening of selected marine algae from the coastal Tamil Nadu, South India for antibacterial activity. *Asian Pacific Journal of Tropical Biomedicine, 2*(1), 140–146.

Sabour, B., Reani, A., Belattmania, Z., & Loudiki, M., (2016). Seaweed and seaweed farming in Morocco: Potential and opportunities for development of the sector. *Conference, Presented at CIAlgues'2016 – 1st International Congress on Algae "Algae: An Opportunity for Sustainable Development"*. Cadi Ayyad University, Marrakech.

Samaraweera, A. M., Vidanarachchi, J. K., & Kurukulasuriya, M. S., (2011). Industrial applications of macroalgae. In: Kim, S. K., (ed.), *Handbook of Marine Macroalgae: Biotechnology and Applied Phycology* (pp. 500–521). United Kingdom: John Wiley and Sons.

Saraswati, Giriwono, P. E., Iskandriati, D., Tan, C. P., & Andarwulan, N., (2019). *Sargassum* seaweed as a source of anti-inflammatory substances and the potential insight of the tropical species: A review. *Marine Drugs, 17*(10), 590.

Sartal, C. G., Barciela, A. M. C., & Barrera, P. B., (2011). Application of seaweeds in the food industry. In: Kim, S. K., (ed.), *Handbook of Marine Macroalgae: Biotechnology and Applied Phycology* (pp. 522–531). United Kingdom: John Wiley and Sons.

Sasidharan, M., Liu, D., Gunawardhana, N., Yoshio, M., & Nakashima, K., (2011). Synthesis, characterization and application for lithium-ion rechargeable batteries of hollow silica nanospheres. *Journal of Materials Chemistry, 21*, 13881–13888.

Satapathy, D., & Natarajan, G. S., (2006). Potassium bromated modification of the granular activated carbon and its effect on nickel adsorption. *Adsorption, 12*, 147–154.

Shahidi, F., Zhong, Y., & Chandrasekara, A., (2012). Antioxidants and human health. In: Yu, L. L., Tsao, R., & Shahidi, F., (eds.), *Cereals and Pulses: Nutraceutical Properties and Health Benefits* (pp. 273–308). Hoboken: John Wiley & Sons, Inc.

Shi, Q., Wang, A., Lu, Z., Qin, C., Hu, J., & Yin, J., (2017). Overview on the antiviral activities and mechanisms of marine polysaccharides from seaweeds. *Carbohydrate Research, 453–454*, 1–9.

Shibata, T., Fujimoto, K., Nagayama, K., Yamaguchi, K., & Nakamura, T., (2002). Inhibitory activity of brown algal phlorotannins against hyaluronidase. *International Journal of Food Science and Technology, 37*, 703–709.

Shiraishi, K., (2005). Dietary intakes of eighteen elements and ^{40}K in eighteen food categories by Japanese subjects. *Journal of Radioanalytical and Nuclear Chemistry, 266*, 61–69.
Shiraishi, K., (2006). Dietary intakes of eighteen elements and ^{40}K in eighteen food categories by Japanese subjects. [Erratum to document cited in CA143:477119]. *Journal of Radioanalytical and Nuclear Chemistry, 267*, 261–264.
Simopoulos, A. P., (2002). The importance of the ratio of omega-6/omega-3 essential fatty acids. *Biomedicine and Pharmacotherapy, 56*(8), 365–379.
Simopoulos, A. P., (2008). The importance of the omega-6/omega-3 fatty acid ratio in cardiovascular disease and other chronic diseases. *Experimental Biology and Medicine, 233*(6), 674–688.
Singh, S., Kate, B. N., & Banerjee, U. C., (2005). Bioactive compounds from cyanobacteria and microalgae: An overview. *Critical Reviews in Biotechnology, 25*, 73–95.
Siriwardhana, N., Kim, K. N., Lee, K. W., et al., (2008). Optimization of hydrophilic antioxidant extraction from *Hizika fusiformis* by integrating treatments of enzymes, heat and pH control. *International Journal of Food Science and Technology, 43*, 587–596.
Sparidams, R. W., Stokvis, E., Jimeno, J. M., Lopez-Lazaro, L., Schellens, J. H., & Beijnen, J. H., (2001). Chemical and enzymatic stability of a cyclic depsipeptide, the novel, marine-derived, anticancer agent kahahalide F. *Anticancer Drugs, 12*, 575–587.
Spolaore, P., Joannis-Cassan, C., Duran, E., & Isambert, A., (2006). Commercial applications of microalgae. *Journal of Bioscience and Bioengineering, 101*, 87–96.
Stamey, J. A., Shepherd, D. M., De Veth, M. J., & Corl, B. A., (2012). Use of algae or algal oil rich in n-3 fatty acids as a feed supplement for dairy cattle. *Journal of Dairy Science, 95*, 5269–5275.
Stanley, N. F., (2006). Agars. In: Stephen, A. M., Phillips, G. O., & Williams, P. A., (eds.), *Food Polysaccharides and Their Applications* (pp. 217–238). Florida: CRC Press.
Stirk, W. A., Novak, M. S., & Van, S. J., (2003). Cytokinins in macroalgae. *Plant Growth Regulation, 41*, 13–24.
Thomas, N. V., & Kim, S. K., (2011). Health beneficial aspects of phloroglucinol derivatives from marine brown algae. In: Kim, S. K., (ed.), *Handbook of Marine Macroalgae: Biotechnology and Applied Phycology* (pp. 378–386). United Kingdom: John Wiley and Sons.
Tiwari, B. K., & Troy, D. J., (2015). Seaweed sustainability – food and nonfood applications. In: Tiwari, B. K., & Troy, D. J., (eds.), *Seaweed Sustainability Food and Non-Food Applications* (pp. 1–7). United Kingdom: Academic Press.
Vandermeulen, H., & Gordin, H., (1990). Ammonium uptake using *Ulva* (Chlorophyta) in intensive fishpond systems: Mass culture and treatment of effluent. *Journal of Applied Phycology, 2*, 363–374.
Vauchel, P., (2007). *Optimisation de Procédés innovants pour L'obtention de Phycocolloïdes*. PhD Thesis, University of Nantes, France.
Vijayaraghavana, K., Mahadevana, A., Sathishkumara, M., Pavagadhia, S., & Balasubramanian, R., (2011). Biosynthesis of Au(0) from Au(III) via biosorption and bioreduction using brown marine alga *Turbinaria conoides*. *Chemical Engineering Journal, 167*, 223–227.
Volesky, B., Weber, J., & Park, J. M., (2003). Continuous-flow metal biosorption in a regenerable *Sargassum* column. *Water Research, 37*, 297–306.
Wajahatullah, K., Rayirath, U. P., Subramanian, S., et al., (2009). Seaweed extracts as biostimulants of plant growth and development. *Journal of Plant Growth Regulation, 28*, 386–399.

Wan, M. W. O., Al-Fawwaz, A. T., & Surif, M., (2012). Biosorption of copper and zinc by immobilized and free algal biomass, and the effects of metal biosorption on the growth and cellular structure of *Chlorella* sp., and *Chlamydomonas* sp. isolated from rivers in Penang, Malaysia. *Journal of Environmental Sciences, 24*(8), 1386–1393.

Wang, H. M. D., Chen, C. C., Huynh, P., & Chang, J. S., (2015). Exploring the potential of using algae in cosmetics. *Bioresource Technology, 184*, 355–362.

Wang, J., & Chan, C., (2009). Biosorbent for heavy metals removal and their future. *Biotechnology Advances, 27*, 195–226.

Wang, T., Jonsdottir, R., Kristinsson, H. G., et al., (2010). Enzyme-enhanced extraction of antioxidant ingredients from red algae *Palmaria palmata. Food Science and Technology, 43*, 1387–1393.

White, W. L., & Wilson, P., (2015). World seaweed utilization. In: Tiwari, B. K., & Troy, D. J., (eds.), *Seaweed Sustainability: Food and Non-Food Applications* (pp. 7–25). United Kingdom: Academic Press.

Wong, K. H., & Cheung, P. C. K., (2000). Nutritional evaluation of some subtropical red and green seaweeds: Part I: Proximate composition, amino acid profiles and some physicochemical properties. *Food Chemistry, 71*(4), 475–482.

Yamamoto, I., Nagumo, T., Takahashi, M., Fujihara, M., Suzuki, Y., & Iizima, N., (1981). Antitumor effects of seaweeds III. Antitumor effect of an extract from *Sargassum kjellmanianum. The Japanese Journal of Experimental Medicine, 51*, 157–189.

Yang, H., Zeng, M., Dong, S., Liu, Z., & Li, R., (2010). Anti-proliferative activity of phlorotannin extracts from brown algae *Laminaria japonica* Aresch. *Chinese Journal of Oceanology and Limnology, 28*, 122–130.

Yousefzadi, M., Rahimi, Z., & Ghafori, V., (2014). The green synthesis, characterization and antimicrobial activities of silver nanoparticles synthesized from green alga *Enteromorpha flexuosa* (Wulfen) J. Agardh. *Materials Letters, 137*, 1–4.

Yung-Bum, S., Youn-Woo, L., Chun-Han, L., & Hack-Chul, Y., (2010). Red algae and their use in papermaking. *Bioresource Technology, 101*, 2549–2553.

Zakhama, S., Dhaouadi, H., & M'Henni, F., (2011). Nonlinear modelization of heavy metal removal from aqueous solution using *Ulva lactuca* algae. *Bioresource Technology, 102*(2), 786–796.

Zhao, X., Xue, C. H., & Li, B. F., (2008). Study of antioxidant activities of sulfated polysaccharides from *Laminaria japonica. Journal of Applied Phycology, 20*, 431–436.

Zhou, Y., Yang, H., Hu, H., et al., (2006). Bioremediation potential of the macroalga *Gracilaria lemaneiformis* (Rhodophyta) integrated into fed fish culture in coastal waters of north China. *Aquaculture, 252*, 264–276.

CHAPTER 7

SEAWEED AS PLANT BIOSTIMULANTS

TIAGO SOUSA,[1] JOÃO PEDRO NUNES,[2] JOSÉ LOPES,[2] JOÃO COTAS,[1] ANA MARTA MENDES GONÇALVES,[1,3] KIRIL BAHCEVANDZIEV,[2] and LEONEL PEREIRA[1]

[1]*University of Coimbra, MARE – Marine and Environmental Sciences Centre, Department of Life Sciences, Coimbra – 3001-456, Portugal*

[2]*Agricultural College of Coimbra (ESAC/IPC), Research Centre for Natural Resources Environment and Society (CERNAS), Institute of Applied Research (IIA), Coimbra – 3045-601, Portugal*

[3]*Department of Biology and CESAM, University of Aveiro, Aveiro – 3810-193, Portugal*

ABSTRACT

Seaweeds have been used since ancient times as soil fertilizer and to turn infertile soil into fertile soil. Seaweeds' research permitted in the last years to profile various species of seaweeds and their valuable compounds for various fields, such as human food, animal feed, biomedical, and pharmaceutical. In this case, with the pressure to turn agriculture more eco-friendly and lowering the usage of synthetic products, there was a gain in the research for seaweed compounds as plant biostimulants. However, in some countries of the North Atlantic Area, there was a tradition of collecting the seaweeds that came ashore. In Portugal, the most relevant one is named "sargaço." This chapter explores not only the seaweeds as plant biostimulants but also the data already published about "sargaço" species as plant biostimulants to understand the potential of this tradition.

7.1 INTRODUCTION

Seaweeds are multitrophic photosynthetic organisms and one of the main sources of organic matter in aquatic ecosystems, having an important role in the carbon cycle by fixing it during photosynthesis. Seaweeds usually inhabit rocky shores between shallow waters and 90 m of depth, and some macroalgae species inhabit rivers, lakes, and other freshwater sources (Pereira and Correia 2015; Silva et al., 2019). Their growth, reproduction, and survival depend on many factors like temperature, pH, salinity, wave undulation, carbon dioxide concentration, and upwelling nutrients (Harley et al., 2012). There are around 1,000 species of macroalgae on the planet, and they are classified due to their pigmentation in three phyla: Chlorophyta (green algae), Rhodophyta (red algae), and Ochrophyta (brown algae) (Sangha et al., 2014). Besides the pigmentation, they differ in other aspects like structure and chemical characteristics, chloroplast structure, photosynthetic pigments, storage compounds, cell wall composition, flagella (presence or absence), connections between adjacent cells, and ultrastructure of mitosis (Rindi et al., 2012).

The Chlorophyta phylum (green algae) belongs to Kingdom Plantae, having the chlorophyll *a* as the main pigment and in similarity with land plants, also have chloroplasts involved in a double membrane, stacked thylakoids and other accessory pigments as xanthophylls and carotenes. Another characteristic of these algae is the starch as reserve substance that is deposited into the plastids (Friedl, 1997). Besides being very abundant and diverse in aquatic ecosystems and in some terrestrial habitats, green algae have an important role in global ecosystems since hundreds of millions of years ago, when from an ancestor of these algae the evolution to terrestrial plants took place (Robic et al., 2008; Leliaert et al., 2012).

The Rhodophyta phylum (red algae) belongs to Kingdom Plantae too and is one of the most ancient eukaryotic organisms (with the remarkable 1,2 million years old *Bangiomorpha pubescens* fossil) (Saunders and Hommersand, 2004). This phylum has more than 6,135 species (mainly marine species), representing much of the biodiversity of seaweed forests. The Rhodophyta are divided into three subphyla (*Cyanidiophytina, Eurhodophytina*, and *Proteorhodophytina*) and seven classes (*Cyanidiophyceae, Bangiophyceae, Compsopogonophyceae, Florideophyceae, Porphyridiophyceae, Rhodellophyceae*, and *Stylonematophyceae*) (Rindi et al., 2012; Guiry and Guiry, 2020). This group is different from the other eukaryotic algae because the ones belonging to this group are the only ones without flagella and centrioles in their life cycle. The photosynthetic organelle (plastid) of red algae is bounded by two membranes and contains chlorophyll-*a*, phycocyanin, and phycoerythrin as photosynthetic pigments. These pigment complexes, organized in phycobilisomes, are located on the surface of unstacked thylakoid membranes to capture light energy. As a storage product, the red algae produce granulated floridean starch in the cytoplasm that is different from green algal starch (Yoon et al., 2010).

The brown seaweeds are a large group of multicellular algae of the class Phaeophyceae, phylum Ochrophyta (or Heterokontophyta), Kingdom Chromist. They usually live in cold and temperate oceans in the Northern Hemisphere from the North Pacific to the North Atlantic (Sun et al., 2014). The brown seaweeds go from filamentous seaweed (*Ectocarpus*) to more complex species that can grow longer than a blue whale like the giant kelps of the Californian Coast (*Macrocystis*) that can grow up to 60 m (Cavalier-Smith and Chao, 2006). Today, there are more than 1500 known species from brown seaweeds distributed for about 285 genera, of which six are from freshwater (Jia et al., 2014). Their pigments are the chlorophyll-*a* and chlorophyll-*c* (many types) and carotenes (mainly fucoxanthin), responsible for their coloration (Pereira, 2015). These seaweeds do not have chlorophyll-*b*, and their storage product is the laminarin. The cell wall, besides having an internal layer of cellulose, has phlorotannins and a second layer (external) with calcium alginate, that has a major importance due to its economic value in various industries (Maneveldt and Keats, 2003).

The seaweed from Ochrophyta phylum also have four membranes around the plastids (two from the endoplasmic reticulum), which is indicative of a secondary endosymbiosis event, in which green or red algae were assimilated by other eukaryotic hosts, and for this reason they are classified in different Kingdoms. The thylakoids are stacked in groups of three (referred to as lamellae) and these algae have in the sporous and the gamete stage two flagella inserted laterally, one directed forward and another backward (Maneveldt and Keats, 2003; Blackwell, 2009; Jia et al., 2014).

In oriental countries like China and Japan, the seaweeds are harvested to be used as an ingredient in a wide variety of dishes such as sushi, salads or soups. The most important seaweeds in the diet of these countries are: *Undaria* (*Undaria pinnatifida*; Wakame), *Porphyra/Pyropia* (*Pyropia yezoensis, Pyropia tenera, Porphyra umbilicalis* and *Porphyra* sp.; Nori) and *Saccharina* (*Saccharina japonica* (formerly known as *Laminaria japonica*); Kombu) (Gupta et al., 2010). In western countries like the United States, Ireland, or France, the consumption, production, and commercialization of seaweeds have been increasing, although the main use of seaweeds in these countries is as a source of phycocolloids (alginates, carrageenan, and agar) and thickening and gelling agents for application in the food industry and other industries. France was the first European country to regulate the use of seaweeds for human consumption as vegetables and condiments (Ruperez and Saura-Calixto, 2001).

7.2 VALUABLE SEAWEED COMPOUNDS

The chemical composition of seaweeds varies according to species, habitat, maturity, and environmental conditions (Wong and Cheung, 2000). Seaweeds are an excellent source of dietary fiber due to the high content of polysaccharides in the algal cell wall (33%-50% dry weight (DW) of seaweed) that are not digestible by humans (Ruperez and Saura-Calixto, 2001). Polysaccharides can be soluble fibers such as alginates, fucoidans, and carrageenans (Raposo et al., 2016). Seaweeds are also an excellent source of minerals, protein (found in high amounts in red seaweeds, mainly), vitamins, and lipids, thanks to their ability to absorb inorganic substances from aquatic media and store in their tissues (Ratana-arporn and Chirapart, 2006; Cofrades et al., 2010).

For this reason, seaweeds are considered a valuable source of bioactive compounds, as they are capable of producing a wide variety of secondary metabolites responsible for biological activities with potential industrial applications, such as cosmetics, food, and nutrition, where natural products have more market value (Gupta et al., 2010; Sanjeewa et al., 2016).

Bioactive compounds are secondary metabolites that have beneficial or harmful effects on organisms depending on their concentration. From approximately 1 million bioactive compounds produced by terrestrial and marine organisms, more than 20% showed antimicrobial, antifungal, anti-protozoal, anthelmintic, anticancer, antiviral or anti-inflammatory properties (Busetti et al., 2017).

Of all macroalgae, green seaweeds are those with less-known compounds. Only about 300 secondary metabolites are known and very few are discovered each year. Seaweeds are known to produce compounds like those found in red seaweeds, primarily functionalized diterpenoids and sesquiterpenoids, but lacking the extensive halogenation of the red algal compounds (Maschek and Baker, 2008). Red seaweeds are responsible for the production of more than 1,500 secondary metabolites, and mainly produce derivatives of isoprenoids and acetogenin, but also derivatives of amino acids, chiquimates, and nucleic acids. Brown seaweeds also produce several secondary metabolites, of which more than 1,140 are known, including diterpenes, chlorotannins (exclusive to brown seaweeds), and small acetogenins (Busetti et al., 2017).

Seaweeds can be eaten fresh, dried or as ingredients, such as alginates, agar, and carrageenans (Ruperez and Saura-Calixto, 2001). The use of these seaweeds in the diet brings numerous advantages to the humans, such as, growth, and protection of the intestinal flora and the antihypertensive and anti-hypocholesterolemic effects from seaweeds fibers, reducing the risk of atherosclerosis, coronary heart disease, inflammatory diseases and even behavioral disorders through the consumption of lipids from seaweeds as long-chain omega-3 polyunsaturated fatty acids (PUFA), the known anti-oxidant activity of carotenoids, polyphenols, and tocopherols (Ratana-arporn and Chirapart, 2006; Cofrades et al., 2010; Gupta et al., 2010).

Carrageenan is a gelling agent used as an emulsifier, binder, and product stabilizer in the food processing, pharmaceutical, and cosmetic industries (Valderrama et al., 2015). There are three types of carrageenan with great importance: kappa, iota, and lambda. Kappa is more rigid and used as an additive in food and lambda is more used in dairy products with a creamy texture (Ferdouse et al., 2018). Carrageenans are composed of alternative units of β-D-galactopyranose and 3,6-anhydro-galactose (3,6-AG) (sulfated or not) joined by α-1,3 and β-1,4-glycosidic connections. The type of carrageenan varies with the sulfate ester groups it contains and 3,6-AG. The more ester sulfate of carrageenan is present, the lower is the gelation strength and solubilization temperature. The kappa-carrageenan contains the lower concentration of sulfate ester, and the lambda-carrageenan have the higher concentration of sulfate esters (Rodríguez et al., 2009; Pereira, 2013). The additive number of carrageenan is E407 (Featherstone, 2015).

Agar is a biopolymer consisting of agarose and agaropectin. Agarose contains repeated units of 3-linked D-galactose and 4-linked 3,6-anhydro-L-galactose (3,6-AG). An agaropectin is composed of alternative units of D-galactose and L-galactose modified with sulfate, methoxy, and pyruvate groups in the polysaccharide chain. The type, degree, and pattern of substitution determine the gelling properties of the agar. The difference between carrageenans and agar is if 4-linked residue is in the D configuration (carrageenan) or in L configuration (agar). The difference between carrageenans and agar is how β-D-galactose units bind to the D (carrageenan) or L (agar) configurations (Montaño et al., 1999; Rodríguez et al., 2009). Agar gels have medical, pharmaceutical, industrial, and food applications (Villanueva et al., 2010). The agar additive number is E406 (Featherstone, 2015).

Alginates are linear polymers composed of β-D-mannuronic acid and α-L-glucuronate units joined by β-1,4 bonds (Huebbe et al., 2017). Alginates can be used in desserts, ice cream and beverage emulsions. The alginate food additive numbers are as follows: E400, alginic acid; E401, sodium alginate; E402, potassium alginate; E403, ammonium alginate; E404, calcium alginate; E405, propane-1,2-diol alginate (Featherstone, 2015).

Phlorotannins are formed by the polymerization of phloroglucinol monomers (1,3,5-trihydroxybenzene) and synthesized by the acetate-malonate pathway. They are hydrophilic compounds that can be classified into four subclasses according to the bonding methods: halogens and phlorethols (phlorotannins with an ether bond), fucols (a phenyl bond), fucophloroethols (an ether and phenyl bond) and eckols (with an ether bond) dibenzodioxin binding. They are compounds with different activities of pharmacological interest mainly due to their antimicrobial, antiviral, antidiabetic, antioxidant, antitumor, anti-allergic properties, among others (Li et al., 2011; Busetti et al., 2017).

Fucoidans are heteropolymers with sulfated fucose (Huebbe et al., 2017). These polysaccharides, in addition to being constituted by ester groups of L-fucose and sulfate, include other monosaccharides (mannose, galactose, glucose, xylose, etc.), proteins, and acetyl-equivalent groups in the same species. Fucoidans have several biological activities, mainly anticoagulant, antithrombotic, antiviral, antitumor, anti-inflammatory, blood lipid-lowering, antioxidant, and gastric protective effects (Li et al., 2008).

7.3 SEAWEED EXTRACTS AS PLANT BIOSTIMULANTS

The first documented use of seaweed as agricultural fertilizer in the Atlantic coast of the European continent is dated from the ancient Romans and ancient Celtic tribes. L.J.M. Collumella, the most notable Roman writer about agricultural practices, wrote that the roots should be wrapped with seaweeds to maintain the humidity and freshness of the seedlings (Battacharyya et al., 2015). In AD 79, Plínio observed the collection "margo" (considered maerl, a red alga native to the coast of Brittany) by "peoples of Great Britain and Gaul" to fertilize their soils. Seaweeds were used regularly by people from the ancient coastal areas along the Atlantic to fertilize the soil, but only the Romans left written records of this practice (Monagail et al., 2017). In the 14th century, in Portugal, farmers started to use seaweeds as a soil improver and vegetable compost (Pereira, 2019).

The seaweed extract exhibits various biostimulating activities in agricultural production when applied either in a soil or on leaves (Chatzissavvidis and Therios, 2014). Biostimulants are metabolic enhancers that improve plant productivity as a consequence of the novel or emergent properties of the complex of constituents, used in small quantities (Yakhin et al., 2017). Biostimulants have several benefits for plants, such as increased stress tolerance to nutrients, water, and salinity (Khan et al., 2009; Sangha et al., 2014). These beneficial effects on plants are germination and early seed establishment, better crop performance and yield, maximum resistance to abiotic and biotic stress and longer shelf life of the fruits after harvesting (Khan et al., 2009; Sousa et al., 2020). In fruits, vegetables, and flowers, extracts provide better vigor, greater absorption of nutrients from the soil, greater yield, greater resistance to pests and diseases, a longer post-harvest life of fruits and a better tolerance of abiotic stresses (Khan et al., 2009; Ramarajan et al., 2012; Chatzissavvidis and Therios, 2014). Extracts can also stimulate cell division and plant defense mechanisms by polysaccharides (laminarin) involved in the polyamine pollution that stimulates these activities (Chatzissavvidis and Therios, 2014). At the ground level, seaweed also improves the soil structure and its ability to retain moisture due to their carbohydrate content (Thirumaran et al., 2009a; López-Mosquera et al., 2011).

As seaweeds are an organic product, they generate great interest due to the restriction of the use of pesticides and chemical fertilizers in the national times of the various countries of the world (Verkleij, 1992). The use of chemical fertilizers in the European Union (EU) has become more regulated through changes in the availability of fertilizers on the market such as Regulation (EU) 2019/1009 of the European Parliament and the European Council from June 5, which establishes rules for fertilizers on the European market. This regulation aims to encourage the production of large-scale fertilizers from national organic or secondary raw materials, thus aligned with the circular economy model, transforming waste into nutrients for crops. On the other hand, the regulation also sets harmonized limits for several contaminants, such as cadmium, present in mineral fertilizers. In this sense, new rules are defined to stimulate the production and use of organic and phosphate fertilizers with low cadmium content and provide a greater variety of options for those oriented towards more sustainable agriculture (Regulation EU 2019/1009 of the European Parliament and of the Council, 2019).

The application of seaweed as a fertilizer in crops replacing chemical fertilizers and reducing the cost of crop production has been done (Silva et al., 2019; Thirumaran et al., 2009b). These biofertilizers are considered superior to synthetics because they contain high levels of organic matter, macro-, and microelements, vitamins, fatty acids, and growth regulators (Thirumaran et al., 2009b; Sathya et al., 2010).

7.4 SEAWEED EXTRACTS AS PLANTS FOLIAR BIOSTIMULANTS

Liquid seaweed extracts are applied in the form of leaf spray to crops in order to achieve greater productivity (Pise and Sabale, 2010; Ramarajan et al., 2012). These seaweed extracts contain several active organic substances that act in the regulation and growth of plants such as cytokinins, auxins, betaines, gibberellins, carbohydrates, vitamins, polysaccharides, alginates, amino acids, and trace elements (Fe, Cu, Zn, Co, Mo, Mn, and Ni) (Thirumaran et al., 2009a; Pise and Sabale, 2010; Chatzissavvidis and Therios, 2014; Silva et al., 2019).

The mechanisms of action of these seaweed substances are complex, and it is not yet known why they are effective, but it is possible that these substances together exhibit synergistic activity (Khan et al., 2009). These mechanisms are only now being studied through molecular biology, metabolomics, and genomics. As the genomes of several plants are being sequenced, then it will be possible to observe the effects of seaweed extracts on the plant genome to better understand the mechanisms of action of the extracts (Chatzissavvidis and Therios, 2014).

A practical example of the application of an aqueous extract and its benefits in plants is the use of *Sargassum wightii* (Phaeophyceae), which when applied as a foliar spray on *Ziziphus mauritiana* (Ber Tree or Jujube) increased the productivity and quality of its fruits (Rama Rao, 1991). In addition to this case, countless beneficial cases have been reported in which seaweed extracts have provided plants with foliar applications with increased production, such as a 25% increase in bean production, the early production of a variety of cucumbers in the greenhouse or an increase in total weight of greenhouse tomatoes by 17% (Csizinszky, 1984; Crouch and Van Staden, 1992).

Seaweed extracts have been marketed for several years as fertilizer additives. Some of these commercial products available on the market are Maxicrop, Algifert, Goemar, Kelpak, Seaspray, Seasol, SM3, Cytex, and Seacrop (Pise and Sabale, 2010). Most commercial extracts for application in agriculture are based on brown seaweeds such as *Ascophyllum nodosum*, *Fucus* spp., *Laminaria* spp., *Sargassum* spp., *Ecklonia maxima* and *Durvillaea* spp. (Craigie, 2011; Sangha et al., 2014). The commercialized extracts are liquids with different colors (from colorless to dark brown) derived from the high content of polyphenols or phlorotannins present in brown seaweeds in large quantities, but also present in red seaweeds

(Craigie, 2011). Although the methods of producing the extracts are rarely published, remaining as restricted information of the owners, the methods of manufacture are generally made with water, acids or alkalis, physical methods to break cell walls at low temperatures to obtain a suspension of fine greenish to greenish-brown particles that are slightly acidic (Craigie, 2011; Rodriguez-Jasso et al., 2011). Alternatively, the cell walls can be disrupted by high pressure devices, and the soluble cytosol components are recovered in the filtered liquid. Physical processes avoid the use of organic, acidic or alkaline solvents, and make them more ecological, economical, and compatible with food (Craigie, 2011; Godlewska et al., 2016; Fradinho et al., 2020).

Other alternative processes are fluid extraction with CO_2 as a solvent, extraction by Soxhlet (older), or microwave-assisted extraction. CO_2 extraction is used in the food and pharmaceutical industries as it is a non-toxic, non-flammable, inexpensive solvent, easily separable from extracts and allows the extraction of seaweed compounds without degradation because the process is not carried out at very high temperatures. The efficiency of extractions is reduced due to the presence of the cell wall, which can be affected by the solvent used, temperature, time, and pH. There are several solvents that can be used, such as acetone, ethanol, methanol, butanol, petroleum ether, and dichloromethane, but the use of these solvents makes the extraction process more expensive in addition to their toxicity (Cheung et al., 1998; Godlewska et al., 2016).

7.4.1 SCIENTIFIC ACTUALIZATION OF PORTUGUESE SEAWEED TRADITION

The collection or harvesting of the seaweeds from the sea or the sea throws to the beach was a tradition in central and northern Portugal, to fertilize the agricultural areas near the maritime shore, from the middle-ages until the mid-20th century (Figure 7.1a). This activity played a key role in the soil fertilization, and it was economically relevant. In Ria de Aveiro, harvest of seaweeds (mainly red and green seaweeds) with seagrasses in the estuarine zone is denominated "Moliço." In the North of Portugal, the mixture of seaweeds collected is "Sargaço." These seaweeds were first washed by the rain to remove the salts derived from the seawater and then they were used as soil fertilizer (Pereira and Cotas, 2019) (Figure 7.1b).

FIGURE 7.1 The various stage of the "sargaço" and targeted seaweeds for plant biostimulants assays; 7.1a – North of Portugal Beach (Praia near Viana do Castelo, Portugal) with some seaweeds washed-off in the tidal pools; 7.1b – Identification of the seaweeds washed-off in the laboratory; 7.1c – *Sargassum muticum* crude extract; 7.1d and 7.1e – lettuce assay with seaweed extracts and visual check for plant damages from abiotic and biotic factors; 7.1f – underexploited biomass, from extraction method, germination assay.

More recently, there have been studies to understand the scientific potential of these Portuguese traditions for plant foliar biostimulants applications (Queiroz, 2016; Lopes, 2018; Nunes, 2018; Silva, 2019), using the extracts of one seaweed species to understand the potential of the specific species and the optimization of the protocol (Figure 7.1c). Queiroz (2016) studied the properties of fresh *Saccorhiza polyschides* and *Fucus vesiculosus* (two seaweeds that are present nowadays in the "Sargaço" composition) liquid fertilizer in germination and growth of *Solanum lycopersicum* (Tomato plant). This work revealed that the seaweed extract has curve response where the best results achieved with a concentration between 10 and 20% (m:v), over that there is an inhibition effect that was prejudicial for the seed germination and plant growth parameters. In the germination assay, maximum seedling vigor index was obtained with 15% of seaweed extract concentration (1:1 v:v *S. polyschides* and *F. vesiculosus* extract), while 5 and 25% concentrations showed lower results, proving that the extract effect is not dose-dependent. In the tomato plants, with the same extract (1:1 v:v *S. polyschides* and *F. vesiculosus* extract), higher dosages were demonstrated to be more efficient. Thus, the best results of plant growth and fruit production were obtained with extract concentration between 20 and 25% with an accentuated decrease to the other concentrations assayed. The impact of the extract extended to the number of leaves and the plant length but not to flower number. This proves that the liquid extracts of these seaweed species have promising potential and can be tested for other agricultural crops.

Silva et al. (2019) studied the effects of *Ascophyllum nodosum* and *Sargassum muticum* extracts as biofertilizer in two commercial sub-variants of *Lactuca sativa* (Lettuce) (Figure 7.1d and 7.1e). The germination assay proved that a high seaweed concentration inhibits the seed germination. The concentration of 25% gave the best results. With lettuce plant, the Aqueous Extract Processed (AEP) of *S. muticum* (C=25%) demonstrated to be the most effective treatment in the overall assay when compared with the crude extract. The AEP of *S. muticum* treatment also demonstrated, after soil analysis, that the lettuce was able to assimilate more minerals (P, K, Ca, Mg) from the soil, when compared with the control or the other extracts. These results show that seaweed treatment can have a multirole effect in the plant. Silva et al. (2019) also tested the seaweed crude extracts with the hydroponically grown vegetable crops (fertirrigation solution) from two seaweed species, with some success. This experiment shows that seaweed extract can be also used in hydroponic systems.

More recent studies have been carried out in terms of the optimization of seaweed extraction methods, namely temperature, with differences when the extraction is carried out at 25 and 80°C. Extract of the seaweed *S. muticum* carried out at 25°C induced higher average production of total biomass and the aerial part of lettuce, significantly favoring the crop. In root biomass, the *A. nodosum* extract was the one who favored better root growth. *S. polyschides* (25 and 80°C) extracts also presented better results than the control. When chemically analyzed, *S. muticum* extract carried out at 25°C, showed significant differences in the plant nutrients such as calcium, potassium, and phosphorus (Lopes, 2018). In another study with the application of seaweed extracts in Peppers (*Capsicum annuum*), the use of *S. muticum* extract at 25°C presented significantly higher mean seed and fruit production (pepper weight) than the other treatments, with a positive influence on photosynthesis levels. However, when all other treatments (*S. muticum* at 80°C; *S. polyschides* at 25°C; *S. polyschides* at 80°C and *A. nodosum* at 25°C) were used, plant production was higher than in the control plants. When compared to the control, *S. muticum* extract induced higher carbohydrate and phosphorus amounts in pepper fruits (Nunes, 2018). There is also a need to study the discharges from the extraction process (Figure 7.1f) to understand the total value of the seaweeds for agricultural field.

7.5 A MARKET VIEW

The world production of seaweeds has started to grow over the years. In 2005, world production of seaweeds was 14.7 million tons. Ten years later, in 2015, production has doubled the value to 30.4 million tons (Ferdouse et al., 2018). Seaweed industry is valued at more than $7 million (USD). Food products for human consumption represent $6 million (USD), and the remaining value is associated with the production of phycocolloids, bioactive compounds, food for animal consumption, and fertilizers (Nayar and Bott, 2014; Ferdouse et al., 2019).

The largest global seaweed producers are Chile, China, and Norway for wild species. For cultured species, the main producers are China, Indonesia, the Republic of Korea, and the Philippines. The most important species cultivated are *Eucheuma* spp., *Saccharina japonica*, *Gracilaria* spp., *Undaria pinnatifida* and *Kappaphycus* spp. (Ferdouse et al., 2019). China, which is the largest consumer, importer, and producer of seaweeds, produces mainly seaweeds for food such as kelps (*Saccharina japonica*, *Undaria pinnatifida*) and red seaweeds (of the genus *Gracilaria* and *Pyropia*). Indonesia produces more carrageenophytes (of the genus *Kappaphycus* and *Eucheuma*) (Titlyanov and Titlyanova, 2010; Nayar and Bott, 2014; Buschmann et al., 2017). In addition to production, there is a seaweed harvest from the seacoast, where Chile leads with 42% of this harvest followed by China with 28%. The most harvested wild species are Chileans (*Lessonia nigrescens* and *Lessonia trabeculata*). Out of the Asian continent, South Africa and Madagascar also produce seaweeds. There are 221 seaweed species exploited globally (32 green seaweeds, 125 red seaweeds and 64 brown seaweeds), 145 species for food and 101 species for phycocolloid production (Nayar and Bott, 2014). The most cultivated species globally are *Saccharina japonica* and *Undaria pinnatifida* (brown seaweeds); and *Porphyra/Pyropia* spp., *Eucheuma* sp., *Kappaphycus alvarezii* and *Gracilaria* spp. (red seaweeds) (Xiu-geng et al., 1999; Hafting et al., 2012; Buschmann et al., 2017).

Current agriculture needs more sustainable nutrients to increase the crop production for human food needs. Developed countries use growth hormones, regulators, and promoters in crops to enhance yield attributes (Thirumaran et al., 2009a). In agriculture and horticulture, seaweed has found application as animal food, as soil conditioner and manure, and, in the form of liquid extracts, as growth promoter and crop protectant against pests and diseases (Verkleij, 1992). Biostimulants are metabolic enhancers, which can stimulate plant growth and development when used in small quantities (Khan et al., 2009; Sangha et al., 2014).

7.6 CONCLUSION

Seaweeds provide an important ecosystem service due to their photosynthetic capacity they are natural habitat for fishes and marine wildlife and can express bioaccumulation properties for metal ions. However, human beings must change their natural resources conception because the increasing pressure on the environment such as water and air pollution and the depletion of natural resources are threatening the life support systems on Earth.

Seaweeds also represent an organic, healthy, and nutritious source. They demonstrate benefits for world agriculture. Seaweeds compounds as agar, carrageenan, fucoidan, and alginates already play an important role in the industry of cosmetics, food, and nutrition. Seaweed extracts combine all these bioactive compounds and minerals in a liquid biostimulant solution that improves the plants, so it is very important to study this natural resource in order to take advantage of it and to make our agricultural crops more efficient, sustainable, cheaper, and healthier. The use of the macroalgae also favors the environment by reducing the use of synthetic fertilizers, betting more on organic farming.

Seaweeds have been used as plant biostimulants since ancient times, and, for instance, the work of evaluating the Portuguese historical use of seaweeds applying scientific and modern approaches clarifies that these traditions have the potential to be useful in the future in the plant biostimulants area. However, more research needs to be done to really understand this seaweed extract mechanism of action in the plants and how this historical use as plant biostimulants can be commercially explored.

KEYWORDS

- **aqueous extract processed**
- **Chlorophyta**
- **European Union**
- **macroalgae**
- **Ochrophyta**
- **Rhodophyta**
- **xanthophylls**

ACKNOWLEDGMENTS

This work was financed through Foundation for Science and Technology (FCT), within the scope of the projects UIDB/04292/2020 – Marine and Environmental Sciences Centre (MARE) and UIDP/50017/2020+UIDB/50017/2020 (by FCT/MCTES) granted to Centre for Environmental and Marine Studies (CESAM). João Cotas shows his gratitude to the European Regional Development Fund through the Interreg Atlantic Area Program, under the project NASPA (EAPA_451/2016) and Ana M. M. Gonçalves acknowledges University of Coimbra for the contract IT057–18–7253.

REFERENCES

Battacharyya, D., Babgohari, M. Z., Rathor, P., & Prithiviraj, B., (2015). Seaweed extracts as biostimulants in horticulture. *Scientia Horticulturae 196*, 39–48. https://doi.org/10.1016/J.SCIENTA.2015.09.012 format like this everywhere.

Blackwell, W., (2009). Chromista revisited: A dilemma of overlapping putative kingdoms, and the attempted application of the botanical code of nomenclature. *Phytologia, 91*(2), 191–225.

Buschmann, A. H., Camus, C., Infante, J., Neori, A., Israel, Á., Hernández-González, M. C., Pereda, S. V., et al., (2017). Seaweed production: Overview of the global state of exploitation, farming and emerging research activity. *European Journal of Phycology, 52*(4), 391–406.

Busetti, A., Maggs, C. A., & Gilmore, B. F., (2017). Marine macroalgae and their associated microbiomes as a source of antimicrobial chemical diversity. *European Journal of Phycology, 52*(4), 452–465.

Cavalier-Smith, T., & Chao, E. E., (2006). Phylogeny and mega systematics of phagotrophic heterokonts (kingdom Chromista). *Journal of Molecular Evolution, 62*(4), 388–420.

Chatzissavvidis, C., & Therios, I., (2014). Role of algae in agriculture. In: Pomin, V. H., (ed.), *Seaweeds* (pp. 1–37). Nova Science Publishers, Inc.

Cheung, P. C., Leung, A. Y., & Ang, P. O., (1998). Comparison of supercritical carbon dioxide and Soxhlet extraction of lipids from a brown seaweed, *Sargassum hemiphyllum* (Turn.) C. Ag. *Journal of Agricultural and Food Chemistry, 46*(10), 4228–4232.

Cofrades, S., López-Lopez, I., Bravo, L., Ruiz-Capillas, C., Bastida, S., Larrea, M. T., & Jiménez-Colmenero, F., (2010). Nutritional and antioxidant properties of different brown and red Spanish edible seaweeds. *Food Science and Technology International, 16*(5), 361–370.

Craigie, J. S., (2011). Seaweed extract stimuli in plant science and agriculture. *Journal of Applied Phycology, 23*(3), 371–393.

Crouch, I. J., & Van, S. J., (1992). Effect of seaweed concentrate on the establishment and yield of greenhouse tomato plants. *Journal of Applied Phycology, 4*(4), 291–296.

Csizinszky, A. A., (1984). Response of tomatoes to seaweed based nutrient sprays. *Proc. Fla. State Hort. Soc., 97*,151–157.
Featherstone, S., (2015). Ingredients used in the preparation of canned foods. In: *A Complete Course in Canning and Related Processes. Microbiology, Packaging, HACCP and Ingredients* (pp. 147–211).
Ferdouse, F., Holdt, S. L., Smith, R., Murua, P., & Yang, Z., (2018). *The Global Status of Seaweed Production, Trade and Utilization.* Food and Agriculture Organization of the United Nations.
Fradinho, P., Flórez-Fernández, N., Sousa, I., Raymundo, A., Domínguez, H., & Torres, M. D., (2020). Environmentally friendly processing of *Laminaria ochroleuca* for soft food applications with bioactive properties. *Journal of Applied Phycology,* 1–11.
Friedl, T., (1997). The evolution of the green algae. In: *Origins of Algae and Their Plastids* (pp. 87–101). Springer, Vienna.
Godlewska, K., Michalak, I., Tuhy, Ł., & Chojnacka, K., (2016). Plant growth biostimulants based on different methods of seaweed extraction with water. *BioMed Research International, 2016,* 5973760. https://doi.org/10.1155/2016/5973760.
Guiry, M. D., & Guiry, G. M., (2020). *AlgaeBase*. World-wide electronic publication, National University of Ireland, Galway. Available on: https://www.algaebase.org/pub_taxonomy/?id=97240 (accessed on 13 October 2021).
Gupta, S., Rajauria, G., & Abu-Ghannam, N., (2010). Study of the microbial diversity and antimicrobial properties of Irish edible brown seaweeds. *International Journal of Food Science and Technology, 45*(3), 482–489.
Hafting, J. T., Critchley, A. T., Cornish, M. L., Hubley, S. A., & Archibald, A. F., (2012). On-land cultivation of functional seaweed products for human usage. *Journal of Applied Phycology, 24*(3), 385–392.
Harley, C. D., Anderson, K. M., Demes, K. W., Jorve, J. P., Kordas, R. L., Coyle, T. A., & Graham, M. H., (2012). Effects of climate change on global seaweed communities. *Journal of Phycology, 48*(5), 1064–1078.
Huebbe, P., Nikolai, S., Schloesser, A., Herebian, D., Campbell, G., Glüer, C. C., et al., (2017). An extract from the Atlantic brown algae *Saccorhiza polyschides* counteracts diet-induced obesity in mice via a gutrelated multi-factorial mechanisms. *Oncotarget, 8*(43), 73501.
Jia, S., Wang, X., Li, T., Qian, H., Sun, J., Wang, L., et al., (2014). Phylogenomic analysis of transcriptomic sequences of mitochondria and chloroplasts of essential brown algae (Phaeophyceae) in China. *Acta Oceanologica Sinica, 33*(2), 94–101.
Khan, W., Rayirath, U. P., Subramanian, S., Jithesh, M. N., Rayorath, P., Hodges, D. M., et al., (2009). Seaweed extracts as biostimulants of plant growth and development. *Journal of Plant Growth Regulation, 28*(4), 386–399.
Leliaert, F., Smith, D. R., Moreau, H., Herron, M. D., Verbruggen, H., Delwiche, C. F., & De Clerck, O., (2012). Phylogeny and molecular evolution of the green algae. *Critical Reviews in Plant Sciences, 31*(1), 1–46.
Li, B., Lu, F., Wei, X., & Zhao, R., (2008). Fucoidan: Structure and bioactivity. *Molecules, 13*(8), 1671–1695.
Li, Y. X., Wijesekara, I., Li, Y., & Kim, S. K., (2011). Phlorotannins as bioactive agents from brown algae. *Process Biochemistry, 46*(12), 2219–2224.
Lopes, J., (2018). Estudo da influência do uso de extratos de macroalgas na produção de alface (*Lactuca sativa*). Relatório de estágio profissionalizante em Licenciatura em Engenharia Agro-Pecuária. Polytechnic Institute of Coimbra, College of Agriculture, pp. 41–53.

López-Mosquera, M. E., Fernández-Lema, E., Villares, R., Corral, R., Alonso, B., & Blanco, C., (2011). Composting fish waste and seaweed to produce a fertilizer for use in organic agriculture. *Procedia Environmental Sciences, 9*, 113–117.

MacMonagail, M., Cornish, L., Morrison, L., Araújo, R., & Critchley, A. T., (2017). Sustainable harvesting of wild seaweed resources. *European Journal of Phycology, 52*(4), 371–390.

Maneveldt, G. W., & Keats, D. W., (2003). Chromista. In: *ELS*. John Wiley & Sons, Ltd.

Maschek, J. A., & Baker, B. J., (2008). The chemistry of algal secondary metabolism. In: *Algal Chemical Ecology* (pp. 1–24). Springer, Berlin, Heidelberg.

Montaño, N. E., Villanueva, R. D., & Romero, J. B., (1999). Chemical characteristics and gelling properties of agar from two Philippine *Gracilaria* spp. (Gracilariales, Rhodophyta). *Journal of Applied Phycology, 11*(1), 27.

Nayar, S., & Bott, K., (2014). Current status of global cultivated seaweed production and markets. *World Aquaculture, 45*(2), 32–37.

Nunes, J., (2018). Influência do uso de extratos de macroalgas na produção de pimento (*Capsicum annuum*). Relatório de estágio profissionalizante em Licenciatura em Biotecnologia. Polytechnic Institute of Coimbra, College of Agriculture, pp. 23–32.

Pereira, L., & Correia, F., (2015). *Algas Marinhas da Costa Portuguesa – Ecologia, Biodiversidade e Utilizações* (p. 340). Nota de Rodapé Edições, Paris.

Pereira, L., & Cotas, J., (2019). Historical use of seaweed as an agricultural fertilizer in the European Atlantic area. In: Pereira, L., Bahcevandziev, K., & Joshi, N. H., (eds.), *Seaweeds as Plant Fertilizer, Agricultural Biostimulants and Animal Fodder* (pp. 1–22). Boca Raton, FL: CRC Press. https://doi.org/10.1201/9780429487156.

Pereira, L., (2013). Population studies and carrageenan properties in eight gigartinales (Rhodophyta) from the western coast of Portugal. *The Scientific World Journal, 2013*.

Pereira, L., (2015). Seaweed flora of the European North Atlantic and Mediterranean. In: Kim, S. K., (ed.), *Springer Handbook of Marine Biotechnology*. Springer, Berlin, Heidelberg.

Pereira, L., (2019). Traditional use of seaweed in agriculture in Mainland Portugal. *The Silk Road Universities Network Web Magazine 1*, 49–54.

Pise, N. M., & Sabale, A. B., (2010). Effect of seaweed concentrates on the growth and biochemical constituents of *Trigonella foenum-graecum* L. *Journal of Phytology, 2*(4), 50–56.

Queiroz, P., (2016). *Biotechnological Evaluation of Seaweeds as Bio-fertilizer*. MSc Thesis, University of Coimbra, Coimbra, Portugal.

Rama, R. K., (1991). Effect of seaweed extract on *Zyziphus mauratiana* Lamk. *Journal of Indian Botanical Society, 71*, 19–21.

Ramarajan, S., Joseph, L. H., & Ganthi, A. S., (2012). Effect of seaweed liquid fertilizer on the germination and pigment concentration of soybean. *Journal of Crop Science and Technology, 1*(2), 1–5.

Raposo, J. M., De Morais, A., & De Morais, R., (2016). Emergent sources of prebiotics: Seaweeds and microalgae. *Marine Drugs, 14*(2), 27.

Ratana-arporn, P., & Chirapart, A., (2006). Nutritional evaluation of tropical green seaweeds *Caulerpa lentillifera* and *Ulva reticulata*. *Kasetsart J. (Nat. Sci.), 40*(Suppl.), 75–83.

Regulation (EU) 2019/1009 of the European Parliament and of the Council of 5 June 2019 laying down rules on the making available on the market of EU fertilizing products and amending Regulations (EC) No 1069/2009 and (EC) No 1107/2009 and repealing Regulation (EC) No 2003/2003 (Text with EEA relevance). PE/76/2018/REV/1. OJ L 170,

25.6.2019, pp. 1–114. http://data.europa.eu/eli/reg/2019/1009/oj (accessed on 13 October 2021).

Rindi, F., Soler-Vila, A., & Guiry, M. D., (2012). Taxonomy of marine macroalgae used as sources of bioactive compounds. In: *Marine Bioactive Compounds* (pp. 1–53). Springer, Boston, MA.

Robic, A., Sassi, J. F., & Lahaye, M., (2008). Impact of stabilization treatments of the green seaweed *Ulva rotundata* (Chlorophyta) on the extraction yield, the physicochemical and rheological properties of ulvan. *Carbohydrate Polymers, 74*(3), 344–352.

Rodríguez, M. C., Matulewicz, M. C., Noseda, M. D., Ducatti, D. R. B., & Leonardi, P. I., (2009). Agar from *Gracilaria gracilis* (Gracilariales, Rhodophyta) of the patagonic coast of Argentina – content, structure and physical properties. *Bioresource Technology, 100*(3), 1435–1441.

Rodriguez-Jasso, R. M., Mussatto, S. I., Pastrana, L., Aguilar, C. N., & Teixeira, J. A., (2011). Microwave-assisted extraction of sulfated polysaccharides (fucoidan) from brown seaweed. *Carbohydrate Polymers, 86*(3), 1137–1144.

Ruperez, P., & Saura-Calixto, F., (2001). Dietary fibre and physicochemical properties of edible Spanish seaweeds. *European Food Research and Technology, 212*(3), 349–354.

Sangha, J. S., Kelloway, S., Critchley, A. T., & Prithiviraj, B., (2014). Seaweeds (macroalgae) and their extracts as contributors of plant productivity and quality: The current status of our understanding. In: *Advances in Botanical Research* (Vol. 71, pp. 189–219). Academic Press.

Sanjeewa, K. K. A., Kim, E. A., Son, K. T., & Jeon, Y. J., (2016). Bioactive properties and potentials cosmeceutical applications of phlorotannins isolated from brown seaweeds: A review. *Journal of Photochemistry and Photobiology B: Biology, 162*, 100–105.

Sathya, B., Indu, H., Seenivasan, R., & Geetha, S., (2010). Influence of seaweed liquid fertilizer on the growth and biochemical composition of legume crop, *Cajanus cajan* (L.) Mill sp. *Journal of Phytology, 2*, 50–63.

Saunders, G. W., & Hommersand, M. H., (2004). Assessing red algal supraordinal diversity and taxonomy in the context of contemporary systematic data. *American Journal of Botany, 91*(10), 1494–1507.

Silva, L. D., Bahcevandziev, K., & Pereira, L., (2019). Production of bio-fertilizer from *Ascophyllum nodosum* and *Sargassum muticum* (Phaeophyceae). *Journal of Oceanology and Limnology, 37*(3), 918–927.

Sousa, T., Cotas, J., Bahcevandziev, K., & Pereira, L., (2020). Atualização científica da tradição: Revisitação científica aos métodos tradicionais de fertilização com macroalgas em Portugal. *Associação Portuguesa de Horticultura, 136*, 38–41.

Sun, J., Wang, L., Wu, S., Wang, X., Xiao, J., Chi, S., et al., (2014). Transcriptome-wide evolutionary analysis on essential brown algae (Phaeophyceae) in China. *Acta Oceanologica Sinica, 33*(2), 13–19.

Thirumaran, G., Arumugam, M., Arumugam, R., & Anantharaman, P., (2009). Effect of seaweed liquid fertilizer on growth and pigment concentration of *Abelmoschus esculentus* (L) medikus. *American-Eurasian Journal of Agronomy, 2*(2), 57–66.

Thirumaran, G., Arumugam, M., Arumugam, R., & Anantharaman, P., (2009). Effect of seaweed liquid fertilizer on growth and pigment concentration of *Cyamopsis tetrogonolaba* (L) Taub. *American-Eurasian Journal of Agronomy, 2*(2), 50–56.

Titlyanov, E. A., & Titlyanova, T. V., (2010). Seaweed cultivation: Methods and problems. *Russian Journal of Marine Biology, 36*(4), 227–242.

Verkleij, F. N., (1992). Seaweed extracts in agriculture and horticulture: A review. *Biological Agriculture and Horticulture, 8*(4), 309–324.

Villanueva, R. D., Sousa, A. M. M., Gonçalves, M. P., Nilsson, M., & Hilliou, L., (2010). Production and properties of agar from the invasive marine alga, *Gracilaria vermiculophylla* (Gracilariales, Rhodophyta). *Journal of Applied Phycology, 22*(2), 211–220.

Wong, K. H., & Cheung, P. C., (2000). Nutritional evaluation of some subtropical red and green seaweeds: Part I – proximate composition, amino acid profiles and some physicochemical properties. *Food Chemistry, 71*(4), 475–482.

Xiu-geng, F., Ying, B., & Shan, L., (1999). Seaweed cultivation: Traditional way and its reformation. *Chinese Journal of Oceanology and Limnology, 17*(3), 193–199.

Yakhin, O. I., Lubyanov, A. A., Yakhin, I. A., & Brown, P. H., (2017). Biostimulants in plant science: A global perspective. *Front. Plant Sci., 7*, 2049.

Yoon, H. S., Zuccarello, G. C., & Bhattacharya, D., (2010). Evolutionary history and taxonomy of red algae. In: *Red Algae in the Genomic Age* (pp. 25–42). Springer, Dordrecht.

CHAPTER 8

SEAWEEDS AND THEIR DERIVATES AS A MULTIROLE TOOL IN AGRICULTURE

PAULO CESAR DE MELO,[1,2] TIAGO SOUSA,[2] RITA TEIXEIRA,[3] JOÃO COTAS,[2] DIANA PACHECO,[2] ANA MARTA MENDES GONÇALVES,[2,4] KIRIL BAHCEVANDZIEV,[3] and LEONEL PEREIRA[2]

[1]*Department of Agriculture, Federal University of Lavras/UFLA, Lavras, Brazil*

[2]*University of Coimbra, MARE-Marine, and Environmental Sciences Centre, Department of Life Sciences, Coimbra, Portugal*

[3]*Agricultural College of Coimbra (ESAC/IPC), Research Centre for Natural Resources Environment and Society (CERNAS), Institute of Applied Research (IIA), Coimbra, Portugal*

[4]*Department of Biology and CESAM, University of Aveiro, Aveiro, Portugal*

ABSTRACT

Seaweeds have been explored since pre-historic times as manure and to fertilize coastal areas and turn less fertile soil into fertile soil to sustain important crops for human and animal consumption. With the search for new products, seaweeds are regaining attention because of the restrain and reduction of synthetic products in agriculture. Thus, seaweeds have a multi-role purpose in agriculture, not only as fertilizer but also as soil conditioner and plant health promoter. This chapter presents a comprehensive review of the utilization of seaweeds in agriculture: from the soil fertilization usage in the past to the new applications of seaweeds and their extracts as new anti-pathogen solutions.

8.1 INTRODUCTION

Historically, seaweeds (also known as marine macroalgae) were frequently used in agriculture as plant growth promoters. However, this natural product has fallen into disuse due to the governmental encouragement for the application of synthetic fertilizers, aiming the increase of crops productivity (Traon et al., 2014). Therefore, the environmental and ecological harmful effects of these synthetic products led to the reform of the Common Agricultural Policy regulated by the Regulation (EU) N° 1308/2013 (European Parliament and of the Council, 2013) prioritizing natural alternatives to synthetic products and establishing a common organization of the markets with agricultural products.

Regarding biodiversity and soil protection, the European Union (EU) has encouraged Circular Economy through nutrient recycling and by-products exploitation. In this context, according to the Regulation (EU) 2019/1242 (Council of European Parliament, 2019), fertilizers manufacture should be standardized, and privilege recycled organic matter, previously selected.

It is also in the EU that pesticides legislation is stricter with the purpose of reducing synthetic compounds, and also toxic and harmful products, being regulated by Regulation (EC) N° 396/2005 (European Commission, 2005), Regulation (EC) N° 1107/2009 (European Commission, 2009a), Regulation (EC) N° 1185/2009 (European Commission, 2009b), and Regulation (EU) N° 2017/625 (European Commission, 2017).

The pressure under land resources is exponentially growing, in order to provide food and feed for a growing world population. According to the report of the UN Convention to combat desertification, it is estimated that 1.3 billion people have agriculture properties degraded (United Nations 2018). This is a global-scale problem, thus being included in the 2030 Agenda and in the sustainable development goals (SDG). In this context, SDG # 2 ("End hunger, achieve food security and improved nutrition and promote sustainable agriculture") defines objectives to guarantee more resilient and sustainable agricultural practices while simultaneously promoting an adaptive capacity to climate change and soil characteristics improvement (United Nations, 2015). Due to the farming intensity nowadays, it has a negative impact on the surrounding ecosystems, and it is creating food insecurity. Thus, there is the need to modify the products applied in the crops without compromising the yield of the crop. It is required to shift the agriculture to be more resilient and sustainable (United Nations, 2015). The current challenge to shift for greener products urges in developing new effective approaches for intensive and small agriculture scales (Timilsena et al., 2015).

Seaweeds and their derivatives are rich in polysaccharides, minerals, vitamins, fatty acids, antioxidants, and hormones, which can stimulate plant growth and enhance their nutritional profile (Khan et al., 2009; Craigie, 2011; Michalak and Chojnacka, 2014). Nevertheless, this profile differs in the same proportion as the macroalgal biodiversity, which taxonomically is divided in three phyla, according to the pigmentation they produce: Chlorophyta (green seaweeds), Rhodophyta (red seaweeds) and Ochrophyta, Phaeophyceae (brown seaweeds).

Green seaweeds, those belonging to the *Ulva* genus, are a rich source of minerals (sodium, potassium, and manganese) and essential amino acids (alanine, glycine, and threonine) (Raj et al., 2018). The red seaweed *Pyropia perforata* (formerly *Porphyra perforata*) exhibits a low concentration of auxins (Zhang et al., 1993). For instance, the brown seaweeds *Ascophyllum nodosum*, *Fucus vesiculosus*, and *Saccharina longicruris* synthesize the polysaccharides laminarin, fucoidan, and alginate. Hence, the polysaccharide laminarin has antimicrobial properties and stimulates plant natural defense mechanisms (Fritig et al., 1998; Van Loon and Van Strien, 1999).

Brown seaweeds are also rich in phytohormones, such as cytokinins and auxin compounds (Stirk et al., 2003; Stirk et al., 2004). Another interesting compound found in the brown macroalgae *A. nodosum* extracts are betaines, which is a compound with a pivotal role in plant resilience and adaption capacity under stress conditions (Blunden et al., 1986; Blunden et al., 1996; Akula et al., 2000). Besides being a plant growth promoter, seaweeds or their extracts have a beneficial effect on soil properties by stimulating rhizosphere microbes and improving soil retention, moisture, and nourishment (Kuwada et al., 2006; Cardozo et al., 2007; Craigie et al., 2008). All these seaweeds' compounds and properties have application in agriculture and horticulture industries. Consequently, there are numerous commercially available products based on seaweeds and their derivatives as plant biostimulants and health promotors (Khan et al., 2009). These products can be applied directly in the soil or on the plants as a spray with foliar application with positive effects (du Jardin, 2015; El Boukhari et al., 2020).

This chapter reviews seaweeds and their derivative bioactivities in soil and in important crops, to understand their role in agriculture and horticultural industry, as also to the current challenges, in order to tackle the need to mitigate desertification impacts and develop novel and natural fertilizers and phytopharmaceuticals.

8.2 SEAWEED AND AGRICULTURE: HISTORY OVERVIEW

Seaweeds have been used in agriculture for centuries in various coastal countries. In Europe, there are several historical usages of seaweed as fertilizers in agriculture. For example, there are documents or artefacts from the ancient Greeks, Britons, and Celts tribes supporting that seaweeds were harvested or collected to enrich the soils near the shore. Thus, it is demonstrated the importance of this practice in villages near the sea and also their use as human food and animal feed sources (Craigie, 2011; Chatzissavvidis and Therios, 2014). The most reported cases are from the Roman writers, such as L.J.M. Columella, who reported that cabbage roots should be covered and also manured with seaweeds, in order to retain the greenness of the seedlings and plants, to produce better and lower cabbage losses. Other Roman writer, Pliny, highlights the collection of "Margo" (possibly related to maerl, a red seaweed from Ireland and Britain), by tribes of Britain and the Gaul to fertilize their soils. Also, Palladius described the harvest/collection of marine algae in March to apply as manure in pomegranate and citron orchards (Aitken and Senn, 1965; Dillehay et al., 2008; Battacharyya et al., 2015; Mac Monagail et al., 2017; Pereira and Cotas, 2019).

Depending on the regional techniques and traditions, seaweeds can be directly mixed with the soil, or decompose for some period. This technique was used to join other organic wastes, such as peat or straw to make a manure. But in most of the cases, the seaweeds were stacked in the field first, and overturned periodically to reduce the production of toxic sulfhydryl compounds. This procedure was normal in various countries from the Atlantic region of Europe, such as Portugal, North of Spain, France, Ireland, and Great Britain. This occurred for various centuries, it was possible from the ancient times until the 20th Century (Craigie, 2011; Chatzissavvidis and Therios, 2014; Pereira and Cotas, 2019). The seaweeds were the main source of organic matter in various soil types for multiple crops and plants. The negative point was their weight, incapacity of growers to transport the seaweeds very far inland. However, in Portugal and Ireland, there are reports of seaweed transport systems to take the dried seaweed to at least 10 km inland (McHugh et al., 2003; Pereira and Cotas, 2019). Also, there was a technique to gather seaweeds, by collecting the seaweed that drifted or beach-washed or cutting the seaweeds that were exposed to the sun. In Great Britain and Scotland, farmers cut mainly brown seaweed exposed during the low tide to use as fertilizers, as in Portugal with "moliço" (a mixture of seaweeds and marine herbs) that was harvested (Chatzissavvidis and Therios, 2014; Pereira and

Cotas, 2019) and used for the same purpose. However, the most common technique was the collection of seaweed that was washed and drifted to the beaches (McHugh et al., 2003; Chatzissavvidis and Therios, 2014). There are some reports of similar activities around the world, but with lesser information available (McHugh et al., 2003). These traditional activities were so relevant that were regulated and taxed during various centuries, as in Portugal (Pereira and Cotas, 2019).

Before the industrial revolution, this tradition was studied in more detail, since the first official seaweed application as fertilizer, reported in the mid-17th Century (Basak, 2008). The first alkaline seaweed extract was produced in the first decades of the 20th Century, in the tentative to turn economically feasible the use of seaweeds as manure/fertilizer in inland agricultural soils (Craigie, 2011).

The Second World War was very important for the evolution of seaweed extracts as fertilizers. The British government expressed the need of new sources of fibers. Thus, R.F. Milton, a Scottish Biochemist, tackled this need with the extraction of alginate fibers from kelps (brown seaweeds). However, this option was not validated due to alginate dissolution with rain (Craigie, 2011). Milton dedicated part of his research to the development of fertilizers based on kelp extracts. The hot pressurized alkaline extraction method was the technique with better results, and it allowed the creation of a patent in 1947. The fertilizer based on seaweeds' extracts was commercially available by Maxicrop in 1949 (Craigie, 2011). This was the "revolution" in the seaweed extracts for agricultural application, which has been under investigation. Nowadays, there are several companies and a whole range of novel products in the fertilizer market for specific or non-specific plant cultivars, for example, seaweed's extract specific for corn crop but also seaweed fertilizer that can be applied on diverse types of crops and trees (Craigie, 2011).

The use of marine algae as fertilizers is known in several European North Atlantic and Eastern Adriatic Mediterranean countries, such as Iceland, Norway, Denmark, Ireland, Scotland, Spain, Scandinavia, Baltic countries, Great-Britain, Orkney, Hebrides, Aran, Channel, and Kirk Islands, as well as in Azores and Cape Verde Archipelagos (Pereira and Cotas, 2019). In the North of Portugal, soil fertilization with seaweeds was a regular and economical pivotal activity along the coastline areas. On the South of Douro River, seaweed application with the mentioned purpose was applied at a small-scale in Peniche, Ericeira, and Cabo Espichel (Pereira and Cotas, 2019). Still, in the Algarve and Alentejo coast, seaweeds were never used with an agricultural aim.

8.3 APPLICATIONS OF SEAWEEDS IN AGRICULTURE

Since the last centuries, algae, and agriculture have been directly related. Biofertilizers based on seaweeds can promote biological fixation of nitrogen in soil and provide beneficial compounds to the plants. However, the increase of agricultural production allied with healthy (nutraceutical) food products demand led to the need of phytopharmaceuticals crop application reduction. For this reason, researchers, and the industrial sector have been working together to develop products based on natural bioactive compounds (Derner, 2020). In precision agriculture, one of the main concepts is the application of a needed product in the right place, time, and concentration, homogenously, and progressively in smaller areas. The market of products based on algae extracts is in growing expansion, being already worldwide commercialized since laboratory and commercial tests have shown positive results in crop production by increasing yields (Derner, 2020). Among the beneficial effects of seaweed extracts on plants (through soil or foliar application), it is highlighted some aspects such as: rooting, budding, and flowering promotion, better quality and longer-lasting fruits; biotic (caused by bacteria, virus, fungi or insects) and abiotic (caused by salinity or temperature variations and water availability) stress mitigation (Chojnacka et al., 2012).

Besides the minerals and nutrients that seaweed-based biofertilizers can provide (Table 8.1), these extracts contain other bioactive compounds, such as plant hormones (i.e., auxins, gibberellins, cytokinins or abscisic acid), amino acids, carbohydrates, vitamins, as well as other biomolecules that are not yet identified, but have a biostimulant effect on plants (Yokoya et al., 2010; Pereira, 2018). These extracts are manufactured, almost exclusively, from coast stranded seaweeds or harvested from their natural habitat (Pereira et al., 2019). However, seaweeds harvested for agriculture industry supply could have destructive impacts on ecosystems. Hereby, seaweed aquaculture could be a long-term solution to support this industry (Pereira, 2020).

Seaweed bioextracts were firstly manufactured with the purpose of being a concentrated product with a long shelf-life. Through seaweed liquefaction (solid-liquid extraction) it is possible, in some cases, to extract bioactive compounds with a 100% yield (Gupta and Abu-Ghannam, 2011). Among the different valuable biomolecules that could be found in seaweed extracts, hereby is highlighted sulfated polysaccharides (PLS), pigments (e.g., carotenoids), phytosterols, peptides, polyunsaturated fatty acids (PUFA), betaines, taurine, polyphenols (particularly, phlorotannins), minerals, and vitamins (Amano et al., 2005). Some of the bioactive substances, such as phlorotannins, are only found in algae. Nevertheless, the biochemical composition of marine algae extracts differs seasonally and according to seaweeds' geo-localization and the extraction method selected (Pereira et al., 2019).

Also, between taxonomic groups there are species more suitable for agriculture application. Seaweeds are a rich source of different metabolites (Table 8.1) that enhance soil properties and are involved in plant defense mechanisms against hostile conditions (El Boukhari et al., 2020). The bioextract has shown differences between the mineral and amino acids concentrations between the green, red, and brown seaweeds. The green seaweeds present a high content of sodium, potassium, and manganese minerals and, in the amino acids, alanine, glycine, threonine, and aspartate. The red seaweeds have a lower concentration of minerals, to be noted the concentration of iron. However, the amino acid profile is identical to the green seaweeds. The brown seaweeds are very rich in calcium, nickel, chromium, and zinc. However, the brown seaweeds' amino acid profile is considerable poorer when compared to the green and red seaweeds' content.

TABLE 8.1 Seaweed Chemical Composition According to the Taxonomic Group (Adapted from Nabti, 2019 and El Boukhari et al., 2020).

Mineral compounds (μg/g of bioextract)	Green seaweeds	Red seaweeds	Brown seaweeds
Copper	0.38	4.89	8.641
Manganese	62	57.5	8.75
Zinc	1.01	15.8	19.92
Iron	0.37	915	858.5
Potassium	113	5.18	29.65
Magnesium	1.83	25.8	9.6
Cobalt	0.06	0.082	3.47
Chromium	n.d.	0.82	16.6
Lead	n.d.	0.158	0.4
Nickcl	10.4	1.84	25.2
Cadmium	2	0.07	5.9
Sodium	185	4.15	39.11
Calcium	195.26	351.5	2053.4
Organic compounds	Amylase, Amylopectin Cellulose, Laminaran Pectin	Agar, Carrageenan, Cellulose, Xylan	Alginates, Cellulose Complex sulfated heteroglucans Lichenan-like glucan

TABLE 8.1 *(Continued)*

Mineral compounds (μg/g of bioextract)	Green seaweeds	Red seaweeds	Brown seaweeds
Alanine	+++	+++	++
Glycine	+++	+++	++
Valine	++	++	++
Leucine	++	++	+
Isoleucine	++	++	+
Serine	++	++	+
Threonine	+++	+++	+
Cystein	+	+	+
Methionine	+	+	+
Aspartate	+++	+++	++

n.d. – not determined.

Seaweeds' growth depends on certain conditions, such as location, temperature or salinity, whereas shallow waters movement through tides or storms could stimulate seaweed development (Pereira et al., 2019). Despite their presence on several coastal countries, some of them never took advantage of seaweeds' beneficial properties in soil and plant growth, although, it was already demonstrated by several researchers that marine seaweeds are a rich source of many bioactive substances, macro-, and micronutrients (Fe, Mn, Co, Cu) that affect positively plant cellular metabolism and development (Khan et al., 2009). These nutrients are pivotal for the plant life cycle, occurring in active sites of metalloproteins. Besides that, these elements enhance crops quality and yields through their impact on soil conditioning (Dobromilska and Gubarewicz, 2008). Seaweed bioextracts could act as plant biostimulants, increasing plant vitality due to the presence of many biomolecules essential for superior plants (Khan et al., 2009) and through the increment of nutrient soil absorption capacity (Pereira et al., 2019). Among the several advantages already proven by seaweed extracts application, many commercially available products based on these natural sources are applied with the aim to promote seed germination and improve root development (Melo et al., 2020b).

In this context, seaweeds exhibit a high potential for the development of novel agricultural tools and technologies, highlighting the three main roles of seaweed application in agriculture: (a) direct control of phytopathogens through the antimicrobial activity of seaweed extracts; (b) induction of natural plant defense mechanisms and (c) plant growth promoter (Chatzissavvidis and Therios, 2014). According to the literature (Figure 8.1), 45% of seaweeds used in agriculture are brown seaweeds (Phaeophyceae), mainly *Ascophyllum nodosum*, followed by green seaweeds (Chlorophyta) with 36%, particularly the genus *Ulva*, whereas red seaweeds (Rhodophyta) assume only 19%.

Most of the articles (40.4%) describe the use of seaweed extract in induction of plant disease resistance (reducing the severity of the symptoms from fungal diseases, mainly) and 34% of this activity result of the foliar application of bioactive polysaccharides, such as ulvan, carrageenan or laminarin. On the other hand, 36.2% of the articles described the biostimulating effect of seaweed extracts (increase in leaves, roots, seeds/grains, favoring germination) and, only 12.8% of articles report antimicrobial activity (antiviral, antifungal or antibacterial) of marine extracts or polysaccharides (Torres et al., 2020). The main species used as agricultural fertilizer produce phycocolloids, such as agar and carrageenan (Table 8.2). These polysaccharides are also important for other industries, namely food, pharmaceutical, and cosmetic. Still, the types of polysaccharides produced by seaweeds are distinctive between taxonomic groups (Pereira et al., 2019).

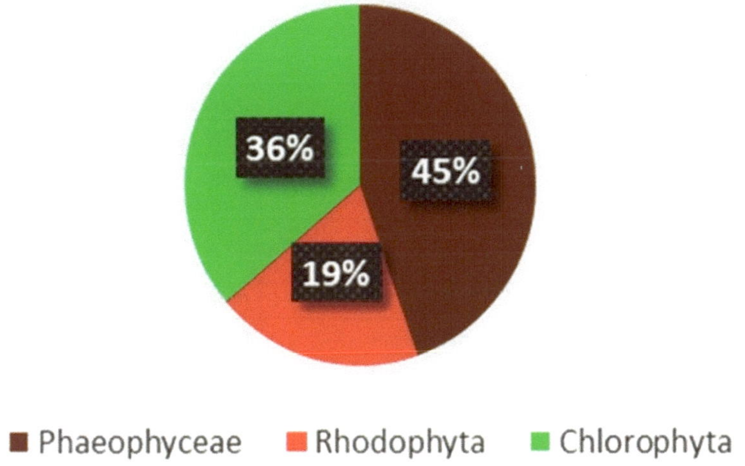

FIGURE 8.1 Published information related with seaweed application in agriculture according to their taxonomic group (brown, red, and green).

TABLE 8.2 Commercially Available Products Based on Seaweeds and Their Derivatives (Adapted from Nabti, 2019 and Melo et al., 2020b).

Commercial product name	Seaweed	Company
Acadian	*Ascophyllum nodosum*	Acadian Agritech
ACM-MAR 20	*A. nodosum*	Epagro – Serviços Agrícolas, Ltda.
Agri-Gro Ultra	*A. nodosum*	Agri-Gro Marketing Inc.
AgroKelp	*Macrocystis pyrifera*	Algas y Bioderivados Marinos S.A.
Alg-A-Mic	*A. nodosum*	BioBizz Worldwide N.V.
Biof-Nathure®	*Kappaphycus alvarezii*	Ceres Technologic Ltd.
Bio-Genesis™/High Tide™	*A. nodosum*	Green Air Products, Inc.
Biovita	*A. nodosum*	PI Industries Ltd.
Calmar®	*Lithothamnion brasilienses*	Acquamar Oceanologia Ltda.
Espoma	*A. nodosum*	The Espoma Company
FitoAlgas Green	-	Trade Corporation International S.A.
Garantee®	*A. nodosum*	Maine Stream Organics
Kelp Meal	*A. nodosum*	Acadian Seaplants Ltd.
Kelpak	*Ecklonia maxima*	BASF
Kelpro	*A. nodosum*	Tecniprocesos Biologicos, S.A.
Kelprosoil	*A. nodosum*	Productos del Pacifico, S.A.
Maxicrop	*A. nodosum*	Maxicrop USA, Inc.
Nitrozime	*A. nodosum*	Hydrodynamics International Inc.
Profert®	*Durvillaea potatorum*	Antártica BASF
Profertil®	*A. nodosum*	ADP Fertilizantes
Reabilit Algas®	*K. alvarezii* and *Sargassum vulgare*	Nutri-Fertilizantes Indústria e Comércio Agropecuário – Ltda.
Sea Winner	-	China Ocean University Product Development Co., Ltd.
Seanure	-	Farmura Ltd.
Seasol	*D. potatorum*	International Pty Ltd.
Soluble Seaweed Extract	*A. nodosum*	Technaflora Plant Products, Ltd.
Stimplex®	*A. nodosum*	Acadian Agritech
Synergy	*A. nodosum*	Green Air Products, Inc.
Tasco®	*A. nodosum*	Acadian Agritech

This field of research has raised interest from the agriculture industry, and this is why currently biostimulants based on seaweed extracts are already worldwide commercially available (Table 8.2) (Nabti, 2019; El Boukhari et al., 2020). The commercial products are based mainly in *A. nodosum*, a brown seaweed traditionally cultivated in the North Atlantic, such as Canada and Ireland, and was the first seaweed used in the development of commercial seaweed fertilizers in 1970 (Craigie, 2011). These commercial products are plant biofertilizers and biostimulants. The companies presented have a worldwide distribution; however, the north hemisphere has a greater impact in this industry due to the *A. nodosum* and the knowledge to produce the extract, even though globalization is happening, and south hemisphere is growing in this field, with other seaweeds being commercially explored (Craigie, 2011; Chatzissavvidis and Therios, 2014).

8.4 SEAWEED AS SOIL CONDITIONER

The seaweed *Lithothamnion* (Rhodophyta) is a calcareous seaweed that is typically used for pH soil correction or to tackle the lack of calcium in the soil. There are studies suggesting that the application of this seaweed in the soil dates from 1018 (Le Beu, 1983). Nowadays, this technique is widely used, particularly in coastal areas of France and the United Kingdom. In these regions, we can find Calcified Seaweed or maërl, being composed by *Phymatolithon calcareum* and *Lithothamnion corallioides*.

In Brazil, it can be observed *Lithothamnion* as deposit from Amazonian region to the South of Rio de Janeiro. Some deposit locations are still unknown (de Macedo Dias and Villaça, 2012), but usually they can be observed in the coastal areas and could be explored for agriculture industry (Kempf, 1974). Maranhão continental shelf is explored by several companies, for example, the Oceana Minerals®, which extracts this seaweed species for agriculture application, from the marine sediment and afterwards it is milled, hot air dried and cold micro-sprayed. Due to this, seaweed increases soil porosity which is beneficial for the plants (Melo and Furtini Neto, 2003).

Miranda (1985) studied the application of two commercially available products from France, Lithothamne C and Lithothamne 400 (Melo and Furtini Neto, 2003), their effect on lactosoils pH and how they can affect maize plants. Miranda (1985) concluded the application of these products has positive effects and their application was feasible to correct soil pH. However, the raw product extracted from marine sediment usually needs to be enriched with micronutrients for soil application. Melo and Furtini Neto (2003) evaluated *Lithothamnion* powder as a source of chemical compounds and nutrients and their beneficial effects on bean plant (*Phaseolus vulgaris*) cultivated on three different types of soils, highlighting the beneficial properties of this process as a plant nitrogen uptake promoter. From their study, they found out that a neutralization power of 99.26% is an indicator that the sample analyzed shows a fast and intensive corrective action of soil acidity (Table 8.3).

TABLE 8.3 *Lithothamnion brasiliensis* Chemical Characterization (Adapted from Melo and Furtini Neto, 2003)

CaO ($g\ kg^{-1}$)	Ca ($g\ kg^{-1}$)	MgO ($g\ kg^{-1}$)	Mg ($g\ kg^{-1}$)	PN (%)	Reactivity (%)	PRNT (%)
462.7	330.5	42.3	25.4	99.26	92.62	93.31

CaO – calcium oxide; Ca – calcium; MgO – magnesium oxide; Mg – magnesium; PN – neutralization power; PRNT – Relative total neutralization power.

This result shows the feasibility of the application of the seaweed *Lithothamnion* as a corrective material for acid and deficient in Ca and Mg, soils. In this way, it is possible to increase soil pH and neutralize the toxic effect of some elements by providing Ca and Mg as nutrients (Souza et al., 2009). Nevertheless, according to Turan and Köse (2004), it is necessary to consider that Ca content is one of the major drivers to better perform culture establishment in the soil after germination.

When compared with the method of liming, *Lithothamnion* application advantages are evidenced as plant biostimulators contributing to a higher nutrient uptake, thus promoting plant development and higher production (Moreira et al., 2015). Literature demonstrates that the coupled application with vinasse can promote soil conditioning in sugar cane cultures. With this method, it was verified a higher nutrient availability for plant absorption and an increase of soil pH. Besides that, it also enhances soil remineralization, increasing the Ca, Mg, and Fe content in 224, 40 and 88%, respectively (Rodriguez, 2017). For this reason, researchers observed a yield increment of 50% in sugar and alcohol production, while in maize, soybean, and tomato, the increase was between 20 and 35% (Crouch et al., 1990; Lima et al., 2002; Rodriguez, 2017). However, there are other studies that evidence higher yield on fruit crops (i.e., orange, guava, and passion fruit) and a better radicular development in citrus fruits with the application of calcareous seaweeds (Moreira et al. 2015). Besides all these positive effects already mentioned, the content of marine seaweeds needed to achieve these results is lower than from the conventional products used (e.g., lime) (López-Benito, 1963; Melo and Furtini Neto, 2003).

8.5 SEAWEED AS PLANT BIOSTIMULANT

Seaweed extracts exhibit various biostimulating activities in agricultural production, also when applied on leaves (foliar spray) (Chatzissavvidis and Therios, 2014). Biostimulants are defined as "materials, other than fertilizers, that promote plant growth when applied in small quantities" and are also referred to as "metabolic enhancers" (Khan et al., 2009).

Seaweed species are still often regarded as an underutilized bioresource, although seaweeds have been used as a source of food, industrial raw materials, and in therapeutic and botanical applications. Moreover, seaweed extracts are widely used in horticulture crops due to their plant growth-promoting effects in the improvement of flowering, yield enhanced product quality, improve nutritional content of edible product, as well as shelf life and for their ameliorating effect on crop tolerance to abiotic stresses, such as salinity, extreme temperatures, nutrient deficiency and drought (Khan et al., 2009; Zodape et al., 2011; Battacharyya et al., 2015; Nabti et al., 2017; Yakhin et al., 2017).

The chemical constituents of seaweeds' extracts include complex polysaccharides (e.g., galactan, fucoidan, alginate, and laminarin), proteins (e.g., lectins), fatty acids, pigments (e.g., chlorophylls, carotenoids, and phycobiliproteins), polyphenols (e.g., phenolic acids, flavonoids, cinnamic acid, isoflavones, benzoic acid, lignans, quercetin), minerals (e.g., K, Mg, Ca, and Na), vitamins, and phytohormones (e.g., cytokinins, auxins, gibberellins, and abscisic acid) (Khan et al., 2009; Battacharyya et al., 2015; Godlewska et al., 2016; Yakhin et al., 2017). However, the biostimulatory potential of many of these products has not been fully exploited due to the lack of scientific data on growth factors present in seaweeds and their mode of action in affecting plant growth, although plant biostimulants have received more attention by both the scientific community and commercial enterprises (Table 8.2) in the last two and a half decades (Khan et al., 2009; Battacharyya et al., 2015; Godlewska et al., 2016; Yakhin et al., 2017).

Several experiments have already been carried out with seaweed extracts. One example was from *Kappaphycus alvarezii* and *Hydropuntia edulis* (formerly *Gracilaria edulis*) extracts applied in rice (*Oryza sativa*). The rice seeds were soaked with lower concentrations (2.5 and 5%) of both seaweed extracts, and it was recorded a higher germination percentage, root, and shoot length and seedling vigor index when compared to water soaking.

Silva et al. (2019) applied *A. nodosum* and *Sargassum muticum* extracts to improve the lettuce (*Lactuca sativa*) seed germination where it was demonstrated that the seaweed extract at a low concentration (25%) potentiated the seed germination, but higher dosages inhibited or had adverse effects in the seed germination and the seedling formations. Thus, the seaweed concentration needs to be controlled for the full potential, still if the dosage is not correct, there is a high possibility of fully damaging of the crop. These results are comparable to the lettuce plants where the ones in which 25% of the seaweed extract was applied had a better growth than the control. However, in the hydroponics it did not show an improvement when compared with the commercial fertilizer, due to the main problem detected related with the high electrical conductivity, which is very relevant to the plants (Silva et al., 2019). The aqueous extract of *S. muticum* (foliar extract) showed a beneficial impact in soil minerals assimilation by the lettuce, for example, in phosphorous, potassium, calcium, and magnesium (Silva et al., 2019). These effects show that *S. muticum* extract could have a multirole influence in the crops.

Foliar spray at 5% and above concentrations significantly increased plant height, dry matter accumulation, chlorophyll index, crop growth rate, yield attributes and yield of rice as compared to control (Layek et al., 2018). Another experiment was made with Polysaccharide Enriched Extracts (PEEs) obtained from 17 Moroccan seaweeds, on tomato seed germination and plant growth. Results indicated an increase of germination rate and germination speed of tomato seeds treated with 0.02 mg mL^{-1} of PEEs obtained from the Moroccan seaweeds. The PEEs also enhanced plant dry weight (DW) and chlorophyll-*a* content (Mzibra et al., 2020).

Scientific research has proven that some algal metabolites show potential antioxidant, antiproliferative, antidiabetic, antitumor, anti-inflammatory, antiallergic, and anti-HIV properties. Because of their composition and the functional properties, they are used as human food, especially in Asia (China, Japan, and Korea) and as animal feed (Godlewska et al., 2016).

8.6 SEAWEED AS PLANT PROTECTION FACTOR

Seaweeds are known as a rich source of bioactive compounds that have a broad range of biological activities spectrum, such as antibacterial and antifungal (Vallinayagam et al., 2009). Although several industries use seaweeds as a valuable source of bioactive compounds, a long road is needed in phytopharmacology to apply seaweed extracts and compounds in this field (Righini et al., 2019). However, there are studies demonstrating the efficacy of seaweeds and their extracts against pathogens in various types of crops.

Plant diseases are considered to be one of the main factors leading to decreased food production worldwide (Savary et al., 2017). Synthetic pesticides are often used to control these bacterial and fungal diseases, but the toxic environmental risks caused by the overuse of these fungicides' present health problems as modern society is becoming more aware, but also due to the resistance of pathogens against fungicides (Soliman et al., 2018). The European Directive 2009/128/EC came to convert this ideology trying to develop and implement sustainable management of the protection of agriculture, using other means, including natural products (Righini et al., 2019).

8.6.1 ANTIBACTERIAL

The treatment of bacterial infections is increasingly hampered by the ability of bacteria to develop resistance to antimicrobial agents. Continuous discovery and development of new compounds is necessary to ensure a sustainable future (Palma, 2011). The ability of algae to inhibit growth and toxicity of pathogens has been known for a long time (Ibrahim and Lim, 2015). However, to be effective against bacterial pathogens, it must exhibit anti-biofilm activity, as the predominant mode of growth of bacteria is through the formation of biofilm since the possibility of forming biofilms also provides enhanced protection against chemical antimicrobial treatments, making it difficult to eradicate such infections (Righini et al., 2019).

The effect of extracts from brown seaweeds *Cystoseira humilis* var. *myriophylloides* (formerly *Cystoseira myriophylloides*), *Laminaria digitata* and *Fucus spiralis* against tomato pathogens *Agrobacterium tumefaciens* (bacteria) and *Verticillium dahliae* (fungi) was studied by Esserti et al. (2017), in which a significant inhibition of the pathogen growth with the methanol extracts from seaweed was observed. The application of spray and aqueous extracts of seaweed of *C. myriophylloides* and *F. spiralis* also resulted in a significant reduction of crown gall disease caused by the bacterial pathogen *A. tumefaciens* (Esserti et al., 2017).

Three strains of Gram-positive bacteria, namely *Clavibacter michiganensis* subsp. *nebraskensis* infects corn (*Zea mays*) provoking leaf lesions and symptoms of vascular wilt, for example, orange internal discoloration of vascular bundles and the irregular external appearance of the stem, presenting a viscous and soaked in water like stem. *C. michiganensis* subsp. *michiganensis* has as main host a tomato plant (*Solanum lycopersicum*), causing on it a bacterial disease that causes the desiccation of the leaves and, eventually, of the whole plant. In the field, the bacteria dehydrate the edge of the tomato plant leaflets and the plant slowly and completely dehydrates. As *C. michiganensis* subsp. *tessellarius* infects wheat (*Triticum aestivum*), causing bacterial mosaic in the form of freckles and leaf spots. Whereas *Polysiphonia lanosa* (currently *Vertebrata lanosa*) methanolic extract has an intermedium effect against these strains compared to the positive control. However, the *F. serratus* and *A. nodosum* methanolic or aqueous extract do not appear to have an antibacterial effect (O'Keefe et al., 2019). The *V. lanosa* methanolic and aqueous extract have antibacterial activity against *Xanthomonas arboricola* which attacks predominantly species of *Prunus*, but also fruit trees, such as peach, apricot, cherry, and plum, causing damage on leaves, branches, and fruits (O'Keefe et al., 2019). Lately, a variety of studies has been carried out for a wide range of bacterial pathogens, including *Xanthomonas* sp., *Clavibacter michiganensis* subsp. *sepedonicus* and *Pseudomonas syringe*, with promising results, demonstrating the potential of *Ecklonia* sp., and *Jania* sp. polysaccharides as antibacterial agents (Righini et al., 2019).

A research work perceived that the halogenated polyphenols present in seaweeds, mainly bromophenols, are the compounds predominantly responsible for the expressive antimicrobial activity; however, the seaweed chemical composition with acids, alkaloids, and amines, can also give a strong antimicrobial activity response (González-Castro et al., 2019). The research showed countless potential compounds that can be extracted from the *V. lanosa*, which have an enormous antibacterial capacity, consequently being a good opportunity to tackle antibiotic resistance driven by the normally used phytopharmaceuticals.

8.6.2 ANTIFUNGAL

Several studies suggest that seaweeds' extracts could be a key driver to develop new tools to plant diseases control due to their antifungal activity (Cardozo et al., 2007). Currently, seaweeds metabolites are known for their capability of stimulating crops growth. Besides that, macroalgae compounds with antifungal bioactivity are seen as an unexplored wealth (De Corato et al., 2017).

In 2017, *in vivo* and *in vitro* activity of two brown seaweed extracts was evaluated (*L. digitata* and *Undaria pinnatifida*) and extracts from three red seaweeds (*Porphyra umbilicalis*, *Eucheuma denticulatum* and *Gelidium pusillum*) against three post-harvest pathogens (*Botrytis cinerea*, *Monilinia laxa* and *Penicillium digitatum*). In this study, *L. digitata*, *U. pinnatifida* and *P. umbilicalis* showed a higher efficiency in *in vitro* tests, inhibiting significantly the mycelium and conidia growth of all the pathogens tested. In opposition, *E. denticulatum* and *G. pusillum* extracts demonstrated a lower inhibition of mycelium growth and conidia germination in all the pathogens tested (de Corato et al., 2017).

Soliman et al. (2018) studied the antifungal effect of the red seaweed *Gracilariopsis longissima* (formerly *Gracilaria confervoides*) in three fungi, namely *Rhizoctonia solani*, *Fusarium solani* and *Macrophomina faseolina*. The extracts caused a deformed fungal growth. However, the extract of *G. confervoides* with chloroform and ethyl acetate provoked the highest inhibition of *R. solani*. These results indicated that the extracts have potential as a novel product to control soil pathogenic organisms (Soliman et al., 2018).

In vivo assays with seaweed extract application (*L. digitata*, *U. pinnatifida* and *P. umbilicalis*) performed in damaged fruits showed a reduction of gray mold in strawberries, brown mold in peach and green mold in lemons. In contrast, *E. denticulatum* and *G. pusillum* extracts could delay and reduce these diseases' development. Antifungal bioactivity could be a result of the high fatty acids content in *L. digitata*, *U. pinnatifida* and *P. umbilicalis*. While polysaccharides found in seaweed extracts could be possibly related to systemic resistance mediated by peroxidase (Khan et al., 2017).

In fact, seaweeds' extracts exhibit bioactivities against several phytopathogens. Recent studies showed that extracts from the macroalgae *L. digitata*, *U. pinnatifida* and *P. umbilicalis* also inhibit the mycelial growth of *B. cinerea* and the germination of their spores (Soliman et al., 2018). Another brown seaweed extract showed an effective protector effect against *B. cinerea* in tomato leaves (Soliman et al., 2018). Besides that, these extracts were also capable of increasing the transcription of certain plant enzymes involved in fungal resistance and promote their enzymatic activity (Soliman et al., 2018).

Between the five extracts tested from the brown seaweed *Sargassum wightii*, the highest concentration was considered the most promising extract to inhibit *R. solani* spore germination (fungistatic activity). Also, foliar extracts demonstrated direct impact, by doing the fungicidal *in vitro* methods, demonstrating that seaweed extracts can control the *R. solani*, not only the propagation but can also be a recovery tool for infected plants (Raj et al., 2018). Raj et al. (2018) hypothesizes that seaweeds' extracts antifungal bioactivity efficiency is due to their high content of phenolic compounds and phytoalexines.

8.6.3 IMMUNOMODULATORY EFFECTS

Immunomodulation effects in plants occur through the potentiation or suppression of elements of the plant's innate immune system. Mainly, though the compound will modulate the defense response from the plant and not interact directly with the pathogen. Several studies are being carried out in search of substances capable of assisting the immune system in the fight against pathogenic microorganisms (Machado et al., 2019). There are different types of immunomodulators and depending on their origin, they can be divided into natural or synthetic. They can be derived from fatty acids, flavonoids, polysaccharides, and terpenes that can be found in plants (Corrêa et al., 2006).

Many studies which refer immunomodulatory effect of plants have used crude extract, in others, combinations of various herbs or herbs in combination with minerals (Corrêa et al., 2006). Polysaccharides obtained from natural sources are considered biological response modulators and have been shown to increase plant immune responses (Wijesinghe and Jeon, 2012). Regarding studies of seaweeds' extracts with immunomodulatory bioactivity, there was a lack of information, which reveals an interesting field of investigation for researchers, since macroalgae have bioactive compounds with different physiological and biological activities (Wijesinghe and Jeon, 2012).

Jayaraj et al. (2008) studied, in a greenhouse, carrot plants cultivated with fungal pathogens (*Alternaria radicina* and *B. cinerea*) and sprayed with an extract of *A. nodosum*. The plants treated with the seaweed extract showed a lower severity of the disease compared with the extract with salicylic acid. In this study, it was observed that the activity of enzymes related to plant defense mechanisms (including peroxidase, polyphenol oxidase, phenylalanine ammonia-lyase, chitinase, and β-1,3-glucanase) increased significantly in plants treated with the seaweed extract, thus demonstrating that this extract increases disease resistance in carrots through induction of defense genes or proteins (Jayaraj et al., 2008). In 2018, in a study performed with four different seaweeds, *S. wightii*, *Dictyota bartyrensiana*, *Ulva reticulata*, *Gelidiella acerosa* were applied in powdered fish extract (*Odonus niger*), in order to control *Rhizoctonia solani*, which is a pathogenic agent of the rice sheath pest, which has become a major problem in rice production. Seaweed provides a rich source of structurally diverse and biologically active secondary metabolites and has proven their efficiency in reducing leaf fungal diseases, which ultimately promotes plant fertility and growth (Raj et al., 2018).

8.6.4 INSECTS PROTECTION

Insecticides are widely used for pest control and have diverse origins, including plants, invertebrates or more recurring synthetic compounds. For decades, there has been a strong dependence on synthetic insecticides for the control of insects, due to their efficacy and convenience. However, the excessive use of synthetic insecticides can cause ecological adverse and dangerous effects, such as the development of insect tolerance and resistance to insecticide compounds, environmental pollution, accidental poisoning of other organisms and the growing threat of infectious diseases (Yu et al., 2015). Consequently, it is indispensable to search for more sustainable alternatives to these synthetic products (Yu et al., 2015).

New natural insecticide solutions require a full understanding of the potential activity against insects, rendering to each stage of their life. In addition, more information is needed about insecticide toxic effect, in order to develop safer and more selective products (Hamed et al., 2018), thus reducing the hazard and noxious effects of insecticides application in ecosystems.

Seaweeds are a great natural resource and could be a solution for this problem, since they present a wide range of insecticidal and anti-herbivory substances (Machado et al., 2019). Seaweeds are becoming the focus of studies to discover new molecules with insecticidal effect. Nowadays, there are already studies showing that seaweeds have secondary metabolites with insecticidal effect (Sahayaraj and Kalidas, 2011; Kitherian Sahayaraj and Jeeva, 2012; Hamed et al., 2018).

The insecticide capability of *Caulerpa scalpelliformis* has been successfully tested against *Dysdercus cingulatus*. This insect causes severe destruction in cotton crops, by feeding on developing cotton capsules and on mature cotton seeds, or by the transmission of pathogenic fungi, becoming an economically important problem (Kombiah and Sahayaraj, 2012). In the study of Sahayaraj and Kalidas (2011), the impact of the seaweed extract from *Padina pavonica* against *D. cingulatus* was evaluated. It was proven that *P. pavonica* extract was more effective in provoking the nymph death and in reducing the capability of the insect hatching, concluding that *P. pavonica* has a partial insecticidal activity. The presence of saponin, sterols, and phenolic compounds in the *P. pavonica* extract can be the reason for these activities, but further investigation is needed (Sahayaraj and Kalidas, 2011). Also, *Sargassum tenerrimum* extract reduced the period of nymph development of the *D. cingulatus*. In this specific case, several extracts were tested, all reducing the period of nymph development. The extract effect suffered variations due to the different solvents used. For example, the chloroform extract of *S. tenerrimum* reduced the periods of oviposition, pre-oviposition, and post-oviposition of *D. cingulatus*, as well as fertility. With the *S. tenerrimum* benzene extract, the emerging females were unable to lay eggs. Consequently, *S. tenerrimum* extracts have proven to be a potential candidate to develop an insecticide against *D. cingulatus*. However, there is a long road until its commercial application (Asaraja and Sahayaraj, 2013).

In other case, extracts from *Caulerpa sertularioides, Laurencia johnstonii* and *Sargassum horridum* showed potential activity against adults of *Diforina citri*. This is the insect vector of the bacteria *Candidatus Liberibacter* spp. that causes the huanglongbing (citrus greening disease) provoking significant yield losses in citrus production worldwide, and so, it is considered the worst disease in citriculture nowadays. The three seaweeds' extracts showed various compounds of interest, such as alkaloids, terpenes, phenols, tannins, flavonoids, anthraquinones, and saponins, which have been associated with the insecticidal and repellent activity. With these results, these seaweeds' extracts and compounds can be considered as alternative in the development of novel agrochemicals for control of this and other pests (Machado et al., 2019).

8.7 CONCLUSION AND FUTURE PERSPECTIVES

Farmers and agriculture technicians recognize the need of reducing phytopharmaceuticals in soil and plants. Thus, seaweeds are a natural source to tackle this need by promoting plant health and resistance to diseases and pathogens. Moreover, seaweed application as soil conditioner and fertilizer enhances fruit and vegetables flavor and shelf-life. Within seaweed soil decomposition, it is verified an increase of aerobic fermentation temperature that enhances organic compounds biodegradation and improves soil quality through pH correction, soil moisture and is also capable of turning sandy soils into more interesting ones for crop production. Nevertheless, there are areas where more research is needed, mainly more information about seaweeds and their extracts' impacts on the plant, such as anti-pathogenic and immunomodulatory mechanisms.

ACKNOWLEDGMENTS

This work is financed by national funds through FCT – Foundation for Science and Technology, I.P., within the scope of the projects UIDB/04292/2020 – MARE – Marine and Environmental Sciences Centre and UIDP/50017/2020+UIDB/50017/2020 (by FCT/MTCES) granted to CESAM – Centre for Environmental and Marine Studies. João Cotas thanks MARE, Universidade de Coimbra and Universidade Nova de Lisboa, through its Chemistry Department of the Faculty of Sciences and Technology within the scope of the research project "Algae in Gastronomy – Development of innovative techniques for conservation and use – Alga4Food" (MAR-01.03.01-FEAMP-0016). Ana M.M. Gonçalves acknowledges the University of Coimbra for the contract IT057–18–7253.

KEYWORDS

- **abscisic acid**
- **antioxidants**
- ***Ascophyllum nodosum***
- **auxins**
- **cultivation trials**
- **fatty acids**
- **fertilizers**
- **food security**
- ***Fucus vesiculosus***
- **gibberellins**
- **macroalgae**
- **polysaccharide enriched extracts**
- **pyropia perforate**
- ***Saccharina longicruris***
- **sustainable development goals**

REFERENCES

Aitken, J. B., & Senn, T. L., (1965). Seaweed products as a fertilizer and soil conditioner for horticultural crops. *Bot. Mar., 8*(1), 144–147. doi: https://doi.org/10.1515/botm.1965.8.1.144.

Akula, A. C., & Bateson, M., (2000) Betaine a novel candidate for rapid induction of somatic embryogenesis in tea (*Camellia sinensis* (L.) O. Kuntze). *J. Plant Growth Regul., 30*(3), 241–246. doi: https://doi.org/10.1023/A:1006323213621.

Amano, H., Kakinuma, M., Coury, D. A., Ohno, H., & Hara, T., (2005). Effect of a seaweed mixture on serum lipid level and platelet aggregation in rats. *Fish. Sci., 71*(5), 1160–1166. doi: https://doi.org/10.1111/j.1444-2906.2005.01076.x.

Araújo, P. O. L. C., Gonçalves, F. C., Ramos, J. D., Chalfun, N. N. J., & Carvalho, G. J., (2007). Crescimento e percentual de emergência de plântulas de citrumeleiro "Swingle" em função dos substratos e das doses de corretivo à base de *Lithothamnion*, após cem dias da semeadura. *Ciênc. Agrotec., 31*(4), 982–988. doi: http://dx.doi.org/10.1590/S1413-70542007000400006.

Asaraja, A., & Sahayaraj, K., (2013). Screening of insecticidal activity of brown macroalgal extracts against *Dysdercus cingulatus* (Fab.) (Hemiptera: Pyrrhocoridae). *J. Biopestic., 6*(2), 193–203.

Basak, A., (2008). Effect of preharvest treatment with seaweed products, Kelpak® and Goëmar BM 86®, on fruit quality in apple. *Int. J. Fruit Sci., 8*(1, 2), 1–14. doi: https://doi.org/10.1080/15538360802365251.

Battacharyya, D., Babgohari, M. Z., Rathor, P., & Prithiviraj, B., (2015). Seaweed extracts as biostimulants in horticulture. *Sci. Hortic., 196*(2015), 39–48. doi: https://doi.org/10.1016/j.scienta.2015.09.012.

Blunden, G., Cripps, A. L., Gordon, S. M., Mason, T. G., & Turnei, C. H., (1986). The characterization and quantitative estimation of betaines in commercial seaweed extracts. *Bot. Mar., 29*(2), 155–160. doi: https://doi.org/10.1515/botm.1986.29.2.155.

Blunden, G., Jenkins, T., & Liu, Y. W., (1996). Enhanced leaf chlorophyll levels in plants treated with seaweed extract. *J. Appl. Phycol., 8*(6), 535–543. doi: https://doi.org/10.1007/BF02186333.

Cardozo, K. H. M., Guaratini, T., Barros, M. P., Falcão, V. R., Tonon, A. P., Lopes, N. P. S., Campos, S., et al., (2007). Metabolites from algae with economical impact. *Comp. Biochem. Physiol. C Toxicol. Pharmacol., 146*(1), 60–78. doi: https://doi.org/10.1016/j.cbpc.2006.05.007.

Chatzissavvidis, C., & Therios, I., (2014). Role of algae in agriculture. In: Pomin, V. H., (ed.), *Seaweeds* (pp. 1–37). Nova Science Publishers Inc, New York.

Chojnacka, K., Saeid, A., Witkowska, Z., & Tuhy, L., (2012). Biologically active compounds in seaweed extracts – the prospects for the application. *The Open Conference Proceedings Journal, 3*(1), *20–28.* doi: http://dx.doi.org/10.2174/1876326X01203020020.

Corrêa, V. S. C., Maynié, J. C., França, E. L., & Honório-França, A. C., (2006). Atividade funcional de fagócitos na presença do fitoterápico "mais vida". *Rev. Bras. Pl. Med., 8*(2), 26–32.

Council of European Parliament, (2019). Regulation (EU) 2019/1242 of the European Parliament and of the Council of 20 June 2019. *Official Journal of European Union, 198*(338), 202–240.

Craigie, J. S., (2011). Seaweed extract stimuli in plant science and agriculture. *J. Appl. Phycol., 23*(3), 371–393. doi: https://doi.org/10.1007/s10811-010–9560-4.

Craigie, J. S., MacKinnon, S. L., & Walter, J. A., (2007). Liquid seaweed extracts identified using ^{1}H NMR profiles. In: Borowitzka, M. A., Critchley, A. T., Kraan, S., Peters, A., Sjøtun, K., & Notoya, M., (eds.), *Nineteenth International Seaweed Symposium* (pp. 215–221). *Developments in Applied Phycology,* Vol. 2. Springer, Dordrecht. doi: https://doi.org/10.1007/978-1-4020-9619-8_27.

Crouch, I. J., Beckett, R. P., & Van, S. J., (1990). Effect of seaweed concentrate on the growth and mineral nutrition of nutrient-stressed lettuce. *J. Appl. Phycol., 2*(3), 269–272. doi: https://doi.org/10.1007/BF02179784.

De Corato, U., Salimbeni, R., De Pretis, A., Avella, N., & Patruno, G., (2017). Antifungal activity of crude extracts from brown and Red Seaweeds by a supercritical carbon dioxide technique against fruit postharvest fungal diseases. *Postharvest Biol. Technol., 131*, 16–30. doi: https://doi.org/10.1016/j.postharvbio.2017.04.011.

De Macedo, D. G. T., & Villaça, R. C., (2012). Coralline algae depositional environments on the Brazilian central-south-eastern shelf. *J. Coast. Res., 28*(1), 270–279. doi: https://doi.org/10.2112/11T-00003.1.

Derner, R. B., (2020). As algas na agricultura – Colunas Biotecnologia de Algas – Aquacultura Brasil. Available online: https://www.aquaculturebrasil.com/coluna/115/as-algas-na-agricultura (accessed on 13 October 2021).

Dillehay, T. D., Ramirez, C., Pino, M., Collins, M. B., Rossen, J., & Pino-Navarro, J. D., (2008). Monte Verde: Seaweed, food, medicine, and the peopling of South America. *Science, 320*(5877), 784–786. doi: https://doi.org/10.1126/science.1156533.

Dobromilska, R., Mikiciuk, M., & Gubarewicz, K., (2008). Evaluation of cherry tomato yielding and fruit mineral composition after using of bio-algeen S-90 preparation. *J. Elementol., 13*(4), 491–199.

Du Jardin, P., (2015). Plant biostimulants: Definition, concept, main categories and regulation. *Sci. Hortic., 196*, 3–14. doi: https://doi.org/10.1016/j.scienta.2015.09.021.

El Boukhari, M., Barakate, M., Bouhia, Y., & Lyamlouli, K., (2020). Trends in seaweed extract-based biostimulants: Manufacturing process and beneficial effect on soil-plant systems. *Plants, 9*(3), 359. doi: https://doi.org/10.3390/plants9030359.

Esserti, S., Smaili, A., Rifai, L. A., Koussa, T., Makroum, K., Belfaiza, M., et al., (2017). Protective effect of three brown seaweed extracts against fungal and bacterial diseases of tomato. *J. Appl. Phycol., 29*(2), 1081–1093. doi: https://doi.org/10.1007/s10811-016-0996-z.

European Commission, (2005). Regulation (EU) N° 396/2005 on maximum residue levels of pesticides in or on food and feed of plant and animal origin, and amending. Council Directive 91/414/EEC. *Official Journal of the European Union, L70*, 1–16.

European Commission, (2009a). (EC) No 1107/2009. *Official Journal of the European Union, 309*(1), 1–50.

European Commission, (2009b). Regulation (EC) No 1185/2009 of the European Parliament and of the Council of 25 November 2009 concerning statistics on pesticides. *Official Journal of the European Union, L 324*(1107), 1–22.

European Commission, (2017). Regulation (EU) 2017/625 of the European Parliament and of the Council. *ERA Forum.*

European Parliament and of the Council, (2013). Regulation (EU) No 1308/2013. *Official Journal of the European Union.*

Fritig, B., Heitz, T., & Legrand, M., (1998). Antimicrobial proteins in induced plant defense. *Curr. Opin. Immunol., 10*(1), 16–22. doi: https://doi.org/10.1016/s0952-7915(98)80025-3.

Godlewska, K., Michalak, I., Tuhy, Ł., & Chojnacka, K., (2016). Plant growth biostimulants based on different methods of seaweed extraction with water. *Biomed Res. Int., 2016*, 5973760. doi: https://doi.org/10.1155/2016/5973760.

González-Castro, A. L., Muñoz-Ochoa, M., Hernández-Carmona, G., & López-Vivas, J. M., (2019). Evaluation of seaweed extracts for the control of the Asian citrus psyllid *Diaphorina citri*. *J. Appl. Phycol., 31*(6), 3815–3821. doi: 10.1007/s10811-019-01896-5.

Gupta, S., & Abu-Ghannam, N., (2011). Recent developments in the application of seaweeds or seaweed extracts as a means for enhancing the safety and quality attributes of foods. *Innov. Food Sci. Emerg. Technol., 12*(4), 600–609. doi: https://doi.org/10.1016/j.ifset.2011.07.004.

Hamed, S. M., El-Rhman, A. A. A., Abdel-Raouf, N., & Ibraheem, I. B. M., (2018). Role of marine macroalgae in plant protection and improvement for sustainable agriculture technology. *Beni-Suef Univ. J. Basic Appl. Sci., 7*(1), 104–110. doi: https://doi.org/10.1016/j.bjbas.2017.08.002.

Ibrahim, D., & Lim, S. H., (2015). *In vitro* antimicrobial activities of methanolic extract from marine alga *Enteromorpha intestinalis*. *Asian Pac. J. Trop. Biomed., 5*(9), 785–788. doi: https://doi.org/10.1016/j.apjtb.2015.07.012.

Jayaraj, J., Wan, A., Rahman, M., & Punja, Z. K., (2008). Seaweed extract reduces foliar fungal diseases on carrot. *Crop Prot., 27*(10), 1360–1366. doi: https://doi.org/10.1016/j.cropro.2008.05.005.

Kempf, M., (1974). Perspectives d'exploitation des fonds de maerl du plateau continental du NE du Brésil. *Exploitation des Oceans, 2*, 1–17.

Khan, S. A., Abid, M., & Hussain, F., (2017). Antifungal activity of aqueous and methanolic extracts of some seaweeds against common soil-borne plant pathogenic fungi. *Pak. J. Bot., 49*(3), 1211–1216.

Khan, W., Rayirath, U. P., Subramanian, S., Jithesh, M. N., Rayorath, P., Hodges, D. M., Critchley, A. T., et al., (2009). Seaweed extracts as biostimulants of plant growth and development. *J. Plant Growth Regul., 28*(4), 386–399. doi: 10.1007/s00344-009-9103-x.

Kombiah, P., & Sahayaraj, K., (2012). Repellent activity of *Caulerpa scalpelliformis* extracts and its formulations against *Spodoptera litura* and *Dysdercus cingulatus* (Fab.). *J. Biopest., 5*, 145–150.

Kuwada, K., Wamocho, L. S., Utamura, M., Matsushita, I., & Ishii, T., (2006). Effect of red and green algal extracts on hyphal growth of arbuscular mycorrhizal fungi, and on mycorrhizal development and growth of papaya and passionfruit. *J. Agron., 98*(5), 1340–1344. doi: https://doi.org/10.2134/agronj2005.0354.

Layek, J., Das, A., Idapuganti, R. G., Sarkar, D., Ghosh, A., Zodape, S. T., et al., (2018). Seaweed extract as organic bio-stimulant improves productivity and quality of rice in eastern Himalayas. *J. Appl. Phycol., 30*(1), 547–558. doi: https://doi.org/10.1007/s10811-017-1225-0.

Le Beu, P., (1983). *Contribution à L'étude des Algues Marines en Bretagne: Bilan de Leur Utilisation en Milieu Agricole*. Thesis, Faculté de Pharmacie de Tours, Tours, France.

Li, B. Y., Zhou, D. M., Cang, L., Zhang, H. L., Fan, X. H., & Qin, S. W., (2007). Soil micronutrient availability to crops as affected by long-term inorganic and organic fertilizer applications. *Soil Tillage Res., 96*(1), 166–173. doi: https://doi.org/10.1016/j.still.2007.05.005.

Lima, L. M. L., Fernandes, D. L., Souza, M. W. R., Melo, P. C., Teodoro, R. E. F., Luz, J. M. Q., Almeida, F. Q., & Carvalho, J. O. M., (2002). Doses de *Lithothamnion calcareum* sobre o desenvolvimento de plantas de pimentão. *Hortic. Bras., 20*(2), 347.

López-Benito, M., (1963). Estudio de la composición química del *Lithothamnion calcareum* (Aresch) y su aplicación como corrector de terrenos de cultivo. *Investigación Pesquera, 23*, 53–70.

Machado, L. P., Gasparoto, M. C. G., Filho, N. A. S., & Pavarini, R., (2019). Seaweeds in the control of plant diseases and insects. In: Pereira, L., Bahcevandziev, K., & Joshi, N. H., (eds.), *Seaweeds as Plant Fertilizer, Agricultural Biostimulants and Animal Fodder* (pp. 100–127). CRC Press, Boca Raton, FL.

McHugh, D. J., (2003). A guide to the seaweed industry. *FAO – Fisheries Technical Paper, 441*, 52–72.

Melo, P. C., & Furtini, N. A. E., (2003). Avaliação do *Lithothamnion* como corretivo da acidez do solo e fonte de nutrientes para o feijoeiro. *Ciênc. Agrotec., 27*(3), 508–519.

Melo, P. C., Abreu, C. G., Bahcevandziev, K., & Pereira, L., (2020a). Marine macroalgae bioextract changes the index of reflectance in pepper plants. *Oceanography and Fisheries Open Access Journal, 11*(5), 85–87. doi: https://doi.org/10.19080/OFOAJ.2020.11.555822.

Melo, P., Abreu, C., Bahcevandziev, K., Araujo, G., & Pereira, L., (2020). Biostimulant effect of marine macroalgae bioextract on pepper grown in greenhouse. *Appl. Sci., 10*(11), 4052. doi: https://doi.org/10.3390/app10114052.

Michalak, I., & Chojnacka, K., (2014). Algal extracts: Technology and advances. *Eng. Life Sci., 2*(14), 581–591. doi: https://doi.org/10.1002/elsc.201400139.

Miranda, L. N., (1985). Utilização de calcários marinhos como corretivos de acidez do solo. *Rev. Bras. Ciênc. Solo, 9*(1), 245–248.

Moreira, R. A., Ramos, J. D., Araújo, N. A., Marques, V. B., & Melo, P. C., (2015). Produtividade e teores de nutrientes em cladódios de pitaia vermelha utilizando-se adubação orgânica e granulado bioclástico. *Rev. Bras. Ciênc. Agrár. Recife, 7*(5), 714–719. doi: https://doi.org/10.5039/agraria.v7isa1551.

Mzibra, A., Aasfar, A., Benhima, R., Khouloud, M., Boulif, R., Douira, A., et al., (2021). Biostimulants derived from Moroccan seaweeds: Seed germination metabolomics and growth promotion of tomato plant. *J. Plant Growth Regul., 40*(1), 353–370. doi: https://doi.org/10.1007/s00344-020-10104-5.

Nabti, E., (2019). Why seaweeds can be utilized as biostimulants. *Adv. Biotech. Microl., 13*(3), 555862. doi: https://doi.org/10.19080/AIBM.2019.13.555862.

Nabti, E., Jha, B., & Hartmann, A., (2017). Impact of seaweeds on agricultural crop production as biofertilizer. *Int. J. Plasma Environ. Sci. Technol., 14*(5), 1119–1134. doi: https://doi.org/10.1007/s13762-016-1202-1.

O'Keeffe, E., Hughes, H., McLoughlin, P., Tan, S. P., & McCarthy, N. A., (2019). Antibacterial activity of seaweed extracts against plant pathogenic bacteria. *J. Bacteriol. Mycol., 6*(3), 1–11.

Palma, D. C., (2011). *Avaliação da Atividade Antibacteriana de Extratos Metanólicos de Diferentes Espécies de Algas e Plantas Halófitas* (pp. 1–33). MSc Thesis, University of Algarve, Faro, Portugal.

Pereira, L., & Cotas, J., (2019). Historical use of seaweed as an agricultural fertilizer in the European Atlantic area. In: Pereira, L., Bahcevandziev, K., & Joshi, N. H., (eds.), *Seaweeds as Plant Fertilizer, Agricultural Biostimulants and Animal Fodder* (pp. 1–22). CRC Press, Boca Raton, FL. doi: https://doi.org/10.1201/9780429487156.

Pereira, L., (2018). Biological and therapeutic properties of the seaweed polysaccharides. *International Biology Review, 2*(2), 1–50. doi: https://doi.org/10.18103/ibr.v2i2.1762.

Pereira, L., (2020). Characterization of bioactive components in edible algae. *Mar. Drugs, 18*(1), 1–6. doi: https://doi.org/10.3390/md18010065.

Pereira, L., Bahcevandziev, K., & Joshi, N. H., (2019). *Seaweeds as Plant Fertilizer, Agricultural Biostimulants and Animal Fodder* (p. 232). CRC Press, Boca Raton, FL. doi: https://doi.org/10.1201/9780429487156.

Raj, T. S., Graff, K. H., & Suji, H. A., (2018). Evaluation of various marine products against *Rhizoctonia solani* under *in vitro* condition. *Afr. J. Microbiol. Res., 12*(3), 46–51. doi: https://doi.org/10.5897/AJMR2016.8036.

Righini, H., Baraldi, E., García, F. Y., Martel, Q. A., & Roberti, R., (2019). Different antifungal activity of *Anabaena* sp., *Ecklonia* sp., and *Jania* sp. against *Botrytis cinerea*. *Mar. Drugs, 17*(5), 299. doi: https://doi.org/10.3390/md17050299.

Rodriguez, W. D. M., (2017). *Viabilidade técnica e econômica do uso de Lithothamnion e vinhaça associado à irrigação na cana-de-açúcar*. PhD Thesis, Universidade Federal de Goiás, Goiás, Brazil. Available online: http://repositorio.bc.ufg.br/tede/handle/tede/8212 (accessed on 13 October 2021).

Sahayaraj, K., & Jeeva, Y. M., (2012). Nymphicidal and ovipositional efficacy of seaweed *Sargassum tenerrimum* (J. Agardh) against *Dysdercus cingulatus* (Fab.) (Pyrrhocoridae). *Chil. J. Agric. Res., 72*, 152–156.

Sahayaraj, K., & Kalidas, S., (2011). Evaluation of nymphicidal and ovicidal effect of a seaweed, *Padina pavonica* (Linn.) (Phaeophyceae) on cotton pest, *Dysdercus cingulatus* (Fab.). *Indian J. Pharm. Sci., 40*(1), 5.

Savary, S., Bregaglio, S., Willocquet, L., Gustafson, D., D'Croz, D. M., Sparks, A., et al., (2017). Crop health and its global impacts on the components of food security. *Food Secur., 9*(2), 311–327. doi: https://doi.org/10.1007/s12571-017-0659-1.

Silva, L. D., Bahcevandziev, K., & Pereira, L., (2019). Production of bio-fertilizer from *Ascophyllum nodosum* and *Sargassum muticum* (Phaeophyceae). *J. Oceanol. Limnol., 37*(3), 918–927. doi: https://doi.org/10.1007/s00343-019-8109-x.

Soliman, A. S., Ahmed, A. Y., Abdel-Ghafour, S. E., El-Sheekh, M. M., & Sobhy, H. M., (2018). Antifungal bio-efficacy of the red algae *Gracilaria confervoides* extracts against three pathogenic fungi of cucumber plant. *Middle-East J. Appl. Sci., 8*(3), 727–735.

Souza, H. A. D., Ramos, J. D., Melo, P. C. D., Hafle, O. M., Rodrigues, H. C. D. A., & Santos, V. A. D., (2009). Avaliação de doses e produtos corretores da acidez em variáveis biométricas na produção de mudas de maracujazeiro. *Acta Sci., Agron., 31*, 607–612. doi: https://doi.org/10.1590/S1807-86212009000400009.

Stirk, W. A., & Staden, J. V., (2006). Seaweed products as biostimulants in agriculture. In: Critchley, A. T., Ohno, M., & Danilo, B. L. D. B., (eds.), *World Seaweed Resources – An Authoritative Reference System*. ETI Information Services Ltd., Amsterdam, Netherlands. Available online: https://linkinghub.elsevier.com/retrieve/pii/S0254629906000846 (accessed on 13 October 2021).

Stirk, W. A., Arthur, G. D., Lourens, A. F., Novák, O., Strnad, M., & Van, S. J., (2004). Changes in cytokinin and auxin concentrations in seaweed concentrates when stored at an elevated temperature. *J. Appl. Phycol., 16*(1), 31. doi: https://doi.org/10.1023/B:JAPH.0000019057.45363.f5.

Stirk, W. A., Novák, O., Strnad, M., & Van, S. J., (2003). Cytokinins in macroalgae. *Plant Growth Regul., 41*(1), 13–24. https://doi.org/10.1023/a:1027376507197.

Timilsena, Y. P., Adhikari, R., Casey, P., Muster, T., Gill, H., & Adhikari, B., (2015). Enhanced efficiency fertilizers: A review of formulation and nutrient release patterns. *J. Sci. Food. Agric., 95*(6), 1131–1142. https://doi.org/10.1002/jsfa.6812.

Torres, M. D., Kraan, S., & Dominguez, H., (2020). *Sustainable Seaweed Technologies – Cultivation, Biorefinery, and Applications* (p. 750). Elsevier, Amsterdam, Netherlands.

Traon, D., Amat, L., Zotz, F., & Du Jardin, P., (2014). *A Legal Framework for Plant Biostimulants and Agronomic Fertilizer Additives in the EU* (p. 133). Report to the European Commission, DG Enterprise and Industry. Available online: http://hdl.handle.net/2268/169265 (accessed on 13 October 2021).

Turan, M., & Köse, C., (2004). Seaweed extracts improve copper uptake of grapevine. *Acta Agric. Scand. B Soil Plant Sci., 54*(4), 213–220. doi: https://doi.org/10.1080/09064710410030311.

United Nations, (2015). *Transforming our World: The 2030 Agenda for Sustainable Development*. U. N. General Assembly. NY, US.

United Nations, (2018). *Contribution to the 2018 High Level Political Forum on Sustainable Development Submission from the UNCCD*. Available online: https://sustainabledevelopment.un.org/content/documents/17997UNCCD_contribution_to_HLPF_2018_FINAL.pdf (accessed on 13 October 2021).

Vallinayagam, K., Arumugam, R., Kannan, R. R. R., Thirumaran, G., & Anantharaman, P., (2009). Antibacterial activity of some selected seaweeds from Pudumadam coastal regions. *Glob. J. Pharmacol., 3*(1), 3.

Van, L. L. C., & Van, S. E. A., (1999). The families of pathogenesis-related proteins, their activities, and comparative analysis of PR-1 type proteins. Physiol. *Mol. Plant Pathol., 55*(2), 85–97. doi: https://doi.org/10.1006/pmpp.1999.0213.

Vernieri, P., Borghesi, E., Tognoni, F., Serra, G., Ferrante, A., & Piagessi, A., (2006). Use of biostimulants for reducing nutrient solution concentration in floating system. *Acta Hortic., 718*, 477–484. doi: https://doi.org/10.17660/ActaHortic.2006.718.55.

Wijesinghe, W. A. J. P., & Jeon, Y. J., (2012). Biological activities and potential industrial applications of fucose rich sulfated polysaccharides and fucoidans isolated from brown seaweeds: A review. *Carbohydr. Polym., 88*(1), 13–20. doi: https://doi.org/10.1016/j.carbpol.2011.12.029.

Yakhin, O. I., Lubyanov, A. A., Yakhin, I. A., & Brown, P. H., (2017). Biostimulants in plant science: A global perspective. *Frontiers in Plant Science, 7*.

Yokoya, N. S., Stirk, W. A., Van, S. J., Novák, O., Turečková, V., Pěnčík, A., & Strnad, M., (2010). Endogenous cytokinins, auxins, and abscisic acid in red algae from Brazil. *J. Phycol., 46*(6), 1198–1205. doi: https://doi.org/10.1111/j.1529-8817.2010.00898.x.

Yu, K. X., Wong, C. L., Ahmad, R., & Jantan, I., (2015). Mosquitocidal and oviposition repellent activities of the extracts of seaweed *Bryopsis pennata* on *Aedes aegypti* and *Aedes albopictus*. *Molecules, 20*(8), 14082–14102. doi: https://doi.org/10.3390/molecules200814082.

Zhang, W., Yamane, H., & Chapman, D. J., (1993). The phytohormone profile of the red alga *Porphyra perforata*. *Bot. Mar., 36*(3), 257–266. doi: https://doi.org/doi:10.1515/botm.1993.36.3.257.

Zodape, S. T., Gupta, A., Bhandari, S. C., Rawat, U. S., Chaudhary, D. R., Eswaran, K., & Chikara, J., (2011). Foliar application of seaweed sap as biostimulant for enhancement of yield and quality of tomato (*Lycopersicon esculentum* Mill.). *J. Sci. Ind. Res., 70*, 215–219.

CHAPTER 9

SEAWEEDS AS A NEW SOURCE OF BIOACTIVE COMPOUNDS AND POTENT BIOTECHNOLOGICAL APPLICATIONS

FATIMA EL KHALLOUFI[1] and BRAHIM OUDRA[2]

[1]*Department of Geography, Biology, and Geology, Polydisciplinary Faculty of Khouribga (FPK), Sultan Moulay Slimane University of Beni-Mellal, Khouribga – 25000, Morocco*

[2]*Department of Biology, Faculty of Sciences Semlalia Marrakech, Cadi Ayyad University, Marrakech – 40000, Morocco*

ABSTRACT

Seaweeds are considered a rich source of structurally different bioactive metabolites endowed with different biological activities. They have received more attention because of their potential activities, such as antibacterial, antifungal, antioxidant properties. Besides, they were highly coveted in agricultural, biomedical, and environmental domains. This current view leads to emphasize the importance of bioactive molecules derived from macroalgae. The development of these newly discovered molecules offers a promising future and prospects for biotechnological exploitation regarding their large spectrum of activities. Several research projects to highlight recent activities, in particular anti-obesity, anti-leishmanial, anti-malarial, and antiviral, are currently based on the exploration of seaweeds bioactive metabolites. In fact, it is the responsibility of applied scientific research, to promote sustainable development and to work on raising awareness on the different ways of developing and exploring these biological resources while proposing alternative solutions to the overexploitation of these capitals.

9.1 INTRODUCTION

Marine ecosystems present great biodiversity of significant natural resources that can be exploited. Most of this marine resource diversity is still unexplored despite their richness as a source of new metabolites likely to be the subject of various biotechnological applications. Seaweeds are among the significant currently exploited marine resources. They are very diverse and make up a heterogeneous group. About 9,000 species of algae have been identified and classified into three main divisions of red (Rhodophyta), green (Chlorophyta), and brown (Phaeophyta) based on their pigmentation (Miyashita et al., 2013).

Seaweeds had attracted increasing attention because of the chemical diversity within all taxonomic divisions offering a wide variety of metabolites (e.g., polysaccharides, proteins, and amino acids, phenolic compounds, pigments, and fatty acids) and gaining importance through several applications and biotechnological processes (Wang et al., 2017a; Zhao et al., 2018; Tanna and Mishra, 2019). Major compounds isolated from various seaweed species are reported to possess *in-vitro* and/or *in-vivo* biological properties and pharmacological activities, including antibacterial, antifungal, antiviral, anticancer, and others (Pérez et al., 2016; Wells et al., 2017; Zerrifi et al., 2018). These compounds represent the potential sources of new beneficial and therapeutic agents (Mayer et al., 2020). However, these biological activities could be influenced by different factors such as seaweed growth stages, life habitat, and season of collection and methods of extraction (Mishra et al., 2016). In fact, different other activities of seaweeds are explored, like anti-leishmanial, anti-malarial, and anti-obesity (Lehnhardt Pires et al., 2013; Murugan et al., 2016; Lee et al., 2020).

On the other side, seaweeds have been extensively exploited in the agriculture and environmental sectors. Farmers have used macroalgae for a long time as an amendment to improve soil structure or as a fertilizer to increase soil fertility (Alobwede et al., 2019). Seaweed-based biostimulants have been reported as environment-friendly strategies to improve crop yields and stresses tolerance (Frioni et al., 2018; Kulkarni et al., 2019). Bioactive compounds from seaweeds were also explored for their roles in plant protection, anti-phytopathogen activities, and induction of plants' resistance to biotic stress (Baloch et al., 2013; Ibraheem et al., 2017). As a potent biological eco-friendly process, seaweed bioactive metabolites have been exploited to deal with growing environmental concerns such as harmful algal blooms (HABs) and biofouling occurrence. Natural metabolites constitute a promising approach for controlling HABs and cyanobacteria in aquatic ecosystems (Schrader, 2003). Seaweeds have been shown to produce and release a wide range of allelopathic substances toxic to HABs species (red tides and HABs) (Jeong et al., 2000; Nakai et al., 2000). Considerable research based on the identification and the evaluation of the seaweeds antialgal potentials has been conducted (Sun et al., 2019) to develop an environmentally friendly, natural product-based antialgal agent. In addition, marine biofouling is considered a dominant phenomenon that occurs in the marine environment and causes serious problems worldwide. Bioactive substances from diverse seaweeds exhibit potential antifouling activity (Saha et al., 2018).

This chapter proposes a current review of the main biological applications of seaweeds bioactive metabolites by emphasizing the new molecules elucidated and the new perspectives for applications.

9.2 SEAWEEDS BIODIVERSITY

Algae are primary producers essential for the survival of many ecosystems. They colonize diverse environments and may establish a symbiotic association with fungi (lichens) (Suutari et al., 2010). Algae can be macroscopic or microscopic (from 0.5 microns to more than 60 meters). Macroalgae are macroscopic algae constituting a diverse multicellular polyphyletic group comprising photosynthetic eukaryotic organisms (Lee, 2008). They are structured on thallus with a holdfast, stipe, and blade; without true roots, stems, or leaves (Dhargalkar and Kavlekar, 2004).

Marine macroalgae or Seaweeds are photosynthetic non-flowering plant-like organisms. They are one among the dominant group of flora in the marine ecosystem (Ba-Akdah et al., 2016). These macroorganisms can be epiphytic, endophytic, endozoic or parasitic, but the majority of them are lithophytic. They are widely spread and sited in different zones, which range from tropical warm waters to freezing polar regions (Collins et al., 2016). The distribution and variety of marine algae are related to several environmental characteristics such as light exposure, temperature, depth, and tides. The habitat and location determine their content on different phytopigments because of the light intensity during the photosynthetic process. Therefore, species differ considerably in many ultrastructural and biochemical features.

More than 10000 species of seaweeds have been currently described worldwide (over 6000 of red algal species; approximately 2000 brown species; and around 1500 green algal species) (Collins et al., 2016; West et al., 2017). Classification of algae is based on their specific pigments (other than chlorophyll) besides other properties (cell wall chemistry, specific storage compounds, presence or absence of flagella, structure of their chloroplasts (Bocanegra et al., 2009). According to their specific pigments, they are then divided into three different divisions as Phaeophyceae (brown algae) which brown color is due to their fucoxanthin content, Chlorophyceae (green algae) dominated with Chlorophyll 'a' and 'b' and Rhodophyceae (red algae) comprising phycocyanin and phycoerythrin (Lee, 2008; Bocanegra et al., 2009; O'Sullivan et al., 2010; Collins et al., 2016).

9.3 SEAWEED BIOACTIVE METABOLITES

Nowadays, biological molecules have received more attention due to their biodegradability and their reduced environmental impact (Bilal et al., 2017). Nontoxic residues, short environmental longevity, and fewer environmental consequences are the primary qualities enhancing the importance of such molecules (Grzywacz et al., 2014). During normal growth conditions, primary metabolites of seaweeds like polysaccharides, proteins, and lipids take part in the main physiological functions. However, under stress conditions (ultraviolet radiation, salinity, water pressure) various secondary metabolites are produced (Senthilkumar et al., 2013; Lopes et al., 2016; Rosa et al., 2020). Based on their polyphyletic origin and their different marine environments; seaweeds are considered as a potential source of metabolites with more than 3000 different identified compounds in distinct seaweeds species (Leal et al., 2013; Belghit et al., 2017).

Seaweeds are endowed with a high number of bioactive compounds. They are rich in proteins (high content of essential amino acids), minerals (iodine, magnesium, iron, zinc, and calcium), vitamins, phytochemicals, polyunsaturated fatty acids (PUFA), and polysaccharides (carrageenan and alginate) (Jiao et al., 2011; Mohamed et al., 2012; Miyashita et al., 2013; Lange et al., 2015). To preserve a more adapted healthy lifestyle, they have been reported to produce several structural and chemical biomolecules for their defense mechanisms (Nylund et al., 2013; Sudatti et al., 2018). This production in terms of amount and type is species-dependent (Souza et al., 2011). Secondary metabolites like alkaloids, terpenes, pigments, polyphenols, and phlorotannins are largely distributed in seaweed species (Jormalainen and Honkanen, 2008; Antunes et al., 2011; Lange et al., 2015). Several bioactive metabolites from marine algae have been widely reported and reviewed for their biological activities. Those biomolecules included: Phlorotannins (Sanjeewa et al., 2016), polysaccharides (Wang et al., 2014), proteins, and peptides (Samarakoon and Jeon, 2012; Harnedy and FitzGerald, 2013), lipids, alkaloids (Güven et al., 2010), terpenoids (Wang et al., 2013) and pigments (Dumay et al., 2015).

9.3.1 SEAWEED POLYSACCHARIDES

Polysaccharides represent a very diverse group of molecules isolated from brown, green, or red macroalgae. Carbohydrates are present typically as sulfated and non-sulfated polysaccharides. They are considered as the major compounds in macroalgae; accounting for up to 76% dry weight (DW) (Holdt and Kraan, 2011; Paniagua-Michel et al., 2014). Seaweed's composition on polysaccharides is different depending on macroalgae species. Thereby Phaeophyceae are presenting alginic acid, laminarin, and fucoidan; Rhodophyceae contain mainly agar, carrageenans, xylans; whereas ulvans occur in Chlorophyceae (Kraan, 2012; Senthilkumar et al., 2017). The biological activities of polysaccharides are widely related to their structural properties (molecular type, size, glycosidic linkages, ratio of constituent monosaccharides, degree of sulfation, and distribution pattern of sulfate (Hu et al., 2013).

Seaweeds are a rich source of sulfated polysaccharides (PLS) (Cunha and Grenha, 2016) with several pharmacological activities (Lordan et al., 2011; de Souza et al., 2012; Aquib et al., 2019; Tanna and Mishra, 2019; Venkatesan et al., 2019; Cicinskas et al., 2020).

9.3.1.1 CARRAGEENANS

Carrageenans are PLS present in red seaweeds and account for 30–75% of algae DW (Vera et al., 2011). They are divided into six basic forms according to their structure. Carrageenans were found to contain repeating disaccharide units with 3-linked-β-D-galactopyranose, 4-linked-β-galactopyranose, or 3,6-anhydro-β-galactopyranose (Funami et al., 2007; Vera et al., 2011). The molecules feature a high molecular weight. κ-carrageenan and ι-carrageenan are mostly found in *Kappaphycus alvarezii* and *Eucheuma denticulatum*, respectively, while λ-carrageenan is more common to *Gigartina* and *Chondrus* (Campo et al., 2009).

9.3.1.2 FUCANS (FUCOIDAN)

Fucans are a group of important algal PLS. They constitute the major content of brown seaweed cell walls and occupy 5–20% of the seaweed DW (Cardoso et al., 2014). They are structured on 2-sulfate-O-fucose or 4-sulfate-O-fucose and ramifications at each two or three fucose residues, representing at all PLS with high molecular weight (Ale and Meyer, 2013). Fucans comprise xylofucoglycuronans, glycuronogalactofucans, and fucoidans.

Fucoidans represent the most studied group of fucans. The chemical structure of this compound is complex and different structures appear depending on the originated seaweed species (Zorofchian Moghadamtousi et al., 2014). They are primarily structured as the (1–3)-linked α-L-fucopyranosyl. The secondary structure consists of alternating α-(1–3) and α-(1–4)-linked L-fucopyranosyls (Pérez et al., 2016). In general, the composition of fucoidans varies with species and geographical origin (Cardoso et al., 2016). They are more abundant in *Ecklonia cava*, *Ascophyllum nodosum,* and *Undaria pinnatifida.*

9.3.1.3 ULVANS

Ulvans are the major polysaccharide isolated from green seaweeds (especially in the Ulvaceae family) representing 8–29% of algal DW (Andrieux et al., 1998). They are part of the cell wall composition with molecular weights varying from 189-8200 kDa. Ulvans are made up of repeating units of disaccharides (rhamnose and xylose) and iduronic acid or glucuronic acid (Lahaye and Robic, 2007; Rupérez et al., 2013; Cunha and Grenha, 2016).

9.3.2 SEAWEED LIPIDS AND FATTY ACIDS

Lipids are representing 0.6–4.14% of seaweeds DW (Rodrigues et al., 2015a). They are found as structural components of cellular membranes (polar lipids) and storage compounds (neutral lipids) (Pérez et al., 2016). The ability of algae to survive and proliferate in different environmental conditions is largely related to diversity and patterns of cellular lipid besides the modification of lipids metabolism in response to environmental variations (Guschina and Harwood, 2006). The average fatty acid profile of Chlorophyceae contains a proportion of 41.8% saturated, 14.1% monounsaturated, and 44.1% PUFA; Phaeophyceae shows 40% of saturated fatty acids, 20%

monounsaturated, and 40% PUFA. While Rhodophyceae shows a proportion of 42% of saturated fatty acids, 13% monounsaturated fatty acids, and 37% PUFA (Wielgosz-Collin et al., 2016). Despite the low algal lipid content, their constitution is interesting. Macroalgae receive more attention for their large and high amounts of PUFA (Kendel et al., 2015; Wielgosz-Collin et al., 2016). The PUFA content is similar to or higher than terrestrial plants (Kumari et al., 2010; Torres et al., 2019). α-linolenic acid (ω3), stearidonic acid (ω3), arachidonic acid (ω6) and eicosapentaenoic acid (EPA) (ω3) are among the most common PUFA found in marine macroalgae (Murata and Nakazoe, 2001; Dawczynski et al., 2007).

9.3.3 SEAWEED PROTEINS, PEPTIDES, AND AMINO ACIDS

Protein content and composition of seaweeds are species-dependent (Kumar et al., 2014). Red and green seaweeds have greater protein content (10–47% DW) unlike brown seaweeds (3–15% DW) (Wijesekara and Kim, 2015). The seaweed proteins content is generally a function of the species, season, and environment. Important bioactive proteins from red and green seaweeds including lectin, phycobiliprotein, and bioactive peptides from brown seaweeds have largely been studied.

The majority of seaweeds are an important source of essential and acidic amino acids such as glutamic acid and aspartic acid. Approximately 42% to 48% of the total amino acids in seaweeds are essential amino acids (Wong and Cheung, 2000). In the same way as the other metabolites, the amino acid composition of seaweed proteins is affected by seasonal variations and location (Marinho et al., 2015; Milledge et al., 2016).

9.3.4 SEAWEED PIGMENTS

Pigments in seaweeds comprise three classes: chlorophyll, carotenoid, and phycobiliproteins (phycocyanin and phycoerythrin). Chlorophyll is a green pigment with a fundamental role in the photosynthetic process (Aryee et al., 2018).

Carotenoids are a group of lipid-soluble compounds comprising carotenes (with carbon and hydrogen atoms) and xanthophylls (with at least one oxygen atom) (Tapiero et al., 2004; Krinsky and Johnson, 2005). The most common carotenes are β-carotene, while lutein, fucoxanthin, and violaxanthin belong to the xanthophylls class (Figure 9.1) (Zorofchian Moghadamtousi et al.,

2014; Aryee et al., 2018). The importance of carotenoids such as β-carotene, α-carotene, β-cryptoxanthin, lycopene, and fucoxanthin, have been reported in different seaweeds species (Rodriguez-Amaya, 2016; Susanto et al., 2019). The β-carotene at the same time was found in several species, for example, *Fucus serratus*, *F. vesiculosus*, *Laminaria digitata*, *Ulva* sp., *Chondrus crispus*. Besides, fucoxanthin is considered one of the most abundant carotenoids found in brown algae and contributes over 10% total production of carotenoids in nature (Gullón et al., 2020). This kind of molecule is related to neuroprotective, photoprotective, and hepatoprotective effects; in addition to anti-inflammatory, anticancer, and anti-obesity activities (Bae et al., 2020; Liu et al., 2020; Miyashita et al., 2020).

On the other hand, phycobiliproteins (water-soluble pigment) contain three different structural classes of molecules: phycocyanins (blue pigment), allophycocyanins (light blue pigment), and phycoerythrins (red pigment), this latter being the most abundant (Aryee et al., 2018).

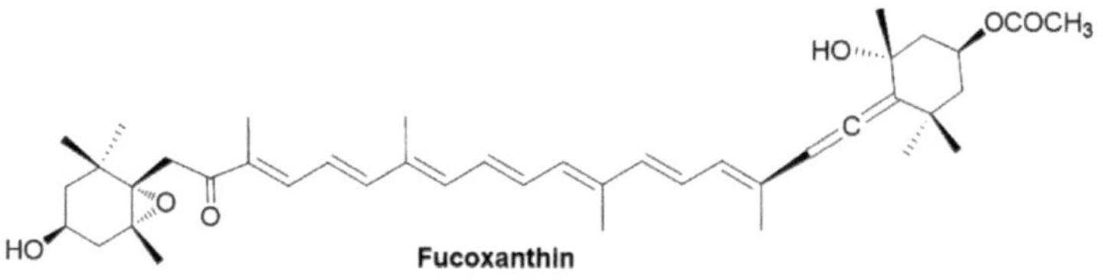

FIGURE 9.1 Chemical structure of fucoxanthin (Example of seaweed pigment) (from Liu et al., 2020).

9.3.5 *PHENOLIC COMPOUNDS AND PHLOROTANNINS*

Phenolic compounds are secondary metabolites widely distributed. They are divided into various groups and have several properties (Giada, 2013). In seaweeds, this group of compounds encompasses phenolic acids, tannins, flavonoids, catechins, and phlorotannins. The extent to which one type or another of phenolic compound is detected depends on the seaweed species.

The most studied algal polyphenolic compound is phlorotannins (Park et al., 2011). They are representing the major phenolic compounds found in brown algae (Steevensz et al., 2012; Machu et al., 2015). Whereas green and red algae contain mainly bromophenols, phenolic acids, and flavonoids (Gómez-Guzmán et al., 2018). Phlorotannins are reported to be extremely variable, heterogenic, and complex. This complexity is due to varying degrees of polymerization of phloroglucinol (1,3,5-trihydroxybenzene: the basic unit) (Kim et al., 2013). Nonetheless, the chemical properties of these marine polyphenols remind similar (Parys et al., 2007). Their molecular weight ranges from 126 Da to 650 kDa and they can be classified into four groups based on their linkages; (1) fuhalols and phlorethols (ether linkage), (2) eckols and carmalols (dibenzodioxin linkage), (3) fucophlorethols (ether and phenyl linkage), and (4) fucols (phenyl linkage) (Li et al., 2011; Martínez and Castañeda, 2013) (Figure 9.2). Phlorotannins were suggested to fulfill significant roles as basic components of the cell wall and function to provide chemical defenses against fluctuating environmental conditions (salinity level, nutrient, and light availability, ultraviolet (UV) radiation, and others) (Creis et al., 2015). To date, these compounds have been identified in many families of brown algae including the Alariaceae, Fucaceae, and Sargassaceae (Ferreres et al., 2012; Sathya et al., 2017). Several species like: *Ecklonia cava, Eisenia bicyclis, Sargassum thunbergii, Undaria pinnatifida,* and *Laminaria japonica;* have been cited for the pharmacological properties of phlorotannins (Rengasamy et al., 2014; Barbosa et al., 2019).

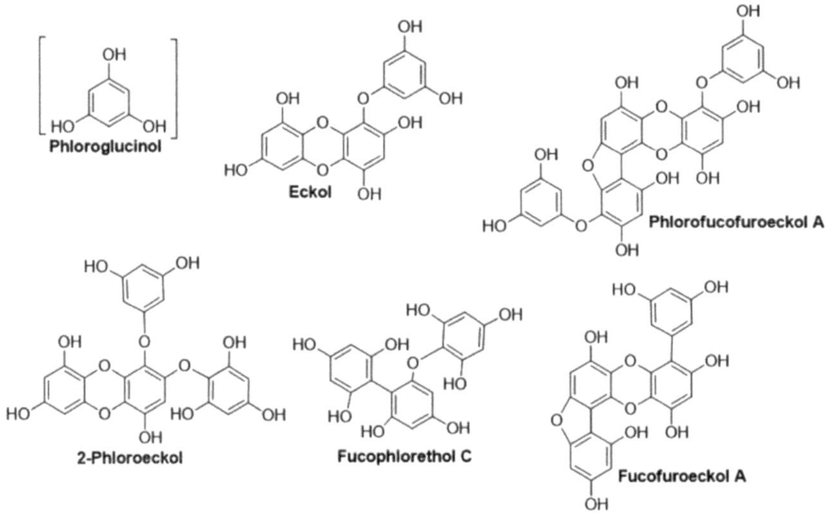

FIGURE 9.2 Chemical structures of Phloroglucinol (basic unit) and some Phlorotannins examples (Li et al., 2011; Barbosa et al., 2019).

Besides phlorotannins, other phenolic compounds have been reported in the composition of macroalgae (even less studied) (Farasat et al., 2014; Mannino et al., 2016). The presence of phenolic compounds such as gallic acid, hydroxybenzoic acid, catechin, epicatechin, has been revealed in different brown, red, and green species of macroalgae (Machu et al., 2015; Montero et al., 2018). Several studies have also reported the content of phenolic acids (Ying-Ying et al., 2015), flavonoids (Yoshie-Stark et al., 2003) and other phenolic compounds (Aigner et al., 2013). In fact, phlorotannins have been associated with different pharmacological and biological properties; among them the antimicrobial, antioxidant, anti-diabetic, anti-inflammatory, anti-allergic, and others (Lee et al., 2008; Heo and Jeon, 2009; Choi et al., 2010; Lee et al., 2012a,b; Wei et al., 2016; Sathya et al., 2017; Sugiura et al., 2018).

9.4 BIOMEDICAL AND PHARMACOLOGICAL SEAWEED EXTRACTS APPLICATIONS

In the last few years, seaweeds received more attention as a new promising source of novel bioactive compounds that can be used for drug development. Seaweeds may produce a wide range of bioactive compounds with a large spectrum of biological activities, such as antibacterial, antifungal, antiviral, antioxidant, anti-inflammatory, and anticancer (Kim et al., 2008; Pádua et al., 2015; Belattmania et al., 2016; Fu et al., 2016; Shi et al., 2017; Aquib et al., 2019; Antony and Chakraborty, 2020).

9.4.1 POTENTIAL ANTIMICROBIAL AND ANTIVIRAL SEAWEED ACTIVITIES

9.4.1.1 ANTIMICROBIAL ACTIVITY OF SEAWEEDS

Antibacterial activity of seaweed bioactive molecules was widely reported, as illustrated in Table 9.1. Extracts from 13 species of seaweeds were tested for their antimicrobial activities. Acetonic extracts of *Sargassum polyceratium* induced a strong antibacterial activity against *Staphylococcus aureus* (Borbón et al., 2012). In another study phlorofucofuroeckol-A, a polyketide from the brown alga *Eisenia bicyclis* has been tested against methicillin-resistant *Staphylococcus aureus* (MRSA). The obtained results revealed the inhibition of the methicillin resistance-associated genes expression at IC_{50} of 32 μg/ml and suppression of penicillin-binding protein 2a (PBP2a) production (Eom et al., 2014). Pereira et al. (2014) reported the inhibition of *Pseudomonas aeruginosa* biofilm formation by furanones (alkaloid) obtained from marine algae *Delisea pulchra* at 1.3 μM. Besides, aqueous extract of the genus *Sargassum* (*S. wightii*) showed moderate antibacterial activity against the Gram-negative bacteria *Escherichia coli* (Kumaresan et al., 2018). Methanol extract of *Laurencia iliformis* presented promising antibacterial activity against *Pseudomonas aeruginosa*, *Staphylococcus aureus*, *Streptococcus pyogenes*, and *Escherichia coli*; while the methanol extract of *Ulva reticulata* exhibited a strong effect against *Pseudomonas aeruginosa* (Begum et al., 2018). Mishra (2018) pointed out the positive results of four seaweeds *Sargassum wightii*, *Gracillaria edulis*, *Gracilaria corticata,* and *Ulva lactuca* from India, against *Pseudomonas aeruginosa, Escherichia coli,* and *Staphylococcus aureus.* It was suggested that the antibacterial activity of seaweeds was related to the ability of the active metabolite to bind with bacterial proteins such as enzymes and cell membranes (Pérez et al., 2016).

Biosynthesized nanoparticles (NPs), recognized as an effective antibiofilm, were tested against *Pseudomonas aeruginosa* biofilms responsible for antibiotic resistance (Khan et al., 2019). Gold NPs stabilized using fucoidan exerted activities in concentration depending manner with a MIC of 512 μg/ml. Antibacterial activity was detected at concentrations above MIC, whereas sub-inhibitory concentration inhibited biofilm formation without effect on bacterial growth (Khan et al., 2019). Most recently, silver nanoparticles (AgNPs) synthesized using the marine seaweed *Gracilaria corticata* have been tested against the biofilm-producing bacteria *Klebsiella pneumonia.* Significant antibacterial activity and biofilm reduction were recorded at 50 μg/mL and 100 μg/mL concentrations of Ag NPs respectively (Rajivgandhi et al., 2020). On the other side, marine natural products isolated from marine bacteria, dinoflagellates, sponges, sea cucumbers, and algae, are described to have a potent antifungal activity (Mayer et al., 2020). Through this section, only antifungal compounds from seaweeds are reported.

Seaweed's antifungal potential was highlighted in several studies (Peres et al., 2012; Arumugam et al., 2019; Arumugam and Rajendran, 2019). The antifungal activity of two new laurane-type sesquiterpenes: laurepoxyene, and 3 β-hydroxyperoxyaplysin, and a new PUFA ethyl ester isolated from the red alga *Laurencia okamurai* caused *Candida glabrata* inhibition with IC50 at 2–4 μg/ml (Yu et al., 2014; Feng et al., 2015). Lopes et al. (2013) screened antifungal activity of purified phlorotannins obtained from brown seaweeds (*Cystoseira nodicaulis, Cystoseira usneoides,* and *Fucus spiralis*). The obtained results showed fungistatic and fungicidal activity against yeast and dermatophytes, respectively. The most sensitive species were *Candida albicans* ATCC 10231, *Epidermophyton floccosum,* and *Trichophyton rubrum* (Lopes et al., 2013). The anti-candidal activity of the seaweeds

TABLE 9.1 Antibacterial Activity of Seaweeds

Seaweeds	Bioactive extract or molecule	Bacteria targeted	Main effects	References
Laurencia nangii	Neolaurene (Terpenoide)	*Salmonella typhi* and *Staphylococcus aureus*	Growth inhibition with IC_{50} of 7.5 µg/ml	Kamada and Vairappan, 2015
Sargassum wightii and *Halimeda gracilis*	Extracts	*Pseudomonas aeruginosa*, *Vibrio parahaemolyticus* and *Escherichia coli*	Antibacterial effect on gram-negative with MIC of 50 µg/ml Inhibition of bacterial pathogens at 25 and 50 µg/ml	Suganya et al., 2019
Sphaerococcus coronopifolius	Bromoditerpenes (Sphaerodactylomelol and 4 other sphaerane bromoditerpenes)	*Escherichia coli*, *Pseudomonas aeruginosa*, *Staphylococcus aureus*, and *Candida albicans*	High antibacterial of Sphaerodactylomelol on *S. aureus* (IC_{50} 96.3 µM) High antibacterial activity of sphaerane bromoditerpene on *S. aureus* (IC_{50} 6.35 µM)	Rodrigues et al., 2015b
Sargassum tenerrimum	Fucoidan	*Enterococcus faecalis* and *Escherichia coli*	Potent antibacterial effect of depolymerized fucoidans	Ashayerizadeh et al., 2020
Dictyopteris polypodioides	Essential oil and volatile fraction	*Staphylococcus aureus*, *Agrobacterium tumefaciens*, *Salmonella typhimurium*, *Bacillus cereus*, *Micrococcus luteus* and *Escherichia coli*	High inhibitory effect of essential oil against all tested bacteria *S. aureus* the most sensitive with MIC of 1519 µg/ml	Riad et al., 2020

TABLE 9.1 *(Continued)*

Seaweeds	Bioactive extract or molecule	Bacteria targeted	Main effects	References
Sargassum latifolium, Sargassum platycarpum and Cladophora socialis	Methanol and acetone extracts	*Escherichia coli, Salmonella* sp., *Staphylococcus xylosus, Staphylococcus aureus, Bacillus subtilis, Enterococcus faecalis* and *Candida albicans*	*S. latifolium* and *S. platycarpum* methanolic extracts more active against gram-positive than gram-negative bacteria *C. socialis* methanolic extract with important effect on methicillin-resistant *Staphylococcus aureus* (MRSA)	Moubayed et al., 2017
Stypocaulon scoparium and *Halopitys incurvus*	Aqueous and Ethanol extract	*Staphylococcus aureus, Bacillus subtilis, Escherichia coli* and *Pseudomonas aeruginosa*	Ethanol extract of *S. scoparium* with remarkable antibacterial activity against all pathogenic bacteria	Khelil-Radji et al., 2017
Ulva lactuca	Methanol extract	*Bacillus subtilis, Corynebacterium diphtheria, Staphylococcus aureus, Escherichia coli, Pseudomonas aeruginosa* and *Salmonella paratyphi*	Potent antibacterial activity against Gram-positive and Gram-negative bacteria strains	Alagan et al., 2017

IC_{50}: Concentration of a compound required for 50% inhibition *in vitro* and MIC: Minimum Inhibitory Concentration.

collected from the Gulf of Mannar (India) was evaluated with well diffusion and micro-dilution techniques. The methanolic extracts of *Dictyota bartayresiana* showed a zone of inhibition of 17.4 ± 0.62 mm against *Candida albicans*. Additionally, all tested extracts showed a fungistatic mechanism of inhibition of *Candida albicans* (Arumugam et al., 2019). In another study, Arumugam and Rajendran (2019) reported that *Stocheospermum marginatum* extracts showed a significant zone of inhibition up to 50 μg/ml against *Candida krusei* and other tested *Candida* strains.

Over all, seaweed's antimicrobial activity is influenced by different factors like the season of collection, the habitat, macroalgae growth stages, besides methods of extraction (Rosaline et al., 2012; Mishra et al., 2016).

9.4.1.2 ANTIVIRAL ACTIVITY OF SEAWEEDS

The marine environment is a rich source of biological and chemical compounds with antiviral activity. Secondary metabolites of algae have been tested as antiviral agents for different viruses of medical and veterinary importance (Aguilar-Briseño et al., 2015; Shi et al., 2017).

Seaweed polysaccharides have been extensively studied for medical antiviral activity as reported in Table 9.2. They have been described as inhibitors of viral attachment, internalization, uncoating, transcription, and replication (Shi et al., 2017). Gheda et al. (2016) evaluated the antiviral effects related to seaweed polysaccharides against Hepatitis C virus. *Sargassum vulgare* extract showed the potentiality to scavenge the free radicals (up to 81.5%) and *Laurencia obtusa* inhibited the Hepatitis C virus by 82.36% (Gheda et al., 2016).

Fucoidan extracted from *Dictyota bartayesiana* and *Turbinaria decurrens* were evaluated for anti-human immunodeficiency virus (anti-HIV) activity. Crude and purified fucoidan samples exhibited inhibitory activity of 90.5% and 89% for *Dictyota bartayesiana* and 89.7% and 92% for *Turbinaria decurrens* (Sanniyasi et al., 2019). Fucoidan was reported to inhibit HIV-1 and HIV-2 as well as further enveloped viruses like Herpes Simplex Virus type-1 (HSV-Type 1), Influenza virus A and B, Measles virus (MeV), and respiratory syncytial virus (RSV). The sulfated polysaccharide inhibited the fusion of the virus and the host cell and viral adsorption to the host cells (Esteves et al., 2011). PLS from *Sargassum fluitans* showed significant cytotoxic and antiviral activities against HSV-Type 1 with EC_{50} of 42.8 mg/ml and antiviral activity at a multiplicity of infection without cytotoxicity.

TABLE 9.2 Examples of Research on Antiviral Activity of Seaweed Compounds

Seaweeds	**Bioactive extract or molecule**	**Virus targeted**	**Main effects**	**References**
Dictyota pfaffii	Dolabelladienol A (Diterpene)	Human Immunodeficiency Virus 1 (HIV-1)	Virus inhibition with IC_{50} of 2.9 μM	Pardo-Vargas et al., 2014
Eisenia bicyclis	Phlorofucofuroeckol-A (Phlorotannin)	Murine norovirus	Virus inhibition with IC_{50} of 0.9 μM	Eom et al., 2015
Dictyota pfaffii	Dolabelladienetriol (Diterpene)	Herpes Simplex Virus-Type 1 (HSV-1)	Similar efficacy on the anti-HSV-1 activity to the Aciclovir drug	Garrido et al., 2017
Osmundaria obtusiloba	Ethanol extract	Chikungunya Virus (CHIKV)	CHIKV inhibition of 1.25 μg/mL	Cirne-Santos et al., 2019
Ecklonia cava	Phlorofucofuroeckol (Phlorotannin)	Influenza A viral strains (H1N1 and H9N2)	Potent antiviral activity with EC_{50} value of 13.48 ± 1.93 μM	Cho et al., 2019
Scytosiphon lomentaria	Galactofucan (Polysaccharide)	Herpes Simplex Virus-Type 1 and 2 (HSV-1 and HSV-2)	Fractions with different viral inhibitory properties	Ponce et al., 2019
Monostroma nitidum	Glucuronorhamnan (Polysaccharide)	Enterovirus 71 (EV71)	Inhibition of EV71 infection and down-regulation of host phosphoinositide 3-kinase / protein kinase B signaling pathway	Wang et al., 2020
Fucus vesiculosus	Fucoidan (Polysaccharide)	Hepatitis B Virus	Reduction of virus replication and activation of the extracellular signal-regulated kinase	Li et al., 2017

The observed activity was linked to the sulfatation, molecular weight, and carbohydrate nature of polysaccharides (Bedoux et al., 2017).

Pholorotannins also exhibited antiviral activity against several viruses like HIV-1, murine norovirus, human papilloma-virus (Ahn et al., 2004; Eom et al., 2015; Kim and Kwak, 2015). The antiviral activity of Polyphenol-rich extracts (PPs) from seaweeds was tested against MeV. Then the best polyphenols were combined with ribavirin and with PLS to discover potent antiviral properties (Morán-Santibañez et al., 2018). In this study, the virucidal activity of PPs extracts was reported by inactivating the viral particle. The observed inhibitory activity reached 83–89% of inhibition for both PPs (*Ecklonia arborea* and *Solieria filiformis*) (Morán-Santibañez et al., 2018).

The situation in 2020 with the global SARS-COV2 pandemic has more than ever emphasized the importance of having an arsenal of effective antiviral molecules to deal with emerging viral pathologies. At least, they are considered the only effective, but not foolproof, way to eradicate viruses, while waiting for the development of a vaccine (Roberts, 2020).

9.4.2 ANTIOXIDANT ACTIVITY OF SEAWEED EXTRACTS

Oxidative stress represents a state of imbalance between the production of radicals or reactive oxygen species (ROS) and the antioxidant cellular capacities in favor of ROS. The latter, are considered toxic products from normal oxygen metabolism and are significant contributors to many pathologies like cancer (Halliwell, 2013) and neurodegenerative diseases, mainly Alzheimer's and Parkinson's diseases (Clark et al., 2010).

Several studies have examined the antioxidant activity of seaweed derived natural products. Antioxidants are the most abundant among all biological compounds in seaweeds; including endogenous antioxidants (enzymes and proteins) and others exogenous (Polyphenols, vitamin C) (Chang and Teo, 2016). Lee et al. (2003) evaluated the antioxidant activity of fucosterol from *Pelvetia siliquosa* (Figure 9.3). They reported a significant increase of free radical scavenging enzyme activities like glutathione peroxidase (GPx), superoxide dismutase (SOD), and catalase (CAT). Zubia et al. (2007) reported the antioxidant potential of different seaweed species. *Avrainvillea longicaulis* portrayed the most potent one with high phenolic content ($3.36 \pm 0.05\%$ DW) and low oxidation index EC_{50} (1.44 ± 0.01 mg/l). Extracts from *Chaetomorpha* were evaluated for their antioxidant activity using the 2, 2-diphenyl-1-picrylhydrazyl radical scavenging activity and total flavonoid, phenolic, and tannin content assays. The results revealed that ethanol extracts of *Chaetomorpha* exhibited higher antioxidant activity IC_{50} (9.41 ± 0.54 mg/ml) and showed totals of 189.14 ± 0.99 mg of Quercetin equivalent QE/g, 21.92 ± 0.43 mg of Gallic acid equivalent (mg GAE/g), and 21.81 ± 0.04 mg GAE/g of flavonoid, phenolic, and tannins contents respectively (Haq et al., 2019).

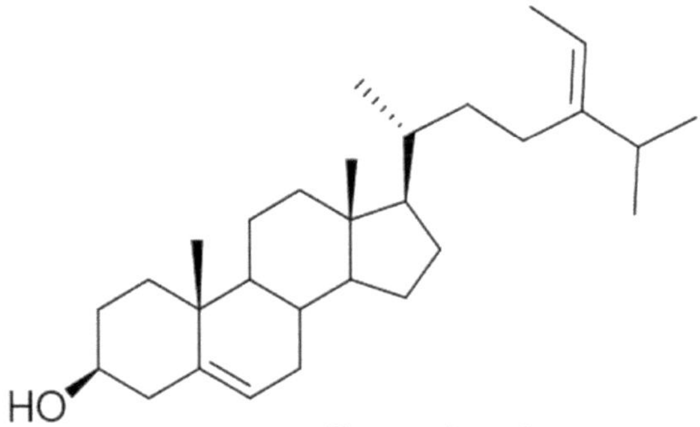

FIGURE 9.3 Chemical structure of antioxidant compound Fucosterol (Lee et al., 2003).

The antioxidant capacities of seaweeds (*Sargassum polycystum, Eucheuma denticulatum, Kappaphycus alvarezzi* var. *Buaya,* and *Kappaphycus alvarezzi* var. *Giant*) were also documented (Fu et al., 2016). Among the tested species *Sargassum polycystum* exhibited the most antioxidant activities (2.00 ± 0.01 μmol TEAC/g DW and 0.84 ± 0.01 μmol TEAC/g DW). Umavandhana and Jayanthi (2018) evaluated the antioxidant activity for *Sargassum swartzi* and *Enteromorpha compressa*. The obtained results showed high free radical scavenging activity of *Sargassum swartzi* aqueous extract. In another work, seven green seaweeds were investigated for their total flavonoid and phenolic content, besides antioxidant, scavenging, and reducing activities (Tanna et al., 2018). *Caulerpa* spp. exhibited potential total antioxidant activity over 50%, high potential scavenging activity above 60%, and maximal reducing activity (>78%) (Tanna et al., 2018). Antioxidant potential of seaweeds was positively correlated to the total phenolic content (Chai et al., 2015; Pinteus et al., 2017), inversely other studies suggested no direct correlation between the antioxidant activity and phenolic content (Lim et al., 2002; Mamelona et al., 2007).

9.4.3 ANTICANCER POTENTIALS OF SEAWEEDS

Cancer is a complex disease in which cells divide uncontrollably, invade tissues, and form tumors. The causes can be infectious, medicinal, linked to environmental toxins, diet, and healthy lifestyle, hereditary genome mutations, hormones, or even diseases or failures of the immune system (Gutiérrez-Rodríguez et al., 2018). The cancer incidence is increasing with 18 million new cases worldwide, while the mortality due to cancer reached 9.6 million people (Bray et al., 2018). Chemotherapy on cancer cells is not specific, so way research on finding novel, efficient, and nontoxic chemotherapeutic drugs based on natural sources (like marine environments) has been of great interest (Vaikundamoorthy et al., 2018).

Several compounds isolated from macroalgae such as terpenes, polysaccharides, and polyphenols showed potent anticancer activity *in vivo* and *in vitro* as noted in Table 9.3 (Peng et al., 2011; Senthilkumar et al., 2013; Atashrazm et al., 2015).

Polysaccharides such as fucoidan, alginate, agarose, and carrageenan, which have immunomodulatory and anticancer effects, have attracted more consideration in the biochemical and medical fields (Ooi and Liu, 2000). Fucoidan and carrageenan obtained respectively from brown or red seaweeds exhibit antitumor and immunomodulating activities (Aquib et al., 2019; Cicinskas et al., 2020). Fucoidan has been reported with promising anticancer and antitumor activities (Lin et al., 2020). Multiple studies have indicated that fucoidan induces cytotoxicity against various cancer cells by preventing cell invasion, angiogenesis, metastasis, and cell cycle arrest; upregulating growth signaling molecules implicated in various cellular regulatory mechanisms; and inducing apoptosis (Senthilkumar et al., 2013).

The anticancer activity has been reported in different types of cancers such as lung cancer, gastric cancer, colon cancer, breast cancer, prostate cancer, and liver cancer (Han et al., 2008; Yamasaki-Miyamoto et al., 2009; Vishchuk et al., 2013; Mak et al., 2014; Choo et al., 2016; Hsu at al., 2017; Yan et al., 2019).

Fucoidan has potent antiproliferative activity against human lung adenocarcinoma cell line A549 (Boo et al., 2011). The sulfated polysaccharide isolated from *Undaria pinnatifida* displayed an anticancer effect against the breast cancer cell line T-47D and the melanoma cancer cell line SK-MEL-28 (Vishchuk et al., 2011). An anti-tumor effect was manifested by inhibition of cell aggregation for T-47D and SK-MEL-28 lines. They also inhibit the viability of B16 murine melanoma cell line by activating apoptosis (IC_{50} 530

TABLE 9.3 Potential Anticancer Activity of Seaweeds Metabolites

Seaweeds	Bioactive extract or molecule	Related disease	Cell lines tested	Main effects	References
Eucheuma cottonii	Polyphenol-rich extract	Breast adenocarcinoma	MCF7 and MDA-MB-231	- Apoptosis induction and downregulation of the endogenous estrogen biosynthesis - Antioxidative improvement	Namvar et al., 2012
Undaria pinnatifda	Fucoxanthin (Carotenoid)	Human gastric adenocarcinoma	SGC-7901 BGC-823	- Proliferation inhibition and apoptosis induction - Arrest of cell cycle at S phase for SGC-7901 cells and at G2/M phase for BGC-823 cells	Yu et al., 2018
Laminaria japonica	Fucoxanthin (Carotenoid)	Lung cancer	Human non-small cell lung cancer cells (NSCLC)	- *In vivo* anti-lung cancer and arrest cell cycle - Apoptosis induction	Mei et al., 2017
Sargassum muticum	Phlorotannins	Adenocarcinoma colon cancer	HT-29 cells	- Antiproliferative activity and cytotoxic potential	Montero et al., 2016
Ecklonia cava	Dieckol (Phlorotannin)	Ovarian cancer	A2780 and SKOV3	- Cytotoxic effect, SKOV3 apoptosis induction and tumor growth suppression - Mitochondrial dysfunction and suppression of anti-apoptotic proteins levels	Ahn et al., 2015

TABLE 9.3 *(Continued)*

Seaweeds	Bioactive extract or molecule	Related disease	Cell lines tested	Main effects	References
Fucus vesiculosus	Fucoidan (Product F5631)	Human leukemia	U937	- Inhibition of growth and induction of apoptosis - Activation of caspases and loss of mitochondria membrane potential	Park et al., 2013
Sphaerococcus coronopifolius	Bromoditerpenes (Sphaerodactylomelol and 4 other sphaerane bromoditerpenes)	Human hepatocellular carcinoma	HepG-2	- Inhibition of cell prolif-eration (IC_{50} 280 μM) and cytotoxicity (IC_{50} 720 μM) by sphaerodactylomelol	Rodrigues et al., 2015b

± 4.3 μg/ml) (Wang et al., 2017b). They seem to be involved in stopping or reducing cell proliferation, cell migration, tumor angiogenesis, invasion, and metastasis (Déléris et al., 2016). Besides, they showed the ability to stop the cell cycle in certain tumor cell lines, suggesting an effect on the cyclines (Déléris et al. 2016). Moreover, an extracted polysaccharide from *Sargassum vulgare* induced a potent inhibition on cells proliferation and antiangiogenic activity (Dore et al., 2013). Anticancer and apoptotic activity of fucoidan from *Sargassum cinereum* have been elucidated also against colon cancer cell line Caco-2 (Narayani et al., 2019). Caco-2 cell growth was inhibited in a dose-dependent manner with an IC_{50} concentration of 250 μg/ml. Besides, fucoidan induced cell apoptosis and increases ROS production and permeability of mitochondrial membrane (Narayani et al., 2019).

Oligo-fucoidan from *Sargassum hemiphyllum* was demonstrated to exhibit hepatocellular carcinoma apoptosis. Anti-hepatoma cell (HepG2 Cells) activity was revealed by cell cycle arrest and activation of caspases as apoptotic pathways inducing cell death (Yan et al., 2019). Carrageenans have an anti-angiogenesis and pro-apoptotic effect. Carrageenans obtained from *Kappaphycus alvarezii* effectively inhibited the growth of different cell lines (breast, colon, liver, and osteosarcoma) (Ariffin et al., 2014; Suganya et al., 2016). However, epidemiological data suggest a link between carrageenans (degraded and non-degraded) and intestinal ulcers and neoplasms (Déléris et al., 2016). Polysaccharides from *Ulva lactuca* exhibited an interesting anticancer activity against breast cancer (Abd-Ellatef et al., 2017).

More recently, seaweed polysaccharides from *Sargassum wightii*, *Enteromorpha compressa,* and *Acanthophora spicifera* (from India) were tested against the HT-29 human colon cancer cell line (Praveen et al., 2020). *Acanthophora spicifera* polysaccharide seaweed extract at 100 μg/ml presented a high *in vitro* cytotoxic effect on HT-29 cells up to 52.13 ± 1.4% (Praveen et al., 2020). Fucoxanthin, a non-polar carotenoid, was also reported to have an antiproliferative effect, inducing apoptosis and anti-angiogenesis and able to stop the cell cycle. This molecule, therefore, has significant potential in the prevention and treatment of cancer. The Fucoxanthin extracts from *Dictyota indica* (at 50 μg/ml) presented effective anticancer activity on the breast cancer cell line, without toxic effects to the normal cells (Karkhane Yousefi et al., 2018).

Phlorotannins from brown algae also show promising effects. It was emphasized that eckol isolated from *Ecklonia maxima* showed important cytotoxic activity against HeLa cells, H157, and MCF7 cell lines (Mwangi et al., 2017). Purified phlorotannin from *Ascophyllum nodosum* inhibited strongly the proliferation of colon cancer cells with IC_{50} values of 33 μg/ml (Nwosu et al., 2011). Furthermore, Alarif et al. (2016) reported the antiproliferative activity of three sesquiterpenoids (eudesma-4(15),7-diene-5,11-diol, teuhetenone, and chabrolidione B) from *Laurencia obtuse* against MCF-7 cell lines. As such, the most promising activity was shown by teuhetenone. Ethanol extracts of *Chaetomorpha* showed the highest anticancer activity through growth inhibition of MDA-MB-231 breast cancer cells. The low IC_{50} obtained was 225.18 ± 0.61 μg/ml. Extracts analysis indicated the presence of active antitumor compounds such as Dichloracetic acid, Oximes, and L-α-Terpinol (Haq et al., 2019).

9.4.4 ADDITIONAL POTENT BIOLOGICAL ACTIVITIES OF SEAWEED METABOLITES

Additionally to the above-cited activities, seaweeds have been explored for properties with potential health benefits in perspective of developing upcoming drug agents. A new path of the biological function of seaweed metabolites has emerged related to their potential applications in the treatment of diseases such as obesity, diabetes, malaria, leishmaniasis, as reported in Table 9.4.

TABLE 9.4 Summary of Potent Biological Activities of Seaweeds Biomolecules

Seaweeds	Bioactive extract or molecule	Biological activity	References
Ecklonia stolonifera	Phlorofucofuroeckol A (Phlorotannin)	Anti-obesity	Jung et al., 2014
Sargassum thunbergii	Indole derivatives	Anti-obesity	Kang et al., 2017
Plocamium telfairiae	Ethanol extracts	Anti-obesity	Lu et al., 2020
Grateloupia elliptica	Ethanol extracts	Anti-obesity	Lee et al., 2020
Ishige okamurae	Diphlorethohydroxycarmalol (Phlorotannin)	Anti-diabetes	Heo et al., 2009
Laminaria japonica and *Hizikia fusiforme*	Extracts	Anti-diabetes	Kang et al., 2018
Undaria pinnatifida	Fucoidan	Anti-diabetes	Sim et al., 2019
Sargassum confusum	Oligosaccharides	Anti-diabetes	Yang et al., 2019b
Undaria pinnatifida	Fucoidan	Anti-plasmodial Anti-malarial	Chen et al., 2009

TABLE 9.4 *(Continued)*

Seaweeds	Bioactive extract or molecule	Biological activity	References
Sargassum wightii	Zinc oxide nanoparticles (ZnO NPs)	Anti-malarial	Murugan et al., 2018
Valoniopsis pachynema	CdS nanoparticles	Anti-malarial Anti-plasmodial	Sujitha et al., 2017
Laurencia dendroidea	Elatol (Sesquiterpene)	Anti-leishmanial	dos Santos et al., 2010
Solieria filiformis, *Botryocladia occidentalis*, *Caulerpa racemosa* and *Gracilaria caudata*	Sulfated polysaccharides	Anti-leishmanial	Lehnhardt Pires et al., 2013
Cystoseira baccata and *Cystoseira barbata*	Extracts	Anti-leishmanial	de Sousa et al., 2018

9.4.4.1 ANTI-OBESITY AND ANTI-DIABETIC ACTIVITIES OF SEAWEED EXTRACTS

Obesity is defined as the excessive or abnormal accumulation of body fat in the adipose tissue, energy imbalance, and lipogenesis (Ojulari et al., 2020). Different studies reported the anti-obesity effects of seaweed metabolites (Seca and Pinto, 2018).

The anti-obesity potential effect of *Ecklonia stolonifera* was related to phlorotannins by reducing lipid accumulation in 3T3-L1 cells. Phlorofucofuroeckol A was the most active with IC_{50} of 17.86 μM. The study reported that phlorotannins suppressed CCAAT/enhancer-binding protein α (C/EBP α) and peroxisome proliferator-activated receptor γ (PPARγ) expressions (Jung et al., 2014). Besides, fucosterol from *Ecklonia stolonifera* exhibited inhibition of adipogenesis preadipocytes of 3T3-L1 cells, downregulation of sterol regulatory element-binding protein 1 (SREBP1), and modulation of multiple signaling pathways (Lee et al., 2017).

Indole derivatives isolated from *Sargassum thunbergii* have been tested for their potential anti-obesity effects (Kang et al., 2017). The results showed inhibition of the 3T3-L1 adipocytes differentiation and lipid accumulation as well as downregulation of the expression of PPARγ and C/EBP α, and sterol regulatory element-binding protein 1c (SREBP-1c) (Kang et al., 2017). Sharma et al. (2017) showed that ethanol extract of *Caulerpa okamurae* induced significant inhibition of lipid accumulation and reduction of the expression of PPAR-γ, SREBP-1c, and C/EBPα in 3T3-L1 adipocytes. The *in vivo* conducted experiment revealed a decrease in body weight, fat weight, and liver weight in HFD-fed mice. Additionally, *Caulerpa okamurae* extracts caused a decrease in free fatty acids, triglyceride, total cholesterol, glucose, and insulin in the plasma (Sharma et al., 2017).

Recently, the anti-obesity effect of *Plocamium telfairiae* was evaluated by the optimal antiadipogenesis on 3T3-L1 cells and the anti-obesity properties in obese C57BL/6 mice (Lu et al., 2020). The obtained results indicated a decrease of fat accumulation and suppression in 3T3-L1 cells of the major adipogenesis factors expression, such as PPAR- γ, SREBP-1, C/EBPα, and phosphorylated ACC (Lu et al., 2020). Ethanolic extracts of *Plocamium telfairiae* resulted in a decrease of triglycerides, total cholesterol, adiponectin, and insulin in high-fat diet-induced obese mice (Lu et al., 2020). Similarly, ethanolic extract of *Grateloupia elliptica* caused inhibition of intracellular lipid accumulation in 3T3-L1 cells, and significant reduction of adipogenic protein expression. Meanwhile, *in vivo* experiments indicated a net reduction in body weight and white adipose tissue weight with downregulation of PPAR-γ and C/EBP-α mRNA (Lee et al., 2020).

Obesity is associated with lipid accumulation, oxidative stress and an increase in insulin resistance that results in diabetes (Yang et al., 2019a). Diabetes is a chronic metabolic disease that could develop with obesity and aging. Diabetes and obesity are closely interrelated disorders. Phlorotannins were reported for their anti-diabetic activity. Diphlorethohydroxycarmalol (DPHC) isolated from *Ishige okamurae* inhibited α-amylase and α-glucosidase (IC_{50} = 0.53 ± 0.08 and 0.16 ± 0.01 mM). The rise in postprandial blood glucose levels in both streptozotocin-induced diabetic and normal mice was also suppressed (Heo et al., 2009). While Lee et al. (2012c) indicated that DPHC treatment protected high glucose-induced damage in RINm5F pancreatic β-cells through the anti-oxidant effects. *Ishige okamurae* extract was shown to prevent insulin resistance and regulate blood glucose levels in hyperglycemia in C57BL/KsJ-db/db mice (Min et al., 2011).

The anti-diabetic effects of *Laminaria japonica* and *Hizikia fusiforme* extracts were assessed both *in vitro* and *in vivo* models (Kang et al., 2018). The study reported significant inhibition of the α-glucosidase activity (with best activities for *Hizikia fusiforme* extracts) and a net increase in glucose uptake. Besides, tested extracts induced phosphorylation of protein kinase B and AMP-activated protein kinase in skeletal muscle models (*in vivo* and *in vitro*) (Kang et al., 2018). Fucoidan isolated from *Undaria pinnatifida* has been reported to reduce blood glucose levels and improve insulin sensitivity in mice (Sim et al., 2019). The authors tested fucoidan anti-diabetic function with direct action on adipocytes in 3T3-L1 cells. The results showed a reduction in lipid accumulation and glycerol-3-phosphate dehydrogenase activity as well as suppression of the PPARγ expression, stimulation of glucose uptake in normal adipocytes, and restoration of insulin-stimulated glucose uptake in obesity-induced insulin-resistant adipocytes (Sim et al., 2019). *Sargassum confusum* oligosaccharides were evaluated for their anti-diabetic effects *in vivo* using high-fat/high-sucrose fed hamsters (Yang et al., 2019b). The oligosaccharides induced a significant decrease in fasting blood glucose levels and protected the cellular architecture of the liver. The main active anti-diabetic effect revealed in this study was the regulation of insulin receptor substrate 1/phosphatidylinositol 3-kinase and c-Jun N-terminal kinase pathways (Yang et al., 2019b).

9.4.4.2 ANTI-MALARIAL AND ANTI-LEISHMANIAL POTENTIALITIES OF SEAWEEDS

Malaria is caused by *Plasmodium* parasites, essentially *Plasmodium falciparum, Plasmodium vivax* and *Plasmodium malariae. Plasmodium* parasites are transmitted to vertebrates through the bites of infected *Anopheles* mosquitoes (Murugan et al., 2016). The treatment and control of malaria are facing different problems like the emergence of artemisinin and chloroquine resistance in *Plasmodium* parasites (Benelli and Mehlhorn, 2016).

Chen et al. (2009) through the evaluation of the anti-plasmodial activity *in vitro* and *in vivo* assessed the anti-malarial activity of fucoidan from *Undaria pinnatifida.* Significant inhibition of erythrocyte invasion by *Plasmodium falciparum* was recorded, the same as for the chloroquine-sensitive (CQ-s) *Plasmodium falciparum* 3D7 strain and the chloroquine-resistant (CQ-r) K1 strain. Besides, fucoidan induced 37% of suppressive effect in *Plasmodium berghei*-infected mice and had no toxic effects on RAW 264.7 cells (Chen et al., 2009). The green-synthesized zinc oxide nanoparticles (ZnO NPs) synthesized using *Sargassum wightii* were tested against female anopheline mosquitoes *Anopheles stephensi* the vector of *Plasmodium* parasites (Murugan et al., 2018). Larvicidal and pupicidal toxicity on *Anopheles stephensi* were obtained with LC_{50} of ZnO NPs ranged from 4.330 (larva

I) to 7.430 ppm (pupa); as well as an important reduction of longevity and fecundity of *Anopheles stephensi* (Murugan et al., 2018). CdS NPs utilizing *Valoniopsis pachynema* extracts showed toxicity on *Anopheles stephensi* and *Anopheles sundaicus* young instars (Sujitha et al., 2017). The mosquitocidal effect of *Valoniopsis pachynema*-synthesized CdS NPs on *Anopheles stephensi* ranged from LC_{50} of 16.856 (larva I) to 30.301 μg/ml (pupa) and for *Anopheles sundaicus* from 13.584 to 22.496 μg/ml respectively. The anti-plasmodial response of the nano-synthesized product against CQ-r and CQ-susceptible (CQ-s) strains of *Plasmodium falciparum* was also evaluated. The results indicated that the IC_{50} of *Valoniopsis pachynema* extract was 58.1 μg/ml (CQ-s) and 71.46 μg/ml (CQ-r). Whereas, the IC_{50} of *Valoniopsis pachynema*-CdS NPs was 76.14 μg/ml (CQ-s) and 89.21 μg/ml (CQ-r) (Sujitha et al., 2017).

Leishmania parasites are the causative agents of Leishmaniasis. These pathologies are vector-borne diseases reported worldwide. Seaweeds, due to their composition of biomolecules, have been the subject of investigations relating to the anti-leishmanial activity. The anti-leishmanial activity of elatol a sesquiterpene from *Laurencia dendroidea* was investigated against *Leishmania amazonensis* (dos Santos et al., 2010). Elatol induced IC50 of 4.0 μM and 0.45 μM respectively, on promastigote and intracellular amastigote forms of *Leishmania amazonensis*. In addition, elatol induced several structural modifications in parasite like pronounced swelling of the mitochondrion, concentric membrane structures inside the organelle, destabilization of the plasma membrane, and formation of an extension of the endoplasmic reticulum (dos Santos et al., 2010).

In another study, PLS from *Solieria filiformis*, *Botryocladia occidentalis*, *Caulerpa racemosa,* and *Gracilaria caudata* were tested *in vitro* against *Leishmania amazonensis* promastigotes (Lehnhardt Pires et al., 2013). A purified polysaccharide from *Caulerpa racemosa* showed the most potent activity with an EC_{50} value of 34.5 μg/ml. Moreover, *in vitro* cytotoxic assays using peritoneal macrophages and J774 macrophages demonstrated a decrease in J774 cell survival with no notable effect on peritoneal macrophages (Lehnhardt Pires et al., 2013). de Sousa et al. (2019) in *in vitro* screening of anti-leishmanial properties of Iberian seaweed showed also cytotoxicity and strong morphological alterations in the parasites. Besides, seaweed extracts of *Cystoseira baccata* and *Cystoseira barbata* were also effective against intracellular amastigotes with respective IC_{50} values of 5.1 and 6.8 μg/ml.

9.5 AGRICULTURAL APPLICATIONS OF SEAWEEDS

Agriculture represents an economic sector of great importance. Global climate changes, environmental stresses and competition raise the evidence of searching for new plant growth-promoting substances to increase crops production.

9.5.1 SEAWEEDS BIOSTIMULANT EFFECTS

Biostimulants include diverse compounds, mainly organic, used in small doses through soil or foliar application to enhance plant growth and development (Schmidt et al., 2003). Plant growth biostimulants should improve plant resistance to stress, increase plant growth, and improve nutrient use efficiency of fertilizers. The major advantage of using this type of product is to reduce fertilizer utilization by increasing crop quality.

Agricultural use of biostimulants comprises obtained products from seaweeds with a significant positive impact on plant development, nutrients uptake, and resistance to abiotic and biotic stresses (Hamed et al., 2017). In agriculture, algae can be used in different forms, of which extracts appear to be the most effective (Khan et al., 2009; Chojnacka et al., 2015). Seaweed extracts could be applied on soil and/or on plants as a foliar spray (Du Jardin, 2015) and stated many potentialities such as plant growth regulation (auxin, cytokinin, gibberellins, betains) (Stirk et al., 2014), fungicides, and pesticides activities (Rengasamy et al., 2015a). Extracts from *Ascophyllum nodosum*, *Fucus* sp., *Laminaria* sp., and *Sargassum* are widely used in agriculture (Khan et al., 2009). The use of seaweed extracts in agriculture promotes improved germination, nutrient uptake, stimulation of root development and shoots elongation, and increased water holding capacity (Craigie, 2011). Algal extracts are involved in increasing plant growth and biomass, the content of chlorophyll, carotenoids, and improving root development (Khan et al., 2009; Michalak and Chojnacka, 2016).

9.5.1.1 SEAWEEDS CROP IMPROVEMENT

Seaweed-based biostimulants play a significant role in developing environment-friendly strategies to improve crop yields, stresses tolerance, and fruit quality (Frioni et al., 2018). Improving local yields through integrated nutrient management is pertinent to sustainable stewardship of agricultural ecosystems.

The bio-stimulating effects of seaweeds may be related to their content on polyphenols, polysaccharides, alginates, polyamines, pigments, free amino acids, betaines, vitamins, micro, and macro-nutrients, and natural phytohormones (cytokinins, auxins, and auxin-like compounds, indole acetic acid, abscisic acid) (Papenfus et al., 2012; Wally et al., 2013; Stirk et al., 2014). This mixture of useful biologically active substances led to different beneficial effects (Khan et al., 2009; Yusuf et al., 2012; Michalak et al., 2017). Several studies related to the ability of seaweed extract on enhancing plant growth, yield, and fruiting have been performed (Abd El-Gawad and Osman, 2014). Brown algae extracts have attracted great interest. They have been used extensively in sustainable and biological agriculture as either natural plant growth stimulants or bio-fertilizers (Hong et al., 2007; Craigie, 2011). *Ascophyllum nodosum* extract enhances growth and yield in many crops and improves quality.

Many studies have reported the positive and beneficial effects of seaweed extracts on the growth and yield of tomato (Demir et al., 2006), soybean (Rathore et al., 2009), wheat (Shah et al., 2013), eggplant (Abd El-Gawad and Osman, 2014), maize (Layek et al., 2015; Rengasamy et al., 2015b), red radish (Mahmoud et al., 2019) and others. Seed germination of *Cyamopsis tetragonoloba* was improved using seaweed fertilizers (Chithra et al., 2016). Godlewska et al. (2016) also emphasized the positive influence of the Baltics seaweeds (*Polysiphonia*, *Ulva,* and *Cladophora*) extract on *Lepidium sativum* seeds germination. An improvement in germination percentage and seedling vigor of rice (*Oryza sativa*) was recorded under *Kappaphycus alvarezii* and *Gracilaria edulis* saps (Layek et al., 2018). The application of seaweed extracts increased root growth and plants development (Khan et al., 2009; Alam et al., 2013; El Miniawy et al., 2014; Arioli et al., 2015).

Seaweed extract from a brown seaweed (*Ecklonia maxima*) has been reported as a plant biostimulant (Rouphael et al., 2017). A phlorotannin 'Eckol' was found to have auxin-like activities, and interesting growth-promoting activity (Rengasamy et al., 2015a, b). Moreover, spinach plants (*Spinacia oleracea*) showed a net increase in growth and biochemical parameters of plants exposed to eckol (from the seaweed *Ecklonia maxima*) a seaweed-based biostimulant (Kulkarni et al., 2019). Foliar applications of *Ascophyllum nodosum* extracts have been reported to affect plant hormone biosynthesis (Wally et al., 2013) and enhance crop tolerance to pathogens (Khan et al. 2009; Battacharyya et al. 2015). Seaweed extract from *Sargassum swartzii* significantly improved shoot length, leaves number, total phenolic content, proteins, and flavonoids of cowpea (*Vigna unguiculata*). The results highlighted that *Sargassum swartzii* application has significantly ameliorated the phytochemical contents and antioxidant activity in *Vigna unguiculata* (Vasantharaja et al., 2019).

More recently, arbuscular mycorrhizal fungus strain (*Rhizophagus irregularis*) combined with seaweed extract derived from *Fucus spiralis* induced significant increases in different growth, physiological, and histological parameters of the date palm (*Phoenix dactylifera* L.) (Anli et al., 2020).

9.5.1.2 SEAWEEDS APPLICATION AS BIO-FERTILIZER FOR SOIL QUALITY IMPROVEMENT

Even though seaweed extracts are mainly applied through foliar spray; they may have as well several effects on soil compartments. Seaweeds have been used as fertilizers and were applied to soils. Metabolites from algae such as polysaccharides (alginate, laminarin) have a positive effect on soil structure (Kraan, 2012) and are considered as perfect metal ion chelators.

The use of algal biomass for the improvement of soil quality for agricultural crop production has been reported (Cole et al., 2016). *Ascophyllum*-based composts used in the form of suspension, have improved the degraded soil structure (Alam et al., 2013). Algal biomass with its content of organic matter and other macronutrients (nitrogen, phosphorus, potassium, etc.); are useful for the improvement of soil quality and rhizosphere nutrients availability (Tabarsa et al., 2012). Seaweed extracts enhance rhizospheric microbes with plant growth-promoting traits and improve interaction with rhizosphere microorganisms and enzyme activity. When seaweed polysaccharides are applied in soils, they contribute to gel formation, water retention and soil aeration (Du Jardin, 2015), whereas the polyanionic compounds contribute to the cations fixation and exchanges closely related to the heavy metals fixation and soil remediation (Du Jardin, 2015). Some positive effects on the promotion of plant growth-promoting bacteria and pathogen antagonists in soils have also been reported. Alginate seaweed-based extracts increased rhizospheric-available phosphorus on *Chrysanthemum* plants, with no effects on microbial community (Ji et al., 2017). Recently, *Macrocystis pyrifera* biostimulant extracts have enhanced the hydrogenase activity of soil-degraded samples by erosion (Onet et al., 2019).

Application of seaweed fertilizers based on *Lessonia nigrescens* and *Lessonia flavicans* improved significantly beneficial microbes and fungi/bacteria ratio of *Malus hupehensis* rhizosphere (Wang et al., 2017c).

9.5.1.3 SEAWEEDS APPLICATION IN REALISTIC FIELD TRIALS

Mattner et al. (2018) showed the potential benefits of the integrated use of seaweed extracts in strawberry production. The biostimulatory effect of the seaweeds *Duvillaea potatorum* and *Ascophyllum nodosum* was manifested by an increase in root growth and fruit yield of strawberry. The extracts of *Ascophyllum nodosum* sprayed on grapevines were found to have positive effects, mainly on the quality of the wine grapes grown (Frioni et al., 2018). Foliar applications of *Ascophyllum nodosum* extracts resulted in a significant increase in tomato and sweet pepper plant growth parameters, including plant height, leaf number, plant dry biomass, root length, and chlorophyll content. The number of flower clusters, flower numbers, fruits per cluster, and total harvested fruit yield were also higher for plants treated with *Ascophyllum nodosum* extract in combination with fungicide treatment (Ali et al., 2019).

Layek et al. (2018) reported a high dry matter accumulation, chlorophyll index, crop growth rate, and rice yield after foliar spray of *Kappaphycus alvarezii* and *Gracilaria edulis* saps. Besides, the seaweed extracts increased the concentration of micronutrients (Fe, Cu, Zn, and Mn) and proteins in rice grains (Layek et al., 2018). On the other hand, seaweed extracts of *Ascophyllum nodosum* and *Sargassum muticum* used as biofertilizer, positively affected rice (*Oryza sativa*) and lettuce (*Lactuca sativa*) seed germination, plant development, and production (Silva et al., 2019). In another study, *Sargassum swartzii* extract enhanced *Vigna unguiculata* nutritional quality and yield (Vasantharaja et al., 2019).

9.5.1.4 SEAWEEDS ENHANCE ABIOTIC STRESS TOLERANCE

Crop productivity is negatively affected by different factors such as salinity, drought, extreme temperatures, and nutrient deficiencies inducing abiotic stresses. Abiotic stresses are associated with oxidative stress (ROS accumulation). ROS are reported to deteriorate biomolecules (DNA, proteins, and lipids) (Das and Roychoudhury, 2014). So, to prevent those negative effects, natural defense mechanisms are used for ROS scavenging. That includes enzymatic and non-enzymatic antioxidants such SOD, peroxidase, phenolic compounds, ascorbic acid, tocopherol, and others (Das and Roychoudhury, 2014). In order to mitigate abiotic stress, several studies were conducted. A positive impact of seaweed extract was also reported under stress conditions (abiotic and biotic) (Khan et al., 2009; Latique et al., 2014; Battacharyya et al., 2015). Their direct and indirect effects on plants under stress raised their role in alleviating stress conditions (Battacharyya et al., 2015). The extracts of *Ascophyllum nodosum, Laminaria digitata*, and *Fucus serratus* have been used as biofertilizers for their high content in betaines, organo-osmolytic compounds that play a crucial role in protecting plants against salt, drought, and extreme temperature stress (Blunden et al., 2010). Liu et al. (2019) reported that *Grateloupia filicina* polysaccharides alleviate salt stress on rice during the seed germination stage, and they stimulated rice seed development under salinity conditions. Polysaccharides extracted from *Lessonia nigrescens* were reported to enhance wheat seedlings adaptability to salt stress. Several plant parameters were improved under stress like shoot, root lengths and dry, fresh matters (Zou et al., 2019). On other study, *Ulva lactuca* and *Enteromorpha intestinalis* extracts significantly promoted diazotrophic bacterial growth and wheat seed germination under salt stress (Rai et al., 2018). Under drought stress, the application of *Gracilaria dura* extract induced an increase of wheat biomass and yield by 57% and 70% respectively (Sharma et al., 2019). Besides, *Ascophyllum nodosum* enhanced fresh biomass of lettuce exposed to potassium deficiency (Chrysargyris et al., 2018). A commercial extract of *Ascophyllum nodosum* evaluated for drought stress alleviation in soybean (*Glycine max*) was reported to induce higher relative water content and higher stomatal conductance under drought stress. The obtained results suggested that applications of *Ascophyllum nodosum* improved soybean drought tolerance through physiological changes and gene expressions (Shukla et al., 2018).

9.5.2 SEAWEEDS AGAINST PHYTOPATHOGENS

Phytopathogens like fungi and bacteria are responsible for serious diseases to humans and lead to great losses in crop productions (Ramaswamy et al., 2007). Natural products are then used as safer alternatives to antibiotics and chemical synthetic pesticides, commonly used for pest control (Derbalah et al., 2012). Several bioactive compounds obtained from seaweeds exhibit different biological activities. They were highlighted for their roles in plant protection, anti-phytopathogen activities, and induction of plant resistance to biotic stress as potential antimicrobial, anti-insecticidal, and anti-nematodal agents (Sahayaraj et al., 2012; Baloch et al., 2013; Ibraheem et al., 2017). Seaweed extracts are organic and cost-effective and offer a non-toxic alternative way for disease biocontrol (Baloch et al., 2013; Kulkarni et al., 2019).

9.5.2.1 ANTIMICROBIAL ACTIVITIES

Kumar et al. (2008) reported the activity of *Sargassum wightii* against *Pseudomonas syringae* responsible for leaf spot disease on *Gymnema sylvestre*. Moreover, *Sargassum swartzii* extracts were reported to inhibit the growth of *Xanthomonas oryzae* that causes bacterial blight of rice (Arunkumar et al., 2005). Also, the use of *Cystoseira myriophylloides* and *Fucus spiralis* extracts markedly decreased the severity of crown gall diseases caused by *Agrobacterium tumefaciens* on tomato (Esserti et al., 2017). Those extracts induced plant resistance against the crown gall diseases due to their high antioxidant content. The *Padina gymnospora* extracts exhibited significant antibacterial activity against the soil pathogenic bacteria *Ralstonia solanacearum* and *Pectobacterium carotovora* (Ibraheem et al., 2017). In addition, a recent study reported the pesticide effects of AgNPs synthesized using aqueous extracts of agar seaweeds (*Gracilaria corticata* and *Gracilaria edulis*) and carrageenan seaweeds (*Hypnea musciformis* and *Spyridia hypnoides*) (Roseline et al., 2019). The AgNPs exhibited effective antibacterial activities against *Xanthomonas axonopodis* and *Xanthomonas oryzae* (Roseline et al., 2019). On the other side, plant virus diseases are responsible for a great loss in agricultural production. Natural anti-viral substances are much sought to limit the development and progression of these diseases. Seaweed extracts, due to their content on proteins and polysaccharides, exhibited several antiviral activities (Zhao et al., 2017). *Durvillaea antarctica* extracts were able to reduce damages caused by tobacco mosaic virus in tobacco leaves (Jiménez et al., 2011).

Seaweeds also received much attention as a source of fungicidal compounds used to protect plants against several pathogenic fungi. Different seaweeds metabolites were reported efficient against common pathogens like *Fusarium solani*, *Rhizoctonia solani*, *Aspergillus* sp., *Fusarium oxysporum*, *Penicillium* sp., and *Botrytis cinerea* (Liu et al., 2014; Khallil et al., 2015; Ibraheem et al., 2017). Algal polysaccharides have been widely studied as a source of agricultural drugs against plant disease (Stadnik and Freitas, 2014). Polysaccharides from brown and red seaweeds can induce defense responses and plant resistance against phytopathogens (Vera et al., 2011). Polysaccharides stimulate pathogen-associated cellular changes, calcium concentration modifications as defense activation and pathogenesis-related proteins activation (Zhao et al., 2012). Jiménez et al. (2011) reported that *Gracillaria chilensis* impeded the growth of *Phytophthora cinnamomi* while *Lessonia trabeculata* suppressed the growth of *Botrytis cinerea* and reduced the necrotic lesion in tomato leaves. Laminarins (polysaccharides from brown seaweeds) elicit defense responses against *Botrytis cinerea* causing gray mold and *Plasmopara viticola* responsible for downy mildew in grapevine plants and suppressed infections (Copping, 2004). Liang et al. (2012) showed the fungicidal activity of laminarin at *Aspergillus flavus*, through reduction of mycelial growth and aflatoxin production. Extracts of *Laminaria digitata* and *Undaria pinnatifia* caused the growth and spore germination inhibition of *Botrytis cinerea.* A net reduction of wilt symptoms caused by *Verticillium dahliae* on olive twigs has been reported by applications of laminarin and alginate (Ben Salah et al., 2018).

On the other hand, carrageenans (red algae polysaccharides) induced protection against different kinds of pathogens like *Botrytis cinerea*, and *Pectobacterium carotovora* on tobacco (Vera et al., 2011). Brown seaweed *Hormophysa cuneiformis* was found to have potential antifungal activity against *Cladosporium herbarum* (Mohamed and Saber, 2019). Furthermore, glucuronan, and oligoglucuronans (saccharides from green algae), showed a reduction of the blue and gray mold (*Penicillium expansum* and *Botrytis cinerea*) severities on postharvest apple fruit (*Malus domestica*); with great effect achieved by oligoglucuronans (Abouraïcha a et al., 2017). Saccharides from green algae showed a protective effect on apple by the activation of related-defense responses such as antioxidant enzyme activities (CAT and SOD) and the levels of lignin and phenolic compounds (Abouraïcha et al., 2017). *Padina gymnospora*, *Sargassum latifolium*, and *Hydroclathrus clathratus* seaweed powders used as soil amendments, diminished the incidence of root rot disease in *Solanum melongena* (eggplant) caused by *Fusarium solani* (Ibraheem et al., 2017). Ammar et al. (2017) reported the fungicidal effect of the *Sargassum vulgare* extract on *Pythium aphanidermatum* by reducing mycelial growth and disease incidence. The sulfated polysaccharide from *Acanthophora spicifera* elicits defense responses on *Hevea brasiliensis* in response to *Phytophthora palmivora* infection (Pettongkhao et al., 2019). Roseline et al. (2019) showed interesting antifungal activity of AgNPs against *Ustilaginoidea virens* under *in vitro* conditions.

9.5.2.2 INSECTICIDAL ACTIVITY

Seaweed extracts have been integrated into pest management. They are considered as an eco-friendly environmental source of novel insecticidal substances. For instance, different seaweed extracts exhibited insecticidal activity against *Dysdercus* spp. (cotton insect pests) (Rajesh et al., 2011; Sahayaraj et al., 2012). *Sargassum swartzii* extracts induced a net reduction in male and female longevity of *Dysdercus cingulatus* (Asharaja and Sahayaraj, 2013). Foliar applications of ι-, κ-, and λ-carrageenans in *Arabidopsis* against the insect *Trichoplusia ni* showed interesting effects (Sangha et al., 2011). Treated plants with ι- and k-carrageenan manifested reduced feeding damage by *Trichoplusia ni* larvae, additionally to the reduction in larval weight under the same treatments (ι- and k-carrageenan) (Sangha et al., 2011).

9.6 SEAWEED-BASED PRODUCTS FOR ENVIRONMENTAL APPLICATIONS

9.6.1 SEAWEED ALGICIDAL ACTIVITY

In recent years, HABs have been reported as a serious hazard to aquatic ecosystems (freshwater and marine), the aquaculture industry and human health (Jeong et al., 2000). HABs are mainly due to the eutrophication of water bodies and climate changes (Paerl and Huisman, 2008; Ying-ying et al., 2015). Red tides or HABs correspond to a massive growth of toxic or non-toxic microalgae in coastal waters. While, harmful cyanobacterial blooms (Cyano HABs) are one of the most serious problems in drinking water resources and freshwater ecosystems (Harke et al., 2016). CyanoHABs are massive blooms of cyanobacteria produced primarily by potentially toxic species such as *Microcystis aeruginosa*, one of the most important freshwater cyanobacteria widely distributed worldwide (Catherine et al., 2013). Due to the undesirable effects caused by these HABs, extensive research in the control or mitigation of HABs has been conducted. Alamsjah et al. (2005) tested several macroalgae species (37 species) and reported that the strongest algicidal effect was obtained by *Ulva fasciata* extract. The growth of the microalgae *Heterosigma akashiwo* and *Alexandrium tamarense* was inhibited by *Sargassum thunbergii, Corallina pilulifera* and *Ulva pertusa* (Wang et al., 2007a). Wang et al. (2012) reported strong growth inhibition of

Heterosigma akashiwo by seaweed species like *Enteromorpha intestinalis, Ulva pertusa* and *Ulva linza.* Besides, the macroalgal extracts of *Porphyra tenera, Ulva pertusa,* and *Enteromorpha clathrata* showed strong growth inhibition on *Skeletonema costatum* (An et al., 2008).

Extract of *Stoechospermum marginatum* containing diethyl phthalate as a major constituent showed a significant algicidal effect on *Oscillatoria* sp. (Manilal et al., 2011). Sesquiterpenoids isolated from *Porphyra yezoensis* had significant selective antialgal activity against test red tide microalgae (Sun et al., 2018a). Growth inhibition or even cell mortality on different bloom-forming dinoflagellates species was significantly recorded by *Ulva pertusa* and *Gracilaria lemaneiformis* (Wang et al., 2007b). *Ulva prolifera* methanol extract also exhibited significant activity mainly against *Heterosigma akashiwo* and *Prorocentrum donghaiense* (Sun et al., 2016), whilst *Gracilaria tenuistipitata* affected the photosynthesis process of *Prorocentrum micans* (Ye and Zhang, 2013). *Ecklonia cava* (brown algae) also caused growth inhibition of *Cochlodinium polykrikide* and *Heterosigma akashiwo* (Chowdhury et al., 2014). Sun et al. (2018b) investigated the antialgal activities of nine novel isolated molecules from *Ulva pertusa* against common red tide microalgae. Selective antialgal effects were obtained for all red tide microalgae tested, suggesting *Ulva pertusa* as a new source of bioactive compounds with antialgal activity. In recent work, methanol extract of *Tricleocarpa jejuensis* was assayed to control the algicidal activity against the phytoplankton *Chattonella antiqua* (Zha et al., 2020). The highest activity was recorded for active compounds identified as (E)-9-hydroxyoctadec-10-enoic acid, (E)-10-hydroxyoctadec-8-enoic acid, (E)-11-hydroxyoctadec-12-enoic acid and (E)-12-hydroxyoctadec-10-enoic acid. The algicidal activity was then related to the presence of the hydroxyl group in the unsaturated fatty acids (Zha et al., 2020). Besides, significant seaweeds effects were reported against Cyano HABs such as *Microcystis aeruginosa* and *Phormidium* sp. Zerrif et al. (2018) reported that the methanol extract of the *Codium elongatum* collected from the coast of Morocco showed a significant progressive reduction in growth of the potential bloom-forming cyanobacteria *Phormidium* sp.

Recently, the potential algicidal activities of 14 species of seaweeds collected from the Moroccan coast were evaluated. Methanol extracts of *Bornetia secundiflora, Laurencia pinnatifida,* and *Gelidium pulchellum* induced a strong inhibitory activity against *Microcystis aeruginosa* on solid medium (27.33 ± 0.33 mm, 17.33 ± 0.33 mm, and 16.67 ± 0.33 mm, respectively) (Zerrifi et al., 2019). While in liquid culture medium, all tested methanol extracts showed a significant inhibitory effect on *Microcystis aeruginosa* compared to the negative control (Zerrifi et al., 2019). Seaweed extracts appear to have a positive impact on monitoring the growth of microalgae and on the development of highly effective and more specific eco-strategies for the biocontrol of HABs.

9.6.2 ANTI-BIOFOULING ACTIVITY OF SEAWEEDS

Biofouling is one of the most serious problems facing worldwide in the marine environment; corresponding to undesirable growth of micro and macroorganisms on submerged surfaces, considering man-made and living structures (Wahl, 1989). Bioactive compounds from diverse marine algae showed potential for unraveling this problem through the inhibition of the fouling process (Hellio et al., 2002). Genera like *Asparagopsis, Laurencia, Sargassum, Ulva,* and others are sources of active antifouling compounds. Different potential substances were identified. Dimethylsulphopropionate and proline qualified as surface-attached antifouling compounds were isolated from *Fucus vesiculosus* (Saha et al., 2012). In another study, acyclic linear diterpenoids from *Bifurcaria bifurcata* exhibited antifouling activity (Muñoz et al., 2013). Several studies reported also different antifouling compounds from *Sargassum* spp. such as phlorotannins (Nakajima et al., 2016). The potential antifouling activity was reported for the 3-bromo-5-(diphenylene)-2(5H)-furanone compound from *Ulva rigida* (Chapman et al., 2014). Whereas, Grosser et al. (2012) related the activity of *Ulva* sp. to β-carotene. Fatty acid derivatives like docosane, hexadecanoic acid, and cholesterol trimethylsilyl ether, were registered as cortical cell products and deposited on the surface of *Laurencia translucida* (Paradas et al., 2016). The same biological activity was reported for *Laurencia* sp. due to the presence of Omaezallene (Figure 9.4) (Umezawa et al., 2014). Greff et al. (2014) reported antimicrobial activity against both marine and terrestrial microorganisms owed to the highly brominated cyclopentenones: mahorone and 5-bromomahorone. The species belonging to *Cystoseira* (Fucales) are known to be a source of bioactive terpenoid derivates, particularly meroterpenoids and diterpenoids. Meroditerpenoid and Cystophloroketals A and B isolated from *Cystoseira tamariscifolia* have shown high potential activity. Potent inhibition activity against fouling microalgae and moderate antimicrobial activity against a diverse range of bacteria and fungi were reported (El Hattab et al., 2015).

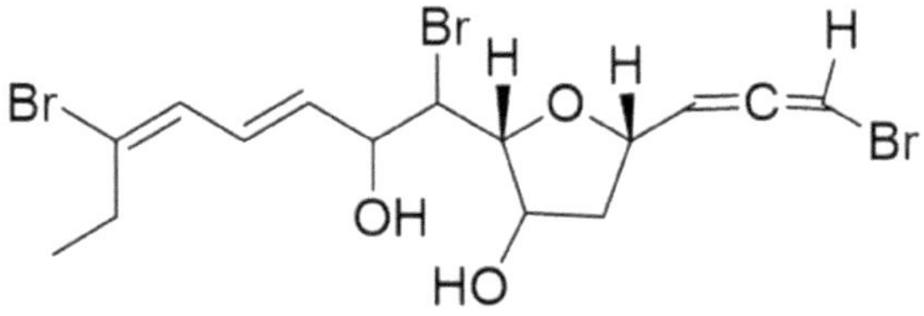

FIGURE 9.4 Omaezallene an anti-fouling compound from *Laurencia* sp. (Umezawa et al., 2014).

9.7 CONCLUSIONS AND FUTURE TRENDS

This chapter reviewed mainly recent studies related to seaweeds biodiversity and their richness on diverse and novel bioactive compounds with a wide large spectrum of biological properties. This suggests that macroalgae, within marine organisms, are the most coveted for possible biotechnological applications. The biological activities assigned to these seaweeds bioactive compounds are multiple like antibacterial, antifungal, antiviral, and anti-cancer activities.

Thereby, these various activities with different potent properties were investigated and involved in different domains of applications such as medicine, pharmacology, agricultural practice and environmental processes. Although there is much fundamental knowledge about the potential of seaweeds to play a key role as a new source of biomolecules with very interesting activities, there is a need for more investigations focused on the structural identification of novel substances and their mode of action to improve their biotechnological applications.

Seaweed bioactive molecules could gain importance and have greater value through the formulation of the natural product-based and environmentally friendly agents (designs of novel drugs and formulation of natural bio-control agents) to benefit the most from the biotechnological advantages offered by these bioactive compounds. The scientific communities must not only focus on the exploration of algal bio-compounds and the search for new biotechnological activities but also adopt alternative solutions for the efficient exploitation of these resources.

KEYWORDS

- **agricultural applications**
- **bioactive metabolites**
- **biological activities**
- **environmental uses of seaweeds**
- **minimum inhibitory concentration**
- **pharmacological potentialities**
- **seaweeds biotechnology**
- **seaweeds biodiversity**
- **seaweeds**
- **nanoparticles**
- **seaweeds extracts**

REFERENCES

Abd-Ellatef, G. E. F., Ahmed, O. M., Abdel-Reheim, E. S., & Abdel-Hamid, A. H. Z., (2017). *Ulva lactuca* polysaccharides prevent Wistar rat breast carcinogenesis through the augmentation of apoptosis, enhancement of antioxidant defense system, and suppression of inflammation. *Breast Cancer, 9,* 67–83.

Abouraïcha, E. F., El Alaoui-Talibi, Z., Tadlaoui-Ouafi, A., El Boutachfaiti, R., Petit, E., Douira, A., Courtois, B., et al., (2017). Glucuronan and oligoglucuronans isolated from green algae activate natural defense responses in apple fruit and reduce postharvest blue and gray mold decay. *J. Appl. Phycol., 29,* 471–480.

Aguilar-Briseño, J. A., Cruz-Suarez, L. E., Sassi, J. F., Ricque-Marie, D., Zapata-Benavides, P., Mendoza-Gamboa, E., Rodríguez-Padilla, C., & Trejo-Avila, L. M., (2015). Sulfated polysaccharides from *Ulva clathrata* and *Cladosiphon okamuranus* seaweeds both inhibit viral attachment/entry and cell-cell fusion, in NDV infection. *Marine Drugs, 13,* 697–712.

Ahn, J. H., Yang, Y. I., Lee, K. T., & Choi, J. H., (2015). Dieckol, isolated from the edible brown algae *Ecklonia cava*, induces apoptosis of ovarian cancer cells and inhibits tumor xenograft growth. *J. Cancer. Res. Clin. Oncology, 141,* 255–268.

Ahn, M. J., Yoon, K. D., Min, S. Y., Lee, J. S., Kim, J. H., Kim, T. G., Kim, S. H., Kim, N. G., Huh, H., & Kim, J., (2004). Inhibition of HIV-1 reverse transcriptase and protease by phlorotannins from the brown alga *Ecklonia cava*. *Biol. Pharm. Bull., 27,* 544–547.

Aigner, S., Remias, D., Karsten, U., & Holzinger, A., (2013). Unusual phenolic compounds contribute to ecophysiological performance in the purple-colored green alga *Zygogonium ericetorum* (Zygnematophyceae, Streptophyta) from a high-alpine habitat. *J. Phycol., 49,* 648–660.

Alagan, V., Valsala, R., & Rajesh, K., (2017). Bioactive chemical constituent analysis, *in vitro* antioxidant and antimicrobial activity of whole plant methanol extracts of *Ulva lactuca* Linn. *Br. J. Pharm. Res., 15,* 1–14.
Alam, M. Z., Brown, G., Norrie, J., & Hodges, D. M., (2013). Effect of *Ascophyllum* extract application on plant growth, fruit yield and soil microbial communities of strawberry. *Can. J. Plant. Sci., 93,* 23–36.
Alamsjah, M. A., Hirao, S., Ishibashi, F., & Fujita, Y., (2005). Isolation and structure determination of algicidal compounds from *Ulva fasciata. Biosci. Biotechnol. Biochem., 69,* 2186–2192.
Alarif, W. M., Al-Footy, K. O., Zubair, M. S., Halid, P. H. M., Ghandourah, M. A., Basaif, S. A., & Badria, F. A., (2016). The role of new eudesmane-type sesquiterpenoid and known eudesmane derivatives from the red alga *Laurencia obtusa* as potential antifungal-antitumor agents. *Nat. Prod. Res., 30,* 1150–1155.
Ale, M. T., & Meyer, A. S., (2013). Fucoidans from brown seaweeds: An update on structures, extraction techniques and use of enzymes as tools for structural elucidation. *RSC Adv., 3,* 8131–8141.
Ali, O., Ramsubhag, A., & Jayaraman, J., (2019). Biostimulatory activities of *Ascophyllum nodosum* extract in tomato and sweet pepper crops in a tropical environment. *PLoS One, 14*(5), e0216710.
Alobwede, E., Leake, J. R., & Pandhal, J., (2019). Circular economy fertilization: Testing micro and macro algal species as soil improvers and nutrient sources for crop production in greenhouse and field conditions. *Geoderma., 334,* 113–123.
Ammar, N., Jabnoun-Khiareddine, H., Mejdoub-Trabelsibi, B., Nefzi, A., Mahjoub, M. A., & Daami-Remadi, M., (2017). Pythium leak control in potato using aqueous and organic extracts from the brown alga *Sargassum vulgare* (C. Agardh, 1820). *Postharvest Biol. Technol., 130,* 81–93.
An, Z., Wang, Z., Li, F., Tian, Z., & Hu, H., (2008). Allelopathic inhibition on red tide microalgae *Skeletonema costatum* by five macroalgal extracts. *Front. Environ. Sci. Eng. China, 2,* 297–305.
Andrieux, C., Hibert, A., Houari, A. M., Bensaada, M., Popot, F., & Szylit, O., (1998). *Ulva lactuca* is poorly fermented but alters bacterial metabolism in rats inoculated with human faecal flora from methane and non-methane producers. *J. Sci. Food Agric., 77,* 25–30.
Anli, M., El Kaoua, M., Ait-el-Mokhtar, M., Boutasknit, A., Ben-Laouane, R., Toubali, S., Baslam, M., et al., (2020). Seaweed extract application and arbuscular mycorrhizal fungal inoculation: A tool for promoting growth and development of date palm (*Phoenix dactylifera* L.) *cv* Boufgous. *South African Journal of Botany, 132,* 15–21.
Antony, T., & Chakraborty, K., (2020). Anti-inflammatory polyether triterpenoids from the marine macroalga *Gracilaria salicornia*: Newly described natural leads attenuate prointlammatory 5-lipoxygenase and cyclooxygenase-2. *Algal Research, 47,* 101791.
Antunes, E. M., Afolayan, A. F., Chiwakata, M. T., Fakee, J., Knott, M. G., Whibley, C. E., Hendricks, D. T., et al., (2011). Identification and *in vitro* anti-esophageal cancer activity of a series of halogenated monoterpenes isolated from the South African seaweeds *Plocamium suhrii* and *Plocamium cornutum. Phytochemistry, 72,* 769–772.
Aquib, M., Farooq, M. A., Filli, M. S., Boakye-Yiadom, K. O., Kesse, S., Maviah, M. B. J., Mavlyanova, R., & Wang, B., (2019). A review on the chemotherapeutic role of fucoidan in cancer as nanomedicine. *Research Journal of Life Sciences, Bioinformatics, Pharmaceutical and Chemical Sciences, 5*(1), 512–539.

Ariffin, S. H. Z., Yeen, W. W., Abidin, I. Z. Z., Wahab, R. M. A., Ariffin, Z. Z., & Senafi, S., (2014). Cytotoxicity effect of degraded and undegraded kappa and iota carrageenan in human intestine and liver cell lines. *BMC Complementary and Alternative Medicine, 14,* 508.

Arioli, T., Mattner, S. W., & Winberg, P. C., (2015). Applications of seaweed extracts in Australian agriculture: Past, present and future. *J. Appl. Phycol., 27,* 2007–2015.

Arumugam, G., & Rajendran, R., (2019). Anti-candidal activity and synergetic interaction of antifungal drugs with differential extract of brown algae *Stocheospermum marginatum. Biocatal Agric. Biotechnol., 19,* 101145.

Arumugam, G., Rajendran, R., Khaleelullah, N. S., & Ramanathan, S., (2019). Anti-candidal and anti-virulence efficiency of selected seaweeds against azole resistance *Candida albicans. Biocatalysis and Agricultural Biotechnology, 20,* 101195.

Arunkumar, K., Selvapalam, N., & Rengasamy, R., (2005). The antibacterial compound sulphoglycerolipid 1-0 palmitoyl-3-0(6-sulpho-α-quinovopyranosyl)-glycerol from *Sargassum wightii* Greville (Phaeophyceae). *Botanica Marina, 40,* 441–445.

Aryee, A. N., Agyei, D., & Akanbi, T. O., (2018). Recovery and utilization of seaweed pigments in food processing. *Current Opinion in Food Science, 19,* 113–119.

Asharaja, A., & Sahayaraj, K., (2013). Screening of insecticidal activity of brown macroalgal extracts against *Dysdercus cingulatus* (Fab.) (Hemiptera: Pyrrhocoridae). *J. Biopest., 6,* 193–203.

Ashayerizadeh, O., Dastar, B., & Pourashouri, P., (2020). Study of antioxidant and antibacterial activities of depolymerized fucoidans extracted from *Sargassum tenerrimum. International Journal of Biological Macromolecules, 151,* 1259–1266.

Atashrazm, F., Lowenthal, R. M., Woods, G. M., Holloway, A. F., & Dickinson, J. L., (2015). Fucoidan and cancer: A multifunctional molecule with anti-tumor potential. *Marine Drugs, 13*(4), 2327–2346.

Ba-Akdah, M., Satheesh, S., & Al-Sofyani, A., (2016). Habitat preference and seasonal variability of epifaunal assemblages associated with macroalgal beds on the central Red Sea coast, Saudi Arabia. *J. Mar. Biol. Assoc. U.K., 96,* 1457–1467.

Bae, M., Kim, M. B., Park, Y. K., & Lee, J. Y., (2020). Health benefits of fucoxanthin in the prevention of chronic diseases. *Biochimica et Biophysica Acta (BBA) – Molecular and Cell Biology of Lipids, 1865*(11), 158618.

Baloch, G. N., Tariq, S., Ehteshamul-Haque, S., Athar, M., Sultana, V., & Ara, J., (2013). Management of root diseases of eggplant and watermelon with the application of asafoetida and seaweeds. *J. Appl. Bot. Food Qual., 86,* 138–142.

Barbosa, M., Lopes, G., Andrade, P. B., & Valentão, P., (2019). Bioprospecting of brown seaweeds for biotechnological applications: Phlorotannin actions in inflammation and allergy network. *Trends in Food Science and Technology, 86,* 153–171.

Battacharyya, D., Babgohari, M. Z., Rathor, P., & Prithiviraj, B., (2015). Seaweed extracts as biostimulants in horticulture. *Sci. Horticult., 196,* 39–48.

Bedoux, G., Caamal-Fuentes, E., Boulho, R., Marty, C., Bourgougnon, N., Freile-Pelegrín, Y., & Robledo, D., (2017). Antiviral and cytotoxic activities of polysaccharides extracted from four tropical seaweed species. *Natural Product Communications, 12*(6), 807–811.

Begum, S., Nyandoro, S., Buriyo, A., Makangara, J., Munissi, J., Duffy, S., Avery, V., & Erdelyi, M., (2018). Bioactivities of extracts, debromolaurinterol and fucosterol from macroalgae species. *Tanzania J. Sci., 44,* 104–116.

Belattmania, Z., Engelen, A. H., Pereira, H., Serrão, E. A., Barakate, M., Elatouani, S., Zrid, R., et al., (2016). Potential uses of the brown seaweed *Cystoseira humilis* biomass: 2-fatty acid composition, antioxidant and antibacterial activities. *J. Mater. Environ. Sci., 7,* 2074–2081.

Belghit, I., Rasinger, J. D., Heesch, S., Biancarosa, I., Liland, N., Torstensen, B., Waagbø, R., Lock, E. J., & Bruckner, C. G., (2017). In-depth metabolic profiling of marine macroalgae confirms strong biochemical differences between brown, red and green algae. *Algal Res., 26,* 240–249.

Ben, S. I., Aghrouss, S., Douira, A., Aissam, S., El Alaoui-Talibi, Z., Filali-Maltouf, A., & El Modafar, C., (2018). Seaweed polysaccharides as bio-elicitors of natural defenses in olive trees against Verticillium wilt of olive. *J. Plant Interact., 13,* 248–255.

Benelli, G., & Mehlhorn, H., (2016). Declining malaria, rising dengue and Zika virus: Insights for mosquito vector control. *Parasitol. Res., 115*(5), 1747–1754.

Bilal, H., Sahar, S., & Din, S., (2017). Bio-pesticides: New tool for the control of *Aedes (Stegomyia) albopictus* (Culicidae: Diptera) in Pakistan. *J. Arthropod. Borne. Dis., 11,* 278–285.

Blunden, G., Morse, P. F., Mathe, I., Hohmann, J., Critchleye, A. T., & Morrell, S., (2010). Betaine yields from marine algal species utilized in the preparation of seaweed extracts used in agriculture. *Nat. Prod. Commun., 5,* 581–585.

Bocanegra, A., Bastida, S., Benedí, J., Ródenas, S., & Sánchez-Muniz, F. J., (2009). Characteristics and nutritional and cardiovascular-health properties of seaweeds. *J. Med. Food, 12,* 236–258.

Boo, H. J., Hyun, J. H., Kim, S. C., Kang, J. I., Kim, M. K., Kim, S. Y., Cho, H., et al., (2011). Fucoidan from *Undaria pinnatifida* induces apoptosis in A549 human lung carcinoma cells. *Phytother. Res., 25,* 1082–1086.

Borbón, H., Herrera, J. M., Calvo, M., Sierra, H. T. L., Soto, R., & Vega, I., (2012). Antimicrobial activity of most abundant marine macroalgae of the Caribbean coast of Costa Rica. *Journal of Asian Scientific Research, 2,* 292–299.

Bray, F., Ferlay, J., Soerjomataram, I., Siegel, R. L., Torre, L. A., & Jemal, A., (2018). Global cancer statistics 2018: GLOBOCAN estimates of incidence and mortality worldwide for 36 cancers in 185 countries. *CA Cancer Journal for Clinicians, 68,* 394–424.

Campo, V. L., Kawano, D. F., Da Silva, D. B., & Carvalho, I., (2009). Carrageenans: Biological properties, chemical modifications and structural analysis – A review. *Carbohydrate Polymers, 77,* 167–180.

Cardoso, M. J., Costa, R. R., & Mano, J. F., (2016). Marine origin polysaccharides in drug delivery systems. *Marine Drugs, 14,* 34.

Cardoso, S. M., Carvalho, L. G., Silva, P. J., Rodrigues, M. S., Pereira, O. R., & Pereira, L., (2014). Bioproducts from seaweeds: A review with special focus on the Iberian Peninsula. *Current Organic Chemistry, 18,* 896–917.

Catherine, Q., Susanna, W., Isidora, E. S., Mark, H., Aurélie, V., & Jean-François, H., (2013). A review of current knowledge on toxic benthic freshwater cyanobacteria – Ecology, toxin production and risk management. *Water Res., 47,* 5464–5479.

Chai, T. T., Kwek, M. T., Ismail, N. I. M., Ooi, J. L. S., Amri, A. Y., Manan, F. A., Law, Y. C., & Wong, F. C., (2015). Antioxidant activities of methanol extract and solvent fractions of marine macroalga, *Avrainvillea erecta* (Berkeley) A. Gepp and E.S. Gepp (Dichotomosiphonaceae). *Trop. J. Pharm. Res., 14,* 503–509.

Chang, V. S., & Teo, S. S., (2016). Evaluation of heavy metal, antioxidant and anti-tyrosinase activities of Red Seaweed (*Eucheuma cottonii*). *Int. Food Res. J., 23*, 2370–2373.

Chapman, J., Hellio, C., Sullivan, T., Brown, R., Russell, S., Kiterringham, E., Le Nor, L., & Regan, F., (2014). Bioinspired synthetic macroalgae: Examples from nature for antifouling applications. *Int. Biodeterior. Biodegrad., 86,* 6–13.

Chen, J. H., Lim, J. D., Sohn, E. H., Choi, Y. S., & Han, E. T., (2009). Growth-inhibitory effect of a fucoidan from brown seaweed *Undaria pinnatifida* on *Plasmodium* parasites. *Parasitology Research, 104,* 245–250.

Chithra, K., Abraham, L. S., Thirugnanasambandham, R., & Prakash, P., (2016). A comparative analysis of the effect of seaweed fertilizers (Liquid and Paste) and vermiwash on seed germination of *Cyamopsis tetragonoloba. Biosciences Biotechnology Research Asia, 13,* 1089–1093.

Cho, H., Doan, T., Ha, T., Kim, H., Lee, B., Pham, H., Cho, T. O., & Oh, W. K., (2019). Dereplication by high-performance liquid chromatography (HPLC) with quadrupole-time-of-flight mass spectroscopy (qTOF-MS) and antiviral activities of phlorotannins from *Ecklonia cava. Marine Drugs, 17,* 149.

Choi, J. G., Kang, O. H., Brice, O. O., Lee, Y. S., Chae, H. S., Oh, Y. C., Sohn, D. H., et al., (2010). Antibacterial activity of *Ecklonia cava* against methicillin-resistant *Staphylococcus aureus* and *Salmonella* spp. *Foodborne Pathogens and Disease, 7,* 435–441.

Chojnacka, K., Michalak, I., Dmytryk, A., Gramza, M., Słowinski, A., & Górecki, H., (2015). Algal extracts as plant growth biostimulators. In: Kim, S. K., & Chojnacka, K., (eds.), *Marine Algae Extracts: Processes, Products, Applications* (pp. 189–211). Wiley-VCH Verlag GmbH and Co. KGaA: Weinheim, Germany.

Choo, G. S., Lee, H. N., Shin, S. A., Kim, H. J., & Jung, J. Y., (2016). Anticancer effect of fucoidan on DU-145 prostate cancer cells through inhibition of PI3K/Akt and MAPK pathway expression. *Marine Drugs, 14*, 126.

Chowdhury, M. T. H., Bangoura, I., Kang, J. Y., Cho, J. Y., Joo, J., Choi, Y. S., Hwang, D. S., & Hong, Y. K., (2014). Comparison of *Ecklonia cava, Ecklonia stolonifera* and *Eisenia bicyclis* for phlorotannin extraction. *J. Environ. Biol., 35,* 713–719.

Chrysargyris, A., Xylia, P., Anastasiou, M., Pantelides, I., & Tzortzakis, N., (2018). Effects of *Ascophyllum nodosum* seaweed extracts on lettuce growth, physiology and fresh-cut salad storage under potassium deficiency. *J. Sci. Food Agric., 98,* 5861–5872.

Cicinskas, E., Begun, M. A., Tiasto, V. A., Belousov, A. S., Vikhareva, V. V., Mikhailova, V. A., & Kalitnik, A. A., (2020). *In vitro* antitumor and immunotropic activity of carrageenans from red algae *Chondrus armatus* and their low-molecular weight degradation products. *Journal of Biomedical Materials Research Part A, 108*, 254–266.

Cirne-Santos, C. C., Barros, C. D. S., Nogueira, C. C. R., Azevedo, R. C., Yamamoto, K. A., Meira, G. L. S., Meira De, V. Z. F., et al., (2019). Inhibition by marine algae of Chikungunya virus isolated from patients in a recent disease outbreak in Rio de Janeiro. *Front. Microbiol., 10,* 2426.

Clark, T. A., Lee, H. P., Rolston, R. K., Zhu, X., Marlatt, M. W., Castellani, R. J., Nunomura, A., et al., (2010). Oxidative stress and its implications for future treatments and management of Alzheimer disease. *Int. J. Biomed. Sci., 6,* 225–227.

Cole, A. J., Roberts, D. A., Garside, A. L., De Nys, R., & Paul, N. A., (2016). Seaweed compost for agricultural crop production. *J. Appl. Phycol., 28*, 629–642.

Collins, K. G., Fitzgerald, G. F., Stanton, C., & Ross, R. P., (2016). Looking beyond the terrestrial: The potential of seaweed derived bioactives to treat non-communicable diseases. *Marine Drugs, 14,* 60.

Copping, L. G., (2004). *The Manual of Biocontrol Agents* (3rd edn., p. 702). BCPC Publications, Alton.

Craigie, J. S., (2011). Seaweed extract stimuli in plant science and agriculture. *J. Appl. Phycol., 23,* 371–393.

Creis, E., Delage, L., Charton, S., Goulitquer, S., Leblanc, C., Potin, P., & Ar Gall, E., (2015). Constitutive or inducible protective mechanisms against UV-B radiation in the brown alga *Fucus vesiculosus*? A study of gene expression and phlorotannin content responses. *PLoS One, 10*, e0128003.

Cunha, L., & Grenha, A., (2016). Sulfated seaweed polysaccharides as multifunctional materials in drug delivery applications. *Marine Drugs, 14,* 42.

Das, K., & Roychoudhury, A., (2014). Reactive oxygen species (ROS) and response of antioxidants as ROS-scavengers during environmental stress in plants. *Front. Environ. Sci., 2,* 53.

Dawczynski, C., Schubert, R., & Jahreis, G., (2007). Amino acids, fatty acids, and dietary fibre in edible seaweed products. *Food Chemistry, 103,* 891–899.

De Sousa, C. B., Lago, J. H. G., Macridachis, J., Oliveira, M., Brito, L., Vizetto-Duarte, C., Florindo, C., et al., (2019). Report of *in vitro* antileishmanial properties of Iberian macroalgae. *Natural Product Research, 33*, 1778–1782.

De Souza, L. A. R., Dore, C. M. P. G., Castro, A. J. G., De Azevedo, T. C. G., De Oliveira, M. T. B., Moura, M. F. V., Benevides, N. M. B., & Leite, E. L., (2012). Galactans from the Red Seaweed *Amansia multifida* and their effects on inflammation, angiogenesis, coagulation and cell viability. *Biomed. Prevent. Nutr., 2,* 154–162.

Déléris, P., Nazih, H., & Bard, J. M., (2016). Seaweeds in human health. In: Fleurence, J., & Levine, I., (eds.), *Seaweed in Health and Disease Prevention* (pp. 319–367). San Diego, Academic Press.

Demir, N., Dural, B., & Yildirim, K., (2006). Effect of seaweed suspensions on seed germination of tomato, pepper and aubergine. *J. Biol. Sci., 6,* 1130–1133.

Derbalah, A. S., Elkot, G. A. E., & Hamza, A. M., (2012). Laboratory evaluation of botanical extracts, microbial culture filtrates and silver nanoparticles against *Botrytis cinerea. Ann. Microbiol., 62,* 1331–1337.

Dhargalkar, V. K., & Kavlekar, D. P., (2004). *Seaweeds – A Field Manual* (p. 42). National Institute of Oceanography. Dona Paula, Goa..

Dore, C. M. P. G., Alves, M. G. C. F., Santos, N. D., Cruz, A. K. M., Câmara, R. B. G., Castro, A. J. G., Alves, L. G., et al., (2013). Antiangiogenic activity and direct antitumor effect from a sulfated polysaccharide isolated from seaweed. *Microvasc. Res., 88,* 12–18.

Dos, S. A. O., Veiga-Santos, P., Ueda-Nakamura, T., Filho, B. P. D., Sudatti, D. B., Bianco, É. M., Pereira, R. C., & Nakamura, C. V., (2010). Effect of elatol, isolated from Red Seaweed *Laurencia dendroidea,* on *Leishmania amazonensis. Marine Drugs, 8,* 2733–2743.

Du Jardin, P., (2015). Plant biostimulants: Definition, concept, main categories and regulation. *Sci. Hortic., 196,* 3–14.

Dumay, J., Morançais, M., Nguyen, H. P. T., & Fleurence, J., (2015). Extraction and purification of R-phycoerythrin from marine red algae. *Nat. Prod. Mar. Algae.: Method. Protocol, 1308*, 109–117.

El Hattab, M., Genta-Jouve, G., Bouzidi, N., Ortalo-Magné, A., Hellio, C., Maréchal, J. P., Piovetti, L., et al., (2015). Cystophloroketals A–E, unusual phloroglucinol-meroterpenoid hybrids from the brown alga *Cystoseira tamariscifolia*. *J. Nat. Prod., 78,* 1663–1670.

El Miniawy, S. M., Ragab, M. E., Youssef, S. M., & Metwally, A. A., (2014). Influence of foliar spraying of seaweed extract on growth, yield and quality of strawberry plants. *J. Appl. Sci. Res., 10,* 88–94.

El-Gawad, H. G. A., & Osman, H. S., (2014). Effect of exogenous application of boric acid and seaweed extract on growth, biochemical content and yield of eggplant. *J. Hortic. Sci. Ornam. Plants, 6,* 133–143.

Eom, S. H., Lee, D. S., Jung, Y. J., Park, J. H., Choi, J. I., Yim, M. J., Jeon, J. M., et al., (2014). The mechanism of antibacterial activity of phlorofucofuroeckol-A against methicillin-resistant *Staphylococcus aureus*. *Appl. Microbiol. Biotechnol., 98,* 9795–9804.

Eom, S. H., Moon, S. Y., Lee, D. S., Kim, H. J., Park, K., Lee, E. W., Kim, T. H., et al., (2015). *In vitro* antiviral activity of dieckol and phlorofucofuroeckol-A isolated from edible brown alga *Eisenia bicyclis* against murine norovirus. *Algae, 30,* 241–246.

Esserti, S., Smaili, A., Rifai, L. A., Koussa, T., Makroum, K., Belfaiza, M., Kabil, E. M., et al., (2017). Protective effect of three brown seaweed extracts against fungal and bacterial diseases of tomato. *J. Appl. Phycol., 29*, 1081–1093.

Esteves, A. I. S., Nicolai, M., Humanes, M., & Goncalves, J., (2011). Sulfated polysaccharides in marine sponges: Extraction and anti-HIV activity. *Marine Drugs, 9,* 139–153.

Farasat, M., Khavari-Nejad, R. A., Nabavi, S. M. B., & Namjooyan, F., (2014). Antioxidant activity, total phenolics and flavonoid contents of some edible green seaweeds from northern coasts of the Persian Gulf. *Iran. J. Pharm. Res., 13,* 163–170.

Feng, M. T., Yu, X. Q., Yang, P., Yang, H., Lin, K., & Mao, S. C., (2015). Two new antifungal polyunsaturated fatty acid ethyl esters from the red alga *Laurencia okamurai*. *Chem. Nat. Compd., 51,* 418–422.

Ferreres, F., Lopes, G., Gil-Izquierdo, A., Andrade, P. B., Sousa, C., Mouga, T., & Valentao, P., (2012). Phlorotannin extracts from fucales characterized by HPLC-DAD-ESI-MSn: Approaches to hyaluronidase inhibitory capacity and antioxidant properties. *Marine Drugs, 10*, 2766–2781.

Frioni, T., Sabbatini, P., Tombesi, S., Norrie, J., Poni, S., Gatti, M., & Palliotti, A., (2018). Effects of a biostimulant derived from the brown seaweed *Ascophyllum nodosum* on ripening dynamics and fruit quality of grapevines. *Scientia Horticulturae, 232*, 97–106.

Fu, C. W. F., Ho, C. W., Yong, W. T. L., Abas, F., Tan, T. B., & Tan, C. P., (2016). Effects of phenolic antioxidants extraction from four selected seaweeds obtained from Sabah. *International Food Research Journal, 23*, 2363–2369.

Funami, T., Hiroe, M., Noda, S., Asai, I., Ikeda, S., & Nishinari, K., (2007). Influence of molecular structure imaged with atomic force microscopy on the rheological behavior of carrageenan aqueous systems in the presence or absence of cations. *Food Hydrocolloids, 21*, 617–629.

Garrido, V., Barros, C., Melchiades, V. A., Fonseca, R. R., Pinheiro, S., Ocampo, P., Teixeira, V. L., et al., (2017). Subchronic toxicity and anti-HSV-1 activity in experimental animal of dolabelladienetriol from the seaweed, *Dictyota pfaffii*. *Regulatory Toxicology and Pharmacology 86*, 193–198.

Gheda, S. F., El-Adawi, H. I., & EL-Deeb, N. M., (2016). Antiviral profile of brown and Red Seaweed polysaccharides against hepatitis C virus. *Iranian Journal of Pharmaceutical Research, 15*, 483–491.

Giada, M. D. L. R., (2013). Food phenolic compounds: Main classes, sources and their antioxidant power. In: Morales-Gonzalez, J. A., (ed.), *Oxidative Stress and Chronic Degenerative Diseases – A Role for Antioxidants* (pp. 87–112). Rijeka, Croatia: InTech.

Godlewska, K., Michalak, I., Tuhy, A., & Chojnacka, K., (2016). Plant growth biostimulants based on different methods of seaweed extraction with water. *BioMed Research International, 2016,* 5973760.

Gómez-Guzmán, M., Rodríguez-Nogales, A., Algieri, F., & Gálvez, J., (2018). Potential role of seaweed polyphenols in cardiovascular-associated disorders. *Marine Drugs, 16,* 250.

Greff, S., Zubia, M., Genta-Jouve, G., Massi, L., Perez, T., & Thomas, O. P., (2014). Mahorones, highly brominated cyclopentenones from the red alga *Asparagopsis taxiformis. J. Nat. Prod., 77,* 1150–1155.

Grosser, K., Zedler, L., Schmitt, M., Dietzek, B., Popp, J., & Pohnert, G., (2012). Disruption-free imaging by Raman spectroscopy reveals a chemical sphere with antifouling metabolites around macroalgae. *Biofouling, 28,* 687–696.

Grzywacz, D., Stevenson, P. C., Mushobozi, L. M., Belmain, S., & Wilson, K., (2014). The use of indigenous ecological resources for pest control in Africa. *Food Sec., 6,* 71–86.

Gullón, B., Gagaoua, M., Barbac, F. J., Gullón, P., Zhang, W., & Lorenzo, J. M., (2020). Seaweeds as promising resource of bioactive compounds: Overview of novel extraction strategies and design of tailored meat products. *Trends in Food Science and Technology, 100,* 1–18.

Guschina, I. A., & Harwood, J. L., (2006). Lipids and lipid metabolism in eukaryotic algae. *Progress in Lipid Research, 45,* 160–186.

Gutiérrez-Rodríguez, A. G., Juárez-Portilla, C., Olivares-Bañuelos, T., & Zepeda, R. C., (2018). Anticancer activity of seaweeds. *Drug Discovery Today, 23,* 434–447.

Güven, K. C., Percot, A., & Sezik, E., (2010). Alkaloids in marine algae. *Marine Drugs, 8,* 269–284.

Halliwell, B., (2013). The antioxidant paradox: Less paradoxical now? *Br. J. Clin. Pharmacol., 75,* 637–644.

Hamed, S. M., Abd, E. A. A., Abdel-Raouf, N., & Ibraheem, B. M. I., (2017). Role of marine macroalgae in plant protection and improvement for sustainable agriculture technology. *Beni-Suef University Journal of Basic and Applied Sciences, 7,* 104–110.

Han, J. G., Syed, A. Q., Kwon, M., Ha, J. H., & Lee, H. Y., (2008). Antioxidant, immunomodulatory and anticancer activity of fucoidan isolated from *Fucus vesiculosus. J. Biotechnol., 136,* S571.

Haq, S. H., Al-Ruwaished, G., Al-Mutlaq, M. A., Naji, S. A., Al-Mogren, M., Al-Rashed, S., Tul, A. Q., et al., (2019). Antioxidant, anticancer activity and phytochemical analysis of green algae, *Chaetomorpha* collected from the Arabian Gulf. *Sci. Rep., 9,* 18906.

Harke, M. J., Steffn, M. M., Gobler, C. J., Otten, T. G., Wilhelm, S. W., Wood, S. A., & Paerl, H. W., (2016). A review of the global ecology, genomics and biogeography of the toxic cyanobacterium, *Microcystis* spp. *Harmful Algae, 54,* 4–20.

Harnedy, P. A., & FitzGerald, R. J., (2013). Cardioprotective peptides from marine sources. *Curr. Protein Pept. Sci., 14,* 162–172.

Hellio, C., Pascal, B. J., Beaupoil, C., Le Gal, Y., & Bougougnon, N., (2002). Screening of marine algal extracts for anti-settlement activities against microalgae and macroalgae. *Biofouling, 18,* 205–215.

Heo, S. J., & Jeon, Y. J., (2009). Evaluation of diphlorethohydroxycarmalol isolated from *Ishige okamurae* for radical scavenging activity and its protective effect against H_2O_2-induced cell damage. *Process Biochem., 44,* 412–418.

Heo, S. J., Hwang, J. Y., Choi, J. I., Han, J. S., Kim, H. J., & Jeon, Y. J., (2009). Diphlorethohydroxycarmalol isolated from *Ishige okamurae*, a brown algae, a potent α-glucosidase and α-amylase inhibitor, alleviates postprandial hyperglycemia in diabetic mice. *Eur. J. Pharmacol., 615,* 252–256.

Holdt, S. L., & Kraan, S., (2011). Bioactive compounds in seaweed: Functional food applications and legislation. *J. Appl. Phycol., 23,* 543–597.

Hong, D. D., Hien, H. M., & Son, P. N., (2007). Seaweeds from Vietnam used for functional food, medicine and biofertilizer. *J. Appl. Phycol., 19,* 817–826.

Hsu, H. Y., Lin, T. Y., Lu, M. K., Leng, P. J., Tsao, S. M., & Wu, Y. C., (2017). Fucoidan induces Toll-like receptor 4-regulated reactive oxygen species and promotes endoplasmic reticulum stress-mediated apoptosis in lung cancer. *Sci. Rep., 7,* 44990.

Hu, D. J., Cheong, K. L., Zhao, J., & Li, S. P., (2013). Chromatography in characterization of polysaccharides from medicinal plants and fungi. *J. Sep. Sci., 36,* 1–19.

Ibraheem, B. M. I., Hamed, S. M., Abd, E. A. A., Farag, F. M., & Abdel-Raouf, N., (2017). Antimicrobial activities of some brown macroalgae against some soil borne plant pathogens and *in vivo* management of *Solanum melongena* root diseases. *Aust. J. Basic Appl. Sci., 11,* 157–168.

Jeong, J. H., Jin, H. J., Sohn, C. H., Suh, K. H., & Hong, Y. K., (2000). Algicidal activity of the seaweed *Corallina pilulifera* against red tide microalgae. *J. Appl. Phycol., 12,* 37–43.

Ji, R., Dong, G., Shi, W., & Min, J., (2017). Effects of liquid organic fertilizers on plant growth and rhizosphere soil characteristics of *Chrysanthemum. Sustainability, 9,* 841.

Jiao, G., Yu, G., Zhang, J., & Ewart, H. S., (2011). Chemical structures and bioactivities of sulfated polysaccharides from marine algae. *Marine Drugs, 9,* 196–223.

Jiménez, E., Dorta, F., Medina, C., Ramírez, A., Ramírez, I., & Peña-Cortés, H., (2011). Anti-phytopathogenic activities of macroalgae extracts. *Marine Drugs, 9,* 739–756.

Jormalainen, V., & Honkanen, T., (2008). Macroalgal chemical defenses and their roles in structuring temperate marine communities. In: Amsler, C. D., (ed.), *Algal Chemical Ecology* (pp. 57–90). Springer-Verlag, Berlin.

Jung, H. A., Jung, H. J., Jeong, H. Y., Kwon, H. J., Ali, M. Y., & Choi, J. S., (2014). Phlorotannins isolated from the edible brown alga *Ecklonia stolonifera* exert anti-adipogenic activity on 3T3 L1 adipocytes by downregulating C/EBP α and PPARγ. *Fitoterapia, 92,* 260–269.

Kamada, T., & Vairappan, C. S., (2015). New Laurene-type sesquiterpene from Bornean *Laurencia nangii. Nat. Prod. Commun., 10,* 843–844.

Kang, M. C., Ding, Y., Kim, E. A., Choi, Y. K., De Araujo, T., Heo, S. J., & Lee, S. H., (2017). Indole derivatives isolated from brown alga *Sargassum thunbergii* inhibit adipogenesis through AMPK activation in 3T3-L1 preadipocytes. *Marine Drugs, 15,* 119.

Kang, S. Y., Kim, E., Kang, I., Lee, M., & Lee, Y., (2018). Anti-diabetic effects and anti-inflammatory effects of *Laminaria japonica* and *Hizikia fusiforme* in skeletal muscle: *In vitro* and *in vivo* model. *Nutrients, 10,* 491.

Karkhane, Y. M., Seyed, H. M., Mashinchian, M. A., & Ghasempour, A. R., (2018). *In vitro* investigating of anticancer activity of focuxanthin from marine brown seaweed species. *Global J. Environ. Sci. Manage., 4,* 81–90.

Kendel, M., Wielgosz-Collin, G., Bertrand, S., Roussakis, C., Bourgougnon, N., & Bedoux, G., (2015). Lipid composition, fatty acids and sterols in the seaweeds *Ulva armoricana*, and

Solieria chordalis from Brittany (France): An analysis from nutritional, chemotaxonomic, and antiproliferative activity perspectives. *Marine Drugs, 13,* 5606–5628.

Khallil, A. M., Daghman, I. M., & Fady, A. A., (2015). Antifungal potential in crude extracts of five selected brown seaweeds collected from the western Libya coast. *J. Microbiol. Mod. Tech., 1,* 103.

Khan, F., Manivasagan, P., Lee, J. W., Pham, D. T. N., Oh, J., & Kim, Y. M., (2019). Fucoidan-stabilized gold nanoparticle-mediated biofilm inhibition, attenuation of virulence and motility properties in *Pseudomonas aeruginosa* PAO1. *Marine Drugs, 17,* 208.

Khan, W., Rayirath, U. P., Subramanian, S., Jithesh, M. N., Rayorath, P., Hodges, D. M., Critchley, A. T., et al., (2009). Seaweed extracts as biostimulants of plant growth and development. *J. Plant Growth Regul., 28,* 386–399.

Khelil-Radji, F., Belhouari, M. Y., Chemlal-Kherraz, D., Matallah-Boutiba, A., & Boutiba, Z., (2017). Antimicrobial activity of aqueous and ethanol extracts of two marine algae collected from Algerian west coast. *Int. J. Biosci., 10,* 217–222.

Kim, E., & Kwak, J., (2015). Antiviral phlorotannin from *Eisenia bicyclis* against human papilloma virus *in vitro*. *Planta Med., 81.*

Kim, M. S., Kim, J. Y., Choi, W. H., & Lee, S. S., (2008). Effects of seaweed supplementation on blood glucose concentration, lipid profile, and antioxidant enzyme activities in patients with type 2 diabetes mellitus. *Nutr. Res. Pract., 2,* 62–67.

Kim, S. M., Kang, S. W., Jeon, J. S., Jung, Y. J., Kim, W. R., Kim, C. Y., & Um, B. H., (2013). Determination of major phlorotannins in *Eisenia bicyclis* using hydrophilic interaction chromatography: Seasonal variation and extraction characteristics. *Food Chemistry, 138,* 2399–2406.

Kraan, S., (2012). Algal polysaccharides, novel applications and outlook. In: Chang, C. F., (ed.), *Carbohydrates – Comprehensive Studies on Glycobiology and Glycotechnology* (Vol. 22, pp. 489–532).

Krinsky, N. I., & Johnson, E. J., (2005). Carotenoid actions and their relation to health and disease. *Mol. Aspects Med., 26,* 459–516.

Kulkarni, M. G., Rengasamy, K. R. R., Pendota, S. C., Gruz, J., Plačková, L., Novák, O., Doležal, K., & Van, S. J., (2019). Bioactive molecules derived from smoke and seaweed *Ecklonia maxima* showing phytohormone-like activity in *Spinacia oleracea* L. *N. Biotechnol., 48,* 83–89.

Kumar, C. S., Raju, D., Sarada, V. L., & Rengasamy, R., (2008). Seaweed extracts control the leaf spot disease of the medicinal plant *Gymnema sylvestre*. *Indian J. Sci. Technol., 1,* 1–5.

Kumar, K. S., Ganesan, K., Selvaraj, K., & Rao, P. S., (2014). Studies on the functional properties of protein concentrate of *Kappaphycus alvarezii* (Doty) Doty – An edible seaweed. *Food Chemistry, 153,* 353–360.

Kumaresan, M., Vijai, A. K., Govindaraju, K., Tamilselvan, S., & Ganesh, K. V., (2018). Seaweed *Sargassum wightii* mediated preparation of zirconia (ZrO_2) nanoparticles and their antibacterial activity against gram positive and gram negative bacteria. *Microb. Pathog., 124,* 311–315.

Kumari, P., Kumar, M., Gupta, V., Reddy, C. R. K., & Jha, B., (2010). Tropical marine macroalgae as potential sources of nutritionally important PUFAs. *Food Chemistry, 120,* 749–757.

Lahaye, M., & Robic, A., (2007). Structure and functional properties of ulvan, a polysaccharide from green seaweeds. *Biomacromolecules, 8,* 1765–1774.

Lange, K. W., Hauser, J., Nakamura, Y., & Kanaya, S., (2015). Dietary seaweeds and obesity. *Food Sci. Human Wellness, 4,* 87–96.

Latique, S., Elouaer, M. A., Chernane, H., Hannachi, C., & Elkaoua, M., (2014). Effect of seaweed liquid extract of *Sargassum vulgare* on growth of durum wheat seedlings (*Triticum durum* L.) under salt stress. *Int. J. Innov. Appl. Stud., 7,* 1430–1435.

Layek, J., Das, A., Ramkrushna, G. I., Sarkar, D., Ghosh, A., Tukaram, Z. S., Lal, R., et al., (2018). Seaweed extract as organic bio-stimulant improves productivity and quality of rice in eastern Himalayas. *J. Appl. Phycol., 30,* 547–558.

Layek, J., Das, A., Ramkrushna, G. I., Trivedi, K., Yesuraj, D., Chandramohan, M., Kubavat, D., et al., (2015). Seaweed sap potential towards sustainable improvement of maize productivity: A dominant staple food crop of the North-east India. *Int. J. Environ. Stud., 72,* 305–315.

Leal, M. C., Munro, M. H. G., Blunt, J. W., Puga, J., Jesus, B., Calado, R., Rosa, R., & Madeira, C., (2013). Biogeography and biodiscovery hotspots of macroalgal marine natural products. *Nat. Prod. Rep., 30,* 1380–1390.

Lee, D. S., Kang, M. S., Hwang, H. J., Eom, S. H., Yang, J. Y., Lee, M. S., Lee, W. J., et al., (2008). Synergistic effect between dieckol from *Ecklonia stolonifera* and β-lactams against methicillin-resistant *Staphylococcus aureus*. *Biotechnol. Bioproc. Eng., 13,* 758–764.

Lee, H. G., Lu, Y. A., Li, X., Hyun, J. M., Kim, H. S., Lee, J. J., Kim, T. H., et al., (2020). Anti-obesity effects of *Grateloupia elliptica*, a Red Seaweed, in mice with high-fat diet-induced obesity via suppression of adipogenic factors in white adipose tissue and increased thermogenic factors in brown adipose tissue. *Nutrients, 12*, 308.

Lee, J. H., Jung, H. A., Kang, M. J., Choi, J. S., & Kim, G. D., (2017). Fucosterol, isolated from *Ecklonia stolonifera*, inhibits adipogenesis through modulation of FoxO1 pathway in 3T3-L1 adipocytes. *J. Pharm. Pharmacol., 69,* 325–333.

Lee, M. S., Shin, T., Utsuki, T., Choi, J. S., Byun, D. S., & Kim, H. R., (2012a). Isolation and identification of phlorotannins from *Ecklonia stolonifera* with antioxidant and hepatoprotective properties in tacrine-treated HepG2 cells. *J. Agric. Food Chem., 60,* 5340–5349.

Lee, R. E., (2008). *Phycology,* (4th edn., p. 560). Cambridge University Press: Cambridge, UK.

Lee, S. H., Choi, J. I., Heo, S. J., Park, M. H., Park, P. J., Jeon, B. T., Kim, S. K., et al., (2012b). Diphlorethohydroxycarmalol isolated from Pae (*Ishige okamurae*) protects high glucose-induced damage in RINm5F pancreatic β cells via its antioxidant effects. *Food Sci. Biotechnol., 21,* 239–246.

Lee, S. H., Choi, J. I., Heo, S. J., Park, M. H., Park, P. J., Jeon, B. T., Kim, S. K., et al., (2012c). Diphlorethohydroxycarmalol isolated from Pae (*Ishige okamurae*) protects high glucose-induced damage in RINm5F pancreatic β cells via its antioxidant effects. *Food Sci. Biotechnol., 21,* 239–246.

Lee, S., Lee, Y. S., Jung, S. H., Kang, S. S., & Shin, K. H., (2003). Anti-oxidant activities of fucosterol from the marine algae *Pelvetia siliquosa*. *Arch. Pharm. Res., 26,* 719–722.

Lehnhardt, P. C., Rodrigues, S. D., Bristot, D., Gaeta, H. H., De Oliveira, T. D., Lobo, F. W. R., & Toyama, M. H., (2013). Evaluation of macroalgae sulfated polysaccharides on the *Leishmania amazonensis* (L.) promastigote. *Marine Drugs, 11*, 934–943.

Li, H., Li, J., Tang, Y., Lin, L., Xie, Z., Zhou, J., Zhang, L., et al., (2017). Fucoidan from *Fucus vesiculosus* suppresses hepatitis B virus replication by enhancing extracellular signal-regulated Kinase activation. *Virology Journal, 14*, 178.

Li, Y. X., Wijesekara, I., Li, Y., & Kim, S. K., (2011). Phlorotannins as bioactive agents from brown algae. *Process Biochemistry, 46,* 2219–2224.

Liang, B. H., Hong, B. L., Jun, L. S., & Zeng, J., (2012). Effect of laminarin on *Aspergillus flavus* growth and aflatoxin production. *Adv. Mater. Res., 343, 344*, 1168–1171.

Lim, S. N., Cheung, P. C. K., Ooi, V. E. C., & Ang, P. O., (2002). Evaluation of antioxidative activity of extracts from a brown seaweed, *Sargassum siliquastrum*. *J. Agric. Food Chem., 50,* 3862–3866.

Lin, Y., Qi, X., Liu, H., Xue, K., Xu, S., & Tian, Z., (2020). The anti-cancer effects of fucoidan: A review of both *in vivo* and *in vitro* investigations. *Cancer Cell International., 20,* 154.

Liu, H., Chen, X., Song, L., Li, K., Zhang, X., Liu, S., Qin, Y., & Li, P., (2019). Polysaccharides from *Grateloupia filicina* enhance tolerance of rice seeds (*Oryza sativa* L.) under salt stress. *Int. J. Biol. Macromol., 124,* 1197–1204.

Liu, M., Li, W., Chen, Y., Wan, X., & Wang, J., (2020). Fucoxanthin: A promising compound for human inflammation-related diseases. *Life Sciences, 255,* 117850.

Liu, M., Wang, G., Xiao, L., Xu, A., Liu, X., Xu, P., & Lin, X., (2014). Bis (2,3-dibromo-4,5 dihydroxybenzyl) ether, a marine algae derived bromophenol, inhibits the growth of *Botrytis cinerea* and interacts with DNA molecules. *Marine Drugs, 12,* 3838–3851.

Lopes, G., Andrade, P. B., & Valentão, P., (2016). Phlorotannins: Towards new pharmacological interventions for diabetes mellitus type 2. *Molecules, 22,* 56.

Lopes, G., Pinto, E., Andrade, P. B., & Valentão, P., (2013). Antifungal activity of phlorotannins against dermatophytes and yeasts: Approaches to the mechanism of action and influence on *Candida albicans* virulence factor. *PLoS One, 8*(8), e72203.

Lordan, S., Ross, R. P., & Stanton, C., (2011). Marine bioactives as functional food ingredients: Potential to reduce the incidence of chronic diseases. *Marine Drugs, 9,* 1056–1100.

Lu, Y. A., Lee, H. G., Li, X., Hyun, J. M., Kim, H. S., Kim, T. H., Kim, H. M., et al., (2020). Anti-obesity effects of Red Seaweed, *Plocamium telfairiae*, in C57BL/6 mice fed a high-fat diet. *Food and Function, 11*, 2299–2308.

Machu, L., Misurcova, L., Vavra, A. J., Orsavova, J., Mlcek, J., Sochor, J., & Jurikova, T., (2015). Phenolic content and antioxidant capacity in algal food products. *Molecules, 20,* 1118–1133.

Mahmoud, S. H., Salama, D. M., El-Tanahy, A. M. M., & El-Samad, E. H. A., (2019). Utilization of seaweed (*Sargassum vulgare*) extract to enhance growth, yield and nutritional quality of red radish plants. *Annals of Agricultural Sciences, 64*, 167–175.

Mak, W., Wang, S. K., Liu, T., Hamid, N., Li, Y., Lu, J., & White, W. L., (2014). Anti-proliferation potential and content of fucoidan extracted from sporophyll of New Zealand *Undaria pinnatifida*. *Front. Nutr., 1,* 9.

Mamelona, J., Pelletier, É., Girard-Lalancette, K., Legault, J., Karboune, S., & Kermasha, S., (2007). Quantification of phenolic contents and antioxidant capacity of Atlantic sea cucumber, *Cucumaria frondosa*. *Food Chem., 104*, 1040–1047.

Manilal, A., Sujith, S., Seghal, K. G., Selvin, J., & Panikkar, M. V. N., (2011). Evaluation of seaweed bioactives on common aquatic floral and faunal weeds of shrimp ponds. *Thalassas, 27,* 47–56.

Mannino, A. M., Vaglica, V., Cammarata, M., & Oddo, E., (2016). Effects of temperature on total phenolic compounds in *Cystoseira amentacea* (C. Agardh). Bory. (Fucales, Phaeophyceae) from southern Mediterranean Sea. *Plant. Biosyst., 150,* 152–160.

Marinho, G. S., Holdt, S. L., & Angelidaki, I., (2015). Seasonal variations in the amino acid profile and protein nutritional value of *Saccharina latissima* cultivated in a commercial IMTA system. *J. Appl. Phycol., 27*, 1991–2000.

Martínez, J. H. I., & Castañeda, H. G. T., (2013). Preparation and chromatographic analysis of phlorotannins. *J. Chromatogr. Sci., 51*, 825–838.

Mattner, S. W., Milinkovic, M., & Arioli, T., (2018). Increased growth response of strawberry roots to a commercial extract from *Durvillaea potatorum* and *Ascophyllum nodosum. J. Appl. Phycol., 30,* 2943–2951.

Mayer, A. M. S., Guerrero, A. J., Rodríguez, A. D., Taglialatela-Scafati, O., Nakamura, F., & Fusetani, N., (2020). Marine pharmacology in 2014–2015: Marine compounds with antibacterial, antidiabetic, antifungal, anti-inflammatory, antiprotozoal, antituberculosis, antiviral, and anthelmintic activities; affecting the immune and nervous systems, and other miscellaneous mechanisms of action. *Marine Drugs, 18*, 5.

Mei, C. H., Zhou, S. C., Zhu, L., Ming, J. X., Zeng, F. D., & Xu, R., (2017). Antitumor effects of *Laminaria* extract fucoxanthin on lung cancer. *Marine Drugs, 15*, 39.

Michalak, I., & Chojnacka, K., (2016). The potential usefulness of a new generation of agro-products based on raw materials of biological origin. *Acta. Sci. Pol. Hortorum Cultus., 15*, 97–120.

Michalak, I., Chojnacka, K., & Saeid, A., (2017). Plant growth biostimulants, dietary feed supplements and cosmetics formulated with supercritical CO_2 algal extracts. *Molecules, 22*, 66.

Milledge, J. J., Nielsen, B. V., & Bailey, D., (2016). High-value products from macroalgae: The potential uses of the invasive brown seaweed, *Sargassum muticum. Reviews in Environmental Science and Bio/Technology, 15*, 67–88.

Min, K. H., Kim, H. J., Jeon, Y. J., & Han, J. S., (2011). *Ishige okamurae* ameliorates hyperglycemia and insulin resistance in C57BL/KsJ-db/db mice. *Diabetes Res. Clin. Pr., 93,* 70–76.

Mishra, A. K., (2018). *Sargassum*, *Gracilaria* and *Ulva* exhibit positive antimicrobial activity against human pathogens. *OALib., 5,* 1–11.

Mishra, J. K., Srinivas, T., & Sawhney, S., (2016). Antibacterial activity of seaweed *Halimeda opuntia* from the coasts of South Andaman. *Glob. J. Biosci. Biotechnol., 5,* 345–348.

Miyashita, K., Beppu, F., Hosokawa, M., Liu, X., & Wang S., (2020). Nutraceutical characteristics of the brown seaweed carotenoid fucoxanthin. *Archives of Biochemistry and Biophysics, 686*, 108364.

Miyashita, K., Mikami, N., & Hosokawa, M., (2013). Chemical and nutritional characteristics of brown seaweed lipids: A review. *J. Funct. Foods, 5,* 1507–1517.

Mohamed, S. S., & Saber, A. A., (2019). Antifungal potential of the bioactive constituents in extracts of the mostly untapped brown seaweed *Hormophysa cuneiformis* from the Egyptian coastal waters. *Egypt. J. Bot., 59*, 695–708.

Mohamed, S., Hashim, S. N., & Abdul, R. H., (2012). Seaweeds: A sustainable functional food for complementary and alternative therapy. *Trends Food Sci. Technol., 23,* 83–96.

Montero, L., Sánchez-Camargo, A. D. P., Ibáñez, E., & Gilbert-López, B., (2018). Phenolic compounds from edible algae: Bioactivity and health benefits. *Current Medicinal Chemistry, 25,* 4808–4826.

Montero, L., Sánchez-Camargo, A. P., García-Cañas, V., Tanniou, A., Stiger-Pouvreau, V., Russo, M., Rastrelli, L., et al., (2016). Anti-proliferative activity and chemical characterization by comprehensive two-dimensional liquid chromatography coupled to mass spectrometry of phlorotannins from the brown macroalga *Sargassum muticum* collected on North-Atlantic coasts. *J. Chromatogr. A, 1428*, 115–125.

Morán-Santibañez, K., Peña-Hernández, M. A., Cruz-Suárez, L. E., Ricque-Marie, D., Skouta, R., Vasquez, A. H., Rodríguez-Padilla, C., & Trejo-Avila, L. M., (2018). Virucidal and synergistic activity of polyphenol-rich extracts of seaweeds against Measles virus. *Viruses, 10,* 465.

Moubayed, N. M. S., Al Houri, H. J., Al Khulaifi, M. M., & Al Farraj, D. A., (2017). Antimicrobial, antioxidant properties and chemical composition of seaweeds collected from Saudi Arabia (Red Sea and Arabian Gulf). *Saudi J. Biol. Sci., 24,* 162–169.

Muñoz, J., Culioli, G., & Köck, M., (2013). Linear diterpenes from the marine brown alga *Bifurcaria bifurcata*: A chemical perspective. *Phytochem. Rev., 12,* 407–424.

Murata, M., & Nakazoe, J., (2001). Production and use of marine algae in Japan. *Japan Agricultural Research Quarterly, 35*, 281–290.

Murugan, K., Panneerselvam, C., Subramaniam, J., Madhiyazhagan, P., Hwang, J. S., Wang, L., Dinesh, D., et al., (2016). Eco-friendly drugs from the marine environment: Spongeweed-synthesized silver nanoparticles are highly effective on *Plasmodium falciparum* and its vector *Anopheles stephensi,* with little non-target effects on predatory copepods. *Environ Sci. Pollut. Res. Int., 23*, 16671–16685.

Murugan, K., Roni, M., Panneerselvam, C., Aziz, A. T., Suresh, U., Rajaganesh, R., Aruliah, R., et al., (2018). *Sargassum wightii* – synthesized ZnO nanoparticles reduce the fitness and reproduction of the malaria vector *Anopheles stephensi* and cotton bollworm *Helicoverpa armigera. Physiological and Molecular Plant Pathology, 101*, 202–213.

Mwangi, M. H., Njue, W. E., Onani, M. O., Thovhoghi, N., & Mabusela, W. T., (2017). Phlorotannins and a sterol isolated from a brown alga *Ecklonia maxima*, and their cytotoxic activity against selected cancer cell lines HeLa, H157 and MCF7. *Interdisciplinary Journal of Chemistry, 2,* 1–6.

Nakai, S., Inoue, Y., Hosomi, M., & Murakami, A., (2000). *Myriophyllum spicatum*-released allelopathic polyphenols inhibiting growth of blue-green algae *Microcystis aeruginosa. Water Res., 34,* 3026–3032.

Nakajima, N., Sugimoto, N., Ohki, K., & Kamiya, M., (2016). Diversity of phlorotannin profiles among sargassasacean species affecting variation and abundance of epiphytes. *Eur. J. Phycol., 51,* 307–316.

Namvar, F., Mohamed, S., Fard, S. G., Behravan, J., Mustapha, N. M., Alitheen, N. B. M., & Othman, F., (2012). Polyphenol-rich seaweed (*Eucheuma cottonii*) extract suppresses breast tumor via hormone modulation and apoptosis induction. *Food Chem., 130,* 376–382.

Narayani, S. S., Saravanan, S., Ravindran, J., Ramasamy, M. S., & Chitra, J., (2019). *In vitro* anticancer activity of fucoidan extracted from *Sargassum cinereum* against Caco-2 cells. *Int. J. Biol. Macromol., 1*, 618–628.

Nwosu, F., Morris, J., Lund, V. A., Stewart, D., Ross, H. A., & McDougall, G. J., (2011). Anti-proliferative and potential anti-diabetic effects of phenolic-rich extracts from edible marine algae. *Food Chem., 126,* 1006–1012.

Nylund, G. M., Enge, S., & Pavia, H., (2013). Costs and benefits of chemical defense in the red alga *Bonnemaisonia hamifera. PLoS One, 8*(4), e61291.

O'Sullivan, L., Murphy, B., McLoughlin, P., Duggan, P., Lawlor, P. G., Hughes, H., & Gardiner, G. E., (2010). Prebiotics from marine macroalgae for human and animal health applications. *Marine Drugs, 8,* 2038–2064.

Ojulari, O. V., Lee, S. G., & Nam, J. O., (2020). Therapeutic effect of seaweed derived xanthophyl carotenoid on obesity management: Overview of the last decade. *Int. J. Mol. Sci., 21*, 2502.

Onet, A., Dincua, L. C., Grenni, P., Laslo, V., Teusdea, A. C., Vasile, D. L., Enescu, R. E., & Crisan, V. E., (2019). Biological indicators for evaluating soil quality improvement in a soil degraded by erosion processes. *J. Soils Sediments, 19,* 2393–2404.

Ooi, V. E., & Liu, F., (2000). Immunomodulation and anti-cancer activity of polysaccharide-protein complexes. *Curr. Med. Chem., 7*, 715–729.

Pádua, D., Rocha, E., Gargiulo, D., & Ramos, A. A., (2015). Bioactive compounds from brown seaweeds: Phloroglucinol, fucoxanthin and fucoidan as promising therapeutic agents against breast cancer. *Phytochem. Lett., 14,* 91–98.

Paerl, H. W., & Huisman, J., (2008). Climate: Blooms like it hot. *Science, 320,* 57–58.

Paniagua-Michel, J. D. J., Olmos-Soto, J., & Morales-Guerrero, E. R., (2014). Algal and microbial exopolysaccharides: New insights as biosurfactants and bioemulsifiers. In: Kim, S. K., (ed.), *Advances in Food and Nutrition Research* (pp. 221–257). Academic Press, San Diego, California, USA.

Papenfus, H. B., Stirk, W. A., Finnie, J. F., & Van, S. J., (2012). Seasonal variation in the polyamines of *Ecklonia maxima. Bot. Mar., 55,* 539–546.

Paradas, W. C., Salgado, L. T., Pereira, R. C., Hellio, C., Atella, G. C., De Lima, M. D., Do Carmo, A. P. B., et al., (2016). A novel antifouling defense strategy from Red Seaweed: Exocytosis and deposition of fatty acid derivatives at the cell wall surface. *Plant Cell Physiol., 57,* 1008–1019.

Pardo-Vargas, A., De Barcelos, O. I., Stephens, P. R., Cirne-Santos, C. C., De Palmer, P. I. C., Ramos, F. A., Jiménez, C., et al., (2014). Dolabelladienols A-C, new diterpenes isolated from Brazilian brown alga *Dictyota pfaffii. Marine Drugs, 12,* 4247–4259.

Park, H. S., Hwang, H. J., Kim, G. Y., Cha, H. J., Kim, W. J., Kim, N. D., Yoo, Y. H., & Choi, Y. H., (2013). Induction of apoptosis by fucoidan in human leukemia U937 cells through activation of p38 MAPK and modulation of Bcl-2 family. *Marine Drugs, 11*, 2347–2364.

Park, H. Y., Han, M. H., Park, C., Jin, C. Y., Kim, G. Y., Choi, I. W., Kim, N. D., et al., (2011). Anti-inflammatory effects of fucoidan through inhibition of NF-IB, MAPK and Akt activation in lipopolysaccharide-induced BV2 microglia cells. *Food and Chemical Toxicology, 49*, 1745–1752.

Parys, S., Rosenbaum, A., Kehraus, S., Reher, G., Glombitza, K. W., & König, G. M., (2007). Evaluation of quantitative methods for the determination of polyphenols in algal extracts. *J. Nat. Prod., 70,* 1865–1870.

Peng, J., Yuan, J. P., Wu, C. F., & Wang, J. H., (2011). Fucoxanthin, a marine carotenoid present in brown seaweeds and diatoms: Metabolism and bioactivities relevant to human health. *Marine Drugs, 9*, 1806–1828.

Pereira, U. A., Barbosa, L. C., Maltha, C. R., Demuner, A. J., Masood, M. A., & Pimenta, A. L., (2014). Gamma-alkylidene-gamma-lactones and isobutylpyrrol-2(5H)-ones analogues to rubrolides as inhibitors of biofilm formation by gram-positive and gram-negative bacteria. *Bioorg. Med. Chem. Lett., 24,* 1052–1056.

Peres, J. C. F., De Carvalho, L. R., Goncalez, E., Berian, L. O. S., & Felicio, J. D., (2012). Evaluation of antifungal activity of seaweed extracts. *Cienc. Agrotec., 36*, 294–299.

Pérez, M. J., Falqué, E., & Domínguez, H., (2016). Antimicrobial action of compounds from marine seaweed. *Marine Drugs, 14*, 52.

Pettongkhao, S., Bilanglod, A., Khompatara, K., & Churngchow, N., (2019). Sulfated polysaccharide from *Acanthophora spicifera* induced *Hevea brasiliensis* defense responses against *Phytophthora palmivora* infection. *Plants, 8,* 73.

Pinteus, S., Silva, J., Alves, C., Horta, A., Fino, N., Rodrigues, A. I., Mendes, S., & Pedrosa, R., (2017). Cytoprotective effect of seaweeds with high antioxidant activity from the Peniche coast (Portugal). *Food Chem., 218,* 591–599.

Ponce, N. M. A., Flores, M. L., Pujol, C. A., Becerra, M. B., Navarro, D. A., Córdoba, O., Damonte, E. B., & Stortz, C. A., (2019). Fucoidans from the phaeophyta *Scytosiphon lomentaria*: Chemical analysis and antiviral activity of the galactofucan component. *Carbohydrate Research, 478,* 18–24.

Praveen, A. M., Parvathy, K. R. K., Patra, S., Khan, I., Natarajan, P., & Balasubramanian, P., (2020). Cytotoxic and pharmacokinetic studies of Indian seaweed polysaccharides for formulating raindrop synbiotic candy. *Int. J. Biol. Macromol., 154,* 557–566.

Rai, A., Cherif, A., Cruz, C., & Nabti, E., (2018). Extracts from seaweeds and *Opuntia ficus-indica* cladodes enhance diazotrophic-PGPR halotolerance, their enzymatic potential, and their impact on wheat germination under salt stress. *Pedosphere, 28,* 241–254.

Rajesh, S., Asha, A., Kombiah, P., & Sahayaraj, K., (2011). Biocidal activity of algal seaweeds on insect pest and fungal plant pathogen. *Proceedings of the National Seminar on Harmful/ Beneficial Insects of Agricultural Importance with Special Reference to the Nuisance Pest Luprops tristis in Rubber Plantations,* 86–91.

Rajivgandhi, G. N., Ramachandran, G., Maruthupandy, M., Manoharan, N., Alharbi, N. S., Kadaikunnan, S., Khaled, J. M., et al., (2020). Anti-oxidant, anti-bacterial and anti-biofilm activity of biosynthesized silver nanoparticles using *Gracilaria corticata* against biofilm producing *K. pneumoniae. Colloids and Surfaces A: Physicochemical and Engineering Aspects, 600,* 124830.

Ramaswamy, V., Cresence, V. M., Rejitha, J. S., Lekshmi, M. U., Dharsana, K. S., Prasad, S. P., & Vijila, H. M., (2007). *Listeria* – review of epidemiology and pathogenesis. *J. Microbiol. Immunol. Infect., 40,* 4–13.

Rathore, S. S., Chaudhary, D. R., Boricha, G. N., Ghosh, A., Bhatt, B. P., Zodape, S. T., & Patolia, J. S., (2009). Effect of seaweed extract on the growth, yield and nutrient uptake of soybean (*Glycine max*) under rainfed conditions. *S. Afr. J. Bot., 75,* 351–355.

Rengasamy, K. R., Kulkarni, M. G., Stirk, W. A., & Van, S. J., (2015a). Eckol – a new plant growth stimulant from the brown seaweed *Ecklonia maxima*. *J. Appl. Phycol., 27,* 581–587.

Rengasamy, K. R., Kulkarni, M. G., Stirk, W. A., & Van, S. J., (2014). Advances in algal drug research with emphasis on enzyme inhibitors. *Biotechnol. Adv., 32,* 1364–1381.

Rengasamy, K. R., Kulkarni, M. G., Stirk, W. A., & Van, S. J., (2015b). Eckol improves growth, enzyme activities, and secondary metabolite content in maize (*Zea mays* cv. Border King). *J. Plant Growth Regul., 34,* 410–416.

Riad, N., Zahi, M. R., Trovato, E., Bouzidi, N., Daghbouche, Y., Utczás, M., Mondello, L., & El Hattab, M., (2020). Chemical screening and antibacterial activity of essential oil and volatile fraction of *Dictyopteris polypodioides*. *Microchemical Journal, 152,* 104415.

Roberts, L., (2020). Global polio eradication falters in the final stretch. *Science, 367*, 14, 15.

Rodrigues, D., Alves, C., Horta, A., Pinteus, S., Silva, J., Culioli, G., Thomas, O. P., & Pedrosa, R., (2015b). Antitumor and antimicrobial potential of bromoditerpenes isolated from the red alga *Sphaerococcus coronopifolius*. *Marine Drugs, 13,* 713–726.

Rodrigues, D., Freitas, A. C., Pereira, L., Rocha-Santos, T. A., Vasconcelos, M. W., Roriz, M., Rodríguez-Alcalá, L. M., et al., (2015a). Chemical composition of red, brown and green macroalgae from Buarcos bay in central west coast of Portugal. *Food Chemistry, 183,* 197–207.

Rodriguez-Amaya, D. B., (2016). Natural food pigments and colorants. *Curr. Opin. Food Sci., 7,* 20–26.

Rosa, G. P., Tavares, W. R., Sousa, P. M. C., Pagès, A. K., Seca, A. M. L., & Pinto, D. C. G. A., (2020). Seaweed secondary metabolites with beneficial health effects: An overview of successes *in vivo* studies and clinical trials. *Marine Drugs, 18*, 8.

Rosaline, X. D., Sakthivelkumar, S., Rajendran, K., & Janarthanan, S., (2012). Screening of selected marine algae from the coastal Tamil Nadu, South India for antibacterial activity. *Asian Pacific Journal Tropical of Biomedicine, 2*, S140–S146.

Roseline, T. A., Murugan, M., Sudhakar, M. P., & Arunkumar, K., (2019). Nanopesticidal potential of silver nanocomposites synthesized from the aqueous extracts of Red Seaweeds. *Environmental Technology and Innovation, 13,* 82–93.

Rouphael, Y., De Micco, V., Arena, C., Raimondi, G., Colla, G., & De Pascale, S., (2017). Effect of *Ecklonia maxima* seaweed extract on yield, mineral composition, gas exchange, and leaf anatomy of zucchini squash grown under saline conditions. *J. Appl. Phycol., 29,* 459–470.

Rupérez, P., Gómez-Ordòñez, E., & Jiménez-Escrig, A., (2013). Biological activity of algal sulfated and non-sulfated polysaccharides. In: Hernandez-Ledesma, B., & Herrero, M., (eds.), *Bioactive Compounds from Marine Foods: Plant and Animal Sources* (pp. 219–247). John Wiley and Sons.

Saha, M., Goecke, F., & Bhadury, P., (2018). Algal natural compounds and extracts as antifoulants. *J. Appl. Phycol., 30,* 1859–1874.

Saha, M., Rempt, M., Gebser, B., Grueneberg, J., Pohnert, G., & Weinberger, F., (2012). Biofouling dimethylsulphopropionate (DMSP) and proline from the surface of the brown alga *Fucus vesiculosus* inhibit bacterial attachment. *Biofouling, 28,* 593–604.

Sahayaraj, K., Rajesh, S., Asha, A., & Rathi, J. M., (2012). Marine algae for the cotton pest and disease management. *Proceedings of International Conference on Agricultural Science and Engineering, Nigeria, 1,* 49–62.

Samarakoon, K., & Jeon, Y. J., (2012). Bio-functionalities of proteins derived from marine algae: A review. *Food Res. Int., 48,* 948–960.

Sangha, J. S., Khan, W., Ji, X., Zhang, J., Mills, A. A. S., Critchley, A. T., & Prithiviraj, B., (2011). Carrageenans, sulfated polysaccharides of Red Seaweeds, differentially affect *Arabidopsis thaliana* resistance to *Trichoplusia ni* (Cabbage Looper). *PLoS One, 6,* e26834.

Sanjeewa, K. K. A., Kim, E. A., Son, K. T., & Jeon, Y. J., (2016). Bioactive properties and potentials cosmeceutical applications of phlorotannins isolated from brown seaweeds: A review. *J. Photochem. Photobiol. B Biol., 162,* 100–105.

Sanniyasi, E., Venkatasubramanian, G., Anbalagan, M. M., Raj, P. P., & Gopal, R. K., (2019). *In vitro* anti-HIV-1 activity of the bioactive compound extracted and purified from two different marine macroalgae (seaweeds) (*Dictyota bartayesiana* J.V.Lamouroux and *Turbinaria decurrens* Bory). *Scientific Reports, 9,* 12185.

Sathya, R., Kanaga, N., Sankar, P., & Jeeva, S., (2017). Antioxidant properties of phlorotannins from brown seaweed *Cystoseira trinodis* (Forsskal) C. Agardh. *Arabian Journal of Chemistry, 10*(Suppl 2), S2608–S2614.

Schmidt, R. E., Ervin, E. H., & Zhang, X., (2003). Questions and answers about biostimulants. *Golf Course Manag., 71,* 91–94.

Schrader, K. K., (2003). *Natural Algicides for the Control of Cyanobacterial-Related Off-Flavor in Catfish Aquaculture.* American Chemical Society: Washington, DC, USA.
Seca, A., & Pinto, D., (2018). Overview on the antihypertensive and anti-obesity effects of secondary metabolites from seaweeds. *Marine Drugs, 16*, 237.
Senthilkumar, K., Manivasagan, P., Venkatesan, J., & Kim, S. K., (2013). Brown seaweed fucoidan: Biological activity and apoptosis, growth signaling mechanism in cancer. *Int. J. Biol. Macromol., 60,* 366–374.
Senthilkumar, K., Ramajayam, G., Venkatesan, J., Kim, S. K., & Ahn, B. C., (2017). Biomedical applications of fucoidan, seaweed polysaccharides. In: *Seaweed Polysaccharides* (pp. 269–281).
Shah, M. T., Zodape, S. T., Chaudhary, D. R., Eswaran, K., & Chikara, J., (2013). Seaweed sap as an alternative liquid fertilizer for yield and quality improvement of wheat. *J. Plant Nutr., 36,* 192–200.
Sharma, B. R., Kim, H. J., Kim, M. S., Park, C. M., & Rhyu, D. Y., (2017). *Caulerpa okamurae* extract inhibits adipogenesis in 3T3-L1 adipocytes and prevents high-fat diet-induced obesity in C57BL/6 mice. *Nutrition Research, 47,* 44–52.
Sharma, S., Chen, C., Khatri, K., Rathore, M. S., & Pandey, S. P., (2019). *Gracilaria dura* extract confers drought tolerance in wheat by modulating abscisic acid homeostasis. *Plant Physiol. Biochem., 136,* 143–154.
Shi, Q., Wang, A., Lu, Z., Qin, C., Hu, J., & Yin, J., (2017). Overview on the antiviral activities and mechanisms of marine polysaccharides from seaweeds. *Carbohydr. Res., 453, 454*, 1–9.
Shukla, P. S., Shotton, K., Norman, E., Neily, W., Critchley, A. T., & Prithiviraj, B., (2018). Seaweed extract improve drought tolerance of soybean by regulating stress-response genes. *AoB Plants, 10*(1), plx051.
Silva, L. D., Bahcevandziev, K., & Pereira, L., (2019). Production of bio-fertilizer from *Ascophyllum nodosum* and *Sargassum muticum* (Phaeophyceae). *Journal of Oceanology and Limnology, 37*, 918–927.
Sim, S. Y., Shin, Y. E., & Kim, H. K., (2019). Fucoidan from *Undaria pinnatifida* has anti-diabetic effects by stimulation of glucose uptake and reduction of basal lipolysis in 3T3-L1 adipocytes. *Nutrition Research, 65,* 54–62.
Souza, B. W., Cerqueira, M. A., Martins, J. T., Quintas, M. A., Ferreira, A. C., Teixeira, J. A., & Vicente, A. A., (2011). Antioxidant potential of two Red Seaweeds from the Brazilian coasts. *J. Agric. Food Chem., 59,* 5589–5594.
Stadnik, M. J., & Freitas, M. B., (2014). Algal polysaccharides as source of plant resistance inducers. *Trop. Plant Pathol., 39,* 111–118.
Steevensz, A. J., MacKinnon, S. L., Hankinson, R., Craft, C., Connan, S., Stengel, D. B., & Melanson, J. E., (2012). Profiling phlorotannins in brown macroalgae by liquid chromatography high resolution mass spectrometry. *Phytochem. Anal., 23*, 547–553.
Stirk, W. A., Tarkowská, D., Turečová, V., Strnad, M., & Van, S. J., (2014). Abscisic acid, gibberellins and brassinosteroids in Kelpak, a commercial seaweed extract made from *Ecklonia maxima*. *J. Appl. Phycol., 26,* 561–567.
Sudatti, D. B., Fujii, M. T., Rodrigues, S. V., Turra, A., & Pereira, R. C., (2018). Prompt induction of chemical defenses in the Red Seaweed *Laurencia dendroidea*: The role of herbivory and epibiosis. *J. Sea Res., 138,* 48–55.
Suganya, A. M., Sanjivkumar, M., Chandran, M. N., Palavesam, A., & Immanuel, G., (2016). Pharmacological importance of sulfated polysaccharide carrageenan from Red

Seaweed *Kappaphycus alvarezii* in comparison with commercial carrageenan. *Biomed. Pharmacother., 84,* 1300–1312.

Suganya, S., Ishwarya, R., Jayakumar, R., Govindarajan, M., Alharbi, N. S., Kadaikunnan, S., Khaled, J. M., et al., (2019). New insecticides and antimicrobials derived from *Sargassum wightii* and *Halimeda gracillis* seaweeds: Toxicity against mosquito vectors and antibiofilm activity against microbial pathogens. *South African Journal of Botany, 125,* 466–480.

Sugiura, Y., Usui, M., Katsuzaki, H., Imai, K., Kakinuma, M., Amano, H., & Miyata, M., (2018). Orally administered phlorotannins from *Eisenia arborea* suppress chemical mediator release and cyclooxygenase-2 signaling to alleviate mouse ear swelling. *Marine Drugs, 16,* 267.

Sujitha, V., Murugan, K., Dinesh, D., Pandiyan, A., Aruliah, R., Hwang, J. S., Kalimuthu, K., et al., (2017). Green-synthesized CdS nano-pesticides: Toxicity on young instars of malaria vectors and impact on enzymatic activities of the non-target mud crab *Scylla serrata. Aquatic Toxicology, 188,* 100–108.

Sun, Y. Y., Xing, J. Z., Zhang, J. S., Zhou, W. J., & Pu, Y. F., (2018a). Sesquiterpenoids with antialgal activity against the common red tide microalgae from marine macroalga *Porphyra yezoensis*. *Environ. Sci. Pollut. Res., 25,* 7844–7859.

Sun, Y., Dong, S., Zhou, W., Guo, L., Guo, G. L., & Zhang, X., (2019). A comprehensive review of secondary metabolites with antialgal activity from marine macroalgae against red tide microalgae. *J. Coast. Res., 93*(S1), 475–488.

Sun, Y., Wang, H., Guo, G., Pu, Y., Yan, B., & Wang, C., (2016). Isolation, purification, and identification of antialgal substances in green alga *Ulva prolifera* for antialgal activity against the common harmful red tide microalgae. *Environ. Sci. Pollut. Res., 23,* 1449–1459.

Sun, Y., Zhou, W., Wang, H., Guo, G., Sua, Z., & Pu, Y., (2018b). Antialgal compounds with antialgal activity against the common red tide microalgae from a green algae *Ulva pertusa. Ecotoxicology and Environmental Safety, 157,* 61–66.

Susanto, E., Fahmi, A. S., Hosokawa, M., & Miyashita, K., (2019). Variation in lipid components from 15 species of tropical and temperate seaweeds. *Marine Drugs, 17,* 630.

Suutari, M., Majaneva, M., Fewer, D. P., Voirin, B., Aiello, A., Friedl, T., Chiarello, A. G., & Blomster, J., (2010). Molecular evidence for a diverse green algal community growing in the hair of sloths and a specific association with *Trichophilus welckeri* (Chlorophyta, Ulvophyceae). *BMC Evolutionary Biology, 10,* 86.

Tabarsa, M., Rezaei, M., Ramezanpour, Z., & Waaland, J. R., (2012). Chemical compositions of the marine algae *Gracilaria salicornia* (Rhodophyta) and *Ulva lactuca* (Chlorophyta) as a potential food source. *J. Sci. Food Agric., 92,* 2500–2506.

Tanna, B., & Mishra, A., (2019). Nutraceutical potential of seaweed polysaccharides: Structure, bioactivity, safety, and toxicity. *Compr. Rev. Food Sci. Food Saf., 18,* 817–831.

Tanna, B., Choudhary, B., & Mishra, A., (2018). Metabolite profiling, antioxidant, scavenging and anti-proliferative activities of selected tropical green seaweeds reveal the nutraceutical potential of *Caulerpa* spp. *Algal Research, 36,* 96–105.

Tapiero, H., Townsend, D. M., & Tew, K. D., (2004). The role of carotenoids in the prevention of human pathologies. *Biomed. Pharmacother., 58,* 100–110.

Torres, M. D., Flórez-Fernández, N., & Domínguez, H., (2019). Integral utilization of Red Seaweed for bioactive production. *Marine Drugs, 19,* 314.

Umavandhana, R., & Jayanthi, S., (2018). Phytochemical screening and free radical scavenging activity on some selected seaweeds from Gulf of Mannar, India. *Res. J. Pharm. Tech., 11,* 3385–3388.

Umezawa, T., Oguri, Y., Matsuura, H., Yamazaki, S., Suzuki, M., Yoshimura, E., Furuta, T., et al., (2014). Omaezallene from red alga *Laurencia* sp.: Structure elucidation, total synthesis, and antifouling activity. *Angew. Chem. Int. Ed., 53*, 3909–3912.

Vaikundamoorthy, R., Krishnamoorthy, V., Vilwanathan, R., & Rajendran, R., (2018). Structural characterization and anticancer activity (MCF7 and MDA-MB-231) of polysaccharides fractionated from brown seaweed *Sargassum wightii*. *Int. J. Biol. Macromol., 111,* 1229–1237.

Vasantharaja, R., Abraham, L. S., Inbakandan, R. D., Thirugnanasambandam, R., Senthilvelana, T., Ayesha, J. S. K., & Prakash, P., (2019). Influence of seaweed extracts on growth, phytochemical contents and antioxidant capacity of cowpea (*Vigna unguiculata* L. Walp). *Biocatalysis and Agricultural Biotechnology, 17*, 589–594.

Venkatesan, M., Arumugam, V., Pugalendi, R., Ramachandran, K., Sengodan, K., Vijayan, S. R., Sundaresan, U., et al., (2019). Antioxidant, anticoagulant and mosquitocidal properties of water soluble polysaccharides (WSPs) from Indian seaweeds. *Process Biochemistry, 84,* 196–204.

Vera, J., Castro, J., Gonzalez, A., & Moenne, A., (2011). Seaweed polysaccharides and derived oligosaccharides stimulate defense responses and protection against pathogens in plants. *Marine Drugs, 9,* 2514–2525.

Vishchuk, O. S., Ermakova, S. P., & Zvyagintseva, T. N., (2011). Sulfated polysaccharides from brown seaweeds *Saccharina japonica* and *Undaria pinnatifida*: Isolation, structural characteristics, and antitumor activity. *Carbohydr. Res., 346,* 2769–2776.

Vishchuk, O. S., Ermakova, S. P., & Zvyagintseva, T. N., (2013). The fucoidans from brown algae of Far-Eastern seas: Anti-tumor activity and structure-function relationship. *Food Chem., 141*, 1211–1217.

Wahl, M., (1989). Marine epibiosis. I. Fouling and antifouling: Some basic aspects. *Mar. Ecol. Prog. Ser., 58,* 175–189.

Wally, O. S. D., Critchley, A. T., Hiltz, D., Craigie, J. S., Han, X., Zaharia, L. I., Abrams, S. R., & Prithiviraj, B., (2013). Regulation of phytohormone biosynthesis and accumulation in *Arabidopsis* following treatment with commercial extract from the marine macroalgae *Ascophyllum nodosum*. *J. Plant Growth Regul., 32,* 324–339.

Wang, B. G., Gloer, J. B., Ji, N. Y., & Zhao, J. C., (2013). Halogenated organic molecules of Rhodomelaceae origin: Chemistry and biology. *Chem. Rev., 113,* 3632–3685.

Wang, H. M. D., Li, X. C., Lee, D. J., & Chang, J. S., (2017a). Potential biomedical applications of marine algae. *Bioresour. Technol., 244,* 1407–1415.

Wang, L., Wang, X., Wu, H., & Liu, R., (2014). Overview on biological activities and molecular characteristics of sulfated polysaccharides from marine green algae in recent years. *Marine Drugs, 12,* 4984–5020.

Wang, R., Feng, L., Tang, X., Wang, J., & Dong, S., (2012). Allelopathic growth inhibition of *Heterosigma akashiwo* by the three *Ulva* species (*Ulva pertusa, Ulva linza, Enteromorpha intestinalis*) under laboratory conditions. *Acta Oceanol. Sin., 31,* 138–144.

Wang, R., Xiao, H., Wang, Y., Zhou, W., & Tang, X., (2007a). Effects of three macroalgae, *Ulva linza* (Chlorophyta), *Corallina pilulifera* (Rhodophyta) and *Sargassum thunbergii* (Phaeophyta) on the growth of the red tide microalga *Prorocentrum donghaiense* under laboratory conditions. *J. Sea Res., 58,* 189–197.

Wang, S., Wang, W., Hou, L., Qin, L., He, M., Li, W., & Mao, W., (2020). A sulfated glucuronorhamnan from the green seaweed *Monostroma nitidum*: Characteristics of its structure and antiviral activity. *Carbohydr. Polym., 227,* 115280.

Wang, Y., Xiang, L., Wang, S., Wang, X., Chen, X., & Mao, Z., (2017b). Effects of seaweed fertilizer on the *Malus hupehensis* Rehd. seedlings growth and soil microbial numbers under continue cropping. *Acta Ecol. Sin., 37,* 180–186.

Wang, Y., Yu, Z., Song, X., Tang, X., & Zhang, S., (2007b). Effects of macroalgae *Ulva pertusa* (Chlorophyta) and *Gracilaria lemaneiformis* (Rhodophyta) on growth of four species of bloom-forming dinoflagellates. *Aquatic Botany, 86,* 139–147.

Wang, Z. J., Xub, W., Liang, J. W., Wang, C. S., & Kang, Y., (2017c). Effect of fucoidan on B16 murine melanoma cell melanin formation and apoptosis. *Afr. J. Tradit. Complement Altern. Med., 14,* 149–155.

Wei, R., Lee, M. S., Lee, B., Oh, C. W., Choi, C. G., & Kim, H. R., (2016). Isolation and identification of anti-inflammatory compounds from ethyl acetate fraction of *Ecklonia stolonifera* and their anti-inflammatory action. *J. Appl. Phycol., 28,* 3535–3545.

Wells, M. L., Potin, P., Craigie, J. S., Raven, J. A., Merchant, S. S., Helliwell, K. E., Smith, A. G., et al., (2017). Algae as nutritional and functional food sources: Revisiting our understanding. *J. Appl. Phycol., 29,* 949–982.

West, J., Calumpong, H. P., & Martin, G., (2017). Seaweeds. In: *United Nations the First Global Integrated Marine Assessment* (pp. 223–228). Cambridge University Press.

Wielgosz-Collin, G., Kendel, M., & Couzinet-Mossion, A., (2016). Lipids, fatty acids, glycolipids, and phospholipids. In: Fleurence, J., & Levine, I., (eds.), *Seaweed in Health and Disease Prevention* (pp. 185–221). San Diego: Academic Press.

Wijesekara, I., & Kim, S. K., (2015). Application of marine algae derived nutraceuticals in the food industry. In: Kim, S. K., & Chojnacka, K., (eds.), *Marine Algae Extracts: Processes, Products and Applications* (pp. 627–638). Weinheim: Wiley-VCH Verlag GmbH and Co.

Wong, K., & Cheung, P. C., (2000). Nutritional evaluation of some subtropical red and green seaweeds: Part I – proximate composition, amino acid profiles and some physico-chemical properties. *Food Chemistry, 71,* 475–482.

Yamasaki-Miyamoto, Y., Yamasaki, M., Tachibana, H., & Yamada, K., (2009). Fucoidan induces apoptosis through activation of caspase-8 on human breast cancer MCF-7 cells. *J. Agric. Food Chem., 57,* 8677–8682.

Yan, M. D., Lin, H. Y., & Hwang, P. A., (2019). The anti-tumor activity of brown seaweed oligo-fucoidan via lncRNA expression modulation in HepG2 cells. *Cytotechnology, 71,* 363–374.

Yang, C., Lai, S., Chen, Y., Liu, D., Liu, B., Ai, C., Wan, X. Z., et al., (2019b). Anti-diabetic effect of oligosaccharides from seaweed *Sargassum confusum* via JNK-IRS1/PI3K signaling pathways and regulation of gut microbiota. *Food and Chemical Toxicology, 131,* 110562.

Yang, H. W., Fernando, K. H. N., Oh, J. Y., Li, X., Jeon, Y. J., & Ryu, B. M., (2019a). Anti-obesity and anti-diabetic effects of *Ishige okamurae. Marine Drugs, 17,* 202.

Ye, C., & Zhang, M., (2013). Allelopathic effect of macroalga *Gracilaria tenuistipitata* (Rhodophyta) on the photosynthetic apparatus of red-tide causing microalga *Prorocentrum micans. IERI Procedia, 5,* 209–215.

Ying-Ying, S., Hui, W., Gan-Lin, G., Yin-Fang, P., Bin-Lun, Y., & Chang-Hai, W., (2015). Green alga *Ulva pertusa* – a new source of bioactive compounds with antialgal activity. *Environ. Sci. Pollut. Res., 22,* 10351–10359.

Yoshie-Stark, Y., Hsieh, Y. P., & Suzuki, T., (2003). Distribution of flavonoids and related compounds from seaweeds in Japan. *Journal – Tokyo University of Fisheries, 89,* 1–6.

Yu, R. X., Yu, R. T., & Liu, Z., (2018). Inhibition of two gastric cancer cell lines induced by fucoxanthin involves down regulation of Mcl-1 and STAT3. *Human Cell, 31,* 50–63.

Yu, X. Q., He, W. F., Liu, D. Q., Feng, M. T., Fang, Y., Wang, B., Feng, L. H., Guo, Y. W., & Mao, S. C., (2014). A seco-laurane sesquiterpene and related laurane derivatives from the red alga *Laurencia okamurai* Yamada. *Phytochemistry, 103*, 162–170.

Yusuf, R., Kristiansen, P., & Warwick, N., (2012). Potential effect of plant growth regulators on two seaweed products. *Acta Hortic., 958,* 133–138.

Zerrif, S. E. A., El Ghazi, N., Douma, M., El Khalloufi, F., & Oudra, B., (2018). Potential uses of seaweed bioactive compounds for harmful microalgae blooms control: Algicidal effects and algal growth inhibition of *Phormidium* sp. (freshwater toxic cyanobacteria). *Smetox. J., 1,* 59–62.

Zerrifi, A. S., Tazart, Z., El Khalloufi, F., Oudra, B., Campos, A., & Vasconcelos, V., (2019). Potential control of toxic cyanobacteria blooms with Moroccan seaweed extracts. *Environmental Science and Pollution Research, 26*, 15218–15228.

Zerrifi, S. E. A., El Khalloufi, F., Oudra, B., & Vasconcelos, V., (2018). Seaweed bioactive compounds against pathogens and microalgae: Potential uses on pharmacology and harmful algae bloom control. *Marine Drugs, 16*, 55.

Zha, S., Kuwano, K., Shibahara, T., & Ishibashi, F., (2020). Algicidal hydroxylated C18 unsaturated fatty acids from the red alga *Tricleocarpa jejuensis*: Identification, synthesis and biological activity. *Fitoterapia, 145,* 104639.

Zhao, C., Yang, C., Liu, B., Lin, L., Sarker, S. D., Nahar, L., Yu, H., et al., (2018). Bioactive compounds from marine macroalgae and their hypoglycemic benefits. *Trends Food Sci. Technol., 72,* 1–12.

Zhao, G., Zhao, J., Peng, L., Zou, L., Wang, J., Zhong, L., & Xiang, D., (2012). Effects of yeast polysaccharide on growth and flavonoid accumulation in *Fagopyrum tataricum* sprout cultures. *Molecules, 17*, 11335–11345.

Zhao, L., Feng, C., Wu, K., Chen, W., Chen, Y., Hao, X., & Wu, Y., (2017). Advances and prospects in biogenic substances against plant virus: A review. *Pest. Biochem. Physiol., 135*, 15–26.

Zorofchian, M. S., Karimian, H., Khanabdali, R., Razavi, M., Firoozinia, M., Zandi, K., & Abdul, K. H., (2014). Anticancer and antitumor potential of fucoidan and fucoxanthin, two main metabolites isolated from brown algae. *The Sci. World J., 3,* 768323.

Zou, P., Lu, X., Zhao, H., Yuan, Y., Meng, L., Zhang, C., & Li, Y., (2019). Polysaccharides derived from the brown algae *Lessonia nigrescens* enhance salt stress tolerance to wheat seedlings by enhancing the antioxidant system and modulating intracellular ion concentration. *Front. Plant Sci., 10,* 48.

Zubia, M., Robledo, D., & Freile-Pelegrin, Y., (2007). Antioxidant activities in tropical marine macroalgae from the Yucatan Peninsula, Mexico. *J. Appl. Phycol., 19,* 449–458.

CHAPTER 10

ROLES OF THE INGREDIENTS CONTAINED IN EDIBLE SEAWEEDS ON PARTICULATE MATTER (PM)-INDUCED SKIN DAMAGES

KALU KAPUGE ASANKA SANJEEWA[1,2*] and YOU-JIN JEON[2,3*]

[1]*Department of Biosystems Technology, Faculty of Technology, University of Sri Jayewardenepura,Pitipana, Homagama 10206, Sri Lanka*

[2]*Laboratory of Marine Bioresource Technology, Department of Marine Life Science, Jeju National University, Jeju 63243, Republic of Korea*

[3]*Marine Science Institute, Jeju National University, Jeju Self-Governing Province 63333, Republic of Korea*

ABSTRACT

According to the literature reviews, most published studies on particulate matter (PM) concerning less than PM10 reported in urban areas mainly focused on PM-induced effects on the respiratory and cardiovascular systems. However, epidemiological and mechanistic studies suggest that PM less than 10 μm are a potential threat to skin health and exacerbates pre-existing skin diseases (atopic dermatitis). This chapter first discusses the biological and toxicological effects of PM on skin, such as oxidative stress-related factors (apoptosis), inflammatory responses, aryl hydrocarbon receptor (Ahr) activations, and matrix metalloproteinases expression activities and then summarizes the protective mechanisms reported from seaweeds metabolites against PM-induced skin damages. In addition, this chapter reports potential seaweed metabolites to develop as functional and medicinal ingredient against PM-induced skin damages. Specifically, the functional properties of seaweeds have been identified long ago, and for the promising bioactive properties, seaweed metabolites are applied as active ingredients in the number of commercial products. Thus, skin protective products developed from seaweeds against PM-induced skin damage will acquire great market share in the future. In addition, the possible protective mechanism of seaweed metabolites was determined to inhibit PM-induced oxidative stress, Ahr activation, and inflammatory responses, which can provide useful information for future researchers attempting to evaluate protective mechanisms of seaweed metabolites against PM-induced skin damages.

10.1 INTRODUCTION

Particulate matter (PM), also known as aerosol, refers to condensed phases suspended in the air. Other than molecules, PM contains considerable amounts of atoms, radicals, and precipitations (Liang, 2013). The particles, with aero dimension less than 2.5 μm (PM2.5), mainly produced by fuel combustion of motor vehicles, industrial facilities, power generation plants, and wood stoves. The particles in-between 2.5–10 μm (PM10), are mainly produced by non-combustion sources. Additionally, gases including NO_x, SO_2, and volatile organic compounds reacted with other particles in the atmosphere and form fine particles (Miller, 2017). Besides anthropogenic reasons, dust storms originated in the Chinese and Mongolian deserts also responsible for the elevated levels of PM in the countries like China, Korea, and Japan. According the previous studies, tiny dust particles originating from these Asian dust storms has a possibility to reach the west coast of North America moving through the Pacific Ocean. Thus, these tiny Asian dust particles also can affect the health of people who lives in North American territory. Other than Asia, dust storms are common in arid and semi-arid regions around the world, such as the Sahara Desert (Choi et al., 2011). According to the World Health Organization, during 2016, ambient air pollution (outdoor air pollution) was responsible for 4.2 million premature deaths. Furthermore, worldwide, ambient air pollution is estimated to cause about 16% of the lung cancer deaths, 25% of chronic obstructive pulmonary disease deaths, about 17% of ischemic heart disease and strokes, and about 26% of respiratory infection deaths (WHO, 2019). Due to the known correlation of increased PM concentrations in the atmosphere with increased clinical visits for respiratory system-related completions, toxicological effects of PM on the respiratory tract have been extensively evaluated. Besides, the aforementioned health hazards, recent epidemiological studies pointed out that PM is a potential threat to the skin health and exacerbates pre-existing skin diseases (Park et al., 2018).

Skin (Figure 10.1) is the largest and complex organ of the human body and accounts for 15% of the total body weight. Skin is act as the exterior interface of the human body with the outer environment, and contains three main layers as epidermis, dermis, and sub-cutis (Lai-Cheong and McGrath, 2017) (Figure 10.1). In general, skin functioning as a physical barrier to avoid excessive water losses from internal organs, and provide protection from pathogens attacks, harmful chemical substances, physical damages. Additionally, skin cells actively support for immune defense, support for thermo-regulation, and endocrine metabolic mechanisms to support normal body functions and keep optimal environmental conditions for internal organs (Byrd et al., 2018). It has been demonstrated that PM deposited on the skin responsible for scab, itching, hives, and skin allergy reported in some patients (Kim et al., 2013; Park et al., 2018). The best prevention and alleviation of PM-induced diseases is reducing the chance of exposure to PM as much as possible, minimize time use for outdoor activities, and wearing long sleeves are common preventive methods to avoid PM-induced health complications. Due to the size and reactivity of PM, which are capable to clog skin pores and reaching to the bottom layers of the skin. Thus, cosmetic dermatological approaches are required to attenuate complications arise from PM exposure (Lee et al., 2018). Specifically, upregulated ROS productions, activation of aryl hydrocarbon receptor (Ahr), activation of inflammatory signal pathways, upregulated MMPs expression in skin cells, alter skin microflora are some common reasons for PM-induced skin cell damages and skin aging. However, to the date, very few data are available regarding the impact of PM on the skin and protective effects of seaweed extracts against PM-induced skin damages. Thus, for better understanding, PM-induced skin damage-related mechanisms and application of seaweed as a functional material against PM-induced skin damages require to collect available data into one piece. Therefore, this chapter reviews all available literature published over the last 25 years against PM-induced skin damages and protective effect of seaweeds against PM-induced skin damages.

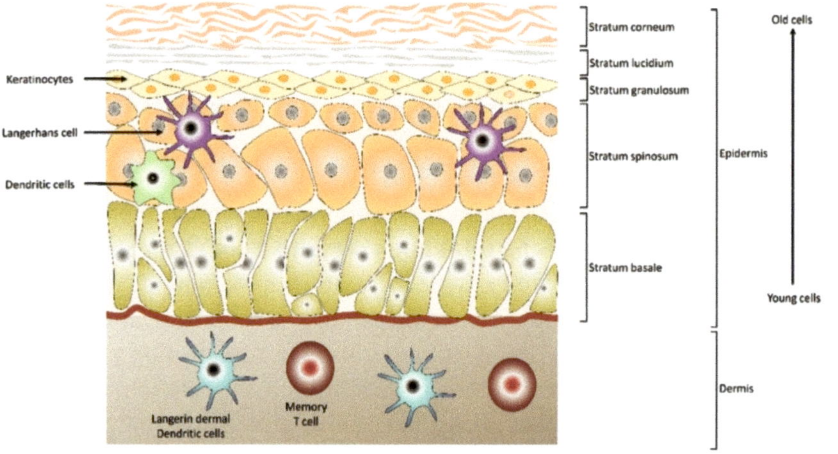

FIGURE 10.1 General structure of skin and cell layer.

10.2 PROTECTIVE EFFECTS OF SEAWEEDS AGAINST PM-INDUCED SKIN DAMAGE AND ROS GENERATION

Seaweed species are often regarded as an underutilized bio-resource, many have been used as sea vegetable for human and animal consumptions, industrial raw materials, in traditional medicine, and as fertilizers for land plants (Khan et al., 2009; Buschmann et al., 2017; Sanjeewa et al., 2017). The use of seaweeds has a long history in East Asian countries such as Korea, China, and Japan for the medicinal and food applications (Sanjeewa et al., 2017). In addition to Asian countries, seaweeds have traditionally been used by the Western food industry to extract polysaccharides such as agar, alginate, carrageenan for a wide range of food products (Mabeau and Fleurence, 1993). Other than the polysaccharides, seaweeds contain substantial amounts of essential nutrients such as proteins, minerals, vitamins, and dietary fiber. Furthermore, seaweeds are also rich in bioactive compounds such as phlorotannins, peptides, phytosterols, pigments, fatty acids, and mycosporine-like amino acids (MAAs) (Wang et al., 2018). Even though seaweeds are rich in bioactive secondary metabolites, most of the studies were limited to basic biological and hydrocolloid related studies. However, during the last decade, the research interest towards the identification and characterization of bioactive metabolites present in seaweeds and identification of specific bioactive compounds and their effects on human health applications has been gradually increased (Hafting et al., 2015). As a result of this trend changes, several thousands of novel bioactive secondary metabolites have been identified from seaweeds and deposited them in databases such as MarinLit (http://pubs.rsc.org/marinlit/) (Tanna and Mishra, 2019). From those identified compounds, some compounds have shown excellent bioactive properties against PM-induced skin damages. In addition to the pure compounds, some crude extracts also demonstrated protective mechanisms against PM-induced skin cell damages. Specifically, bioactive properties of seaweeds in relation to PM-induced skin damages such as anti-oxidant, anti-inflammation, MMP inhibition, and Ahr inhibition reported are discussed in detail in this chapter.

The skin is the biggest organ of the human body and the outermost layer of skin is called stratum corneum (SC); the outer layer of the epidermis (Matsui and Amagai, 2015). Epidermal keratinocytes in the basal layer of the epidermis move upward and ultimately differentiate into cornified cells in the epidermal SC. Keratinocytes are the main constituents of the epidermis, and it has been reported that more than 95% of epidermis cells are keratinocytes (Figure 10.1). Furthermore, keratinocytes in epidermis produce variety of cytokeratin's and mucopolysaccharides to keep biochemical and physical integrity of skin (Barker et al., 1991). However, it has been reported that disruption of the epidermal due to various environmental pollution through the stratum corneum cause to trigger reactive oxygen species (ROS) production in skin cells. Many environmental pollutants, including PM are capable to act as oxidants or catalyze the production of ROS directly or indirectly. The upregulated production of ROS are believed to activate proliferative and cell survival signaling that can induce pathogenesis of skin disorders (Bickers and Athar, 2006). Specifically, Romani et al. (2018) reported that PM has a potential to trigger oxidative stress-related factors in reconstituted human epidermis and the ability of PM to penetrate the skin tissue (Romani et al., 2018). However, several studies reported the potential of seaweed extracts to inhibit PM-induced oxidative stress in keratinocytes. Specifically, high level of DNA damage, lipid peroxidation, and protein carbonylation, endoplasmic reticulum stress, mitochondrial swelling, autophagy, and apoptosis are some characteristics features of oxidative stress in PM-exposed keratinocytes (Figure 10.2) (Hu et al., 2017; Piao et al., 2018; Mihindukulasooriya et al., 2019).

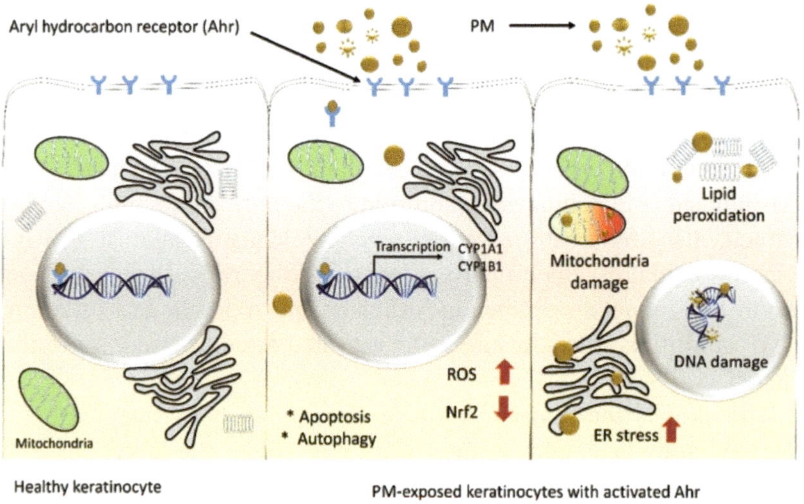

FIGURE 10.2 Effects of particulate matter on ROS induction in skin cells.

Lee et al. (2018) evaluated the protective effects of phlorotannins isolated from *Ecklonia cava* against PM10-induced cellular lipid peroxidation in keratinocytes. According to the authors, PM10 concentrations between 0–100 µg/ml significantly upregulated the lipid peroxidation and cell death rates in keratinocytes. However, the co-treatment of phlorotannins shown a protective effect against PM-induced cell death and lipid peroxidation in keratinocytes (Lee et al., 2018). Diphlorethohydroxycarmalol (DPHC) is another well-known brown algal phlorotannin with interesting bioactive properties such as antioxidant and anti-inflammatory. Recently, Zhen et al. (2019) demonstrated the protective effects of DPHC isolated from *Ishige okamurae* against PM2.5-induced oxidative stress in human keratinocytes. The authors observed that, in response to PM2.5 cell death rates, were increased in the keratinocytes through the upregulated DNA damage, endoplasmic reticulum stress, and autophagy. However, co-treatment of DPHC found to reduce PM2.5-induced oxidative stress related symptoms in the stimulated cells (Zhen et al., 2019). In addition, Ha et al. (2019) reported that dieckol isolated from *E. cava* has a potential to inhibit PM10-induced cell death rates in HaCaT Keratinocytes via blocking Prostaglandin E2 (PGE2). Specifically, in this study the authors found that inhibition of PGE2 in PM-exposed HaCaT cells were achieved by dieckol via blocking several mRNAs important to conversion of arachidonic acid to PGE2 such as Cyclooxygenase (COX)-1, COX-2, mPGES-1, and/or mPGES-2 (AlKahtane et al., 2019). Additionally, Mihindukulasooriya et al. (2019) and Jee et al. (2019) reported ethanolic extract separated from *Sargassum horneri* has a potential to reversed the PM-induced oxidative stress and apoptosis in MLE-12 cells (lung epithelial cells) and splenocytes, respectively (Jee et al., 2019; Mihindukulasooriya et al., 2019). Even though both studies were performed with immune cells and lung epithelial cells, the results demonstrate the potential of *S. horneri* extract to develop as a protective agent against PM-induced oxidative stress and apoptosis against skin cells. However, further studies are required to confirm this hypothesis.

10.2.1 PM REPRESS NUCLEAR FACTOR ERYTHROID 2-RELATED FACTOR 2 (Nrf2) EXPRESSION IN SKIN CELLS

Nrf2 is an important antioxidant protein which regulate the induction of genes encoding antioxidant proteins and phase 2 detoxifying enzymes. Under the normal conditions, Nrf2-dependent transcription is repressed by a negative regulator Kelch-like ECH-associated protein 1 (Keap1). Under the oxidative stress conditions, Nrf2 escapes Keap1-mediated repression and activates antioxidant responsive element (ARE)-dependent gene expression to keep cellular redox homeostasis (Zhang, 2006). Recently a number of studies highlighted that the high levels of ROS in PM-exposed skin cells are co-related with suppression of Nrf2/Keap1 signal transduction (Ranouille et al., 2018). In addition to the ROS inhibition, Nrf2 also repressed inflammatory signal cascades such as nuclear factor kappa B (NF-κB) (Magnani et al., 2016). Thus, restorations of Nrf2 activities in PM-exposed skin cells might be a possible approach to reduce oxidative stress in skin cells. Several compounds isolated from seaweeds such as fucoxanthin (Zheng et al., 2013; Zheng et al., 2014), 3-bromo-4,5-dihydroxybenzaldehyde (Kim et al., 2017), and fucoidan (Ryu and Chung, 2016) found to induce activities of Nrf2 under oxidative stress conditions (Table 10.1). Recently, Jayawardena et al. (2018) reported that ethanolic extract separated from *S. horneri* has a potential to upregulate Nrf2/HO-1 mediated antioxidant protein activity in PM-exposed RAW 264.7 macrophages. The results in Jayawardena et al. (2018) showing the possibility of seaweed metabolites to protect cells against oxidative stress-induced by PM. Unfortunately, till now there are no studies reported the protective

TABLE 10.1 Potential Antioxidant Activities Reported from Seaweeds to Develop as Functional Products against PM-Induced Oxidative Stress.

Effect	Cell type	Source	Compound	References
Lipid peroxidation	Keratinocytes	*Ishige okamurae*	Diphlorethohydroxycarmalol	Zhen et al., 2019
Lipid Peroxidation	Keratinocytes	*Ecklonia cava*	Dieckol	Lee et al., 2018
Cell death	Keratinocytes	*Ecklonia cava*	Dieckol	AlKahtane et al., 2019
Cell death	MLE-12	*Sargassum horneri*	Ethanol extract	Mihindukulasooriya et al., 2019
Cell death	Splenocyte	*Sargassum horneri*	Ethanol extract	Jee et al., 2019
Nrf2/Keap1 signal transduction	RAW 264.7	*Sargassum horneri*	70% ethanol extract	Jayawardena et al., 2018
Nrf2/GSH signal transduction	Keratinocytes	Red algae	3-bromo-4,5-dihydroxybenzaldehyde	Kim et al., 2017
Anti-apoptosis	Keratinocytes	Brown seaweeds	Fucoxanthin	Zheng et al., 2013; Zheng et al., 2014
HO-1/Nrf2 transduction	A549 cells	*Sargassum binderi*	Fucosterol	Fernando et al., 2019
Stimulate CAT and SOD activities	Hairless Kun Ming mice	*Sargassum fusiforme*	Polysaccharide	Ye et al., 2018
Nrf2/ERK signal transduction	Keratinocytes	Brown seaweeds	Fucoidan	Ryu and Chung, 2016

effects of seaweed metabolites against PM-induced Nrf2 suppression in keratinocytes or skin related studies. Therefore, mechanistic studies urgently require to the understand protective effects of seaweed metabolites against PM-induced oxidative stress in the skin, in order to develop cosmeceuticals and functional products from seaweeds.

10.2.2 PM INDUCED ARYL HYDROCARBON RECEPTOR (Ahr) ACTIVATION IN SKIN CELLS

The aryl hydrocarbon receptor (Ahr) is a transcription factor involved in skin barrier adaptations in response to environmental stresses. In the un-liganded state, Ahr is retained in the cytoplasm as an inactive protein complex. When Ahr ligands engage the receptor, a conformational change occurs in it, which leads to its nuclear translocation and subsequently activates the gene transcriptions related to the inflammation and immune modulations (Justiniano and Wondrak, 2016). Activation of Ahr was found to induce the expression of cytochrome P450 genes, including CYP1A1. The upregulated expression of CYP1A1 cause to worsening disease symptoms of atopic dermatitis (AD) conditions (Hidaka et al., 2017). Other than the pollen and sand particles PM contains considerable amounts of organic compounds, heavy metals, particularly polycyclic aromatic hydrocarbons (PAHs), sulfates, and nitrites formed by combustion (Juliano and Magrini, 2018). Among them, PAHs in the PM penetrate into the stratum corneum due to the lipophilic nature of PAHs and retain in the skin for long times. Recently, it has been reported that PAHs exert their biological effect in keratinocytes via binding to the Ahr ligands and activating subsequent gene transcriptions of Ahr (Hidaka et al., 2017). A number of studies have demonstrated the effect of Ahr in the disease conditions like AD (Ji and Li, 2016). Specifically, Hidaka et al. (2016) demonstrated, the overexpression of Ahr in response to the PAHs, could lead to the development of AD in mice. Thus, compounds with Ahr inhibitions from seaweeds might have the potential to develop as Ahr inhibitors against PM-induced skin damages. Therefore, authors attempted to find previous studies reported about Ahr inhibitors identified from seaweed using keywords "Aryl hydrocarbon receptor/Ahr; seaweeds." However, to date there are no studies recorded in the SCOPUS or Google scholar databases in related to the Ahr inhibitors and seaweed keyword searchers.

10.3 PROTECTIVE EFFECTS OF SEAWEED METABOLITES AGAINST PM-INDUCED INFLAMMATORY RESPONSES IN SKIN CELLS

According to the epidemiological research studies carried out to investigate the effects of ambient air pollution, on common skin diseases (AD, acne, and psoriasis) reported that PM has a potential to induce progression of inflammatory skin diseases and worsening the disease conditions (PM10 and lower sizes) (Kim et al., 2016). AD is one of the most common chronic and recurrent inflammatory skin diseases in children and is often an early sign of other subsequent allergic diseases. Majority of children suffered from AD are infants. However, the symptoms of AD can persist in school-age children with concomitant involvement of mucosal surfaces such as asthma and rhinitis (Song et al., 2011). According to Song et al. (2011) the concentrations of PM and the itchiness symptom in children with AD had a positive relationship. Psoriasis is another chronic inflammatory skin disease, found to associate with altered PM levels in the atmosphere. According to the physical appearance, abnormal differentiation in epidermis together with dermal inflammation and thick bright red to pink plaques with silvery scale are characteristics features of psoriasis (Dinulos, 2007). Recently, a number of studies reported that PM-exposed keratinocytes had upregulated expressions of psoriasis-related genes, such as S100A7 and S100A9, chemokines, and pro-inflammatory cytokines (Cheng et al., 2020).

Specifically, human epidermal keratinocytes respond to PM by secreting various pro-inflammatory autacoids and cytokines which can, in turn, affect the integrity of the epidermal permeability barrier (Choi et al., 2011). In response to the inflammatory stimuli keratinocytes secrete interleukins (IL-1, IL-6, and IL-8), chemokines (CCL20, CXCL1, CXCL2, CXCL3, and CXCL8) cytokines (IFN-y), tumor necrosis factor-alpha (TNF-α), transforming growth factor-alpha (TGF-α), and monocyte chemotactic and activating factor (MCAF) through the activation of different inflammatory pathways (Figure 10.3) (Cheng et al., 2020). The elevated production of aforementioned cellular products increases the infiltration of mononuclear cells, either by release of pro-inflammatory cytokines or via intercellular adhesion reactions (Barker et al., 1991; Choi et al., 2011). A number of studies reported that compounds and crude extracts collected from seaweeds have the potential to inhibit PM-induced inflammation via blocking pro-inflammatory secretion, Nuclear factor kappa-light-chain-enhancer of activated B-cells (NF-κB) activation/translocation, and MAPKs expressions in keratinocytes. Thus, in the present section authors reviewed the anti-inflammatory properties reported from different seaweed metabolites against PM-induced inflammatory responses.

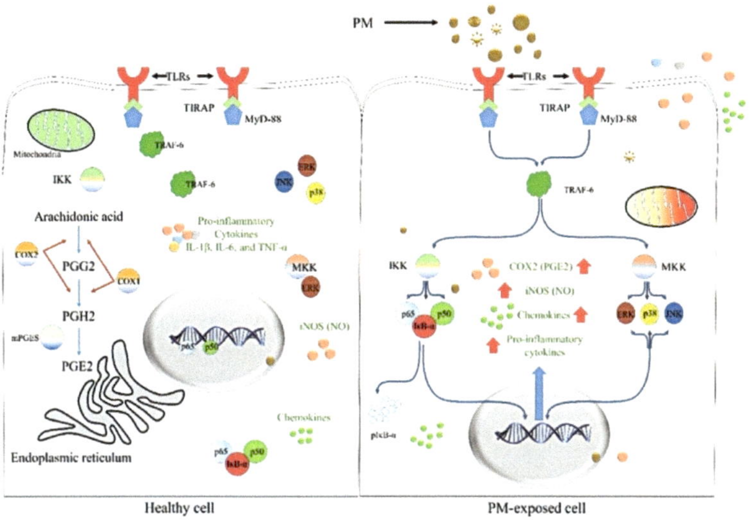

FIGURE 10.3 Possible inflammatory cascades, activated with particulate matter exposure in skin cells.

10.3.1 SEAWEED METABOLITES REPRESS PM-INDUCED INFLAMMATORY RESPONSES IN MACROPHAGES

Recently, Fernando et al. (2017) attempted to evaluate the protective effects of DPHC isolated from *I. okamurae* using urban fine dust induced keratinocytes. According to the results, in response to PM, keratinocytes had upregulated production of COX-2, PGE_2, IL-1β, and IL-6 in HaCaT keratinocytes. In addition, the authors pointed out that incubation of RAW 264.7 macrophages with HaCaT culture medium exposed to PM caused to trigger inflammatory responses in macrophages.

Mitogen-Activated Protein Kinases (MAPK) is a group of proteins involved in signal transduction from the cell membrane to the nucleus. Extracellular signal-regulated kinases (ERK), Jun amino-terminal kinases (JNKs), and stress-activated protein kinases (p38) are the members of MAPKs and involved in cell proliferation, cell growth, and cell death (Thimraj et al., 2018). Recently, several studies reported that PM have a potential to activate MAPKs in keratinocytes, which lead to epithelium injury and endothelial dysfunction (Bae et al., 2019).

Besides, NF-κB is a well-known transcription factor, which regulates adaptive and innate immune responses. In addition to the immune responses, NF-κB transcription is also linked to many biological processes, such as inflammatory responses, cell differentiation, cell proliferation and survival, embryonic development, and cancer cell pathogenesis and progression (Reale et al., 2018). NF-κB proteins are retained in the cytoplasm as an inactive complex form through their association with the inhibitor IκBα; which inhibit the nuclear localization signal of NF-κB. NF-κB is activated through activation of the IκB kinase complex (IKK). IKK degrade IκBα and releases NF-κB from the complex, exposing the nuclear localization signal (Figure 10.3). The translocation of NF-κB encodes a number of genes which are involved in inflammatory responses (Shroff and Abbruzzese, 2015). Specifically, NF-κB triggers the gene expressions related to the various pro-inflammatory genes (IL- 1β, IL-6, IL-12, and TNF-α), including those encoding cytokines and chemokines (MCP-1, IL-18, CXCL1, and CXCL2), and also regulate the activities of inflammasome (Liu et al., 2017). Several studies recently demonstrated PM-exposed skin cells had upregulated NF-κB activations which linked to the malfunctions of skin tissues (Zhang et al., 2017; Romani et al., 2018).

Fernando et al. (2018) compared FD-induced NF-κB and MAPKs activation in HaCaT keratinocytes using urban fine dust collected from China. According to the authors, FD-exposed HaCaT cells had elevated levels of NF-κB and MAPKs activations. However, the authors reported that alginic acid isolated from *S. horneri* has a potential to repress urban aerosol-induced inflammatory responses in keratinocytes. In addition, the same research team recently reported the protective fucosterol isolated from *S. binderi* against PM-induced inflammatory responses in HaCaT and HDF cells. According to the authors, fucosterol treated HaCaT and HDF cells had reduced NF-κB and MAPKs phosphorylation levels in PM-activated skin cells. Additionally, the authors found that fucosterol has the potential to inhibit MMP-1, and 2 secretion levels from tested skin cells (Fernando et al., 2019). Zhen et al. (2019) studied the MAPKs phosphorylation levels in HaCaT cells using standard PM (SRM 1650b) purchased from Sigma-Aldrich (St. Louis, MO, USA). The authors reported PM2.5 increased the formation of phosphor-ERK1/2, phosphor-p38, and phosphor-JNK compared to the unstimulated group. However, the results further demonstrated, the phosphorylation of MAPKs was reversed by DPHC pre-treatment. In addition to these specific studies, several hundreds of studies were demonstrated the ability of seaweed extracts (polyunsaturated fatty acids (PUFA), phlorotannins, and polysaccharides) to inhibit inflammation in skin cells induced by different inflammatory stimuli (Khan et al., 2007; Berthon et al., 2017). In addition, Mihindukulasooriya et al. (2019) reported the MAPKs inhibitory effect of ethanolic extract separated from *S. horneri* in PM-activated MLE-12. According to the authors, the concentration between 0 ~125 μg/mL of *S. horneri* reversed the MAPKs phosphorylation levels in PM-exposed MLE-12 cells compared to the vehicle-treated control group. However, anti-inflammatory mechanisms of seaweed extracts against PM-induced skin damages were not studied in detail. The compounds isolated from seaweeds with anti-inflammatory properties might also have anti-inflammatory properties against PM-induced inflammatory responses in skin cells. Thus, further research studies with isolated compounds might be a possible approach to fill the knowledge gap between seaweed metabolites and PM-induced inflammation in the skin.

10.4 PROTECTIVE EFFECTS OF SEAWEED METABOLITES AGAINST PM-INDUCED MMPs EXPRESSION IN SKIN CELLS

Matrix metalloproteinases (MMPs) are zinc-containing (Zn^{2+}) endopeptidases that comprise a family of structurally related enzymes that are capable of promoting physiological turnover of extracellular matrix components (ECM). According to the substrate specificity and structure, MMPs are categorized into five groups as collagenases (MMP1, 8, and 13), gelatinases (MMP2 and 9), stromelysins (MMP3, 10, and MMP11), matrilysins (MMP7 and 26), and membrane-type (MT) MMPs (MMP14, 15, and 16) (Pittayapruek et al., 2016). According to the previous studies, high levels of MMPs are caused to accelerate the formation of coarse wrinkles and sagging skin via degrading ECM such as collagen and elastin fibers. Specifically, the upregulated production of MMP-1 and MMP-3 is considered as the major reason for the characteristic wrinkling and sagging of photo-damaged skins (Bugel et al., 2016). In a recent study, Wang et al. (2019) reported that PM-exposed human dermal fibroblasts (HDF) cells have an elevated level of MMPs expression and low levels of collagen synthesis in HDF cells. Even though there are no studies reported about PM-induced MMPs expression and protective effects of seaweed metabolites, a large number of studies highlighted the potential of compounds isolated from seaweeds to act as MMPs inhibitors in skin cells. Kim et al. (2013) and Hwang et al. (2014)

demonstrated that fucosterol found in seaweeds have the potential to inhibit MMPs expression and promote procollagen production in skin cells. In addition, sulfated polysaccharides (PLS) (fucoidans) and phlorotannins also found demonstrated promising MMPs inhibitory activities in different skin cells (Joe et al., 2006; Sanjeewa et al., 2016; Ye et al., 2018). According to Joe et al. (2006) out of 29 seaweed species eckol and dieckol isolated from *E. stolonifera* found to possess MMP1 inhibitory effect on HDF cells. Besides that, Kim et al. (2018) recently found that a low molecular weight fucoidan (LMWF) prepared from *E. cava* has an inhibitory effect on the expression of several MMPs (1, 9, and 13) in six-week female HR-1 hairless mice (Kim et al., 2018). Taken together, these research studies providing the possibility of secondary metabolites present in seaweeds to act as MMPs inhibitors against PM-induced MMPs expressions (Table 10.2). However, specific studies require in future to conclude whether aforementioned compounds also have a possibility to act against PM-induced MMPs production.

TABLE 10.2 Potential Anti-Inflammatory and MMPs Inhibitory Activities Reported from Seaweeds to Develop as Functional Products against PM-Induced Oxidative Stress

	Effect	Cell type	Source	Compound	References
Anti-inflammation	Pro-inflammatory cytokine inhibition	Keratinocytes	*Ishige okamurae*	Diphlorethohy-droxycarmalol	Fernando et al., 2017
	PGE2 inhibition	Keratinocytes	*Ecklonia cava*	Dieckol	AlKahtane et al., 2019
	IL-6, p-c-Jun, and p-c-Fos inhibitor	Human dermal fibroblasts	*Hizikia fusiformis*	Fucosterol	Hwang et al., 2014
	NF-κB and MAPKs inhibition	Keratinocytes	*Sargassum horneri*	Alginic acid	Fernando et al., 2018
	Inhibit MAPKs phosphorylation	Keratinocytes	Commercial extract purchased from Sigma Chemicals	Fucosterol	Kim et al., 2013
	Inhibit MAPKs phosphorylation	Keratinocytes	Brown seaweeds	Diphlorethohy-droxycarmalol	Zhen et al., 2019
	NF-κB and MAPKs inhibition	Fibroblast cells	*Sargassum binderi*	Fucosterol	Fernando et al., 2019
	MAPKs inhibition	MLE-12 cells	*Sargassum horneri*	Ethanol extract	Mihindukula-sooriya et al., 2019
	NF-κB and MAPKs inhibition	A549 cells	*Sargassum binderi*	Fucosterol	Fernando et al., 2019

TABLE 10.2 *(Continued)*

	Effect	Cell type	Source	Compound	References
MMPs inhibition	MMP-1 inhibitor	Human dermal fibroblasts	*Hizikia fusiformis*	Fucosterol	Hwang et al., 2014
	MMP-1, 2, and 9 inhibitors	Keratinocytes	Commercial extract purchased from Sigma Chemicals	Fucosterol	Kim et al., 2013
	MMP-1, 9, and 13 inhibitors	HR-1 hairless mice	*Ecklonia cava*	Fucoidan	Kim et al., 2018
	MMP-1 inhibitor	Human dermal fibroblasts	*Ecklonia stolonifera*	Eckol and dieckol	Joe et al., 2006
	MMP-1 and 2 inhibitors	Human dermal fibroblasts	*Sargassum binderi*	Fucosterol	Fernando et al., 2019
	MMP-1 and 9 inhibitors	Hairless kun ming mice	*Sargassum fusiforme*	Polysaccharide	Ye et al., 2018

10.5 CONCLUSION

In summary, this chapter shows that exposure to particulate matter (PM) to skin cells has the potential to affect skin integrity and its natural biological functions. Specifically, a number of studies reported that upregulated pro-inflammatory cytokine production, NF-κB and MAPK phosphorylation, Ahr activation, MMPs expression, pro-collagen inhibition, and oxidative stress are some characteristic features of PM exposed skin cells. Altogether, the protein and gene expression changes observed in skin cells provide new evidences to elucidate the mechanisms underlying to trigger skin diseases and photo-aging of skin in response to PM. Taken together, these results indicate that impaired air quality affects adversely the skin health and the importance of preventive measure to protect skin from PM exposure. This chapter focused on the protective effect of seaweed metabolites against PM-induced skin damages and potential of seaweed metabolites to develop them as functional materials against PM skin damages. According to the literature search polyphenols (eckol, dieckol, and DPHC), polysaccharides (fucoidan and alginic acid), sterols (fucosterol), and crude organic solvent extracts separated from seaweeds have a potential to act against PM-induced skin dysfunctions. However, further studies urgently require to identify Ahr inhibitors, MMP inhibitors, antioxidant, and anti-inflammatory compounds from seaweeds to develop them as functional or medicinal ingredients against PM-induced skin damages.

KEYWORDS

- **antioxidant responsive element**
- **aryl hydrocarbon receptor**
- **bio-active secondary metabolites**
- **cyclooxygenase**
- **diphlorethohydroxycarmalol**
- **edible seaweeds**
- **inflammation**
- **keratinocytes**
- **mycosporine-like amino acids**
- **oxidative stress**
- **particulate matter**
- **polycyclic aromatic hydrocarbons**
- **reactive oxygen species**
- **tumour-necrosis factor-alpha**

REFERENCES

Al Kahtane, A. A., Abushouk, A. I., Mohammed, E. T., Alarifi, S., Ali, D., Alessia, M. S., et al., (2019). Fucoidan alleviates microcystin-LR-induced hepatic, renal, and cardiac oxidative stress and inflammatory injuries in mice. *Environ. Sci. Pollut. Res. Int.* doi: https://doi.org/10.1007/s11356-019-06931-z.

Bae, J. E., Choi, H., Shin, D. W., Na, H. W., Park, N. Y., Kim, J. B., Jo, D. S., et al., (2019). Fine particulate matter (PM2.5) inhibits ciliogenesis by increasing SPRR3 expression via c-Jun activation in RPE cells and skin keratinocytes. *Sci. Rep., 9*(1), 3994. doi: https://doi.org/10.1038/s41598-019-40670-y.

Barker, J. N. W. N., Griffiths, C. E. M., Nickoloff, B. J., Mitra, R. S., Dixit, V. M., & Nickoloff, B. J., (1991). Keratinocytes as initiators of inflammation. *The Lancet, 337*(8735), 211–214. doi: https://doi.org/10.1016/0140-6736(91)92168-2.

Berthon, J. Y., Nachat-Kappes, R., Bey, M., Cadoret, J. P., Renimel, I., & Filaire, E., (2017). Marine algae as attractive source to skin care. *Free Radical Research, 51*(6), 555–567.

Bickers, D. R., & Athar, M., (2006). Oxidative stress in the pathogenesis of skin disease. *J. Invest. Dermatol., 126*(12), 2565–2575. doi: https://doi.org/10.1038/sj.jid.5700340.

Bugel, S. M., Wehmas, L. C., La Du, J. K., & Tanguay, R. L., (2016). Phenotype anchoring in zebrafish reveals a potential role for matrix metalloproteinases (MMPs) in tamoxifen's effects on skin epithelium. *Toxicol. Appl. Pharmacol., 296*, 31–41. doi: https://doi.org/10.1016/j.taap.2016.02.013.

Buschmann, A. H., Camus, C., Infante, J., Neori, A., Israel, Á., Hernández-González, M. C., Pereda, S. V., et al., (2017). Seaweed production: Overview of the global state of exploitation, farming and emerging research activity. *European Journal of Phycology, 52*(4), 391–406. doi: https://doi.org/10.1080/09670262.2017.1365175.

Byrd, A. L., Belkaid, Y., & Segre, J. A., (2018). The human skin microbiome. *Nat. Rev. Microbiol., 16*(3), 143–155. doi: https://doi.org/10.1038/nrmicro.2017.157.

Cheng, Z., Liang, X., Liang, S., Yin, N., & Faiola, F., (2020). A human embryonic stem cell-based *in vitro* model revealed that ultrafine carbon particles may cause skin inflammation and psoriasis. *Journal of Environmental Sciences, 87*, 194–204. doi: https://doi.org/10.1016/j.jes.2019.06.016.

Choi, H., Shin, D. W., Kim, W., Doh, S. J., Lee, S. H., & Noh, M., (2011). Asian dust storm particles induce a broad toxicological transcriptional program in human epidermal keratinocytes. *Toxicol. Lett., 200*(1, 2), 92–99. doi: https://doi.org/10.1016/j.toxlet.2010.10.019.

Dinulos, J. G. H., (2007). Psoriasis. In: Lisa, B. Z., & Vincent, W. C., (eds.), *Comprehensive Pediatric Hospital Medicine* (pp. 967–970). Philadelphia: Mosby.

Fernando, I. P. S., Jayawardena, T. U., Kim, H. S., Lee, W. W., Vaas, A., De Silva, H. I. C., Abayaweera, G. S., et al., (2019). Beijing urban particulate matter-induced injury and inflammation in human lung epithelial cells and the protective effects of fucosterol from *Sargassum binderi* (Sonder ex J. Agardh). *Environ. Res., 172*, 150–158. doi: https://doi.org/10.1016/j.envres.2019.02.016.

Fernando, I. P. S., Jayawardena, T. U., Kim, H. S., Vaas, A. P. J. P., De Silva, H. I. C., Nanayakkara, C. M., Abeytunga, D. T. U., et al., (2019). A keratinocyte and integrated fibroblast culture modal for studying particulate matter-induced skin lesions and therapeutic intervention of fucosterol. *Life Sciences,* 116714. doi: https://doi.org/10.1016/j.lfs.2019.116714.

Fernando, I. P. S., Jayawardena, T. U., Sanjeewa, K. K. A., Wang, L., Jeon, Y. J., & Lee, W. W., (2018). Anti-inflammatory potential of alginic acid from *Sargassum horneri* against urban aerosol-induced inflammatory responses in keratinocytes and macrophages. *Ecotoxicol. Environ. Saf., 160*, 24–31. doi: https://doi.org/10.1016/j.ecoenv.2018.05.024.

Fernando, I. P. S., Kim, H. S., Sanjeewa, K. K. A., Oh, J. Y., Jeon, Y. J., & Lee, W. W., (2017). Inhibition of inflammatory responses elicited by urban fine dust particles in keratinocytes and macrophages by diphlorethohydroxycarmalol isolated from a brown alga *Ishige okamurae. Algae, 32*(3), 261–273. doi: https://doi.org/10.4490/algae.2017.32.8.14.

Hafting, J. T., Craigie, J. S., Stengel, D. B., Loureiro, R. R., Buschmann, A. H., Yarish, C., Edwards, M. D., & Critchley, A. T., (2015). Prospects and challenges for industrial production of seaweed bioactives. *J. Phycol., 51*(5), 821–837. doi: https://doi.org/10.1111/jpy.12326.

Hidaka, T., Ogawa, E., Kobayashi, E. H., Suzuki, T., Funayama, R., Nagashima, T., Fujimura, T., et al., (2017). The aryl hydrocarbon receptor AhR links atopic dermatitis and air pollution via induction of the neurotrophic factor artemin. *Nat. Immunol., 18*(1), 64–73. doi: https://doi.org/10.1038/ni.3614.

Hu, R., Xie, X. Y., Xu, S. K., Wang, Y. N., Jiang, M., Wen, L. R., Lai, W., & Guan, L., (2017). PM2.5 exposure elicits oxidative stress responses and mitochondrial apoptosis pathway activation in HaCaT keratinocytes. *Chin. Med. J., 130*(18), 2205–2214. doi: https://doi.org/10.4103/0366-6999.212942.

Hwang, E., Park, S. Y., Sun, Z. W., Shin, H. S., Lee, D. G., & Yi, T. H., (2014). The protective effects of fucosterol against skin damage in UVB-irradiated human dermal fibroblasts. *Mar. Biotechnol., 16*(3), 361–370. doi: https://doi.org/10.1007/s10126-013-9554-8.

Jayawardena, T. U., Sanjeewa, K. K. A., Fernando, I. P. S., Ryu, B. M., Kang, M. C., Jee, Y., Lee, W. W., & Jeon, Y. J., (2018). *Sargassum horneri* (Turner) C. Agardh ethanol extract inhibits the fine dust inflammation response via activating Nrf2/HO-1 signaling in RAW 264.7 cells. *BMC Complement. Altern. Med., 18*(1), 249. doi: https://doi.org/10.1186/s12906-018-2314-6.

Jee, Y., Kim, A., Herath, K. H. I. N. M., Kim, H. J., Mihindukulasooriya, S. P., Ko, M. O., & Jeon, Y. J., (2019). *Sargassum horneri* inhibits particulate matter-induced apoptotic cell death in splenocytes. *The Journal of Immunology 202*(1 Supplement), 187.31–187.31.

Ji, H., & Li, X. K., (2016). Oxidative stress in atopic dermatitis. *Oxid. Med. Cell. Longev., 2016*, 2721469. doi: https://doi.org/10.1155/2016/2721469.

Joe, M. J., Kim, S. N., Choi, H. Y., Shin, W. S., Park, G. M., Kang, D. W., & Kim, Y. K., (2006). The inhibitory effects of eckol and dieckol from *Ecklonia stolonifera* on the expression of matrix metalloproteinase-1 in human dermal fibroblasts. *Biol. Pharm. Bull., 29*(8), 1735–1739.

Juliano, C., & Magrini, G. A., (2018). Cosmetic functional ingredients from botanical sources for anti-pollution skincare products. *Cosmetics, 5*(1), 19.

Justiniano, R., & Wondrak, G. T., (2016). The aryl hydrocarbon receptor (AhR) as an environmental stress sensor and regulator of skin barrier function: Molecular mechanisms and therapeutic opportunities. In: Georg, T. W., (ed.), *Skin Stress Response Pathways* (pp. 325–359). Cham: Springer International Publishing.

Khan, M. N., Cho, J. Y., Lee, M. C., Kang, J. Y., Park, N. G., Fujii, H., & Hong, Y. K., (2007). Isolation of two anti-inflammatory and one pro-inflammatory polyunsaturated fatty acids from the brown seaweed *Undaria pinnatifida*. *J. Agric. Food. Chem., 55*(17), 6984–6988. doi: https://doi.org/10.1021/jf071791s.

Khan, W., Rayirath, U. P., Subramanian, S., Jithesh, M. N., Rayorath, P., Hodges, D. M., Critchley, A. T., et al., (2009). Seaweed extracts as biostimulants of plant growth and development. *Journal of Plant Growth Regulation, 28*(4), 386–399. doi: https://doi.org/10.1007/s00344-009-9103-x.

Kim, H. O., Kim, J. H., Cho, S. I., Chung, B. Y., Ahn, I. S., Lee, C. H., & Park, C. W., (2013). Improvement of atopic dermatitis severity after reducing indoor air pollutants. *Ann. Dermatol., 25*(3), 292–297. doi: https://doi.org/10.5021/ad.2013.25.3.292.

Kim, K. C., Hyun, Y. J., Hewage, S., Piao, M. J., Kang, K. A., Kang, H. K., Koh, Y. S., et al., (2017). 3-bromo-4,5-dihydroxybenzaldehyde enhances the level of reduced glutathione via the Nrf2-mediated pathway in human keratinocytes. *Mar. Drugs, 15*(9), 291. doi: https://doi.org/10.3390/md15090291.

Kim, K. E., Cho, D., & Park, H. J., (2016). Air pollution and skin diseases: Adverse effects of airborne particulate matter on various skin diseases. *Life. Sci., 152*, 126–134. doi: https://doi.org/10.1016/j.lfs.2016.03.039.

Kim, M. S., Oh, G. H., Kim, M. J., & Hwang, J. K., (2013). Fucosterol inhibits matrix metalloproteinase expression and promotes type-1 procollagen production in UVB-induced

HaCaT cells. *Photochem. Photobiol., 89*(4), 911–918. doi: https://doi.org/10.1111/php.12061.

Kim, Y. I., Oh, W. S., Song, P. H., Yun, S., Kwon, Y. S., Lee, Y. J., Ku, S. K., et al., (2018). Anti-photoaging effects of low molecular-weight fucoidan on ultraviolet B-irradiated mice. *Mar. Drugs., 16*(8), 286. doi: https://doi.org/10.3390/md16080286.

Lai-Cheong, J. E., & McGrath, J. A., (2017). Structure and function of skin, hair and nails. *Medicine, 45*(6), 347–351. doi: https://doi.org/10.1016/j.mpmed.2017.03.004.

Lee, J. W., Seok, J. K., & Boo, Y. C., (2018). *Ecklonia cava* extract and dieckol attenuate cellular lipid peroxidation in keratinocytes exposed to PM10. *Evid. Based. Complement. Alternat. Med., 2018,* 8248323. doi: https://doi.org/10.1155/2018/8248323.

Liang, J., (2013). Particulate matter. In: Jinyou, L., (ed.), *Chemical Modeling for Air Resources* (pp 189–219). Boston: Academic Press.

Liu, T., Zhang, L., Joo, D., & Sun, S. C., (2017). NF-kappa B signaling in inflammation. *Signal Transduct. Target. Ther., 2*, 17023. doi: https://doi.org/10.1038/sigtrans.2017.23.

Mabeau, S., & Fleurence, J., (1993). Seaweed in food products: Biochemical and nutritional aspects. *Trends in Food Science and Technology, 4*(4), 103–107. doi: https://doi.org/10.1016/0924-2244(93)90091-n.

Magnani, N. D., Muresan, X. M., Belmonte, G., Cervellati, F., Sticozzi, C., Pecorelli, A., Miracco, C., et al., (2016). Skin damage mechanisms related to airborne particulate matter exposure. *Toxicol. Sci., 149*(1), 227–236. doi: https://doi.org/10.1093/toxsci/kfv230.

Matsui, T., & Amagai, M., (2015). Dissecting the formation, structure and barrier function of the stratum corneum. *Int. Immunol., 27*(6), 269–280. doi: https://doi.org/10.1093/intimm/dxv013.

Mihindukulasooriya, S. P., Herath, K. H. I. N. M., Kim, A., Kim, H. J., Ko, M. O., Jeon, Y. J., & Youngheun, J., (2019). *Sargassum horneri* mitigates particulate matter induced inflammatory response in MLE-12 cells via MAPK pathways. *The Journal of Immunology,* 202(Supplement 1), 117.22-117.22.

Miller, B. G., (2017). Particulate formation and control technologies. In: Bruce, G. M., (ed.), *Clean Coal Engineering Technology* (pp. 419–465). Butterworth-Heinemann.

Park, S. Y., Byun, E. J., Lee, J. D., Kim, S., & Kim, H. S., (2018). Air pollution, autophagy, and skin aging: Impact of particulate matter (PM10) on human dermal fibroblasts. *Int. J. Mol. Sci., 19*(9), 2727. doi: https://doi.org/10.3390/ijms19092727.

Piao, M. J., Ahn, M. J., Kang, K. A., Ryu, Y. S., Hyun, Y. J., Shilnikova, K., Zhen, A. X., et al., (2018). Particulate matter 2.5 damages skin cells by inducing oxidative stress, subcellular organelle dysfunction, and apoptosis. *Arch. Toxicol., 92*(6), 2077–2091. doi: 10.1007/s00204-018-2197-9.

Pittayapruek, P., Meephansan, J., Prapapan, O., Komine, M., & Ohtsuki, M., (2016). Role of matrix metalloproteinases in photoaging and photocarcinogenesis. *Int. J. Mol. Sci., 17*(6), 868. doi: https://doi.org/10.3390/ijms17060868.

Ranouille, E., Boutot, C., Bony, E., Bombarde, O., Grosjean, S., Lazewski, A., Berthon, J. Y., & Filaire, E., (2018). *Schisandra chinensis* protects the skin from global pollution by inflammatory and redox balance pathway modulations: An *in vitro* study. *Cosmetics, 5*(2), 36.

Reale, C., Zotti, T., Scudiero, I., Vito, P., & Stilo, R., (2018). The NF-κB family of transcription factors and its role in thyroid physiology. In: Gerald, L., (ed.), *Vitamins and Hormones* (pp. 195–210). Academic Press.

Romani, A., Cervellati, C., Muresan, X. M., Belmonte, G., Pecorelli, A., Cervellati, F., Benedusi, M., et al., (2018). Keratinocytes oxidative damage mechanisms related to airborne particle matter exposure. *Mech. Aging. Dev., 172*, 86–95. doi: https://doi.org/10.1016/j.mad.2017.11.007.

Ryu, M. J., & Chung, H. S., (2016). Fucoidan reduces oxidative stress by regulating the gene expression of HO-1 and SOD-1 through the Nrf2/ERK signaling pathway in HaCaT cells. *Molecular Medicine Reports, 14*(4), 3255–3260.

Sanjeewa, K. K. A., Kim, E. A., Son, K. T., & Jeon, Y. J., (2016). Bioactive properties and potentials cosmeceutical applications of phlorotannins isolated from brown seaweeds: A review. *J. Photochem. Photobiol. B, 162*, 100–105. doi: https://doi.org/10.1016/j.jphotobiol.2016.06.027.

Sanjeewa, K. K. A., Lee, W. W., Kim, J. I., & Jeon, Y. J., (2017). Exploiting biological activities of brown seaweed *Ishige okamurae* Yendo for potential industrial applications: A review. *Journal of Applied Phycology, 29*(6), 3109–3119. doi: https://doi.org/10.1007/s10811-017-1213-4.

Shroff, R. T., & Abbruzzese, J. L., (2015). Molecular pathogenesis of pancreatic adenocarcinoma. In: John, M., Joe, W. G., Peter, M. H., Mark, A. I., & Craig, B. T., (eds.), *The Molecular Basis of Cancer* (4th edn., pp. 515–522). Philadelphia.

Song, S., Lee, K., Lee, Y. M., Lee, J. H., Lee, S. I., Yu, S. D., & Paek, D., (2011). Acute health effects of urban fine and ultrafine particles on children with atopic dermatitis. *Environ. Res., 111*(3), 394–349. doi: https://doi.org/10.1016/j.envres.2010.10.010.

Tanna, B., & Mishra, A., (2019). Nutraceutical potential of seaweed polysaccharides: Structure, bioactivity, safety, and toxicity. *Comprehensive Reviews in Food Science and Food Safety, 18*(3), 817–831. doi: https://doi.org/10.1111/1541-4337.12441.

Thimraj, T. A., George, L., Asrafuzzaman, S., Upadhyay, S., & Ganguly, K., (2018). Oxidative signaling in chronic obstructive airway diseases. In: Shampa, C., Wolfgang, J., & Debasis, B., (eds.), *Immunity and Inflammation in Health and Disease* (pp. 79–98). Academic Press.

Wang, L., Lee, W., Cui, Y. R., Ahn, G., & Jeon, Y. J., (2019). Protective effect of green tea catechin against urban fine dust particle-induced skin aging by regulation of NF-kappaB, AP-1, and MAPKs signaling pathways. *Environ. Pollut., 252*(Pt B), 1318–1324. doi: https://doi.org/10.1016/j.envpol.2019.06.029.

Wang, L., Park, Y. J., Jeon, Y. J., & Ryu, B., (2018). Bioactivities of the edible brown seaweed, *Undaria pinnatifida*: A review. *Aquaculture, 495*, 873–880. doi: https://doi.org/10.1016/j.aquaculture.2018.06.079.

WHO, (2019). *Mortality and Burden of Disease from Ambient Air Pollution.* World Health Organization. https://www.who.int/gho/phe/outdoor_air_pollution/burden_text/en/ (accessed on 13 October 2021).

Ye, Y., Ji, D., You, L., Zhou, L., Zhao, Z., & Brennan, C. J. J. O. T. F., (2018). Structural properties and protective effect of *Sargassum fusiforme* polysaccharides against ultraviolet B radiation in hairless Kun Ming mice. *Journal of Functional Foods, 43*, 8–16.

Zhang, D. D., (2006). Mechanistic studies of the Nrf2-Keap1 signaling pathway. *Drug Metab. Rev., 38*(4), 769–789. doi: https://doi.org/10.1080/03602530600971974.

Zhang, Y., Zheng, L., Tuo, J., Liu, Q., Zhang, X., Xu, Z., Liu, S., & Sui, G., (2017). Analysis of PM2.5-induced cytotoxicity in human HaCaT cells based on a microfluidic system. *Toxicol. In Vitro., 43*, 1–8. doi: https://doi.org/10.1016/j.tiv.2017.04.018.

Zhen, A. X., Piao, M. J., Hyun, Y. J., Kang, K. A., Madushan, F. P. D. S., Cho, S. J., Ahn, M. J., & Hyun, J. W., (2019). Diphlorethohydroxycarmalol attenuates fine particulate matter-induced subcellular skin dysfunction. *Mar. Drugs, 17*(2), 95. doi: https://doi.org/10.3390/md17020095.

Zheng, J., Piao, M. J., Keum, Y. S., Kim, H. S., & Hyun, J. W., (2013). Fucoxanthin protects cultured human keratinocytes against oxidative stress by blocking free radicals and inhibiting apoptosis. *Biomol. Ther (Seoul)., 21*(4), 270–276. doi: https://doi.org/10.4062/biomolther.2013.030.

Zheng, J., Piao, M. J., Kim, K. C., Yao, C. W., Cha, J. W., & Hyun, J. W., (2014). Fucoxanthin enhances the level of reduced glutathione via the Nrf2-mediated pathway in human keratinocytes. *Mar. Drugs, 12*(7), 4214–4230. doi: https://doi.org/10.3390/md12074214.

CHAPTER 11

SEAWEED BIOTECHNOLOGY: AN ANSWER TO ENVIRONMENTAL ISSUES AND HUMAN HEALTH PROBLEMS

ELHAN KHAN and IFFAT ZAREEN AHMAD

Natural Products Laboratory, Department of Bioengineering, Integral University, Lucknow – 226026, Uttar Pradesh, India

ABSTRACT

This chapter provides information relating to seaweeds' pharmacological, clinical, industrial, and environmental researches that have been carried out. Compared to terrestrial foodstuff, marine alga provides unique compounds that are only found in marine life along with the high concentrations of trace elements, minerals, enzymes, vitamins, protein, polysaccharides, and various secondary metabolites. Seaweeds have been used in culinary from earlier times in Asian foods. Advance farming also encourages the farmers to utilize the seaweeds as nature manure and/or fertilizer to promote standing crop yield and increase crop fertility. Recently, seaweed biomasses have gained attention for the production of biofuels as an alternative renewable source, also referred to as "next-generation fuel." Aquacultures are also widely accepting the seaweeds for the process of the wastewater containing dyes, heavy metals, nitrogen, and phosphorous. Agar and alginates are two major compounds extracted from seaweed by industries to be used as raw sources. Thus, this provides an insight to the application of seaweed biotechnology to human and environmental welfare.

11.1 INTRODUCTION

Seaweeds, also known as microalgae, are aquatic species that contribute to the macrophytic marine algae family, belonged to the *Thallophyta* division of the plant kingdom. They are non-flowering, with no true leaves, stem, and root, primitive plants whose whole body consists of stipe, blade, and holdfast which is collectively called thallus. They are the primary producers in the marine ecosystem and commercially important living marine renewable resources (Egan et al., 2013). Generally, seaweeds grow abundantly in the intertidal region due to the presence of the substratum; they also grow extensively along the coastline, particularly on rocky shore regions. The existence of photosynthetic pigments, as well as the morphological, structural, and reproductive arrangements of the cells, categorized seaweeds into four classes namely *Chlorophyceae* (green algae – contain photosynthetic pigments like carotenoids and chlorophyll a and b), *Phaeophyceae*/brown algae (fucoxanthin along with chlorophyll a, c pigments), *Cyanophyceae* (blue-green algae – coloration is due to phycobilins; red color phycoerythrin and blue color phycocyanin presence) and *Rhodophyceae* (Red algae - have chlorophyll a, phycobilins, and carotenoids as photosynthetic pigments) (Dhargalkar and Devanand, 2004).

The use of seaweed has been cited in Chinese and Japanese literature for 2,500 years for its effectiveness in human health. In earlier times, seaweed was used in folk medicines (Tseng, 2004). Seaweed curative power for the treatment of arthritis, colds, influenza, and tuberculosis were used for traditional treatment and in foods. Anecdotal evidence from the Roman and Egyptian periods suggested its application in burns, rashes, wounds treatment, and even in the treatment of breast cancer. Modern studies have also reported the antiviral, antibacterial, anthelmintic, anticoagulant, antiprotozoal, antifungal, anti-inflammatory, and antiplatelet activities of the seaweeds (Mayer and Haman, 2002).

Recently, seaweeds have attracted natural products research. In comparison to the terrestrial ecosystem, marine organisms live in a rather demanding climate, so they have adapted complex properties to support a diverse availability of metabolites both primary and secondary, represents an enormous natural reservoir with many biologically and pharmacologically active substances. Seaweed is the sole origin for the production of certain phytochemicals like carrageenan, laminarin, galactans, alginic acid, agar, and fucoidan, along with certain minerals, protein, trace elements, bromine, iodine, fiber, and vitamins. They are used as a raw product for various factories such as confectionery, food, and dairy, textiles, and paper industries, pharmaceutical, cosmetic, and biotechnology industries. The benefits of seaweed in the agriculture field are used as a soil conditioner for centuries (Pati et al., 2016).

11.2 ECONOMIC IMPORTANCE OF SEAWEED

There are about 11,000 species of seaweeds recorded in AlgaeBase, of which 7,500 are *Rhodophyta*, 2,000 are brown algae, and 1500 are *Chlorophyceae*. While 800 species of *Bryopsidophyceae*, 400 *Siphoncladophyceae*, 250 marine *Ulvophyceae* and 50 species of *Dasycladophyceae* are recorded. In these 291 species have important commercial values and almost ten species are harvested intensively, such as *Enteromorpha clathrata, Porphyra* spp., *Saccharina japonica, Monostroma nitidum, Caulerpa* spp., *Eucheuma* spp., *Undaria pinnatifida, Sargassum fusiforme, Kappaphycus alvarezii* and *Gracilaria* spp. Today, 43 countries represent commercial seaweed activity, harvesting 31.2 million tons of seaweed per year with a worth of 11.7 billion US$, in which 95% account to farming (Tiwari and Troy, 2015). China, France, Chile, Philippines, United Kingdom, Indonesia, Japan, Norway, North Korea, and South Korea are the top ten producers of seaweed worldwide. In total world seaweed production China is leading in the seaweed production (FAO, 2018) with 59%. From 3.2 million tons of production of fresh weight of world seaweed increased to about 7 million tons in between year 1981 and 1994. The brown alga, with around 5.2 million tons (75%) is the most harvested seaweed, followed by 1.73 million tons of red alga production (25%) and a small percentage of about 0.5% of chlorophyte. The red alga *Porphyra yezoensis*, often called Nori is a valuable crop and is utilized as a food and feed in China, Japan, and Pacific regions.

Laminaria japonica, the most widely grown seaweed, accounts for the majority of Chinese harvest (about 3.8 million years). Nori, a red alga, is the most valuable crop (*Porphyra* species, mainly *P. yezoensis*). Japanese production is projected to value about $1 billion, while Korean production is estimated to be worth $0.5 billion. Newer cultivation figures suggests that China alone harvest nearly 9.5 million tons of world seaweed. The products of the seaweed industry contribute about US$ 5.5 to 6 billion total annual value with food products that are used for consumption estimated to possess the largest share of about 5 billion US$ of this followed by hydrocolloids.

Seaweed products like agar is extensively utilized in packaging material, leather industry, plywood, and paper manufacturing, culture media, photography, dairy industry, pharmaceutical industry, cosmetics industry, and also in preservation of foods. Another seaweed product, carrageenan is employed in industry of food for manufacture of goods like meatballs, sausages, ham preparations of fish and poultry meat, corned beef, as a juice concentrates, dessert gels, chocolates, marmalade, ice cream and sardine sauce. It is additionally utilized in the processing of non-edible products such as the production of toothpastes, culture media, fungicides, air fresheners, sanitary napkins and tissues, textiles, hair shampoos, and tissues. Seaweed derived alginates are utilized in frozen and baked food products like in frozen desserts, dessert gels, pastry fillings, cooked puddings, bakery icings, instant pudding, dry mixes and also in meat and flavor sauce (FAO, 2018).

11.3 POTENTIAL APPLICATION OF SEAWEEDS

11.3.1 PHARMACOLOGICAL ACTIVITIES

About 7000 aquatic natural products have been isolated to date, with 25% originating from algae. Over the years, the therapeutic and preventive effects of seaweed have been well established. Researchers have reported its anti-hypertension, antibiotics, antihelminthics, anti-diarrhea, cough remediates and drugs or compounds against tumor and cancer activity. *Bryopsis* spp. of seaweed has shown the in-vitro activity against *Delisea pulchra*, a tubercle bacillus (El Sayed et al., 2000).

DPPH assay was utilized to determine the antioxidant properties of three brown seaweeds from the Mediterranean of genus *Cystoseira* (*Cystoseira sedoides, Cystoseira crinita* and *Cystoseira compressa*). The compounds, carrageenan was believed to induce the activity of anti-proliferation cancer cells (HCT15, MCF7 and A549) in comparison to normal cell line of fibroblast and MDCK and is also studied for anti-inflammatory test by inducing rat paw edema (Mhadhebi et al., 2014). Compounds such as sulfated polysaccharides, phlorotannins, and terpenes that are isolated from brown algae are believed to be responsible for its antiproliferative activities. Neelakandan (2014) have studied the activities against bacterial and inflammation along with anti-tumor and antioxidant activities of *Sargassum wightii*, a marine brown alga in human colon carcinoma cell line. Mozhi et al. (2009) has also reported the in-vitro antioxidant activity of *S. wightii* and *U. lactuca,* methanolic extract due to their high concentration of flavonoid.

Padina boergesenii is a major source of alginic acid having antioxidant properties. Its extract treatment in Streptozotocin-induced diabetes in rat shows a significant difference in liver glycogen level and total protein with increased vital glycolytic enzymes activity (phosphoglucoisomerase, aldolase, and hexokinase) and also significantly lowered the liver gluconeogenesis by reducing the activity of gluconeogenic enzymes (fructose-1,6-diphosphatase and flucose-6-phosphatase) (Senthilkumar et al., 2014).

Ulva rigida, methanolic extract of this green seaweed was evaluated for its antipyretic activity on albino mice by determining the rectal temperature on yeast-induced pyrexia. This study reveals the extract have marked highest antipyretic activities in dose-dependent manner (doses like 200 mg/kg and 40 mg/kg, after up to 4 hours of administration lowers the body temperature) as compared to the standard-paracetamol (Sheeba and Paul, 2014).

Seafood is one of the essential reservoirs for humans of new therapeutic compounds to limit or eradicate the progression of cancer. The compound extracted from *Fucus* spp., a brown alga, fucoidan has shown activity towards the both human breast and colorectal cancers (Moussavou et al., 2014). *Sargassum siliquosum* features a significant free radical scavenging property towards OH, hydrogen peroxide (H_2O_2) and NO radicals making it usefulness for its antioxidant activity. *Sargassum oligocystum* water extract showed the most effective activity of anticancer towards Daudi and K562 cell lines (Zandi et al., 2010).

Gracilaria crassa, Laurencia papillosa, Turbinaria ornata and *Turbinaria conoides* have been evaluated for their protective ability against hepatic cells and antiulcer wound healing properties (Senthil and Murugan, 2013). Crassa and papillosa has cardiac glycosides, sulfated algal polysaccharides, flavonoids, saponins, alkaloids, and tannins which may believe to restraint stress due to water immersion and mixture of HCl/ethanol, resulting protection against indomethacin-induced gastric ulcer. The presence of fatty acids, vitamins, amino acids, carbohydrates, minerals, and polyphenolic compounds in *Gracilaria* plays a task within the defense against oxidative damage to tissue resulting healing of the wounds and a significant impact on the enzymes of the serum marker and therefore depicts markedly decreased level of SGPT by the extract of all the concentrations compared to the standard drug and thus responsible for prominent hepatoprotective activity (Senthil and Murugan, 2013). Similarly, a wide variety of seaweeds are investigated and many are reported for their pharmaceutical properties. Some of these are mentioned in Table 11.1.

Turbinaria ornata and *Sargassum wightii* methanolic extracts has been evaluated towards the various gram-negative and gram-positive bacteria reportedly are *Enterococcus faecalis, Proteus vulgaris, Bacillus subtilis, Shigella flexneri, Staphylococcus aureus, Klebsiella pneumoniae, Aeromonas hydrophila, Pseudomonas aeruginosa* and *Escherichia coli* (Vijayabaskar and Shiyamala, 2011). The aqueous extract of *Dichotomaria obtusata,* red seaweed shows significant dose dependent anti-inflammatory and analgesic activities by inhibiting tetradecanoylphorbol-13-acetate (TPA)-induced ear edema (Vázquez et al., 2011).

TABLE 11.1 The Various Reported Pharmacological Studies of the Seaweeds

Activities	Reported studies	References
Antibacterial	Alkaloids present in brown algae *Cystoseira barbata* inhibit *Klebsiella* spp. growth.	Karthikaidevi et al. (2009)
	Codium adhaerens, Ulva reticulata and *Halimeda tuna* are reported for their antibacterial activities.	Alghazeer et al. (2013)
	Compound Fucofuroeckol purified from seaweed *Eisenia bicyclis* shows activity against *Propionibacterium,* responsible for acnes.	Lee et al. (2014)
	Sargassum muticum acetone and chloroform extract shows activity against human pathogens.	Moorthi and Balasubramanian (2015)
Antifungal	Polysaccharides extracted from algae *Colpomenia sinuosa* and *Sargassum polycystum* reported to inhibit *Candid albicans.*	Kantachumpoo and Chirapart (2010)
	Compounds extracted from brown alga *Ecklonia cava* like Dieckol shows notable antifungal activity towards *Trichophyton rubrum.*	Lee et al. (2010)
	Padina pavonica, Ulva lactuca, Cystoseira tamariscifolia and *Rhodomela confervoides* show activity towards the *Candida albicans, Aspergillus niger* and *Mucor ramaniannus.*	Saidani et al. (2012)
Antiprotozoal	Algae species like *Codium bursa, Gracilaria verrucosa, Jania rubens, Caulerpa racemosa, Corallina granifera, Cystoseira barbata, Cystoseira crinita, Ceramium rubrum, Dasya pedicellata* and *Gelidium crinale* shows antiprotozoal properties against *Plasmodium falciparum, Trypanosoma brucei rhodesiense, Leishmania donovani* and *T. cruzi.*	Süzgeç-Selçuk et al. (2011)

TABLE 11.1 *(Continued)*

Activities	Reported studies	References
Antiviral	Extract of *Gelidium pulchellum, Hypnea musciformis, Colpomenia bullosa, Undaria pinnatifida, Asparagopsis armata, Ceramium rubrum, Pterosiphonia complanata, Ulva pertusa, Ishige okamurae, Sargassum sagamianum, Carpopeltis affinis, Gelidium spinulosum, Enteromorpha linza, Halopitys incurvus, Codium fragile, Boergeseniella thuyoides, Corallina pilulifera, Grateloupia turuturu, Sphaerococcus coronopifolius, Plocamium cartilagineum, Symphyocladia latiuscula* and *Symphyocladia marchantioides* reported to have ability to prevent the proliferation of poliovirus, Sindbis virus and Herpes simplex virus (HSV).	Hudson et al. (1999); Rhimou et al. (2010)
	Stypopodium zonale contains merditerpenoids aromatic acid, epitaondiol, and peroxylactone which show with high activity against the HSV-1.	Soares et al. (2007)
	Extracts of crude prepared from seaweeds like *Canistrocarpus cervicornis, Palisada perforata, Caulerpa racemosa, Padina gymnospora, Hydroclathrus clathratus* and *Lobophora variegata* are reported to have potential of inhibiting the activity of dengue virus, HSV-1 and HSV-2.	Wang et al. (2008); Koishi et al. (2012)
	Sulfated polysaccharides, namely monogalactosyl diacylglycerides, sulfoquinovosyl diacylglycerides and digalactosyl diacylglycerides showed high activity toward the wide spectrum of viruses such as Human immunodeficiency virus (HIV), Newcastle Disease Virus (NDV), HSV-1 and HSV-2. These are extracted and purified from *Sargassum vulgare,* a brown seaweed and green algae, namely *Caulerpa racemosa, Sphaerococcus coronopifolius, Adenocystis utricularis, Boergeseniella thuyoides* and *Cladosiphon okamuranus*.	Ponce et al. (2003); Trinchero et al. (2009); Bouhlal et al. (2011); Elizondo-Gonzalez et al. (2012)

TABLE 11.1 *(Continued)*

Activities	Reported studies	References
Against cancer	High antitumor activity was reported by two polysaccharide fractions SP-3–1 and SP-3–2 extracted from *Sargassum pallidum* towards the MGC-803, HepG2 and A549 cancer cell lines.	Ye et al. (2008); Pereira and Costa-Lotufo (2012)
	Compounds like kahalalide F, alginate, and kappa-carrageenan found in *Bryopsis*, *Sargassum*, and *Kappaphycus*, known for their impressive anticancer activities.	Kim et al. (2014)
	The marine brown alga *Dictyopteris undulata* were investigated for their cytotoxic and apoptotic effects against human colon adenocarcinoma cells SW480, SNU407 and HT29.	
Antioxidant	Studies have suggested the antioxidant activities of *Cystoseira hakodatensis, Kjellmaniella crassifolia, Alaria crassifolia, Sargassum horneri* and *Eisenia bicyclis* and the activity is attributed by the existence of phenolics compounds and compounds like fucoxanthin.	Airanthi et al. (2011)
	Polysaccharides such as fucoglucoxyloglucuronan, heterofucan, and two glucans present in the brown seaweed *Dictyopteris justii* are reported to provide a notable antioxidant property.	Melo et al. (2013); Farasat et al. (2013); Foon et al. (2013); Indu and Seenivasan (2013); Khairy and El-Sheikh (2015); Tariq et al. (2015)
	Crude extract of seaweeds such as *Dictyota dichotoma, Dictyota indica, Iyengaria stellata, Stokeyia indica, Eucheuma cottonii, Padina* sp., *Chaetomorpha linum, Padina pavonica, Sargassum variegatum, Jolyna laminarioides, Caulerpa taxifolia, Halimeda tuna, Solieria robusta, Melanothamnus afaqhusainii, Stoechospermum marginatum, Grateloupia lithophila, Jania rubens* and *Pterocladia capillacea* are reported for their antioxidant potential.	

TABLE 11.1 *(Continued)*

Activities	Reported studies	References
Anti-inflammatory and Immunomodulatory	The sulfated polysaccharide derived from *Ecklonia cava* and the red algae *Gracilaria cornea* (Gc-TSP) showed significant analgesic, immunomodulatory, and anti-inflammatory effects.	Kang et al. (2011)
	Mixture of *Macrocystis pyrifera, Fucus vesiculosis* and *Laminaria japonica* extract has significant potential as an immune modulator.	Myers et al. (2011)
	Fucans fraction isolated from *Ascophyllum nodosum* has the potential ability to stimulate complement activation, making it an important immunomodulator.	Coura et al. (2012)
Anti-thrombic	Polysaccharides from seaweed species such as *Dictyopteris delicatula, Codium divaricatum, Padina gymnospora, C. adhaerens, C. latum, C. fragile, Sargassum fulvellum* and *Pachymeniopsis elliptica* have been screened for their anticoagulant activity CP2–1 Polysaccharide's that are extracted and purified from *Codium divaricatum* is very efficient as anticoagulant with novel structure	Silva et al. (2005); Ekanayake et al. (2008); Zoysa et al. (2008); Ciancia et al. (2010); Arivuselvan et al. (2011); Magalhaes et al. (2011); Li et al. (2015)
Anti-diabetes	Fucoidan and alginic fractions extracted from *Sargassum duplicatum* and *Turbinaria decurrens* are reported to be an antidiabetic agent.	Hardoko et al. (2014)
	Purified hypoglycemic glycoprotein from the *Undaria pinnatifida* studies suggested for controlling postprandial hyperglycemia.	Rafiquzzaman et al. (2014)
	Methanolic extracts of *Ulva lactuca, Grateloupia lithophila* and *Stoechospermum marginatum* showed the hyperglycemic activity	Murugesan et al. (2015)

The antifungal activities of six important seaweeds *Cladophora glomerata, Gracilaria corticata, Kappaphycus alvarezii, Ulva lactuca, Ulva reticulata* and *Sargassum wightii* have been studied against *Saccharomyces cerevisiae, Aspergillus niger, Mucor indicus, Aspergillus flavus* and *Aspergillus fumigatus* (Aruna et al., 2010).

11.3.2 COSMECEUTICAL APPLICATIONS

Marine-based cosmetic products have attracted many consumers for its natural content. Alginate or carrageenan extracted from seaweed is used in creams and lotions because of its skin moisture retention properties. Seaweed is either cold grinded or freeze crushed to make a paste and is warmed under infrared radiation for utilizing in the thalassotherapy in conjunction with hydrotherapy of seawater and is applied on person's body which is said to offer comfort for rheumatism and osteoporosis (Guiry and Blunden, 1991). A patent was filed in 2009 for *Blidingia minima*, species of the family of *Ulvaceae* by the founder of group of Bioderma-Esthederm, Jean-Noël Thorel, suggests employing *Blidingia minima*, aqueous extract so as to "improve the state of the skin."

It has been observed by Verdy et al. (2011), that *Alaria esculenta* lipophilic extract in Kalpariane® AD has a notable reduction in expression of Pmel17 gene and cutaneous progerin by converting of melanin to melanosomes to stimulate anti-aging and spot correction (Verdy et al., 2012). The serum extract contains *A. esculenta* in Algologie Firming Anti-Dark Spot Serum® as the major active seaweed constituent in their serum. An aqueous extract prepared from a seaweed of *Macrocystis pyrifera,* supposedly stimulates the secretion of mucopolysaccharide (a significant part of the skin extracellular matrix), which is employed in treating a burn victim in earlier days (Price et al., 2007).

The sun protective abilities are the most attractive properties of seaweed for cosmetic purposes. Mycosporine-like amino acids (MAAs), shown to have maximum absorption in ultraviolet A (UVA) regions by a maximum absorption (such as palythine, shinorine, usujirene, etc.) and these characterized MAAs are reportedly marked their presence in many marine species (Pallela et al. 2010). *Porphyra umbilicalis* are already exploited because of their elevated content of MAA by industries. Helionori®, Gelyma is one of the products to be tapped in the future for utilizing the aqueous extract of *P. umbilicalis, Palmaria palmata*, and *Solieria chordalis* (Carreto and Carignan, 2011).

Fucosterol is an interesting molecule that has been shown to greatly reduce expressions of interleukin-6 (IL-6), phospho-c-Jun/c-Fos and matrix metalloproteinase-1 induced by UVB expression (Kim et al., 2013; Hwang et al., 2014). Fucoxanthin extracted from *L. japonica* has been reported as a tyrosinase inhibitor which makes it an active ingredient for skin lightening products (Thomas and Kim, 2013).

11.3.3 IMPACT OF SEAWEEDS IN AGRICULTURE

Seaweeds act as an honest source of bio-stimulants and organic matter, which led to their extensive use in agriculture. Seaweeds are rich in beneficial trace minerals and essential nutrients required for the enhanced seed germination and plant growth and also function as a plausible biocide which in turn enhance the standing crop yield. Additionally, to those, the seaweeds have hormones that stimulate plant growth. The seaweed compost or manure when mixed with the soil, it enriches the fertility and integrity of the soil. Bio-fertilizer enhances the root and shoots development in plants at a faster pace (Arun et al., 2014) and encourages the development of symbiotic soil microbes that aid with nutrient recycling.

Seaweed is also utilized in soil amendments, defense against pests and plant diseases by crystal formation. Seaweed features a natural salt content, which repels unwanted organisms, and within a couple of days of application, it dries and becomes quite crispy. Pests do not like "crispy" surfaces, because the sharp salty edges dig the soft body tissue (Jayaraman et al., 2015). Seaweed is also accounted for the presence of the developmental advancing hormones like auxins, cytokinins, and gibberellins, which triggers root and shoot formation. Various forms of seaweeds, such as *Porphyra perforata, Botryocladia* spp. *Ascophyllum nodosum, Enteromorpha* spp., and including cyanobacteria were found to contain indole-3-acidic acids (IAA). *Fucus vesiculosus* and *Fucus spiralis* are the seaweeds that are reported to contain the seed germinating hormones gibberellins (Khan et al., 2009).

Specific concentration of *Sargassum johnstonii* aqueous extract enhances the rooting and vegetative growth of *Vigna mungo*, while reproductive and conceptive parameters (new weight, blossom number, and natural product number) of tomato (Kumari et al., 2011). Similarly, *Sargassum wightii* gives an improvement of 63% in the development of numbers of horizontal roots with increment in the length of shoots by 46%

and 11% expansion in seed germination of *Triticum aestivum* in contrast with control. *Sargassum wightii* aqueous extract when implemented on *Ziziphus mauritiana* is a foliar splash indicated an elevated productivity and improved nature of organic products (Osman and Olfat, 2011). Likewise, seaweed liquid fertilizer (SLF) prepared by Gandhiyappan and Perumal (2001) by utilizing the seaweed of *Enteromorpha intestinalis* has been accounted for growth-promoting benefits on the crop plant of sesame. Similarly, lower concentration of SLF from *Stoechospermum marginatum* have demonstrated an enhanced growth in the brinjal (Ramya et al., 2015). *Ulva ohnoi* (green seaweed) are reported for soil amelioration and biochar production for carbon sequestration (Nkema and Murto, 2012; Roberts and de Nys, 2016).

11.3.4 SEAWEED AS A BIOFUELS

Bio-fuel such as ethanol, butanol, etc., can be produced from seaweed by converting alginate, mannitol, and fiber. Seaweed is also a proven, nonpartisan carbon-dioxide potential (CO_2) wellspring of second era biofuels (Bhajhaiya et al., 2010). *Laminaria hyperborean* extracts were already reported for the production of ethanol and the yeast *Pichia angophorae* was utilized for evaluation (Hinks et al., 2013). Seaweed extract revealed that both laminarin and mannitol can be employed for ethanol production when provided as a substrate by *Pichia angophorae* seaweed and *Saccharina latissima* provided the best yield of ethanol when used as a substrate (Adams et al., 2009). Numerous scientists announced that kelp may be better for higher biodiesel creation, along these lines' seaweed be fit for utilized as a sustainable power source (Hossain et al., 2008).

11.3.5 SEAWEED'S UTILIZATION AS A BIO-FILTER

Seaweed and other algae have developed different approaches for their tolerance in overwhelming metal-contaminated environments, these life forms are known to create and embrace diverse detoxifying components, for example, bioaccumulation, biotransformation, biosorption, and biomineralization, which can be exploited either *ex situ* or *in situ* for the purpose of bioremediation. Seaweed is successfully incorporated as a biofilter in aquaculture. A certain species such as *Ulva, Gracilaria, Laminaria,* and a few other microalgae have been perceived to incorporate in the framework of bio channel and indicated sensibly high productivity in the evacuation of waste inorganic supplements. Various reported studies of seaweed utilization as a biofilter are mentioned in Table 11.2. In Korea *Porphyra,* also called "Gim" was proposed as an attractive participant with salmonids for the coordinated aquaculture. Several red seaweeds (such as *Gracilaria lemaneiformis, Gracilaria tikvahiae, Gracilaria chouae, Chondrus crispus, Palmaria palmata, Gracilaria vermiculophylla* and many more) are utilized in sewage treatment and to diminish the dyes and absolute level of phosphorous and nitrogen from some rural squanders and it may useful for the expulsion of harmful metals from mechanical wastewater (Arumugam et al., 2018).

TABLE 11.2 Various Reported Species of Seaweed involved in the Treatment of Wastewater

Bioremediation targets	Seaweeds with the activity	References
Phosphorus and nitrogen-containing wastewater treatment	*Chondrus crispus* and *Palmaria palmata* are also reported for its ability of bioremediation and productivity enhancement and removal of nitrogen in a land-based Atlantic halibut farm.	Kim et al. (2013)
	Seaweed of species *Gracilaria* such as *G. lemaneiformis*, *G. tikvahiae*, *G. chouae, G. vermiculophylla* and *G. birdie* are reported to for their efficient decrease in the eutrophication index.	Mariho-Soriano et al. (2009); Abreu et al. (2011); Samocha et al. (2015); Wu et al. (2015); Wei et al. (2017)
Wastewater treatment containing dyes	*Nizamuddin zanardinii* was investigated for the biosorption of Acid Black 1.	Cengiz and Canvas (2008)
	Gracilaria parvispora, Nizamuddin zanardinii and *Ulva fasciata* shows a significant biosorption efficiency and capacity of Methylene Blue dye from aqueous solution.	Daneshvar et al. (2017)
	Gracilaria corticata, Sargassum crassifolium and *Ulva lactuca* aqueous solution were examined to have a potential to remove malachite green dye.	Omar et al. (2018)

TABLE 11.2 *(Continued)*

Bioremediation targets	Seaweeds with the activity	References
Wastewater treatment containing heavy metals	Similarly, *Canthus chamisso* another red alga reported to have the ability lead (II) and cadmium (II) biosorption.	Yipmantin et al. (2011)
	Osmunda pinnatifida, a red alga is also utilized for the biosorption of copper (II) and cadmium (II).	El Hassouni et al. (2014)
	Various *Sargassum* genus alga is reported for their ability to eliminate the heavy metals such as cadmium (II), zinc (II), nickel (II), copper (II), manganese (II) and lead (II) ions from wastewater.	Ghasemi et al. (2016); Verma et al. (2016); Barquilha et al. (2017); Mahmood et al. (2017)

11.3.6 INDUSTRIAL USES

Hydrocolloid-based gums are polymers of variety long chain having the ability to form gels or viscous dispersion when dispersed in water. Agar, carrageenan, and alginate are the most significant seaweed hydrocolloids, and they are distributed as colorless powders. In the last 300 years, agar was the first hydrocolloid to be used in Asian countries as a food additive. Around 90% of the produced agar is used in edible products and with the corresponding 10% used in biotechnology and microbiological applications. Agar was discovered in Japan where it is prevalently known as “kanten” and is utilized as a solid substrate for bacteria and fungi growth. *Ahnfeltiopsis, Gracilaria, Gelidium, Gelidiella, Pterocladiella,* and *Pterocladia* are the seaweed genera which are involved in the production of agar. The highest caliber of agar and its substitute called agarose originates from red-green growth having a place with family *Gelidiaceae*. They are also used in food as in cakes, chocolates, jam, jellies, candies, wafers, liquors salad dressing and also in making of sauces as a stabilizer, emulsifier, and diuretic. Agars are also used in assembling of photographic film, paint, glue, batteries, graphite, and so on. While alginate is being utilized as a potential element in frozen food products, in ice cream and also in numerous pharmaceutical enterprises as a stabilizer and a responsive base in receptive dye printing of textiles. It is also utilized to improve the nature of paper surface. It goes about as an emulsifier for some food items in food ventures as an additive (Pati et al., 2016). Other lower-quality agars can be found in a variety of families, especially the *Gracilariaceae*. As reported, seaweeds contain an appreciable number of polyphenols such as fucoxanthin, a pigment of the carotenoid family, which supplied a new functioning food with an anti-metabolic syndrome activity (anti-diabetes and anti-obesity) as a provider of cosmetic ingredients (Warkoyo and Saati, 2011).

11.3.7 SEAWEED UTILIZATION IN ADDRESSING ENVIRONMENTAL ISSUES

By incorporating the natural seaweed polymer into traditional plastic formulations, *Gracilaria vermiculophylla*, red seaweed, is often used to produce biodegradable plastics (Freile-Pelegrin et al., 2007). Seaweed likewise, an alluring source to use in battling the ecological issues like ameliorates ocean acidification, desertification, and deoxygenation and also break down the environmental pollutants that enters the ocean. Vincent van Ginneken, suggested a special model named "Ulva-Desert" to combat desertification. He utilized the *Ulva* genus seaweed to create a high-temperature-tolerance (HTT) strain (suitable for dessert aquaculture) protoplast. The biomass produced with primary accentuation on unpredictable dimethyl sulfide and its precursor β-dimethyl sulfonium propionate which stimulates the ocean cloud formation via sulfur cycle bringing about rainfall in the deserts and atmosphere cooling. Along these lines' algae seaweed biotechnology in conjunction with characteristic geosphere-biosphere processes may alter the course of a dangerous atmospheric deviation, battle desertification, add to environmental change moderation and increase the worldwide capability of inexhaustible arable zone (van Ginneken, 2018).

Apart from above mentioned uses, seaweeds play a serious role in maintaining and balancing of marine food chain and provide supplements and vitality to marine creatures either directly through the consumption of fronds or indirectly through the decomposition into fine particles. Beds of seaweed provide shelter and habitat of marine creatures for entire or part of their lives. They serve as an important nursery for commercial species of wide diversity.

11.4 CONCLUSION

Seaweeds are the most significant asset for essential nutrients like vitamins, trace elements, lipids, enzymes, polysaccharides, and secondary metabolites. In spite of the fact that seaweeds and its derived products are utilized in regular consumption as a food, by many clinical, pharmaceutical, and industrial sectors may not be safe as it may be contaminated by heavy metals like customary uptake of bladderwrack may lead to toxic to the body by causing thyroid abnormalities, skin-acne lesions, irritation of stomach and may influence the glucose level because of the presence of heavy metals and high iodine. Owing to its complex constituents, marine therapeutic plants have been the potent source of human well-being, which is responsible for its numerous pharmacological activities (Gade et al., 2013).

Seaweed species in the marine world have a wide variety of genetic resources and biodiversity. Researchers have opined that kelp can be used in a totally extraordinary way in the medication business. The restorative medications arranged from seaweeds have gotten an incredible enthusiasm for malignant growth treatment. The utilizations of seaweed and their use have not been completely settled. Because of its optional metabolites, the current restricted use of seaweed should be broadened to different territories of use. Seaweed's multi-functional properties as a plant with a unique architecture and biochemical composition, it could be exploited in the form of medicine and food. The exploration on seaweed should be directed for getting more data about their obscure advantages.

KEYWORDS

- **agar**
- **alga**
- **biofuel**
- **biotechnology**
- **brown algae**
- **carrageenan**
- **Cyanophyceae**
- **environment**
- **fucoxanthin**
- **high-temperature-tolerance**
- **hydrocolloid**
- **indole-3-acetic acid**
- **marine organism**
- **microorganism**

- **Phaeophyceae**
- **pigments**
- **Rhodophyceae**
- **seaweed**
- **seaweed liquid fertilizer**
- **Thallophyta**

REFERENCES

Abreu, M. H., Pereira, R., Yarish, C., Buschmann, A. H., & Sousa-Pinto, I., (2011). IMTA with *Gracilaria vermiculophylla*: Productivity and nutrient removal performance of the seaweed in a land-based pilot-scale system. *Aquaculture, 312*, 77–87.

Adams, J. M., Gallagher, J. A., & Donnison, I. S., (2009). Fermentation study on *Saccharina latissima* for bioethanol production considering variable pre-treatments. *Journal of Applied Phycology, 21*, 569.

Airanthi, M. K. W. A., Hosokawa, M., & Miyashita, K., (2011). Comparative antioxidant activity of edible Japanese brown seaweeds. *Journal of Food Science, 76*, 104–111.

Alghazeer, R., Whida, F., Abduelrhman, E., Gammoudi, F., & Naili, M., (2013). *In vitro* antibacterial activity of alkaloid extracts from green, red and brown macroalgae from western coast of Libya. *African Journal of Biotechnology, 12*(51), 7086–7091.

Arivuselvan, N., Radhiga, M., & Anantharaman, P., (2011). *In vitro* antioxidant and anticoagulant activities of sulfated polysaccharides from brown seaweed (*Turbinaria ornata*) (Turner) J. Agardh. *Asian Journal of Pharmaceutical and Biological Research, 1*, 232–239.

Arumugam, N., Chelliapan, S., Kamyab, H., Thirugnana, S., Othman, N., & Nasri, N. S., (2018). Treatment of wastewater using seaweed: A review. *International Journal of Environmental Research and Public Health, 15*(12), 2851.

Arun, D. V., Gayathri, P. K., Chandran, M., & Dinakarkumar, Y., (2014). Studies on the effect of seaweed extracts on crop plants and microbes. *International Journal of Chemtech Research, 6*.

Aruna, P., Mansuya, P., Sridhar, S., Kumar, J. S., & Babu, S., (2010). Pharmacognostical and antifungal activity of selected seaweeds from Gulf of Mannar region. *Recent Research in Science and Technology, 2*(1), 115–119.

Bajhaiya, A. K., Mandotra, S. K., Suseela, M. R., Toppo, K., & Ranade, S., (2010). Algal biodiesel: The next generation biofuel for India. *Asian Journal of Experimental Biological Science, 1*, 728–739.

Barquilha, C. E. R., Cossich, E. S., Tavares, C. R. G., & Silva, E. A., (2017). Biosorption of nickel (II) and copper (II) ions in batch and fixed-bed columns by free and immobilized marine algae *Sargassum* sp. *Journal of Cleaner Production, 150*, 58–64.

Bouhlal, R., Haslin, C., Chermann, J. C., Colliec-Jouault, S., Sinquin, C., Simon, G., Cerantola, S., et al., (2011). Antiviral activities of sulfated polysaccharides isolated from *Sphaerococcus coronopifolius* (Rhodophyta, Gigartinales) and *Boergeseniella thuyoides* (Rhodophyta, Ceramiales). *Marine Drugs, 9*, 1187–1209.

Carreto, J. I., & Carignan, M. O., (2011). Mycosporine-like amino acids: Relevant secondary metabolites. Chemical and ecological aspects. *Marine Drugs, 9*(3), 387–446.

Cengiz, S., & Cavas, L., (2008). Removal of methylene blue by invasive marine seaweed: *Caulerpa racemosa* var. *cylindracea. Bioresource Technology, 99*, 2357–2363.

Ciancia, M., Quintana, I., & Cerezo, A. S., (2010). Overview of anticoagulant activity of sulfated polysaccharides from seaweeds in relation to their structures, focusing on those of green seaweeds. *Current Medicinal Chemistry, 17*, 2503–2529.

Coura, C. O., De Araffljo, I. W. F., Vanderlei, E. S. O., Rodrigues, J. A., Quindere, A. L., Fontes, B. P., De Queiroz, I. N., et al., (2012). Antinociceptive and anti-inflammatory activities of sulfated polysaccharides from the Red Seaweed *Gracilaria cornea. Basic and Clinical Pharmacology and Toxicology, 110*, 335–341.

Daneshvar, E., Vazirzadeh, A., Niazi, A., Sillanpaa, M., & Bhatnagar, A., (2017). A comparative study of methylene blue biosorption using different modified brown, red and green macroalgae – Effect of pretreatment. *Chemical Engineering Journal, 307*, 435–446.

Dhargalkar, V. K., & Devanand, K., (2004). *Seaweeds — A Field Manual* (p. 36). National Institute of Oceanography.

Egan, S., Harder, T., Burke, C., Steinberg, P., Kjelleberg, S., & Thomas, T., (2013). The seaweed holobiont: Understanding seaweed-bacteria interactions. *FEMS Microbiology Reviews, 37*(3), 462–476. doi: 10.1111/1574-6976.12011.

Ekanayake, P. M., Nikapitiya, C., De Zoysa, M., Whang, I., Kim, S. J., & Lee, J., (2008). Novel anticoagulant compound from fermented red alga *Pachymeniopsis elliptica. European Food Research and Technology, 227*, 897–903.

El Hassouni, H., Abdellaoui, D., El Hani, S., & Bengueddour, R., (2014). Biosorption of cadmium (II) and copper (II) from aqueous solution using red alga (*Osmundea pinnatifida*) biomass. *Journal of Materials and Environmental Science, 5*, 967–974.

El Sayed, K. A., Bartyzel, P., Shen, X., Perry, T. L., Zjawiony, J. K., & Hamann, M. T., (2000). Marine natural products as antituberculosis agents. *Tetrahedron, 56*(7), 949–953. doi: 10.1016/S0040-4020(99)01093-5.

Elizondo-Gonzalez, R., Cruz-Suarez, L. E., Ricque-Marie, D., MendozaGamboa, E., Rodriguez-Padilla, C., & Trejo-Avila, L. M., (2012). *In vitro* characterization of the antiviral activity of fucoidan from *Cladosiphon okamuranus* against Newcastle disease virus. *Virology Journal, 9*, 1–9.

FAO, (2018). *The State of World Fisheries and Aquaculture - Meeting the Sustainable Development Goals*. FAO: Rome, Italy.

Farasat, M., Khavari-Nejad, R. A., Nabavi, S. M. B., & Namjooyan, F., (2013). Antioxidant properties of two edible green seaweeds from northern coasts of the Persian Gulf. *Jundishapur Journal of Natural Pharmaceutical Products, 8*(1), 47–52.

Foon, T. S., Ai, L. A., Kuppusamy, P., Yusoff, M. M., & Govindan, N., (2013). Studies on in-vitro antioxidant activity of marine edible seaweeds from the east coastal region of Peninsular Malaysia using different extraction methods. *Research Journal of Applied Sciences, 9*, 141–146.

Freile-Pelegrin, Y., Madera-Santana, T. J., Robledo, D., Veleva, L., Quintana, P., & Azamar, A., (2007). Degradation of agar films in a humid tropical climate: Thermal, mechanical, morphological and structural changes. *Polymer Degradation and Stability, 92*, 244–252. doi: 10.1016/j.polymdegradstab.2006.11.005.

Gade, R., Tulasi, M., & Bhai, V., (2013). Seaweeds: A novel biomaterial. *International Journal of Pharmacy and Pharmaceutical Sciences, 5*, 40–44.

Gandhiyappan, K., & Perumal, P., (2001). Growth promoting effect of seaweed liquid fertilizer (*Enteromorpha intestinalis*) on the sesame crop plant. *Seaweed Research and Utilisation, 23*, 23–25.
Ghasemi, F. F., Dobaradaran, S., Keshtkar, M., Mohammadi, M. J., Ghaedi, H., & Soleimani, F., (2016). Biosorption of Mn (II) from aqueous solution by *Sargassum hystrix* algae obtained from the Persian Gulf: Biosorption isotherm and kinetic. *International Journal of Pharmacy and Technology, 8*, 18227–18238.
Guiry, M. D., & Blunden, G., (1991). *Seaweed Resources in Europe: Uses and Potential.* Chichester, West Sussex, England: Wiley.
Hardoko, S. T., Eveline, Y. M., & Olivia, S., (2014). An *in vitro* study of antidiabetic activity of *Sargassum duplicatum* and *Turbinaria decurens* seaweed. *International Journal of Pharmaceutical Science Invention, 3*(2), 13–18.
Hinks, J., Edwards, S., Sallis, P. J., & Caldwell, G. S., (2013). The steady-state anaerobic digestion of *Laminaria hyperborea* – effect of hydraulic residence on biogas production and bacterial community composition. *Bioresource Technology, 143*, 221–230.
Hossain, A. B. M. S., Salleh, A., Nasrulhaq, A. B., Partha, C., & Husri, M., (2008). Biodiesel fuel production from algae as renewable energy. *American Journal of Biochemistry and Biotechnology, 4*(3), 250–254. doi: 10.3844/ajbbsp.2008.250.254.
Hudson, J. B., Kim, J. H., Lee, M. K., De Wreede, R. E., & Hong, Y. K., (1999). Antiviral compounds in extracts of Korean seaweeds: Evidence for multiple activities. *Journal of Applied Phycology, 10*, 427.
Hwang, E., Park, S. Y., Sun, Z. W., Shin, H. S., Lee, D. G., & Yi, T. H., (2014). The protective effects of fucosterol against skin damage in UVB-irradiated human dermal fibroblasts. *Marine Biotechnology, 16*(3), 361–370. doi: 10.1007/s10126-013-9554-8.
Indu, H., & Seenivasan, R., (2013). *In vitro* antioxidant activity of selected seaweeds from southeast coast of India. *International Journal of Pharmacy and Pharmaceutical Sciences, 5*, 474–484.
Jayaraman, J., Ali, N., Ganesan, S., & Vadivel, K., (2015). *Use of Seaweed Extracts for Disease Management of Vegetable Crops* (p. 424). CAB International Publishing.
Kang, S. M., Kim, K. N., Lee, S. H., Ahn, G., Cha, S. H., Kim, A. D., Yang, X. D., et al., (2011). Anti-inflammatory activity of polysaccharide purified from AMG-assistant extract of *Ecklonia cava* in LPS-stimulated RAW 264.7 macrophages. *Carbohydrate Polymers, 85*, 80–85.
Kantachumpoo, A., & Chirapart, A., (2010). Components and antimicrobial activity of polysaccharides extracted from Thai brown seaweeds. *Kasetsart Journal – Nature Science, 44*(2), 220–233.
Karthikaidevi, G., Manivannan, K., Thirumaran, G., Anantharaman, P., & Balasubaramanian, T., (2009). Antibacterial properties of selected green seaweeds from Vedalai coastal waters; Gulf of Mannar marine biosphere reserve. *Global Journal of Pharmacology, 3*(2), 107–112.
Khairy, H. M., & El-Sheikh, M. A., (2015). Antioxidant activity and mineral composition of three Mediterranean common seaweeds from Abu-Qir Bay, Egypt. *Saudi Journal of Biological Sciences, 22*(5), 623–630.
Khan, W., Rayirath, U. P., Subramanian, S., Jithesh, M. N., Rayorath, P., Hodges, D. M., Critchley, A. T., et al., (2009). Seaweed extracts as biostimulants of plant growth and development. *Journal of Plant Growth Regulation, 28*(4), 386–399. doi: 10.1007/s00344-009-9103-x.

Kim, A. D., Kang, K. A., Piao, M. J., Kim, K. C., Zheng, J., Yao, C. W., Cha, J. W., et al., (2014). *Dictyopteris undulata* extract induces apoptosis in human colon cancer cells. *Biotechnology and Bioprocess Engineering, 19*, 419–425.

Kim, J. K., Duston, J., Corey, P., & Garbary, D. J., (2013). Marine finfish effluent bioremediation: Effects of stocking density and temperature on nitrogen removal capacity of *Chondrus crispus* and *Palmaria palmata* (Rhodophyta). *Aquaculture, 414, 415*, 210–216.

Kim, M. S., Oh, G. H., Kim, M. J., & Hwang, J. K., (2013). Fucosterol inhibits matrix metalloproteinase expression and promotes type-1 procollagen production in UVB-induced HaCaT cells. *Photochemistry and Photobiology, 89*(4), 911–918.

Koishi, A. C., Zanello, P. R., Bianco, E. M., Bordignon, J., & Santos, C. N. D. D., (2012). Screening of dengue virus antiviral activity of marine seaweeds by an in-situ enzyme-linked immunosorbent assay. *PLoS One, 7*(12), e51089.

Kumara, R., Kaur, I., & Bhatnagar, A. K., (2011). Effect of aqueous extract of *Sargassum johnstonii* Setchell & Gardner on growth, yield and quality of *Lycopersicon esculentum* Mill. *Journal of Applied Phycology, 23*, 623–633.

Lee, J. H., Eom, S. H., Lee, E. H., Jung, Y. J., Kim, H. J., Jo, M. R., Son, K. T., et al., (2014). *In vitro* antibacterial and synergistic effect of phlorotannins isolated from edible brown seaweed *Eisenia bicyclis* against acne-related bacteria. *Algae, 29*(1), 47–55.

Lee, M. H., Lee, K. B., Oh, S. M., Lee, B. H., & Chee, H. Y., (2010). Antifungal activities of dieckol isolated from the marine brown alga *Ecklonia cava* against *Trichophytonrubrum*. *Journal of the Korean Society for Applied Biological Chemistry, 53*(4), 504–507.

Li, N., Mao, W., Yan, M., Liu, X., Xia, Z., Wang, S., Xiao, B., et al., (2015). Structural characterization and anticoagulant activity of a sulfated polysaccharide from the green alga *Codium divaricatum*. *Carbohydrate Polymers, 121*, 175–182.

Magalhaes, K. D., Costa, L. S., Fidelis, G. P., Oliveira, R. M., Nobre, L. T., Dantas-Santos, N., Camara, R. B., et al., (2011). Anticoagulant, antioxidant and antitumor activities of heterofucans from the seaweed *Dictyopteris delicatula*. *International Journal of Molecular Sciences, 12*(5), 3352–3365.

Mahmood, Z., Zahra, S., Iqbal, M., Raza, M. A., & Nasir, S., (2017). Comparative study of natural and modified biomass of *Sargassum* sp. for removal of Cd^{2+} and Zn^{2+} from wastewater. *Applied Water Science, 7*, 3469–3481.

Marinho-Soriano, E., Nunes, S. O., Carneiro, M. A. A., & Pereira, D. C., (2009). Nutrients' removal from aquaculture wastewater using the macroalgae *Gracilaria birdie*. *Biomass Bioenergy, 3*, 327–331.

Mayer, A. M., & Hamann, M. T., (2002). Marine pharmacology in 1999: Compounds with antibacterial, anticoagulant, antifungal, anthelmintic, anti-inflammatory, antiplatelet, antiprotozoal and antiviral activities affecting the cardiovascular, endocrine, immune and nervous systems, and other miscellaneous mechanisms of action. *Comparative Biochemistry and Physiology, Toxicology and Pharmacology, 132*(3), 315–339. doi: 10.1016/s1532-0456(02)00094-7.

Melo, K. R. T., Camara, R. B. G., Queiroz, M. F., Vidal, A. A., Lima, C. R., Melo-Silveira, R. F., Almeida-Lima, J., & Rocha, H. A., (2013). Evaluation of sulfated polysaccharides from the brown seaweed *Dictyopteris justii* as antioxidant agents and as inhibitors of the formation of calcium oxalate crystals. *Molecules, 18*, 14543–14563.

Mhadhebi, L., Mhadhebi, A., Robert, J., & Bouraoui, A., (2014). Antioxidant, anti-inflammatory and antiproliferative effects of aqueous extracts of three Mediterranean

brown seaweeds of the genus *Cystoseira*. *Iranian Journal of Pharmaceutical Research, 13*(1), 207–220.
Moorthi, P. V., & Balasubramanian, C., (2015). Antimicrobial properties of marine seaweed, *Sargassum muticum* against human pathogens. *Journal of Coastal Life Medicine, 3*(2), 122–125.
Moussavou, G., Kwak, D. H., Obiang-Obonou, B. W., Maranguy, C. A., Dinzouna-Boutamba, S. D., Lee, D. H., Pissibanganga, O. G., et al., (2014). Anticancer effects of different seaweeds on human colon and breast cancers. *Marine Drugs, 12*(9), 4898–4911. doi: 10.3390/md12094898.
Mozhi, S., Muthuvel, A., Gnanambigai, D., & Thangavel, B., (2009). Total flavanoid and *in vitro* antioxidant activity of two seaweeds from Rameshwaram coast. *Global Journal of Pharmacology, 3*(2), 59–62.
Murugesan, S., Anand, B. M., Bhuvaneswari, S., Kotteswari, M., & Thennarasan, S., (2015). *In vitro* antidiabetic activity of methanolic extracts of selected marine algae. *European Journal of Pharmaceutical and Medical Research, 2*(6), 256–260.
Myers, S. P., O'Connor, J., Fitton, J. H., Brooks, L., Rolfe, M., Connellan, P., Wohlmuth, H., et al., (2011). A combined Phase I and II open-label study on the immunomodulatory effects of seaweed extract nutrient complex. *Biologics Targets and Therapy, 5*, 45–60.
Neelakandan, Y., (2014). *In vitro* anti-tumor, anti-inflammatory, antioxidant and antibacterial activities of marine brown alga *Sargassum wightii* collected from gulf of Mannar. *Global Journal of Pharmacology, 8*(4), 566–577. doi: 10.5829/idosi.gjp.2014.8.4.84281.
Nkemka, N. K., & Murto, M., (2012). Exploring strategies for seaweed hydrolysis: Effect on methane potential and heavy metal mobilization. *Process Biochemistry, 47*(12), 2523–2526. doi: 10.1016/j.procbio.2012.06.022.
Omar, H., El-Gendy, A., & Al-Ahmary, K., (2018). Bioremoval of toxic dye by using different marine macroalgae. *Turkish Journal of Botany, 42*, 15–27.
Osman, H. E., & Olfat, M. A. S., (2011). Effect of seaweed extracts as foliar spray on sunflower yield and oil content. *Egyptian Journal of Phycology, 12*(1), 57–70.
Pallela, R., Na-Young, Y., & Kim, S. K., (2010). Anti-photoaging and photoprotective compounds derived from marine organisms. *Marine Drugs, 8*(4), 1189–1202. doi: 10.3390/md8041189.
Pati, M. P., Sharma, S. D., Nayak, L., & Panda, C. R., (2016). Uses of seaweed and its application to human welfare: A review. *International Journal of Pharmacy and Pharmaceutical Sciences, 8*(10), 12–20. doi: 10.22159/ijpps.2016v8i10.12740.
Pereira, R. C., & Costa-Lotufo, L. V., (2012). Bioprospecting for bioactives from seaweeds: Potential, obstacles and alternatives. *Brazilian Journal of Pharmacognosy, 22*(4), 894–905.
Ponce, N. M. A., Pujol, C. A., Damonte, E. B., Flores, M. L., & Stortz, C. A., (2003). Fucoidans from the brown seaweed *Adenocystis utricularis*: Extraction methods, antiviral activity and structural studies. *Carbohydrate Research, 338*(2), 153–165.
Price, R. D., Berry, M. G., & Navsaria, H. A., (2007). Hyaluronic acid: The scientific and clinical evidence. *Journal of Plastic, Reconstructive and Aesthetic Surgery, 60*(10), 1110–1119. doi: 10.1016/j.bjps.2007.03.005.
Rafiquzzaman, S. M., Lee, J. M., Ahmed, R., Lee, J. H., Kim, J. M., & Kong, I. S., (2014). Characterisation of the hypoglycaemic activity of glycoprotein purified from the edible brown seaweed, *Undaria pinnatifida*. *International Journal of Food Science and Technology, 50*(1), 143–150. doi: 10.1111/ ijfs.12663.

Ramya, S. S., Vijayanand, N., & Rathinavel, S., (2015). Foliar application of liquid biofertilizer of brown alga *Stoechospermum marginatum* on growth, biochemical and yield of *Solanum melongena*. *International Journal of Recycling of Organic Waste in Agriculture, 4*, 167–173. doi: 10.1007/s40093-015-0096-0.

Rhimou, B., Hassane, R., & Nathalie, B., (2010). Antiviral activity of the extracts of *Rhodophyceae* from Morocco. *African Journal of Biotechnology, 9*, 7968–7975.

Roberts, D. A., & De Nys, R., (2016). The effects of feedstock pre-treatment and pyrolysis temperature on the production of biochar from the green seaweed *Ulva*. *Journal of Environmental Management, 169*, 253–260. doi: 10.1016/j.jenvman.2015.12.023.

Saidani, K., Bedjou, F., Benabdesselam, F., & Touati, N., (2012). Antifungal activity of methanolic extracts of four Algerian marine algae species. *African Journal of Biotechnology, 11*, 9496–9500.

Samocha, T. M., Fricker, J., Ali, A. M., Shpigel, M., & Neori, A., (2015). Growth and nutrient uptake of the macroalga *Gracilaria tikvahiae* cultured with the shrimp *Litopenaeus vannamei* in an integrated multi-trophic aquaculture (IMTA) system. *Aquaculture, 446*, 263–271.

Senthil, K. A., & Murugan, M., (2013). Antiulcer, wound healing and hepatoprotective activities of the seaweeds *Gracilaria crassa, Turbinaria ornata* and *Laurencia papillosa* from the southeast coast of India. *Brazilian Journal of Pharmaceutical Sciences, 49*, 669–678.

Senthilkumar, S., Sudha, S., & Prakash, S., (2014). Antidiabetic activity of aqueous extract of *Padina boergesenii* in streptozotocin-induced diabetic rats. *International Journal of Pharmacy and Pharmaceutical Sciences, 6*(5), 418–422. doi: 10.20959/wjpr20178-9153.

Sheeba, M., & Paul, J. P., (2014). Evaluation of antipyretic activity of methanol extract of *Ulva rigida* C.Ag (green seaweed) in Hare Island, Thoothukudi, Tamil Nadu, India. *International Journal of Experimental Pharmacology, 4*(2), 105–108.

Silva, T. M. A., Alves, L. G., De Queiroz, K. C. S., Santos, M. G. L., Marques, C. T., Chavante, S. F., Rocha, H. A. O., & Leite, E. L., (2005). Partial characterization and anticoagulant activity of a heterofucan from the brown seaweed *Padina gymnospora*. *Brazilian Journal of Medical and Biological Research, 38*(4), 523–533.

Soares, A. R., Abrantes, J. L., Souza, T. M. L., Fontes, C. F. L., Pereira, R. C., Frugulhetti, I. C. D. P. P., & Teixeira, V. L., (2007). *In vitro* antiviral effect of meroditerpenes isolated from the Brazilian seawead *Stylopodium zonale* (Dictyotales). *Planta Medica, 73*(11), 1221–1224.

Süzgeç-Selçuk, S., Meriçli, A. H., Güven, K. C., Kaiser, M., Casey, R., Hingley-Wilson, S., Lalvani, A., & Tasdemir, D., (2011). Evaluation of Turkish seaweeds for antiprotozoal, antimycobacterial and cytotoxic activities. *Phytotherapy Research, 25*(5), 778–783.

Tariq, A., Athar, M., Ara, J., Sultana, V., Ehteshamul-Haque, S., & Ahmad, M., (2015). Biochemical evaluation of antioxidant activity in extracts and polysaccharide fractions of seaweeds. *Global Journal of Environmental Science and Management, 1*(1), 47–62.

Thomas, N. V., & Kim, S. K., (2013). Beneficial effects of marine algal compounds in cosmeceuticals. *Marine Drugs, 11*(1), 146–164. doi: 10.3390/md11010146.

Tiwari, B. K., & Troy, D. J., (2015). *Seaweed Sustainability: Food and Nonfood Applications* (p. 472). Academic Press, London.

Trinchero, J., Ponce, N. M. A., Córdoba, O. L., Flores, M. L., Pampuro, S., Stortz, C. A., Salomón, H., & Turk, G., (2009). Antiretroviral activity of fucoidans extracted from the brown seaweed *Adenocystis utricularis*. *Phytotherapy Research, 23*(5), 707–712.

Tseng, C. K., (2004). The past, present and future of phycology in China. *Hydrobiology, 512*, 11–20.
Van, G. V., (2018). Seaweed biotechnology to combat desertification. *Advances in Biochemistry and Biotechnology, 2*, 156. doi: 10.29011/2574-7258.000056.
Vázquez, A. I. F., Sánchez, C. M. D., Delgado, N. G., Alfonso, A. M. S., Ortega, Y. S., & Sánchez, H. C., (2011). Anti-inflammatory and analgesic activities of Red Seaweed *Dichotomaria obtusata. Brazilian Journal of Pharmaceutical Sciences, 47*(1), 111–118.
Verdy, C., Branka, J. E., & Mekideche, N., (2011). Quantitative assessment of lactate and progerin production in normal human cutaneous cells during normal ageing: Effect of an *Alaria esculenta* extract. *International Journal of Cosmetic Science, 33*(5), 462–466. doi: 10.1111/j.1468-2494.2011.00656.x.
Verdy, C., Branka, J. E., & Mekideche, N., (2012). Melanosome transfer evaluation by quantitative measurement of Pmel 17 in human normal melanocyte-keratinocyte co-cultures: Effect of an *Alaria esculenta* extract. *Journal of Cosmetic Science, 63*(3), 197–203.
Verma, A., Kumar, S., & Kumar, S., (2016). Biosorption of Lead ions from the aqueous solution by *Sargassum filipendula:* Equilibrium and kinetic studies. *Journal of Environmental Chemical Engineering, 4*(4), 4587–4599.
Vijayabaskar, P., & Shiyamala, V., (2011). Antibacterial activities of brown marine algae (*Sargassum wightii* and *Turbinaria ornata*) from the Gulf of Mannar biosphere reserve. *Advances in Biological Research, 5*(2), 99–102.
Wang, H., Ooi, E. V., & Jr, P. O. A., (2008). Antiviral activities of extracts from Hong Kong seaweeds. *Journal of Zhejiang University of Science B, 9*(12), 969–976.
Warkoyo, W., & Saati, E., (2011). The solvent effectiveness on extraction process of seaweed pigment. *Makara Journal of Technology, 15*(1), 5–8. doi: 10.7454/mst.v15i1.850.
Wei, Z., You, J., Wu, H., Yang, F., Long, L., Liu, Q., Huo, Y., & He, P., (2017). Bioremediation using *Gracilaria lemaneiformis* to manage the nitrogen and phosphorous balance in an integrated multi-trophic aquaculture system in Yantian Bay, China. *Marine Pollution Bulletin, 121*, 313–319.
Wu, H., Huo, Y., Han, F., Liu, Y., & He, P., (2015). Bioremediation using *Gracilaria chouae* co-cultured with *Sparus macrocephalus* to manage the nitrogen and phosphorous balance in an IMTA system in Xiangshan Bay, China. *Marine Pollution Bulletin, 91*(1), 272–279.
Ye, H., Wang, K., Zhou, C., Liu, J., & Zeng, X., (2008). Purification, antitumor and antioxidant activities *in vitro* of polysaccharides from the brown seaweed *Sargassum pallidum. Food Chemistry, 111*(2), 428–432.
Yipmantin, A., Maldonado, H. J., Ly, M., Taulemesse, J. M., & Guibal, E., (2011). Pb (II) and Cd (II) biosorption on *Chondracanthus chamissoi* (a red alga). *Journal of Hazardous Materials, 185*(2, 3), 922–929.
Zandi, K., Ahmadzadeh, S., Tajbakhsh, S., Rastian, Z., & Sartavi, K., (2010). Anticancer activity of *Sargassum oligocystum* water extract against human cancer cell lines. *European Review for Medical and Pharmacological Sciences, 14*, 669–673.
Zoysa, M. D., Nikapitiya, C., Jeon, Y. J., Jee, Y., & Lee, J., (2008). Anticoagulant activity of sulfated polysaccharide isolated from fermented brown seaweed *Sargassum fulvellum. Journal of Applied Phycology, 20*(1), 67–74.

CHAPTER 12

SEAWEEDS FOR FOOD SECURITY AND ENVIRONMENTAL MITIGATION

AMBO TUWO,[1,5,6] MITSUTOSHI NAKAJIMA,[2,3] JOEHARNANI TRESNATI,[4–6] INAYAH YASIR,[1,5,6] HAKIKI MELANIE,[7] and RISAL APRIANTO[5]

[1]*Department of Marine Science, Faculty of Marine Science and Fisheries, University of Hasanuddin, Kampus Tamalanrea KM. 10, Makassar – 90245, Sulawesi Selatan, Indonesia*

[2]*Tsukuba Life Science Innovation Program (T-LSI), School of Integrative and Global Majors (SIGMA), University of Tsukuba, 1-1-1 Tennoudai, Tsukuba Ibaraki – 305-8572, Japan*

[3]*Graduate School of Life and Environmental Sciences, University of Tsukuba, 1-1-1 Tennoudai, Tsukuba Ibaraki – 305-8572, Japan*

[4]*Department of Fisheries, Faculty of Marine Science and Fisheries, University of Hasanuddin, Kampus Tamalanrea KM. 10, Makassar – 90245, Sulawesi Selatan, Indonesia*

[5]*Multitrophic Research Group, University of Hasanuddin, Kampus Tamalanrea KM. 10, Makassar – 90245, Sulawesi Selatan, Indonesia*

[6]*Center of Excellence for Development and Utilization of Seaweed, University of Hasanuddin, Kampus Tamalanrea KM. 10, Makassar – 90245, Sulawesi Selatan, Indonesia*

[7]*Research Center for Chemistry, Indonesian Institute of Sciences, Kawasan Puspiptek Serpong, Tangerang Selatan – 15314, Indonesia*

ABSTRACT

Food security is a serious problem. This chapter outlines how seaweed could improve food security and promote environmental mitigation using the holistic-integral approach. Actual conditions that are not expected due to the problems encountered can be resolved if the development paradigm and the existing strategic environmental conditions are considered to formulate the alternative solution. The alternative solution will lead to the expected condition. The expected conditions can be achieved if alternative solutions are successfully implemented. The implementation concept is formulated by using subject, object, and method analysis. The successful implementation of the concept will deliver the expected outputs and outcomes. The actual condition is low food security due to high population growth, limited food resources, limited agricultural lands, water scarcity, land erosion, environmental degradation, and energy scarcity. The related paradigm is sustainable development, livelihood, and equity. The related issues are hunger, global warming, and climate change. The alternative solutions are: develop seaweed as a sea vegetable and other food product; develop seaweed as animal feed; and develop seaweed as raw material for renewable energy production. In the effort to realize the expected conditions and achieve outputs and outcomes, there are twelve implementation concepts have been formulated.

12.1 INTRODUCTION

Food security is a food availability measure and the individual ability to access it. Food security exists when everyone, at all times, has physical and economic access to nutritious food, safe, and sufficient to meet the food needs and preferences for an active and healthy life. Meanwhile, food insecurity is a limited or uncertain situation for the nutritious availability and safe food or uncertain ability or limited to obtain acceptable food in the acceptable way (Bickel et al., 2000). Food security combines measures of resistance to future disruptions or unavailability of essential food supplies due to various risk factors, including drought, socio-economic, and political instability, shipping disruptions, fuel shortages, and war.

The food security pillars are availability, accessibility, utilization, and stability. Food availability is related to production, distribution, and exchange, while access to food is affordability, allocation, and food preferences for individuals and households (Gregory et al., 2005). Food utilization refers to individuals' food metabolism, and food stability refers to obtaining food over time (Tweeten, 1999). Poor food security can be seen in the high chronically malnourished prevalence; in 2010–2012, almost 870 million people (12.5%) of the global population were chronically malnourished. In 2018, the food insecurity prevalence was slightly decreasing for the first time to 11.1% (Coleman-Jensen et al., 2019). Food insecurity is still very high in developing countries, where 852 million people suffer from chronic malnutrition. This chronic malnutrition will become more significant and more prominent in the future if no innovative efforts are found to increase food availability.

Food security could be related to two questions, "Is there food for all of us?" and "is Jacques Cousteau's statement – in order to stabilize the world population, it must be eliminate 250,000 people each day – related food availability?" Both questions relate directly and indirectly to food security and environmental management. Therefore, improving food security and environmental mitigation are two interrelated things that must concern improving human well-being. Environmental degradation is the environmental deterioration through resources depletion (air, water, and soil), ecosystems destruction, habitat destruction, wildlife extinction, and pollution; environmental degradation is the capacity reduction of the environment to meet social and economic needs (Johnson et al., 1997).

Food security and environmental mitigation are described in this chapter using a holistic-integral approach (Tuwo 2013). In this approach, an actual condition (i) that is not expected due to the problems encountered (ii) can be overcome if the developing paradigm (iii) and the existing strategic environmental conditions (iv) are considered in the formulation of alternative solutions (v). A formulated alternative solution will lead to the expected condition (goal) (vi). The common condition can be realized if alternative solutions are implemented successfully. The implementation concept is formulated using S-O-M (Subject, Object, and Method) Analysis (vii). In the medium term, five years, successful in implementing the implementation concept will deliver the expected output (viii); and in the long term, ten years will deliver the expected outcome (ix) (Figure 12.1).

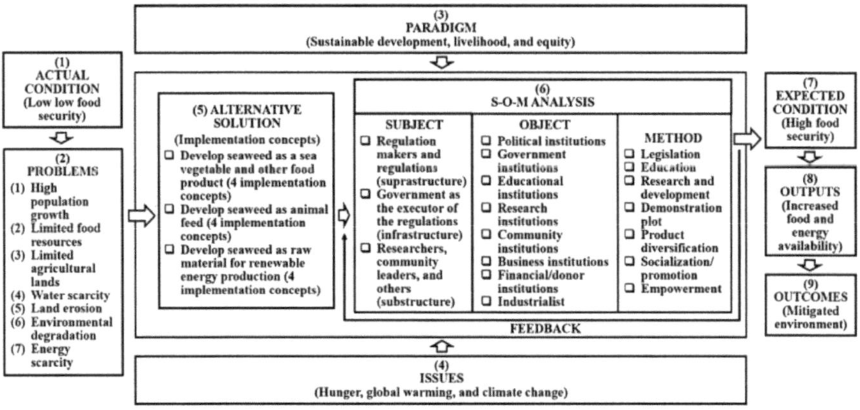

FIGURE 12.1 Holistic-integral approach used for seaweed for food security and environmental mitigation.

12.2 ACTUAL CONDITION AND PROBLEMS

The actual condition is low food security, limited food resources, limited available land for agriculture, water scarcity, land erosion, environmental degradation, and energy scarcity.

12.2.1 HIGH POPULATION GROWTH

The human population grows along with the agriculture development. At the agriculture dawn, about 8000 B.C., the world population was approximately five million. Over the 8,000 years up to one Anno Domini, it grew to 200 million, with a growth rate of less than 0.05% per year. Major changes occurred during the industrial revolution, from the beginning of the history of human civilization, it took about 1800 years to reach first billion, the second billion was reached in just 130 years (in 1930), the third billion was achieved even faster, less than thirty years (in 1959), the fourth billion in fifteen years (in 1974), and the fifth billion was reached in only thirteen years (in 1987). In the 20th century, the world's population grew from 1.65 billion to six billion, and the population reached 7.8 billion people by March 2020 (Anonymous, 2020a). We are adding a billion population every twelve years (Anonymous, 2011). The rapid growth of the human population threatens food security and environmental sustainability (Tuwo, 2018).

12.2.2 LIMITED FOOD RESOURCES

The world's population has grown to more than four billion since the start of the green revolution, but the food production has so far been in line with population growth, but in the future, without a new revolution, there would be more significant levels of hunger and malnutrition than currently documented by the UN. In 2004, a meta-analysis of seventy quantitative studies estimated the sustainable limit for the world population was 7.7 billion people (Anonymous, 2007). The population growth, decreasing energy reserves, and food shortages would create a perfect storm of food shortages by 2030. Our food conditions are paradoxical. On the one hand, our food reserves are at their lowest level in the past fifty years; on the other hand, by 2030, we need 50% more food. Paradoxical conditions also occur in energy and water reserves.

On the one hand, water, and energy reserves continue to decrease; on the other hand, at the same time (2030), it requires 50% more energy and 30% more freshwater. There are dramatic issues out there, particularly food, water, and energy; they are all intimately connected, one cannot think about dealing with one without considering the others. It must deal with all of these together (Anonymous 2009a; Anonymous 2009b). The world must produce a minimum 70% more food by 2050 to feed the projected 2.3 billion additional people (Anonymous, 2009c). Previous data has shown how large the number of malnourished people; there was an absolute increase in the number of malnourished people globally, with 923 million malnourished people in 2007, compared with 832 million in 1995 (Anonymous, 2008a). Food and bioenergy sources must be found that do not require fresh water; therefore, different food and renewable energy must be produced in coastal and marine areas (Tuwo, 2018).

The increasing hunger has created a huge humanitarian crisis. With the economic crisis and high food prices, the number of hunger is expected to grow about 11%. Developing countries suffer from malnutrition. In Asia-Pacific, an estimated 642 million people suffer from chronic hunger; in Sub-Saharan Africa, about 265 million; Latin America and the Caribbean about 53 million; the Near East and North Africa, about 42 million; and developed countries, a total of fifteen million (Anonymous, 2009d).

12.2.3 LIMITED AGRICULTURAL LANDS

Agricultural land is 38.4% of the world land area. The total land used to produce food is 49,116,227 square kilometers. Arable land is 13,963,743 square kilometers, permanent pastures are 33,585,676 square kilometers, and permanent crops are 1,537,338 square kilometers (Anonymous, 2010a). Limited agricultural lands can be described in the case of rice in Indonesia. Rice is the principal food in the Indonesian diet (Taylor, 1944). Indonesia is the world's third-largest rice producer after Tiongkok and India. However, because Indonesia has a large population, rice production is consumed internally. To ensure food security, the Indonesian government fills the gap by importing from neighboring countries such as Thailand, Vietnam, and Cambodia (Anonymous, 2015a). In Indonesia, agricultural land is minimal and unable to produce all the food needed. Indonesia ranks third globally regarding rice production, but also the world's seventh-largest rice importer over the past five years, on average requiring more than 1.1 million tons of rice imports per year (Anonymous, 2012). In 2016, Indonesia's farmland was 39.5 million Ha, consisting of paddy fields 8.1 million Ha, kale yard 11.9 million Ha, farms 5.3 million Ha, and land not cultivated 14.3 million Ha. In 2018, the area of paddy fields fell to 7,105,145 ha (Anonymous, 2019a). In 1989, Indonesian rice fields were 7.08 million ha, and harvested area of about 11.40 million Ha, or harvest index was 1.6. In 2015, Indonesia had 8.00 million Ha, and harvested area of about 14.20 million Ha or harvest index was 1.7. Whereas in 2016, Indonesia had 8.19 million Ha, harvested area of about 15.03 million Ha (harvest index 1.8), and in 2018, officially Indonesia had only 7.11 million Ha, harvested area of about sixteen million Ha (harvest index 2.2). In 2013, Indonesian farmer paddy fields occupation was 0.3 hectares, this area is much smaller than Thailand farmers paddy fields occupation three-hectare. Even if the productivity of paddy fields was increased 1.6 times from 1980, which was 37.4 million tons or 32.93 quintals per hectare to 2015, which was 75.4 million tons or 52.80 quintals per hectare; however, rice food will still be a problem because the increase of rice fields from 1980 to 2016 only 1.11 million Ha, while the population increased by 111.2 million people from 1980 to 2015 (143.7 to 254.9 million people). The conditions in 2018 are critical because the official paddy fields are only 7.11 million Ha (Anonymous, 2020b). Limited agricultural lands must be overcome by opening food production land outside agricultural land, for example, in coastal and marine areas; food from the sea is our future (Tuwo, 2018).

12.2.4 WATER SCARCITY

Environmental problems that arise due to the world population will continue to increase; while the world population is tripled in the twentieth century, renewable water resources have grown six times. This causes competition between agriculture, people, and livestock will increase to get water. Water is clearly physically scarce in densely populated areas in Central and West Asia. In North Africa, water availability is projected to be less than 1000 m^3 $capita^{-1}$ $year^{-1}$. This water scarcity relates to food production, and not water for small household use at this scale (Rijsberman, 2006).

The water covers 79% of the Earth, so it is easy to think that the water will always be abundant. However, fresh water is scarce. The majority of the Earth's water is saltwater, with only 2.5% being freshwater. Approximately 70% of the freshwater available on the planet is frozen in Antarctica and

Greenland's icecaps, leaving the remaining 30% or equal to only 0.7% of total water resources worldwide are available for consumption. From this remaining 0.7%, roughly 87% is allocated to agricultural purposes; consequently, an estimated 1.1 billion people worldwide suffer from a water shortage, and a total of 2.7 billion people finds water shortages one month a year. Sanitation is also a problem, there are 2.4 billion people are exposed to cholera, typhoid fever, and waterborne diseases. Two million people, primarily children, die every year from diarrheal alone. Many water systems that maintain a thriving ecosystem and feed a growing population have become stressed. Rivers, lakes, and aquifers dry up and the water is too polluted to use. When half of the wetlands have been lost, agriculture consumes more water than other sources and wastes a lot through inefficiency. Climate change causing shortages and drought in some areas and flooding in other places. At the current level of water consumption, this situation will get worse, two-thirds of the world population will face water shortages; and ecosystems worldwide will suffer more (Anonymous, 2020c).

Water scarcity is caused by pollution, agriculture, aquaculture, and population growth (Nasir et al., 2016; Suwoyo et al., 2019; Suwoyo et al., 2020a; Suwoyo et al., 2020b). Water pollution comes from agriculture (fertilizers and pesticides), untreated domestic wastewater, and industrial waste. Groundwater is not safe also from pollution because many pollutants can dissolve into underground aquifers. Some immediate effects occur, such as when harmful bacteria from domestic waste pollute the water and make it unfit for drinking. In other cases, it may take years to accumulate in the environment and the food chain before the effects are fully known. Agriculture is inefficient because it uses 70% of the world's accessible freshwater, but wastes around 6% of water due to leaky irrigation systems, improper application methods, and growing crops are too thirsty for water. This wasteful use of water drains rivers, lakes, and underground aquifers. The countries that produce large food amounts have reached or nearly reached their water resource limits. Coupled with this thirsty plant is that agriculture also produces substantial freshwater pollution, both through fertilizers and pesticides, all of which affect humans and other creatures. In the past fifty years, the human population has doubled. The rapid growth has transformed aquatic ecosystems throughout the world and is causing massive biodiversity loss. At present, 41% of the population lives in watersheds suffer from water pressure. The concerns about water availability are increasing when freshwater usage continues at an unsustainable rate (Anonymous, 2020c).

12.2.5 LAND EROSION

Increased land-based food production will increase land cultivation. Increased land cultivation and land use frequency will increase land erosion. Unsustainable agricultural practices could increase erosion rates two times greater than the natural rate and far outweigh replacement by land production (Montgomery, 2007), cultivation of agricultural land, which splits the soil into finer particles, is the main factor. The land erosion has been exacerbated due to mechanical farming equipment that allows deep plowing, increasing the volume of land available for transportation by water erosion (Blanco and Lal, 2008; Lobb, 2009). Nutrient losses from the soil can increase due to soil erosion events. The loss of total phosphorus in finer eroded fractions is more significant than the whole soil (Poirier et al., 2012), the reason is this more transportable material can support lower phosphate solution concentrations than coarser-sized fractions (Scalenghe et al., 2007). Tillage also increases the wind erosion rate by dehydrating and breaking the soil into tiny particles that the wind can pick up. What makes this worse is that most trees are removed from agricultural land, allowing the wind to have long, open trajectories to travel at higher speeds (Whitford, 2002). Heavy grazing can reduce land cover and causes severe soil compaction, that increase the erosion rate (Imeson, 2012).

12.2.6 ENVIRONMENTAL DEGRADATION

One of the main environmental degradation components is freshwater resource depletion on Earth (Warner et al., 2010). Three freshwater uses of now constitute 95% of consumption; about 85% for irrigation of agricultural land, golf courses, and parks; 6% for domestic purposes such as indoor baths and outdoor and courtyard use; and 4% for industrial purposes such as processing and cooling in centers manufacture (Frederick and Major, 1997). An estimated one in three people worldwide faces water shortages, nearly one-fifth of the population lives with physical water scarcity, and in developing countries, only a quarter of the population does not have water shortages. The infrastructure needed to use available water from rivers and other sources. Water scarcity is an increasing problem because many problems are predicted in the future, including population growth, increasing urbanization, higher living standards, and climate change. Climate change affects water supplies in many ways. It is predicted that average global temperatures will rise for the coming years due to a forces number that affects climate.

The atmospheric carbon dioxide amount will rise, and both will affect water resources; evaporation is highly dependent on the availability of temperature and humidity, which can ultimately affect the available water amount to replenish the groundwater supply (Anonymous, 2020d).

Plant respiration can be influenced by an increase of carbon dioxide in the atmosphere, which can reduce water use and increase water use from the possibility of increasing leaf area. Increased temperatures can reduce winter and increase the intensity of melted snow, causing this peak runoff, which affects the risk of soil moisture, flooding, drought, and carrying capacity depending on the area (Ragab and Prudhomme, 2002).

Warmer winter temperatures cause a decrease in the snow amount, which can cause water resources to decrease during the summer. The decrease in water resources significantly affects the mountainous areas that depend on glacier runoff to replenish their river systems and groundwater supplies. The mountainous areas are more vulnerable to water shortages over time; an increase of the temperature will initially result in a rapid increase in melting water from the glacier in the summer, followed by glacier retreats and a decrease in melt and consequently an annual water supply due to smaller and smaller glacier sizes (Anonymous, 2020d). The thermal water expansion and the increase in sea glaciers melting from rising temperatures give way to rising sea levels. Sea level rise can affect the freshwater supply to coastal areas as well. When estuaries and river deltas with higher salinity are pushed further inland, saltwater intrusion results in increased salinity in reservoirs and aquifers (Frederick and Major, 1997). Rising sea levels can also be caused by groundwater depletion (Konikow, 2011) because climate change can affect the hydrological cycle in several ways. An unequal temperature distribution increases and increases rainfall throughout the world, resulting in water surpluses and deficits (Ragab and Prudhomme, 2002). However, a global decline in groundwater shows rising sea levels, even after meltwater and thermal expansion are taken into account (Konikow, 2011), which can provide positive bait feedback for the rising sea levels problem that causes freshwater supply.

Increasing food production to meet food needs could increase environmental pressure if food production only relies on food from agricultural land. Increased land-based food production also encourages the increased use of fresh water, fertilizer, pesticides, and soil erosion, which could negatively impact the environment. Increased land-based food production will increase land cultivation; increased land cultivation will increase land erosion; and increased land erosion will promote ecosystem degradation, especially aquatic ecosystems. For this reason, a new food production area is needed that does not need fresh water, fertilizer, pesticides, and soil tillage. The area that meets the four requirements for new food production is the coastal zone, small islands, and the sea (Tuwo, 2018).

12.2.7 ENERGY SCARCITY

The energy problem is an important global issue that continues to increase in the future. Fossil energy use increases carbon emissions that allegedly cause global warming. Increasing human life comfort led to an increase in energy consumption. It is estimated that in the 22nd Century, there will be a fuel shortage. Meeting energy needs with biofuels can reduce human food availability. Therefore, new bioenergy sources need to be found to avoid competition in meeting human needs (Sari et al., 2017; Wadi et al., 2019). Biofuels production using seaweed can reduce this competition. The use of fossil energy sources could increase carbon emissions that allegedly cause global warming. It's estimated that 22nd Century will be a fuel shortage. Renewable and green energy could cover up this fuel shortage. The biofuels used reduce the rate in carbon emissions.

12.3 PARADIGM

12.3.1 SUSTAINABLE DEVELOPMENT

Sustainable development is realized if the present needs can be met without compromising the future generation's ability to meet their own needs. This definition explains that sustainable development emphasizes the need to balance various interests in development, namely environmental sustainability, economic growth, social progress, and equity, all within the world's natural resources limits (Shah, 2009). Sustainable development is a call to address the significant challenges to ensure all the people have the same opportunities for better life without compromising our planet. There are eight of the millennium development goals that are directly related to a better life, i.e., (i) eradicate poverty and hunger; (ii) guaranteeing a healthy life; (iii) universalize access to essential needs such as water, sanitation, and energy; (iv) support the generation opportunities through education and decent work; (v) foster innovation, infrastructure, and improve communities and cities to produce and consume in sustainable ways; (vi) reduce inequality, especially in gender; (vii) combating climate change and protecting the ecosystems; and (viii) promote collaboration the social agents to create peace environment and sustainable development (Anonymous, 2020e).

Seaweed is one of the best agents for sustainable development. Seaweed is the most considerable unexploited resources on the planet. Seaweed cultivation presents a sustainable and renewable biomass source to meet the growing population demands under climate change pressures, overfishing, and expanding industries. Seaweed and ocean are the human future hope; seaweeds grow faster than any land plant, and 70% of the Earth surface is covered by ocean, and vast coastal areas present a huge potential in ocean farming (Tuwo et al., 2020) for biomass, with no conflict in land use, no need for freshwater, fertilizers, and pesticides. Seaweeds convert sunlight into chemical energy and are a source for several critical applications, presently worth approximately eight billion US$ worldwide. There is an increasing demand for seaweed biomass for high-value applications within food and health products, cosmetics, and animal feed markets. Seaweed can be produced in a controlled cultivation system at a large scale, as proven in Asia, producing thirty million tons cultivated per year. In Europe and the USA, technological advances can replace high labor costs and turn seaweed cultivation into a viable commercial reality. As sustainable development agents, seaweed provides environmental benefits, i.e., renewable biomass and fast growth; CO_2 uptake and storage; increased biodiversity and dissolved oxygen in the sea; increased fish stocks in the vicinity of farms; biofilter ability or bioremediation; assist in reducing coastal eutrophication and improve water quality; integrated aquaculture for cleaner waters and improved resource utilization; sea forestation by planting artificial seaweed forests for the benefit of the environment; and help mitigate climate change and ocean acidification through large scale cultivation (Anonymous, 2020f; Badraeni et al., 2020).

12.3.2 LIVELIHOOD

A livelihood is a series of activities essential to daily life carried out over a person's life span. The essential activities include securing drinking and clean water, food availability, cattle fodder, medicine availability, shelter or housing, and clothing. Individual livelihoods involve the ability to obtain these needs to meet the basic needs and their households. Such activities are usually repeated and repeated in a manner that is sustainable way and provides dignity. For example, the fishers livelihood depends to the fish availability and accessibility. The concept livelihoods go beyond conventional definitions and approaches to poverty reduction. The conventional definition is too narrow because it focuses only on specific poverty manifestations, such as low income, or does not consider poverty vital aspects such as vulnerability and social exclusion. It is recognized that more attention should be focused to various factors and processes that inhibit or enhance the ability of the poor to live in sustainable ways economically, socially, and ecologically. The sustainable livelihoods concept offers a coherent and integrated approach to poverty (Krantz, 2001).

12.3.3 EQUITY

There is a growing recognition of the importance of equity development. The last few decades have seen an increase in inequality and inequities, which causes the world to lag behind in key goals such as the millennium development goals. Inequity rise may be driven mainly by worldwide processes such as globalization and economic integration, rising inequity is a problem that should be tackled by the community development.

Equity comes from the moral equality idea that all people should be treated equally. Thinking about fairness can help decide how to distribute the services throughout society, hold the state accountable for its influence on how the services are distributed in society and use it to ensure fair treatment of all peoples. Applying these ideas in a specific country context involves hard choices, and embedding discussions of distributive justice into domestic political and policy debates is central to national development, but three areas of considerable consensus can be identified. In order of priority, equity is an equal chance of life, there should be: (i) no difference in results based on factors that people cannot be held accountable; (ii) equal attention to the needs of the community, some goods and services are necessary, and must be distributed at the level of need; and (iii) meritocracy, position in society and rewards must reflect differences in effort, ability, and based on fair competition.

There is considerable inequity in developing countries. The access to and interaction with critical institutions is shaped by power balances in the political and socio-economic spheres, often leading to social exclusion. Also, inequality patterns reinforce through intergenerational transmission and various formal and informal institutions, resulting the chronic poverty and inequity between groups, regions, and generations. The evidence confirms a worrying picture of life chances dependent on inequitable access to services, and rising income inequality. This inequity hurts growth, poverty reduction, and social cohesion (Jones, 2009).

12.4 ISSUES

12.4.1 HUNGER

Globally, there are 815 million of the world population who do not have enough food. It is necessary to analyze the hunger causes and consequences, where the situation is worst, and what must be done to overcome hunger. All over the world, each 10 seconds, a child dies caused by hunger. It would seem strange if there are 821.6 million people are hungry, and two billion people in the world suffer from malnutrition when there are enough food and resources to meet the world's food needs. It should be emphasized that food is a human right.

There are three types of hunger; acute, chronic, and hidden hunger. The acute hunger is a form of hunger that most often occurs due to drought, war, and disaster. Drought will exacerbate the suffering of people who are already experiencing chronic hunger. Acute hunger accounts for 8% of all hunger sufferers. The acute hunger is a state of long-term malnutrition because the body absorbs less food. Globally, acute hunger is spreading widely as a result of poverty. The People suffering the chronic hunger do not have enough money to buy healthy food, clean water, and health care. Hidden hunger is chronic hunger. Acute hunger results from an unbalanced diet and lack of essential nutrients, such as iodine, iron, zinc, and vitamin A. Deficiencies in these nutrients lead to severe disease in the long term. Children who do not get enough nutrition will suffer delays in mental and physical development and a high risk of death. Globally, two billion people are suffering from chronic malnutrition, including in industrialized countries. The hidden hunger harms individuals and the overall development of the affected areas as efficiency and public health decline (Anonymous, 2019b).

One sustainable development goal is the hunger end, achieving food security, and promoting sustainable agriculture (Anonymous, 2019c). Various targets and indicators are associated with this objective. The first target addresses hunger and the second malnutrition directly. Other targets are partly instrumental in reducing hunger, such as (1) increasing the agricultural productivity and improving the incomes of small-scale food producers; (2) promoting the sustainable food production systems; and (3) seeds genetic diversity and access to arising benefits from the utilization of genetic resources with traditional knowledge. The objective of sustainable development is to eliminate hunger and malnutrition, not poverty, and this goal must be achieved from the start, i.e., 2025 instead of 2030 (Shenggen and Polman, 2013). The argument is based on Brazil, Thailand, China, and Vietnam's experience that indicated that people suffering from severe hunger face extra barriers to improving their livelihoods, whether through education or work. Three paths to reduce hunger were: (1) agriculture-led; (2) social protection and nutrition intervention-led; and (3) the combination of both of these approaches (Engberg-Pedersen, 2016).

12.4.2 GLOBAL WARMING AND CLIMATE CHANGE

Over the past 50 years, global temperatures have increased at the fastest recorded historical rate. Global warming is the impact of increasing carbon dioxide, air pollutants, and other greenhouse gases that accumulate in the atmosphere. These gases absorb rays and solar radiation that bounce off the Earth's surface. Usually, the radiation escapes into space, but the pollutants, which can last for centuries in the atmosphere, trap heat and cause the planet to become warmer. The rise in Earth's temperature triggers longer and hotter heat waves, and the more frequent droughts, heavier rainfall, and more violent storms (MacMillan, 2016).

The effects of global warming are felt over the world. Extreme heatwaves caused tens of thousands of worldwide deaths in recent years. Also, as a harbinger of alarming events to come, Antarctica has lost ice about 134 billion MT each year since 2002. This rate could increase if fossil fuels continue to burn at current rates, causing sea levels to rise by several meters, for the next 50 to 150 years. Here is just a fraction of what we can look forward to (i) melting glaciers, the onset of melting snow, and severe drought will lead to more dramatic water shortages that can increase the risk of wildfires; (ii) rising sea levels cause the coastal flooding; (ii) forests, agriculture, and cities will face new and troublesome pests, heat waves, heavy rains, and increased flooding, all of these factors can destroy agriculture and fisheries; (iv) habitats disturbance such as coral reefs and alpine grasslands can cause many plants and animals species to become extinct; (v) infectious disease outbreaks, asthma, and allergies will become more frequent due to increase of pollen-producing ragweed, pollution, and conditions favorable for the spread of pathogens and mosquitoes (MacMillan 2016), and the performance of major marine producers, such as seaweed (Arbit et al., 2018).

12.5 ALTERNATIVES SOLUTIONS

Based on the problems, paradigms, and issues above, the expected condition is high food security (goal) by increasing the food availability (output) that mitigates the environment (outcome). Three alternative solutions were formulated to realize the expected conditions, i.e., develop seaweed as sea vegetables and other food products; development of seaweed as animal feed; and development of seaweed as material for renewable energy.

12.5.1 DEVELOP SEAWEED AS SEA VEGETABLE AND OTHER FOOD PRODUCT

The development of seaweed as a sea vegetable and other food product is relevant to people who tend to reduce meat consumption and increase vegetable consumption. Seaweed is very nutritious, contains minerals, essential elements, and vitamins. Edible seaweed is sea vegetables, a food source for humans that could be consumed in various forms. Nutritionally, the seaweed has the ability to absorb the iodine from the sea that the human body itself cannot produce but requires healthy thyroid function. Seaweed is also an excellent micronutrient source, including folate, calcium (Ca), magnesium (Mg), zinc (Zn), iron (Fe), and selenium (Se). Seaweed is low in calories, and nutrient-dense has long been consumed in Asian cuisine, especially Japanese and Korean. An essential part of many well-known Japanese dishes, such as miso soup and sushi rolls, various seaweed varieties are better known throughout the world by their Japanese names (Tan, 2019).

The edible seaweeds are seaweeds that can be used in food preparation. The seaweed contains high amounts of fiber (Wong and Cheung, 2000). The seaweed is multicellular algae groups: red algae, green algae, and brown algae (Garcia-Vaquero and Hayes, 2016; Mala et al., 2016; Mulyani et al., 2018; Arbit et al., 2019). Seaweed is also harvested or cultivated (Aqmal et al., 2016; Irawati et al., 2016; Rima et al., 2016; Armita et al., 2017; Tuwo et al., 2019) for the extraction of polysaccharides (Garcia-Vaquero et al., 2017) such as alginate, agar, carrageenan, and gelatinous substances (Awaluddin et al., 2016; Akmal et al., 2017; Syamsuddin et al., 2019; Mulyani et al., 2021) which are collectively known as hydrocolloids or phycocolloids. The hydrocolloids have reached commercial significance, especially as a food additive (Ma et al., 2019; Melanie et al., 2020). The food industry exploits gels, water retention, emulsifiers, and other physical properties (Garcia-Vaquero et al., 2017).

Seaweed contains high iodine levels compared to other foods (Anonymous, 2020g). In the Philippines created new pancakes or seaweed noodles, which can be cooked into the canton of pancit, luglug pancit, spaghetti, or carbonara and are claimed to have health benefits (Anonymous, 2019d) such as being rich in calcium, magnesium, and iodine (Anonymous, 2008b). A study pointed to certain seaweed species as a biologically active source of Vitamin B12. The study notes that B-12 is found in raw and roasted seaweed. Only four grams of the dried purple laver is considered sufficient to meet the recommended dietary allowances (RDA) for B-12 (Watanabe et al., 2014). Polysaccharides in seaweed can be metabolized by humans through the action of intestinal bacterial enzymes.

Seaweed oil or algae oil is used to make food (Anonymous, 2020h). Seaweed oil can also be used as a source of fatty acid because it contains monounsaturated and polyunsaturated fats, specifically eicosapentaenoic acid (EPA) and docosahexaenoic acid (DHA), both of which are Omega-3 fatty acids (Doughman et al., 2007). Seaweed DHA content is equivalent to that of salmon-based fish oil (Arterburn et al., 2008; Lenihan-Geels et al., 2013). Seaweed oil is also used for biofuels, massage oils, soaps, and lotions.

In some Asia countries, nori (in Japan), zicai (in Tiongkok), and game (in Korea), dried red algae *Porphyra* sheets are used in soups, sushi, or onigiri. *Chondrus crispus* or Irish moss is the red alga used in producing various food additives, along with *Kappaphycus* and various gigartinoid seaweed. There are six most common edible seaweed varieties: nori, kombu, wakame, ogonori, umibudo, and hijiki. Nori is known as zicai in Chinese and gim in Korean. Nori is the most recognizable seaweed. The red seaweed comes into thin dried sheets that are dark green or black, which is eaten as a snack or make sushi rolls. Kombu is an indispensable ingredient for making dashi, the clear yet rich broth that is the cornerstone of Japanese cooking. The brown seaweed is sold dry in wide chunks and used for dashi, shabu-shabu broth, and sushi rice. The powder form kombu is also used to make a Japanese tea (kombucha), not to be confused with the same name's trendy fermented probiotic drink. Wakame is classified as brown seaweed despite its thick and slippery texture, which is dark green. Wakame is slightly sweet and has a creamy texture and is used to make salads and chopped with miso soup. It is usually packed, dried, or salted in the vacuum packs. Ogonori's scientific name is *Gracilaria*, also known in the Western as sea moss. The purplish mossy seaweed is preserved or used in salads and is the popular ingredient in the Caribbean and Hawaii, where poke bowls can be found. Ogonori is also an agar source, a gelatinous substance that is a natural vegetarian counterpart to gelatin. The supporting structures in the cell walls of certain algal species such as ogonori contain a gelatinous component that is released after boiling. It can be used in dry strip or powder form to make jellies and puddings (Tan, 2019).

Umibudo or sea grapes are several seaweed varieties that are sold fresh to preserve the small, bubble-like clusters of leaves that grow on long stems. It looks like caviar; the tiny pearls explode in the mouth when eaten and unleash the beauty of the sea. The sea grapes are also enjoyed in Sabah, Malaysia, where it is known as latok, they are mainly served in Japan, particularly on the island of Okinawa. The umibudo is a staple in Okinawa cuisine valued for its taste and nutritional benefits, earning it the nickname of longevity seaweed. Boiling and electrocuting them in cold water to remove the extra salt, and in Okinawa, they are enjoyed raw with soy sauce. Hijiki is naturally brown when grown along the rocky coastlines of Japan, Korea, and China, and they turn black after they are boiled and dried to share. Dried hijiki has a strong, salty taste and appears like thin strands once reconstructed and prepared to eat. In Japan, hijiki is often boiled with vegetables or fish with spices such as soy sauce and sugar, while in Korea, seaweed is known as a tot and is eaten as a seasoned side dish or cooked with rice (Tan, 2019).

12.5.2 DEVELOPMENT OF SEAWEED AS ANIMAL FEED

Development of seaweed as animal feed is crucial because livestock consumes 40% of world protein production, it takes twelve kg of feed to produce one kg of beef, 6 kg to produce a kg of pork, and 10 to 15kg of feed to produce one kg of tuna (Tuwo, 2018). Some studies have been investigated the effects of using seaweed or macroalgae as part of the diet of ruminants and other livestock. The main feed market area is algae which have potential applications for animal husbandry (poultry, cattle, swine, and others), aquaculture (Salmonidae, Sparidae, bivalves, and others), pets (dogs, cats, birds, tropical fish, and others), and other (racehorses, zoos, aquaria, and others) (Abd-Elrahman, 2020).

The cattle, sheep, and horses have been eating seaweed for a long time living in coastal areas, especially in European countries where the large brown seaweeds have washed ashore. The seaweed availability for animals has been increased by the production of seaweed flour: dried seaweed that has been ground into a powder. Norway was the first producers of seaweed meal using *Ascophyllum nodosum.* Other countries the seaweed meal producers are France using *Laminaria digitata,* Iceland using *Laminaria* and *Ascophyllum*, and the United Kingdom using *Ascophyllum.* As *Ascophyllum* is so easy to reach, it is the main raw material for seaweed flour, and most of the experimental to measure the effectiveness of seaweed flour has been done on this seaweed. Analysis shows that seaweed contains beneficial minerals, vitamins, and trace elements. Trace elements are essential elements for humans and other mammals. Traditionally, the seaweed has been supplied to the livestock as a source of supplemental minerals, trace elements, and vitamins. In trials feeding with poultry shows that seaweed can increase the iodine content of eggs. Trials on giving seaweed to dairy cows showed an increase in milk production of 6.8% and a 13% increase in income. Trials of feeding seaweed to ewes showed stable body weight during winter and resulted in better wool production.

Several companies in Australia, Canada, Ireland, Norway, the UK, and the USA recommend using the seaweed flour as the feed additive for sheep, cattle, horses, goats, poultry, dogs, cats, emus, and alpacas. Horse racing is the main target of giving seaweed powder as a feed additive. One USA university study showed that some animals' immune systems could be improved by giving seaweed as the feed supplement (Abd-Elrahman, 2020). The use of seaweed as supplementary feed for animals is not the new phenomenon for fish breeders and cultivators (Rajauria, 2015).

The growing human population is one reason why livestock needs to look for other feed sources to avoid competition between humans and livestock. Increased consumption of meat due to the increasing human population causes an increase in demand for animal meat. This increased demand will have an impact on increasing the need for animal feed. Currently, livestock plays a significant role in the global food industry (Anonymous, 2010b).

The soybean and corn known as two conventional feedstuffs commonly utilized as the primary source of protein and dietary energy for animal feed. Because the soybean and corn are stapled food crops for humans, their usage for animal feed presents a severe threat to humans and animals' nutrition security (Lum et al., 2013). A long-term vision is needed for exploration and finding alternatives or new food sources, most importantly taken from outside the human food chain (Anonymous, 2010b). There is a growing interest in to use of seaweeds for animal feeding due to their high protein and dietary fibers, relatively well-balanced amino acid profiles, and the rich contents of the minerals and vitamins, along with a unique blend of bioactive compounds (Rajauria, 2015).

The seaweeds are vast renewable resources underutilized in the global economy. Seaweeds do not require the fertilizers, pesticides, and freshwater. The seaweed can grow on one-sixth of the surface needed by terrestrial plants.

Seaweeds constitute 6,000 species with a great diversity of forms and size, and about 5% of them are used as food for either humans or animals (Chojnacka et al., 2012). As the plant of the unique structure and biochemical composition, the seaweed could be exploited for its multifunctional properties for food, feed, fuel, medicine, and cosmetics. Since 1970, seaweed production has been consistently increasing with an annual growth rate of 7.7%. In 2011, 21 million tons of the aquatic plants were produced worldwide, which contributed US$5.5 billion to the world economy, of which 99% of the production comprised of seaweeds only. However, until 2004, only 1% of the global value of seaweed was utilized for the animal feed, wherein *Ascophyllum nodosum* the brown seaweed species constituted the most (US$10 million) (Chopin and Sawhney, 2009), relatively well-balanced amino acid profiles, and rich contents of the minerals and vitamins, along with a unique blend of bioactive compounds (Rajauria, 2015).

In 2014, the world market for animal feed and nutritional supplements was very dynamic and estimated at around US$460 billion (Anonymous, 2015b). In 2013, total world feed production was estimated at 963 million tons, nine million tons higher than the previous year. Asia continues to be the world's leading feed producer region with 348 million tons, followed by Europe with 227 million tons and North America with 189 million tons. Once again, Tiongkok leads the market and produces 189 million tons of total animal feed, while Africa is considered the fastest-growing region where feed output increased 7% in 2013. There is a growing interest in the market potential for available feeds based on the principle of added value linked to livestock's health benefits. The global feed additives market was estimated at US$14.9 billion in 2013, which is expected to grow US$20 billion by 2020 (Rajauria, 2015).

Application of seaweed or its extract as feed additives have many benefits. Apart from the nutritional improvement, the seaweed can add to diets supplemented feed also significantly changed the gastrointestinal flora, improved immunity, enhanced growth performance, and increased the milk and meat quality and the yield in livestock (Carrillo et al., 2008; Bendary et al., 2013). In aquaculture, seaweed-based diets showed improved weight gain, increased the triglyceride and the protein deposition in the muscle, better feed conversion ratios (FCRs), improved resistance to the disease and less mortality, better growth index, higher gutted weight, increased fish digestibility, enhanced natural pigmentation, improve carcass quality, and increase the starvation tolerance, and decreased the nitrogen output to the environment (Becker 2004; Fleurence et al., 2012). Similarly, in the case of pet animals, seaweed meal not only has advantageous effects on health but also affects their external appearance (Pulz and Gross, 2004).

Feed quality is the important exogenous factor influencing the animal health. The animals that consume nutritious feed produce the high-quality food for humans because the quality of animal protein is ultimately a reflection of animal feed's overall nutritional value. The raw materials currently utilized for animal feed products are higher plants such as corn, soybeans, sorghum, oats, and barley. Most of these crops are the short-lived, grow singly in the particular season, and require months to mature. These crops are also a good source of the human diet, and the perception exists that utilizing them for animal feed represents an inefficient or wasteful use of resources, which contributes to rises in food prices (Capper et al., 2013).

Consequently, seaweed application opens space in the animal feed market segment for feed supplements, as they possess many unique and exciting biochemical properties compared to higher plants. They are a sustainable source, reproduce quickly, grow, and harvest in any season, and provide a renewable fodder and fish food choice. Seaweed is the only biomass that allows cultivation with daily harvest. Daily harvest could bring more security to the feed market for a constant supply of raw materials. The demand for a high seaweed production to serve a booming market easily to be fulfilled by increasing the number of ponds or acquiring more sea areas compared to acquiring new agriculture resources. They offer an ideal blend of dietary nutrients, bioactive compounds, and the variety of feed products can be produced using seaweed to grow healthy livestock and aquaculture (Van den Burg et al., 2013).

Seaweed contains beneficial secondary metabolites, and their application can be seen as an almost unlimited field, which cannot be found in other organisms (MacArtain et al., 2007). From a nutritional perspective, seaweeds contain a high concentration of minerals, vitamins, pigments, polysaccharides, fibers, polyphenols, proteins, and lipids (Burtin, 2003). Seaweeds are the source of the vitamins A, B, C, D, E, riboflavin, niacin, pantothenic acid, and folic acid. Seaweeds are also an excellent source of minerals such as Ca, P, Na, and K. Seaweeds have minimal 54 trace elements required for physiological functions (Chapman and Chapman, 1980). The protein content of seaweeds generally varies from five to 35% of its dry weight (DW). These protein levels are comparable to high-protein crops such as soybeans.

Additionally, lipids represent about one to 5% of algal dry and show an important polyunsaturated fatty acid (PUFA) composition, particularly omega-3 and omega 6-acids, which play an essential role in health (MacArtain et al., 2007). The seaweed proteins and lipids are most acceptable for consumption as compared to other crops. The seaweed proteins have a high content of essential amino acids and a relatively high level of unsaturated fatty acids (UFA). Seaweeds are rich in structural and storage types of polysaccharides (Melanie et al., 2020). Their structural polysaccharides are analogous to terrestrial plants and are mainly celluloses, hemicelluloses, and xylans. However, storage polysaccharides, such as carrageenan, laminarin, fucoidan, alginic acid, agar, and alginates, are more specific to the seaweed species and represent the most commercially exploited components in it. Seaweed polyphenols differ from terrestrial plant polyphenols, providing a wide range of the potential biological activity. The highest contents of phlorotannins reported in brown seaweeds, ranging from 5 to 15% of the dried weight (Burtin, 2003). The carotenoid pigment compositions of seaweed are unique. The seaweed contains all kinds of pigments, as can be expected from terrestrial resources. For instance, fucoxanthin, a significant carotenoid of brown seaweed (Ma et al., 2019), contributes more than 10% of the estimated total production of carotenoids in nature (de Quirós et al., 2010).

12.5.3 DEVELOP SEAWEED AS RAW MATERIAL FOR RENEWABLE ENERGY PRODUCTION

The development of seaweed raw material needed for renewable energy production is essential because seaweed can become sustainable energy resources. As of now, the soybean and corn are two conventional feedstuffs commonly utilized as the primary source of dietary energy and protein for animal feed. Because the two crops are stapled food crops for humans, their biofuel production usage presents a severe threat to humans and animals' nutrition security (Lum et al., 2013). This situation requires a long-term vision to find an alternative or new source of biofuels, most importantly taken from outside the human food chain (Anonymous, 2010b).

Among macroalgae, *Laminaria* spp. and *Ulva* spp. is the essential prospect from an energy perspective. Seaweed is usually sold in moderate volumes and shipped fresh to the local mills. It has a moisture content of about eighty to 85% and is expensive to transport. It has a lower negative calorific value, so processes that do not require drying are preferred. The primary energy processes considered for seaweed are fermentation, either anaerobic digestion, to produce biogas, or ethanol fermentation (Sari et al., 2017). The presence of salt, polyphenols, and sulfated polysaccharides (PLS), of course, needs to be managed carefully so as not to inhibit the fermentation process. The biogas production is outdated, and previous experiments have shown that the anaerobic digestion of seaweed is technically feasible. Initially, the seaweed resource should be fed into existing anaerobic digestion plants to allow for smaller quantities. The process is approaching the commercialization process to convert macroalgae into energy, although there is still a need to reduce raw material costs of at least 75% from current levels. Alcohol fermentation is still tricky. The lack of readily fermentable sugar polymers means there is no point in pursuing a standard sugar fermentation process. Existing polysaccharides will require a new commercial process to break down constituent monomers before fermentation. Good work has begun in Ireland to isolate the lyase sea, which will do so efficiently, up to 60% of the dry biomass in *Laminaria* spp. can be fermented by the correct process. *Ulva* spp. also of interest because of their starch content.

The competitiveness of macroalgae biomass for alcoholic fermentation must be seen in other cellulosic biomass such as wood and straw, which also has the potential to become ethanol feed. There is much speculation that an integrated biorefinery solution will allow sufficient scale to enable fuel economy production from macroalgae. The only crucial industrial product of macroalgae is the hydrocolloid, and there is currently no production based in Ireland. Energy extraction from the waste stream is a good commercial biorefinery concept. Suppose the seaweed permit costs, double production of the ethanol and the biogas is also possible. There are many opportunities for the extraction of high-value niche products from seaweed. Each shall be assessed against commercial requirements and demonstrate eligibility for co-production of energy together with higher value products, with particular attention to the appropriate production scale. Ireland has a long track record in macroalgae research. Several research programs were identified in this report where algae as a feedstock for biofuels could be a good complement. Estimates made for this report indicate that up to 447 TJ of the energy could be generated from macroalgae by 2020. This biofuel is about 0.2% of the current national road fuel demand (Bruton et al., 2009).

The US is investing in a new, renewable energy source in the future in the form of seaweed. The US is investing nearly US$1.5 million to help develop the seaweed farms and to explore harvesting methods. Seaweed, with all its slimy beauty, can be processed into biofuel, which can be used for home and vehicle power plants. The US is funding across the country to achieve large-scale seaweed farming, supporting alternative uses of fossil fuels. Recently in Honolulu, US$ 95,978 was awarded to help build a marine simulation model that would assist researchers in designing an offshore seaweed farm, and the US$ 500,000 was awarded to Kampachi Farms to test the method of harvesting seaweed grown on this farm. Kampachi Farms will develop an offshore seaweed farm. Innovators look to seaweed and go further to explore new ways of harvesting renewable energy (Caughill 2017).

The sea has long been a source of wealth in the form of fish or oil. Currently, experts are studying to make the seaweed as the biocrude producer by imitating natural processes to produce oil not on a geological time scale but in minutes – the results of studies using reactor small quartz tubes, which look like small sealed straws. Heating a reactor containing porridge made from seaweed biomass and water up to 350°C at a very high level of 585°C per minute can produce 79% bio-oil. That means that 79% of seaweed biomass in the reactor is converted to biofuel. Biofuel production worldwide is still too small, namely 113 billion liters in 2013, and 140 billion liters in 2018. Biofuel production needs to grow 22-fold by 2025 to produce the amount of biofuel needed by the world to keep global temperatures from rising more than 2°C (Anonymous, 2020i).

Seaweed offers all the benefits of biofuel with the additional benefit of growing at sea. However, turning the big pieces of slippery, salty kelp into the biocrude is a challenge. Several studies have used the catalysts, which are the added chemicals that can help make the process run faster or easier; however, catalysts are usually expensive and require catalyst recovery. To overcome this, a study using heating with high temperatures succeeded in producing biocrude without catalysts. The advantage of this high-temperature process is that it is relatively simple and does not require the catalyst. A high heating rate produces biocrude, which has molecular properties which will make it more easily distilled. However, this research was carried out with a small batch reactor and not suitable for industrial scale. To improve the process to an industrial scale, it must use a flow reactor with the continuous flow of the reactants and the products. Even though the preliminary test gives a 79% result, the percentage can still be increased (Anonymous, 2020i). The development of biocrude from seaweed still requires advanced research and development.

12.6 CONCEPT IMPLEMENTATION

The realization of the expected conditions and achieving outputs and outcomes, the implementation concept involving subjects, objects, and methods is needed. The subject is the regulation makers and regulations (superstructure management), the government as the executor of the regulations (infrastructure management), and researchers, community leaders, and others (substructure management). The management objects are political institutions, government institutions, educational institutions, research institutions, community institutions, business institutions, financial/donor institutions, and industrialists. Management methods are legislation, education, research, and development, demonstration plot, product diversification, socialization/promotion, and empowerment.

Implementation concepts to develop seaweed as sea vegetable and other food product are: (1) identification of edible seaweed potential as sea vegetable and other food product; where educational institutions together with research institutions performing research and development in order to identify potential edible seaweed as sea vegetables and other food products in collaboration with financial/donors institutions and business institutions; (2) domestication of seaweed potential as sea vegetable and other food product; where (i) educational institutions together with research institutions performing research and development in the context of domestication of potential edible seaweed as sea vegetables and other food products in collaboration with financial/donors institutions; and (ii) educational and research institutions together with government institutions to socialize and promote the results of the domestication of potential edible seaweed as sea vegetables and other food products to community institutions, business institutions, and industry; (3) cultivation of edible seaweed potential as sea vegetable and other food product; where (i) educational institutions together with research institutions performing research and development in the context of cultivating new edible seaweed potential as sea vegetables and other food products in collaboration with financial/donors institutions; (ii) educational and research institutions together with government institutions develop demonstration plots of cultivation of new edible seaweed potential as sea vegetables and other food products in collaboration with community institutions, business institutions, and financial/donors institutions; (iii) government institutions together with political institutions develop regulations and incentive that support the development of potential cultivation of edible seaweed as sea vegetables and other food products by involving educational institutions, research institutions, and community institutions; (iv) government institutions together with political institutions develop basic infrastructure that supports the development of potential cultivation of edible seaweed as sea vegetables and other food products by involving financial/donors institutions, business institutions, and industries; and (v) government institutions together with educational and research institutions empower potential cultivators of edible seaweed as sea vegetables and other food products through education, training, demonstration plot by involving community institutions, business institutions, and financial/donors institutions; and (4) diversification product of edible seaweed potential as sea vegetable and other food product, including rice and meat homolog; where educational and research institutions together with the industry diversify potential seaweed as sea vegetables and food products, including rice and meat homologous products by involving financial/donor institutions.

Implementation concepts for development of seaweed as animal feed: (1) identification of seaweed potential as animal feed; where educational institutions together with research institutions performing research and development in order to identify potential seaweed as animal feed in collaboration with financial institutions/donors and business institutions; (2) domestication of seaweed potential as animal feed; where (i) educational institutions together with research institutions performing research and development in order to domesticate potential seaweed as sea vegetables and other food products in collaboration with financial/donors institutions; and (ii) educational and research institutions together with government institutions to socialize and promote the results of domestication of potential seaweed as animal feeds to community institutions, business institutions, and industry; (3) cultivation seaweed potential as animal feed; where (i) educational institutions together with research institutions performing research and development in the context of cultivating new seaweed potential as sea vegetables and other food products in collaboration with financial/donors institutions; (ii) educational and research institutions together with government agencies conduct education and make demonstration plots of cultivation of new edible seaweed potential as sea vegetables and other food products in collaboration with community institutions, business institutions, and financial/donors institutions; (iii) government agencies together with political institutions develop regulations and laws that support the development of potential cultivation edible seaweed as animal feed by involving educational institutions, research institutions, and community institutions; (iv) government agencies together with political institutions develop basic infrastructure that supports the development of potential cultivation edible seaweed as animal feed by involving financial/ donor institutions, business institutions, and industry; and (v) government agencies together with educational and research institutions empower the potential cultivators of edible seaweed as animal feed through education, training, demonstration plot by involving community institutions, business institutions, and financial institutions/donors; and (4) development of seaweed-based animal feeds; where educational and research institutions together with the industry performing research and development in the development of seaweed-based animal feed in collaboration with financial/ donors institutions and business institutions.

Implementation concepts to develop seaweed as raw material for renewable energy production are: (1) identification seaweed potential as raw material for renewable energy; where educational institutions together with research institutions performing research and development in order to identify potential seaweed as raw material for renewable energy production in collaboration with financial/donors institutions, and business institutions; (2) domestication of seaweed potential as raw material for renewable energy production; where (i) educational institutions together with research institutions performing research and development in the context of domestication of potential seaweed as raw material for renewable energy production in collaboration with financial/donors institutions; and (ii) educational and research institutions together with government institutions to socialize and promote the domestication product of potential seaweed as raw material for renewable energy production to community institutions, business institutions, and industry; (3) cultivation of seaweed potential as raw material for renewable energy production; where (i) educational institutions together with research institutions performing research and development in the context of cultivating potential seaweed as raw material for renewable energy production in collaboration with financial/donors institutions; (ii) educational and research institutions together with government agencies performing education and develop demonstration plots of new seaweed potential as raw material for renewable energy production in collaboration with community institutions, business institutions, and financial/donors institutions; (iii) government institutions together with political institutions develop regulations and incentive that support the development of potential cultivation of seaweed as raw material for renewable energy production by involving educational institutions, research institutions, and community institutions; (iv) government institutions together with political institutions develop basic infrastructure that supports the development of potential cultivation of seaweed as raw material for renewable energy production by involving financial /donor institutions, business institutions, and industry; and (v) government institutions together with educational and research institutions empower the potential cultivators of seaweed as raw material for renewable energy production through education, training, demonstration plot by involving community institutions, business institutions, and financial/donor institutions; and (4) renewable energy development by using seaweed as raw materials; where educational and research institutions together with industry performing research and development in the development of seaweed-based renewable energy in collaboration with financial/donors institutions and business institutions.

12.7 CONCLUSION

The holistic-integral approach and S-O-M analysis produced twelve concepts of implementation to make seaweed an agent for strengthening food security and environmental mitigation, i.e., identification edible seaweed potential as sea vegetable and other food product; domestication edible seaweed potential as sea vegetable and other food product; cultivation edible seaweed potential as sea vegetable and other food product; diversification product edible seaweed potential as sea vegetable and other food product, including rice and meat homolog; identification seaweed potential as animal feed; domestication potential seaweed as animal feed; cultivation seaweed potential as animal feed; seaweed-based animal feeds development; identification seaweed potential as raw material for renewable energy production; domestication seaweed potential as raw material for renewable energy production; cultivation seaweed potential as raw material for renewable energy production; and renewable energy development by using seaweed as raw materials.

ACKNOWLEDGMENTS

We would like to thank the Ministry of Research and Technology/ National Research and Innovation Agency of the Republic of Indonesia for providing research and publication funding (contract number 1516/UN4.22/ PT.01.03/2020 dated May 18th, 2020 and 752/UN4.22/PT.01.03/2021 dated March 26th, 2021.

KEYWORDS

- **animal feed**
- **climate change**
- **docosahexaenoic acid**
- **eicosapentaenoic acid**
- **energy scarcity**
- **environmental degradation**
- **environmental mitigation**
- **food security**
- **global warming**
- **high population growth**
- **limited food resources**
- **livelihood**
- **recommended dietary allowances**
- **renewable energy**
- **sustainable development**
- **water scarcity**

REFERENCES

Abd-Elrahman, M. A., (2020). *Seaweed as Animal Feed.* https://www.academia.edu/8732970/seaweed_as_animal_feed (accessed on 13 October 2021).

Akmal, A., Syamsuddin, R., Trijuno, D. D., & Tuwo, A., (2017). Morfologi, kandungan klorofil a, pertumbuhan, produksi, dan kandungan karaginan rumput laut *Kappaphycus alvarezii* yang dibudidayakan pada kedalaman berbeda. *Jurnal Rumput Laut Indonesia, 2*(2), 39–50.

Anonymous, (2007). *The Limits of a Green Revolution?* http://news.bbc.co.uk/2/hi/in_depth/6496585.stm (accessed on 13 October 2021).
Anonymous, (2008a). *The State of Food Insecurity in the World 2008 – High Food Prices and Food Security – Threats and Opportunities*. http://www.fao.org/3/i0291e/i0291e00.htm (accessed on 13 October 2021).
Anonymous, (2008b). *Albay Folk Promote Seaweed 'Pansit'*. https://news.abs-cbn.com/nation/regions/04/08/08/albay-folk-promote-seaweed-pansit (accessed on 13 October 2021).
Anonymous, (2009a). *World Faces 'Perfect Storm' of Problems by 2030, Chief Scientist to Warn*. https://www.theguardian.com/science/2009/mar/18/perfect-storm-john-beddington-energy-food-climate (accessed on 13 October 2021).
Anonymous, (2009b). *Global Crisis to Strike by 2030*. http://news.bbc.co.uk/2/hi/uk_news/7951838.stm (accessed on 13 October 2021).
Anonymous, (2009c). *Global Food Production Will Have to Increase 70% for Additional 2.3 Billion People by 2050*. http://www.finfacts.ie/irishfinancenews/article_1017966.shtml (accessed on 13 October 2021).
Anonymous, (2009d). *1.02 Billion People Hungry*. http://www.fao.org/news/story/en/item/20568/icode/ (accessed on 13 October 2021).
Anonymous, (2010a). *WDI, GDF and ADI Online Databases – Data and Statistics*. https://webarchive.loc.gov/all/20100416035127/http://web.worldbank.org/WBSITE/EXTERNAL/DATASTATISTICS/0,,contentMDK:20398986~menuPK:64133163~pagePK:64133150~piPK:64133175~theSitePK:239419,00.html (accessed on 13 October 2021).
Anonymous, (2010b). *Good practices for the feed industry – Implementing the Codex Alimentarius Code of Practice on Good Animal Feeding – FAO Animal Production and Health Manual No. 9*. http://www.fao.org/3/i1379e/i1379e.pdf (accessed on 13 October 2021).
Anonymous, (2011). *World Population to Hit Milestone with Birth of 7 Billionth Person*. https://web.archive.org/web/20170924090953/http://www.pbs.org/newshour/bb/world-july-dec11-population1_10-27/ (accessed on 13 October 2021).
Anonymous, (2012). *Indonesia: Stagnating Rice Production Ensures Continued Need for Imports*. https://ipad.fas.usda.gov/highlights/2012/03/Indonesia_rice_Mar2012/ (accessed on 13 October 2021).
Anonymous, (2015a). *Jokowi Finally Agrees to Import Rice from Thailand, Vietnam*. https://www.thejakartapost.com/news/2015/10/22/jokowi-finally-agrees-import-rice-thailand-vietnam.html (accessed on 13 October 2021).
Anonymous, (2015b). *Alltech Global Feed Survey*. https://www.alltech.com/sites/default/files/2019-07/global-feed-survey-2015.pdf (accessed on 13 October 2021).
Anonymous, (2019a). *Desember 2019 Ada Data Baru Luas Baku Sawah*. http://agroindonesia.co.id/2019/11/desember-2019-ada-data-baru-luas-baku-sawah/ (accessed on 13 October 2021).
Anonymous, (2019b). *Hunger: Facts and Figures*. Welt Hunger Hilfe. https://www.welthungerhilfe.org/hunger/ (accessed on 13 October 2021).
Anonymous, (2019c). *Progress of Goal 2 in 2019 – Sustainable Development Goals – Knowledge Platform*. https://sustainabledevelopment.un.org/sdg2 (accessed on 13 October 2021).
Anonymous, (2019d). *Cooking with Seaweed and the Health Benefits* (infographic). https://www.meridian-sea.com/cooking-with-seaweed-and-the-health-benefits/ (accessed on 13 October 2021).

Anonymous, (2020a). *World Population Clock: 7.8 Billion People 2020*. https://www.worldometers.info/world-population/#:~:text=7.8%20Billion%20(2020),currently%20living)%20of%20the%20world (accessed on 13 October 2021).

Anonymous, (2020b). *Data Lima Tahun Terakhir*. https://www.pertanian.go.id/home/?show=page&act=view&id=61 (accessed on 13 October 2021).

Anonymous, (2020c). *Water Scarcity*. https://www.worldwildlife.org/threats/water-scarcity (accessed on 13 October 2021).

Anonymous, (2020d). *Water*. http://climate.org/archive/topics/water.html (accessed on 13 October 2021).

Anonymous, (2020e). *What is Sustainable Development?* https://www.acciona.com/sustainable-development/ (accessed on 13 October 2021).

Anonymous, (2020f). *Why Seaweed*. http://www.seaweedenergysolutions.com/en/why-seaweed (accessed on 13 October 2021).

Anonymous, (2020g). *Iodine*. https://lpi.oregonstate.edu/mic/minerals/iodine# food_source (accessed on 13 October 2021).

Anonymous, (2020h). *New Culinary Trend: Cooking with Algae Oil*. https://www.allrecipes.com/article/new-culinary-trend-cooking-with-algae-oil/ (accessed on 13 October 2021).

Anonymous, (2020i). *Turning Humble Seaweed to Biofuel*. https://www.renewableenergyworld.com/2014/10/22/turning-humble-seaweed-to-biofuel/#gref (accessed on 13 October 2021).

Aqmal, A., Tuwo, A., & Haryati, H., (2016). Analisis hubungan antara keberadaan alga filamen kompetitor terhadap pertumbuhan dan kandungan karaginan rumput laut *Kappaphycus* sp. di Provinsi Sulawesi Selatan. *Jurnal Rumput Laut Indonesia, 1*(2), 94–102.

Arbit, N. I. S., Omar, S. B. A., Soekendarsi, E., Yasir, I., Tresnati, J., & Tuwo, A., (2019). Morphological and genetic analysis of *Gracilaria* sp. cultured in ponds and coastal waters. *IOP Conf. Ser.: Earth Environ. Sci., UK, 564*, 012059.

Arbit, N. I. S., Omar, S. B., Tuwo, A., & Soekendarsi, E., (2018). Effect of global warming scenarios on carotenoid pigments *Gracilaria changii*. *International Journal of Environment, Agriculture and Biotechnology, 3*(6), 268–287.

Armita, D., Ali, S. A., Yanuarita, D., & Tuwo, A., (2017). Kualitas air di daerah dengan dan tanpa budidaya rumput laut di Dusun Malelaya, Kecamatan Mangarabombang, Kabupaten Takalar. *Jurnal Rumput Laut Indonesia, 2*(2), 74–80.

Arterburn, L. M., Oken, H. A., Hall, E. B., Hamersley, J., Kuratko, C. N., & Hoffman, J. P., (2008). Algal-oil capsules and cooked salmon: Nutritionally equivalent sources of docosahexaenoic acid. *Journal of the American Dietetic Association, 108*(7), 1204–1209.

Awaluddin, A., Badraeni, B., Azis, H. Y., & Tuwo, A., (2016). Perbedaan kandungan karaginan dan produksi rumput laut *Kappaphycus alvarezii* antara bibit alam dan bibit hasil pengayaan. *Jurnal Rumput Laut Indonesia, 1*(1), 65–70.

Badraeni, B., Azis, H. Y., Tresnati, J., & Tuwo, A., (2020). Seaweed *Gracilaria change* as a bioremediator agent for ammonia, nitrite and nitrate in controlled tanks of white leg shrimp *Litopenaeus vannamei*. *IOP Conf. Ser.: Earth Environ. Sci., UK, 564,* 012059.

Becker, W., (2004). Microalgae in human and animal nutrition. In: Richmond, A., (ed.), *Handbook of Microalgal Culture: Biotechnology and Applied Phycology* (pp. 312–351). Blackwell Science, Victoria.

Bendary, M. M., Bassiouni, M. I., Ali, M. F., Gaafar, H. M., & Shamas, A. S., (2013). Effect of premix and seaweed additives on productive performance of lactating Friesian cows. *Int. Res. J. Agric. Sci. Soil Sci., 3*, 74–181.

Bickel, G., Nord, M., Price, C., Hamilton, W., & Cook, J., (2000). *Guide to Measuring Household Food Security*. https://naldc.nal.usda.gov/download/38369/PDF (accessed on 13 October 2021).

Blanco, H., & Lal, R., (2008). *Principles of Soil Conservation and Management* (p. 617). Springer, Dordrecht.

Bruton, T., Lyons, H., Lerat, Y., Stanley, M., & Rasmussen, M. B., (2009). *A Review of the Potential of Marine Algae as a Source of Biofuel in Ireland.* http://www.fao.org/uploads/media/0902_SEI_-_A_Review_of_the_Potential_of_Marine_Algae.pdf (accessed on 13 October 2021).

Burtin, P., (2003). Nutritional value of seaweeds. *Electronic Journal of Environmental, Agricultural Food Chemistry, 2*(4), 498–503.

Capper, J., Berger, L., Brashears, M., & Jensen, H., (2013). *Animal Feed vs. Human Food: Challenges and Opportunities in Sustaining Animal Agriculture Toward 2050 – Economics Technical Reports and White Papers, Issue.* http://lib.dr.iastate.edu/econ_reportspapers/16 (accessed on 13 October 2021).

Carrillo, S., López, E., Casas, M. M., Avila, E., Castillo, R. M., Carranco, M. E., Calvo, C., & Perez-Gil, F., (2008). Potential use of seaweeds in the laying hen ration to improve the quality of n-3 fatty acid-enriched eggs. *J. Appl. Phycol., 20*(5), 721–728.

Caughill, P., (2017). *Why Seaweed, Not Solar, Maybe the Energy of Tomorrow: It's All About Biofuel!* https://www.nbcnews.com/mach/science/why-seaweed-not-solar-may-be-energy-tomorrow-ncna809426#:~:text=SEAWEED%20ENERGY,That's%20right%20%E2%80%94%20seaweed%20energy.&text=Seaweed%2C%20in%20all%20its%20slimy,power%20our%20homes%20and%20vehicles (accessed on 13 October 2021).

Chapman, V. J., & Chapman, D. J., (1980). Sea vegetables (algae as food for man). In: Chapman, V. J., & Chapman, D. J., (eds.), *Seaweeds and Their Uses* (pp. 62–97). Springer, Dordrecht.

Chojnacka, K., Saeid, A., Witkowska, Z., & Tuhy, L., (2012). Biologically active compounds in seaweed extracts – the prospects for the application. *The Open Conference Proceedings Journal, Netherlands, 3*(1), 20–28.

Chopin, T., & Sawhney, M., (2009). Seaweeds and their mariculture. In Steele, J. H., Thorpe, S. A., & Turekian, K. K., (eds.), *Encyclopedia of Ocean Sciences* (pp. 4477–4487). Elsevier, Amsterdam.

Coleman-Jensen, A., Rabbitt, M. P., Gregory, C. A., & Singh, A., (2019). *Household Food Security in the United States in 2018, ERR-270.* U.S. Department of Agriculture, Economic Research Service. https://www.ers.usda.gov/webdocs/publications/94849/err-270.pdf (accessed on 13 October 2021).

De Quirós, A. R. B., Frecha-Ferreiro, S., Vidal-Perez, A. M., & López-Hernández, J., (2010). Antioxidant compounds in edible brown seaweeds. *European Food Research and Technology, 231*(3), 495–498.

Doughman, S. D., Krupanidhi, S., & Sanjeevi, C. B., (2007). Omega-3 fatty acids for nutrition and medicine: Considering microalgae oil as a vegetarian source of EPA and DHA. *Current Diabetes Reviews, 3*(3), 198–103.

Engberg-Pedersen, P., (2016). *An SDG-Based Results Framework for Development Co-operation.* https://www.oecd.org/dac/peer-reviews/SDG-based-results-framework.pdf (accessed on 13 October 2021).

Fleurence, J., Morançais, M., Dumay, J., Decottignies, P., Turpin, V., Munier, M., Garcia-Bueno, N., & Jaouen, P., (2012). What are the prospects for using seaweed in human nutrition and for marine animals raised through aquaculture? *Trends in Food Science Technology, 27*(1), 57–61.

Frederick, K. D., & Major, D. C., (1997). Climate change and water resources. *Climatic Change, 37*(1), 7–23.

Garcia-Vaquero, M., & Hayes, M., (2016). Red and green macroalgae for fish and animal feed and human functional food development. *Food Reviews International, 32*(1), 15–45.

Garcia-Vaquero, M., Lopez-Alonso, M., & Hayes, M., (2017). Assessment of the functional properties of protein extracted from the brown seaweed *Himanthalia elongata* (Linnaeus) SF Gray. *Food Research International, 99*, 971–978.

Garcia-Vaquero, M., Rajauria, G., O'doherty, J. V., & Sweeney, T., (2017). Polysaccharides from macroalgae: Recent advances, innovative technologies and challenges in extraction and purification. *Food Research International, 99*, 1011–1020.

Gregory, P. J., Ingram, J. S. I., & Brklacich, M., (2005). Climate change and food security. *Philosophical Transactions of the Royal Society B: Biological Sciences, 360*(1463), 2139–2148.

Imeson, A., (2012). *Human Impact on Degradation Processes – Desertification, Land Degradation and Sustainability*. John Wiley & Sons, New Jersey.

Irawati, I., Badraeni, B., Abustang, A., & Tuwo, A., (2016). Pengaruh perbedaan bobot tallus terhadap pertumbuhan rumput laut *Kappaphycus alvarezii* strain coklat yang dikayakan. *Jurnal Rumput Laut Indonesia, 1*(2), 82–87.

Johnson, D. L., Ambrose, S. H., Bassett, T. J., Bowen, M. L., Crummey, D. E., Isaacson, J. S., Johnson, D. N., et al., (1997). Meanings of environmental terms. *Journal of Environmental Quality, 26*, 581–589.

Jones, H., (2009). *Equity in Development: Why it is Important and How to Achieve It.* https://www.odi.org (accessed on 13 October 2021).

Konikow, L. F., (2011). *Contribution of Global Groundwater Depletion Since 1900 to Sea-Level Rise*. https://doi.org/doi:10.1029/2011GL048604.

Krantz, L., (2001). *The Sustainable Livelihood Approach to Poverty Reduction: An Introduction.* https://publikationer.sida.se (accessed on 13 October 2021).

Lenihan-Geels, G., Bishop, K. S., & Ferguson, L. R., (2013). Alternative sources of omega-3 fats: Can we find a sustainable substitute for fish? *Nutrients, 5*(4), 1301–1315.

Lobb, D. A., (2009). Soil movement by tillage and other agricultural activities. In: Jorgensen, S. E., (ed.), *Applications in Ecological Engineering* (pp. 3295–3303). Elsevier, Amsterdam.

Lum, K. K., Kim, J., & Lei, X. G., (2013). Dual potential of microalgae as a sustainable biofuel feedstock and animal feed. *Journal of Animal Science Biotechnology, 4*(1), 1–7.

Ma, Z., Khalid, N., Shu, G., Zhao, Y., Kobayashi, I., Neves, M. A., Tuwo, A., & Nakajima, M., (2019). Fucoxanthin-loaded oil-in-water emulsion-based delivery systems: Effects of natural emulsifiers on the formulation, stability, and bioaccessibility. *ACS Omega, 4*(6), 10502–10509.

MacArtain, P., Gill, C. I. R., Brooks, M., Campbell, R., & Rowland, I. R., (2007). Nutritional value of edible seaweeds. *Nutrition Reviews, 65*(12), 535–543.

MacMillan, A., (2016). *Global Warming 101.* https://www.nrdc.org/stories/global-warming-101 (accessed on 13 October 2021).

Mala, L., Latama, G., Abustang, A., & Tuwo, A., (2016). Analisis perbandingan pertumbuhan rumput laut *Kappaphycus alvarezii* varietas coklat yang terkena epifit di Perairan Libukang, Kabupaten Jeneponto. *Jurnal Rumput Laut Indonesia, 1*(1), 52–56.

Melanie, H., Taarji, N., Zhao, Y., Khalid, N., Neves, M. A., Kobayashi, I., Tuwo, A., & Nakajima, M., (2020). Formulation and characterization of O/W emulsions stabilized with modified seaweed polysaccharides. *International Journal of Food Science and Technology, 55*(1), 211–221.

Montgomery, D. R., (2007). Soil erosion and agricultural sustainability. *Proc. Natl. Acad. Sci. USA, 104*(33), 13268–13272.

Mulyani, S., Tuwo, A., Syamsuddin, R., & Jompa, J., (2018). Effect of seaweed *Kappaphycus alvarezii* aquaculture on growth and survival of coral *Acropora muricata*. *AACL Bioflux, 11*(6), 1792–1798.

Mulyani, S., Tuwo, A., Syamsuddin, R., Jompa, J., & Cahyono, I., (2021). Relationship of the viscosity of carrageenan extracted from *Kappaphycus alvarezii* with seawater physical and chemical properties at different planting distances and depth. *AACL Bioflux, 14*(1), 328–336.

Nasir, A., Lukman, M., Tuwo, A., Hatta, M., & Tambaru, R., (2016). The use of C/N ratio in assessing the influence of land-based material in coastal water of South Sulawesi and spermonde archipelago, Indonesia. *Frontiers in Marine Science, 3*, 1–8.

Poirier, S. C., Whalen, J. K., & Michaud, A. R., (2012). Bioavailable phosphorus in fine-sized sediments transported from agricultural fields. *Soil Science Society of America Journal, 76*(1), 258–267.

Pulz, O., & Gross, W., (2004). Valuable products from biotechnology of microalgae. *Applied Microbiology and Biotechnology, 65*(6), 635–648.

Ragab, R., & Prudhomme, C., (2002). Soil and water: Climate change and water resources management in arid and semi-arid regions: Prospective and challenges for the 21st century. *Biosystems Engineering, 81*(1), 3–34.

Rajauria, G., (2015). Seaweeds: A sustainable feed source for livestock and aquaculture. In: Tiwari, B. K., & Troy, D. J., (eds.), *Seaweed Sustainability* (pp. 389–420). Academic Press, Massachusetts.

Rijsberman, F. R., (2006). Water scarcity: Fact or fiction? *Agricultural Water Management, 80*(1), 5–22.

Rima, R., Yunus, B., Umar, M. T., & Tuwo, A., (2016). Performa rumput laut *Kappaphycus alvarezii* pada habitat berbeda di perairan Kecamatan Arungkeke, Kabupaten Jeneponto. *Jurnal Rumput Laut Indonesia, 1*(1), 17–26.

Sari, A. P., Ahmad, A., Usman, H., & Tuwo, A., (2017). Produksi bioetanol dari selulosa alga merah dengan sistem fermentasi simultan menggunakan bakteri *Clostridium acetobutylicum*. *Jurnal Rumput Laut Indonesia, 2*(2), 58–62.

Scalenghe, R., Edwards, A. C., & Barberis, E., (2007). Phosphorus loss in overfertilized soils: The selective partitioning and redistribution between particle size separates. *European Journal of Agronomy, 27*(1), 72–80.

Shah, A., (2009). *Sustainable Development Introduction*. https://www.globalissues.org/article/408/sustainable-development-introduction (accessed on 13 October 2021).

Shenggen, F., & Polman, P., (2013). An ambitious development goal: Ending hunger and undernutrition by 2025. In: Marble, A., & Fritschel, H., (eds.), *Global Food Policy Report* (pp. 15–28). International Food Policy Research Institute, Washington.

Suwoyo, H. S., Tuwo, A., Haryati, & Anshary, H., (2020a). Applying organic fertilizer from solid waste of super-intensive shrimp pond on production of milkfish fingerlings *Chanos chanos* (Forsskal, 1775). *International Journal on Advanced Science, Engineering and Information Technology, 10*(2), 766–774.

Suwoyo, H. S., Tuwo, A., Haryati, H., & Anshary, H., (2019). Potential, characteristics and utilization of shrimp pond solid waste as organic fertilizer. *International Journal of Environment, Agriculture and Biotechnology, 4*(2), 411–421.

Suwoyo, H. S., Tuwo, A., Haryati, H., & Syah, R., (2020b). The utilization of solid waste originating from super-intensive shrimp farms as organic fertilizers for natural feed productions. *IOP Conf. Ser.: Earth Environ. Sci., UK, 473,* 012110.

Syamsuddin, R., Tuwo, A., & Aswar, N. R., (2019). Weight gain and carrageenan content of *Kappaphycus alvarezii* (Rhodophyta, Solierisceae) polycultured with *Sargassum polycystum* (Phaeophyta, Sargassaceae). *IOP Conf. Ser.: Earth Environ. Sci., UK, 253,* 012005.

Tan, R., (2019). *Most Common Varieties of Edible Seaweed.* https://guide.michelin.com/en/article/dining-in/6-edible-delicious-varieties-of-seaweed (accessed on 13 October 2021).

Taylor, J. G., (1944). *Indonesia: Peoples and Histories.* Yale University Press, New Haven.

Tuwo, A., (2013). *Percepatan Pembangunan Pulau-Pulau Kecil Terluar Guna Meningkatkan Kesejahteraan Masyarakat Dalam Rangka Ketahanan Nasiona* (p. 151). Lemhannas RI, Jakarta.

Tuwo, A., (2018). *Marine farming urgency to improve food security and strengthen national resilience* (p. 34). Asian Agriculture and Food Forum, Jakarta.

Tuwo, A., Auliana, R., Samawi, M. F., Aprianto, R., & Tresnati, J., (2020). Feasibility study of seaweed farming *Kappaphycus alvarezii* in Sub-District North Pulau Laut and Sub-District East Pulau Laut Kota baru Regency, South Borneo, Indonesia. *IOP Conf. Ser.: Earth Environ. Sci., UK, 564*, 012026.

Tuwo, A., Yasir, I., Tresnati, J., Aprianto, R., Yanti, A., Bestari, A. D., & Nakajima, M., (2019a). A microcosm multitrophic aquaculture system. *IOP Conf. Ser.: Earth Environ. Sci., UK, 370*, 012012.

Tweeten, L., (1999). The economics of global food security. *Review of Agricultural Economics, 21*(2), 473–488.

Van, D. B. S. W. K., Stuiver, M., Veenstra, F. A., Bikker, P., Contreras, A. M. L., Palstra, A. P., Broeze, J., et al., (2013). *A Triple Preview of the Feasibility of Sustainable Offshore Seaweed Production in the North Sea* (p. 108). Wageningen UR, Wageningen.

Wadi, A., Ahmad, A., Tompo, M., Hasyim, H., Tuwo, A., Nakajima, M., & Karim, H., (2019). Production of bioethanol from seaweed, *Gracilaria verrucosa* and *Eucheuma cottonii*, by simultaneous saccharification and fermentation methods. *Journal of Physics: Conference Series, UK, 1341*(3), 032031.

Warner, K., Hamza, M., Oliver-Smith, A., Renaud, F., & Julca, A., (2010). Climate change, environmental degradation and migration. *Natural Hazards, 55*(3), 689–715.

Watanabe, F., Yabuta, Y., Bito, T., & Teng, F., (2014). Vitamin B12-containing plant food sources for vegetarians. *Nutrients, 6*(5), 1861–1873.

Whitford, W. G., (2002). *Ecology of Desert Systems.* Academic Press, Massachusetts.

Wong, K. H., & Cheung, P. C. K., (2000). Nutritional evaluation of some subtropical red and green seaweeds: Part I – proximate composition, amino acid profiles and some physicochemical properties. *Food Chemistry, 71*(4), 475–482.

INDEX

1

1,3,5-trihydroxybenzene, 188, 237

2

2,2-diphenyl-1-picrylhydrazyl (DPPH), 73, 114, 125, 244, 314

3

3,6-anhydro-galactose (3,6-AG), 187
3,6-anhydro-L-galactopyranose, 159
3,6-anhydro-β-galactopyranose, 233
3-O-*β*-D Glucopyranosyl-stigmasta-5,25-dien (22), 65
3T3-L1 adipocytes, 251, 252

4

4-hydroxy-4-methyl-2-pentanone, 64
4-linked-β-galactopyranose, 233
4-sulfate-O-fucose, 234

5

5-bromomahorone, 264
5-methyltetrahydrofolic acid, 157

7

7 (*E*)-9-keto-octadec-7-enoic acid, 70

α

α(1→3)-linked L-fucopyranose, 113
α-amylase, 252
α-carotene, 236
α-glucosidase, 56, 57, 70, 252
  activity, 57, 70, 252
α-L-glucuronate, 187
α-tocopherol, 73, 85, 86

β

β-1,3-glucanase, 218
β-carotene, 84, 157, 235, 236, 264
β-cryptoxanthin, 236
β-D-galactopyranose, 187, 233
β-D-glycosidic linkages, 83
β-D-mannuronic acid, 161, 187
β-dimethyl sulfonium propionate, 325
β-glucans, 124
β-hydroxyperoxyaplysin, 239
β-sitosterol, 60, 71

ι

ι-carrageenans, 160

κ

κ-carrageenans, 160

λ

λ-carrageenans, 160, 262

A

Acanthophora spicifera, 54, 64, 67, 71, 249, 261
Alaria, 30, 100, 101, 125, 155, 320
  angusta, 125
  esculenta, 30, 155, 320
  marginata, 125
Ascophyllum, 86–89, 100, 101, 110, 118, 119, 126, 155, 161, 162, 190, 193, 203, 209, 210, 221, 234, 249, 255, 256, 258, 259, 321, 352, 353
  nodosum, 86–89, 101, 110, 113, 119, 126, 127, 155, 161, 190, 193, 203, 209, 210, 211, 214, 216, 218, 221, 234, 249, 255, 256, 258, 259, 321, 352, 353
    extract, 256, 258
Asparagopsis, 58, 62, 264
  taxiformis, 58, 62, 68, 74
Aspergillus, 260
  flavus, 69, 261, 320
  fumigatus, 320
  niger, 65, 69, 320

Abiotic stresses, 189, 213, 258
Abnormal differentiation, 298
Abscisic acid, 163, 206, 213, 221, 256
  growth substances, 163
Acetate-malonate pathway, 188
Acetogenin, 186
Acetone, 66, 67, 191
  extract of *C.spongiosus* AECS, 64
Acetyl-equivalent groups, 188
Acid, 65, 102, 103, 123, 128, 156, 157, 163, 186, 191, 207, 213, 216, 235, 236, 238, 321, 355
  extractable, 131
Active
  compound, 119, 263
  organic substances, 190
Actual condition problems, 18–20, 338
  energy scarcity, 344
  environmental degradation, 342
  high population growth, 338
  land erosion, 342
  limited
    agricultural lands, 339
    domesticated species, 18
    food resources, 338
  low carrying capacity production (brackish water ponds), 19
  low product quality, 20
  low technology input, 19
  water scarcity, 340
Adipogenic protein expression, 252
Aerators, 20
Aerobic
  fermentation temperature, 220
  metabolic pathways, 113
Aeromonas hydrophila, 166, 315
Aerosol-induced inflammatory responses, 300
Affinity chromatography, 84
Affordability, 336
Aflatoxin production, 261
Agar, 6, 8, 30, 82, 98, 120, 159, 170, 185, 187, 195, 207, 209, 233, 260, 292, 311–313, 324, 326, 349, 351, 355
Agaropectin, 159, 187
Agarophyte, 30
Agarose, 159, 187, 246, 324
Agricultural, 2, 26, 27, 32, 63, 100, 154, 171, 183, 188–190, 191, 193–195, 201–209, 211, 213, 215, 220, 229, 230, 255–257, 260, 265, 266, 312, 321, 336, 338–343, 347, 348, 354
  applications, 255, 266
    seaweeds against phytopathogens, 259
    seaweeds biostimulant effects, 255
  lepidopterous pests, 63
Agrobacterium tumefaciens, 215, 260
Ahnfeltia, 160
Ahnfeltiopsis, 324
Air fresheners, 314
Alanine, 203, 207
Alariaceae, 237
Alcoholic, 65, 167
  fermentation, 356
Aldehydes, 65
Alexandrium tamarense, 262
Algae, 52, 56, 118, 133, 221, 231
  bio-compounds, 265
  biomass, 257
  cultivation, 31
  polysaccharides, 166, 260
Alginate, 10, 29, 30, 82, 83, 86–91, 97–109, 120, 122, 126, 128, 132, 161, 162, 185–188, 190, 195, 203, 205, 207, 213, 232, 246, 256, 257, 261, 292, 311, 314, 320, 322, 324, 349, 355
  applications, 104
    biomedical, 104
  chemical structure, 102
    cosmetic, 108
  extraction and isolation methods, 100
    food, 107
  general overview, 99
  hydrogels, 104–107
  inherent biocompatibility, 104
  market, 101, 104
  materials, 106
  matrix, 100, 106
  solutions, 100
  wound dressings, 105, 106
Alginic acid, 60, 83, 100–105, 108, 109, 188, 233, 300, 303, 312, 314, 355
Alginophyte, 30, 170
Alkaline solvents, 191
Alkaloids, 82, 165, 216, 220, 232, 233, 315

Allolobophora caliginosa, 58
Alloxan-induced diabetic rats, 73
Alpha-L-fucosidase, 111
Alternaria radicina, 218
Amino acid, 156, 158, 163, 166, 186, 190, 203, 206, 207, 230, 232, 235, 256, 315, 320, 355
composition, 156, 235
Aminotransferases activity, 72
Amoxycillin, 65
Amphotericin B, 64, 67
Ampicillin, 63–65
Anatheca, 160
Angiogenesis, 246, 249
potential inhibition, 125
Animal feed, 86, 153, 154, 162, 183, 204, 214, 336, 345, 349, 351–355, 359–362
Annual fish consumption per capita, 24
Anopheles
mosquitoes, 253
stephensi, 253, 254
sundaicus, 254
Anoxic hazard, 24
Anteraxanthin, 157
Anthelminthic potentials, 61
Anthelmintic, 53, 186, 312
Anthropogenic
disturbances, 10
GHG emissions, 21
pressure, 1, 7, 8, 11
stresses, 8
Anti-aging, 166, 320
Antialgal, 58, 68, 231, 263
Anti-allergic properties, 188
Antiangiogenic activity, 249
Anti-arthritic, 86
Antibacterial, 53, 58, 66–69, 75, 88, 99, 117, 126, 166, 209, 214, 216, 229, 230, 238–240, 260, 265, 312
activity, 66–69, 75, 99, 216, 238, 239, 260
Anti-biofilm activity, 215
Antibiotic, 90, 259, 314
resistance, 216, 239
Anticancer, 59, 86, 89, 93, 99, 114, 124, 125, 165, 186, 230, 236, 238, 246, 249, 250, 265, 315
activity, 59, 114, 246, 249, 250
properties, 89, 165
Anticandidal activity, 64
Anticoagulant, 87, 89, 98, 99, 117, 119, 121, 132, 161, 188, 312
Antidiabetic, 86, 188, 214, 253
activity, 252
Anti-diarrhea, 314
Antifouling activities, 75
Antifungal, 53, 54, 58, 65, 67–69, 86, 186, 209, 215–218, 229, 230, 238, 239, 261, 265, 312, 320
activity, 54, 58, 65, 67–69, 216, 239, 261, 320
Antihelminthics, 314
Antihepatocarcinoma effects, 61
Anti-hepatoma cell, 249
Anti-herbivory substances, 219
Antihypertensive, 9, 187
Anti-hypocholesterolemic effects, 187
Anti-inflammatory, 52, 54, 57, 71–73, 75, 86, 87, 98, 99, 115–118, 124, 127, 129, 130, 153, 164, 165, 186, 188, 214, 236, 238, 293, 294, 299, 301–303, 312, 314, 315
activity, 52, 54, 57, 71, 73, 75, 115, 116, 124
agents, 115, 165
compounds, 303
effects, 72, 165
properties, 117, 186, 299, 301
Anti-leishmanial, 229, 254
activity, 254
Antimalarial, 55, 63, 71, 86, 229, 230, 253
activities, 71
capacity, 63
Anti-metabolic syndrome activity, 325
Antimicrobial, 47, 52, 54–56, 58, 63–69, 71, 75, 86, 87, 90, 126, 153, 166, 169, 186, 188, 203, 209, 215, 216, 238, 242, 260, 264
activity, 47, 52, 54–56, 58, 63–69, 126, 166, 209, 216, 242, 264
Antinociception, 71, 75
Anti-obesity, 229, 230, 236, 251, 252, 325
properties, 252
Antioxidant, 48, 56, 62, 72–75, 86, 87, 89, 90, 98, 99, 109, 113, 114, 116–118, 121, 123, 124, 153, 164, 166, 187, 188, 203, 214, 221, 229, 238, 244, 245, 252, 257, 258, 260, 261, 293–296, 303, 304, 314, 315
activity, 62, 73, 114, 121, 187, 244, 245, 257, 314, 315

potential, 124, 244
properties, 48, 89, 90, 123, 229, 314
responsive element (ARE), 295, 304
Anti-pathogen solutions, 201
Anti-phytopathogen activities, 230, 259
Anti-plasmodial activity, 253
Anti-proliferation
activities, 60
cancer cells, 314
effect, 165, 249
Antiprotozoal, 86, 186, 312
Antipyretic activities, 315
Antithrombotic, 117, 188
properties, 117
Antitumor, 58, 59, 62, 98, 99, 114, 121, 125, 129, 153, 188, 214, 246, 250
activity, 58, 59, 62, 121, 125
properties, 153
Antiviral, 52–54, 58, 64, 68, 87, 98, 116, 126, 163, 164, 186, 188, 209, 229, 230, 238, 242–244, 260, 265, 312
activity, 243
Apoptosis, 114, 125, 129, 246, 249, 289, 293, 295
activity, 249
cellular death, 125
Apricot, 216
Aquaculture, 2, 16–18, 21–25, 28, 32, 34, 81, 90, 101, 162, 170, 171, 262, 311, 323, 325, 341, 345, 351, 354
business, 34
development, 21
emissions, 22
paradigm, 21
ponds, 34
production, 16, 17
resources, 22, 23
sector, 22, 25, 171
species, 18, 22
Aquatic
animals, 32–34, 90, 162
ecosystems, 183, 184, 231, 262, 341, 344
plant, 16, 17, 353
farming, 154
production, 16
Aqueous
cetylpyridinium chloride, 131
extract processed (AEP), 193, 195
Arabidopsis, 262
Arachidonic acid, 156, 157, 235, 294
Arame, 30
Arsenic, 84
Arthritis, 164, 312
Artificial feed, 20, 23
Aryl
hydrocarbon receptor (Ahr), 289–291, 293, 297, 303, 304
activation, 290, 303
inhibitors, 297, 303
ether linkage, 84
Ascophyllan, 97, 99, 127–130
treated U937 cells, 129
Ascorbic acid, 73, 86, 259
Aspartate, 207
Astaxanthin, 165
Asthma, 298, 348
Atmospheric carbon dioxide, 343
Atopic dermatitis (AD), 188, 289, 297, 298, 320
Autophagy, 293, 294
Autotroph, 19
Auxin, 163, 190, 203, 206, 213, 221, 256, 321
activities, 256
compounds, 256
Avrainvillea longicaulis, 244
Azoxymethane (AOM), 55, 72, 73, 75

B

Bacillus
cereus, 65, 166
subtilis, 65, 66, 315
Botrytis cinerea, 217, 218, 260, 261
Bacteria
apoptosis induction, 117
pathogens, 69, 215, 216
Baculovirus nuclear polyhedrosis virus, 63
Badderlocks, 30
Bangiomorpha pubescens, 184
Bangiophyceae, 3, 6, 9, 11, 184
Barium (Ba), 100
Battle desertification, 325
Benthic algae, 30
Benzoic acid, 213
Beta-amyloid plaques stress, 116
Betaines, 190, 203, 206, 256, 259

Bilirubin, 72
Binding-drying seaweed, 26
Bioaccumulation, 31, 195, 322
Bioactive, 86, 91, 98, 99, 109, 110, 112–115, 117, 118, 129, 132, 203, 217
  compound, 9, 10, 47, 48, 52, 65, 81, 93, 98, 99, 111, 113, 118, 166, 172, 186, 194, 195, 206, 214, 215, 218, 230, 232, 238, 245, 259, 263–265, 292, 353, 354
  constituents, 64
  metabolites, 62, 229–232, 266, 292
  screening equipment, 109
  substances, 231
  terpenoid derivates, 264
Bio-alcohols, 167
Bioavailability, 156
Biochar, 90
  production, 322
Biochemical, 48, 83, 91, 92, 98, 163, 171, 207, 231, 246, 256, 293, 326, 353, 354
Biocompatibility, 86, 89, 104–106
Bio-crude, 90
Biodegradability, 86, 89, 106, 232
Bioenergy, 171, 339, 344
Bioengineering, 90, 92
Biofertilizers, 162, 189, 206, 211, 259
Biofuel, 88, 92, 167, 322, 326, 344, 355–357
  production, 357
Biofunctional characteristics, 158
Biological
  activities, 86, 91, 98, 186, 188, 214, 218, 229, 230, 233, 238, 259, 265, 266
  compounds, 48, 244
  molecules, 232
  properties, 99, 103, 110, 120, 230, 238, 265
Biomass, 29, 31, 32, 84, 90, 98, 100, 110, 118, 121, 154, 161, 162, 167–169, 192, 193, 255, 257–259, 325, 345, 354, 356, 357
Biomaterials, 105
Biomedical, 107
  industries, 99, 104, 105
  products, 115
Biomineralization, 322
Biomolecules, 97, 132, 169, 206, 208, 232, 233, 254, 258, 265
Bionanotechnology, 90
Biopolymers, 89–91
Bioprospecting, 81, 94
Bioremediation, 31, 322, 323, 345
Bioresource, 2, 15, 153, 155, 213
Biosorbent, 168
Biosorption, 168, 322–324
Biostimulants, 163, 172, 183, 188, 189, 192, 194, 195, 206, 211, 213, 255–257
Biostimulating
  activities, 188, 213
  effect, 209, 256
Biostimulators, 212
Biostimulatory
  effect, 258
  potential, 213
Biotechnological, 1, 10, 82, 86, 100, 109, 110, 154, 159, 229, 230, 265, 266, 311, 312, 324–326
  industries, 312
Biotransformation, 322
Bis (2-ethylhexyl) phthalate, 65
Bivalves, 21, 22, 28, 32, 33, 351
Bladderwrack, 30, 155
Blidingia minima, 320
Blood lipid-lowering, 188
Bloom-forming dinoflagellates species, 263
Blue
  belt, 171
  carbon system, 3
Body homeostasis, 164
Bone regeneration, 106
Botryocladia, 321
  occidentalis, 251, 254
Brackish water pond, 16, 18–20, 33
  production, 19
Breast carcinoma cell line, 62
Breeding, 22, 37, 39, 40
Broad-spectrum antibacterial activity, 67
Brominated cyclopentenones, 264
Brown algae, 3, 18, 30, 47, 48, 50–52, 56, 60, 61, 63, 81, 83–87, 90, 92–94, 98–104, 118, 119, 156, 157, 161, 163–168, 184, 232, 236, 237, 249, 256, 263, 312–314, 326, 349
  bio-constituents, 91
Brown seaweeds, 6, 65, 72, 75, 81–84, 89, 90, 92, 93, 97, 109, 110, 114, 126, 132, 133, 157, 161, 167, 185, 186, 190, 194, 203, 205, 207, 209, 215, 235, 239, 261, 314, 352, 355

Bryopsidophyceae, 47, 313
Bryopsis sp., 165, 314
Business
institutions, 36–39, 358–361
management capacities, 26

C

Candida, 53, 63–65, 239, 242
albicans, 65–69, 239, 242
ATCC 10231, 239
albicans
glabrata, 239
krusei, 53, 63, 242
Caulerpa, 8–11, 30, 51, 52, 54, 57, 219, 220, 245, 251, 252, 254, 313
lentillifera, 30
okamurae, 252
production, 10
racemosa, 52, 54, 57, 67, 70, 251, 254
scalpelliformis, 219
Chondrus, 28, 156, 160, 233, 236, 323, 350
crispus, 28, 30, 31, 160, 236, 323, 350
Clavibacter michiganensis, 215, 216
subsp. *sepedonicus*, 216
subsp. *tessellarius*, 216
Codium
arabicum, 53, 64
elongatum, 263
tomentosum, 52, 62, 68, 73, 74
Colpomenia sinuosa, 53, 56, 61, 70, 73, 74
Corallina
mediterranea, 57, 68
officinalis, 53, 58, 61, 63, 69, 73
pilulifera, 262
Cystoseira, 51, 54–56, 64, 66, 67, 87, 100, 119, 122, 215, 239, 251, 254, 260, 264, 314
barbata, 54, 66, 87, 119, 122, 251, 254
compressa, 55, 67, 87, 314
crinita, 314
humilis, 215
myrica, 54–56, 64
myriophylloides, 215, 260
nodicaulis, 239
sedoides, 314
tamariscifolia, 264
trinodis, 51, 55, 66
usneoides, 239
Cystosira trinodis, 60
Cabbage roots, 204
Cadmium, 84, 189, 324
Calcium (Ca), 84, 86, 89, 90, 100–103, 105–108, 128, 158, 185, 188, 193, 207, 211, 212, 214, 232, 261, 349–351, 355
Callophyllis spp., 30
Cancer, 60, 62, 86, 91, 114, 125, 129, 130, 164, 165, 169, 244, 246, 249, 250, 290, 300, 312, 314, 315
Candidatus liberibacter spp., 220
Capillary activity, 163
Capsicum annuum, 193
Caraginophyte, 30
Carbohydrates, 32, 60, 97, 112, 132, 190, 206, 233, 315
Carbon dioxide (CO_2), 21–24, 33, 191, 322, 345
concentration, 184
emissions, 21, 23
Carboxymethylation, 89
Carcinoma cell lines, 58
Cardiovascular, 164, 289
Carmalols, 84, 237
Carotenoids, 73, 84, 99, 157, 164, 165, 187, 206, 213, 235, 236, 249, 255, 312, 325, 355
Carrageenan, 8, 28, 30, 31, 73, 98, 120, 130, 158, 160, 167, 185–187, 195, 207, 209, 232, 233, 246, 249, 260–262, 292, 312–314, 320, 324, 326, 349, 355
Carrageenophytes, 170, 194
Catalase (CAT), 71, 244, 261
Catalysis, 90
hydrothermal gasification, 90
Catechins, 236
Catfishes, 21
Cation exchange chromatography, 112
Caulerpin, 59, 70
CCAAT/enhancer-binding protein α (C/EBP α), 251
mRNA, 252
Cell
cell interactions, 106
cycle arrest, 246, 249
differentiation, 300
migration, 249
proliferation, 125, 249, 299, 300
to-cell communication, 83
wall
composition, 184, 234

integrity, 83
matrix, 98
structural materials, 83
tissue, 104
Cellulose, 83, 98, 99, 131, 167, 185
Centrifugation, 102, 128, 130
Ceramium, 6, 52, 53, 67, 166
rubrum, 53, 67, 166
Cervix cancer cell line, 62
Chabrolidione B, 250
Chaetomorpha, 244, 245, 250
Channeled wrack, 30
Chattonella antiqua, 263
Chemical
fertilizers, 29, 189
modified electrode (CMEs), 90
structure, 97, 98, 102, 103, 105, 107, 109, 110, 112, 113, 119, 132, 160, 234
Chemokines, 298, 300
Chemotherapeutics, 90
Chemotherapy, 246
Cherry, 216
Chitin, 98
Chitinase, 218
Chitosan, 107
Chloroform, 67, 73, 84, 217, 220
extract, 69
Chlorophyceae, 47, 82, 232–234, 312, 313
Chlorophyll, 84, 184, 185, 214, 232, 235, 255, 258, 312
Chlorophyll-α, 184, 185, 214
Chlorophyll-χ, 185
Chlorophyta, 7, 9, 18, 30, 47, 51, 81, 98, 156, 184, 195, 203, 209, 230
Chloroquine
resistant (CQ-r), 253, 254
sensitive (CQ-s), 253, 254
susceptible, 254
Chlorotannins, 186
Chnoospora minima, 118
Chocolates, 313, 324
Cholesterol trimethylsilyl ether, 264
Chondria atropurpurea, 165
Chondriamide A, 165
Chromatographic study, 102
Chromista, 48
Chromium, 84, 207
Chronic
inflammatory skin disease, 298
malnutrition, 336, 337, 347
Chrysanthemum plants, 257
Cinnamic acid, 213
Cladophora, 256, 320
Cladosiphon, 109
okamuranus, 30, 87, 116
Cladosporium herbarum, 261
Cladostephus spongiosus, 53, 64
Climate
change, 1, 2, 8, 17, 22, 23, 171, 202, 262, 336, 341–343, 345, 362
impact, 1, 2
resilience, 6, 171
Coastal
area, 3, 6, 7, 9, 16–19, 26, 39, 47, 71, 188, 201, 211, 343, 345, 352
ecosystems, 97
marine ecosystems, 168
Cobalamin, 157
Cochayuyo, 30
Cochlodinium polykrikide, 263
Coffee, 35
Cold water
insoluble laminarin, 123
soluble laminarin, 123
Collagenases, 301
Colony formation inhibition, 125
Column chromatography, 112, 131
Commercial
alginate, 100
applications, 82
cultivation, 30
species, 6
Community
institutions, 36–39, 358–361
organization, 26
Complementary polymers, 107
Complex
macrothalli, 82
sulfated polysaccharide, 109
Comprehensive assessments, 91
Compsopogonophyceae, 3, 6, 11, 184
Conidia germination, 217
Conventional
applications, 86
extraction methods, 91
products, 213
Cooked puddings, 314
Co-polymerization, 88

Copper, 84, 131, 324
Coral reef, 3, 5, 8, 50, 348
Corynebacterium diphtheria, 64
Cosmetic, 36, 37, 82, 86, 108, 126, 166, 186, 195, 313, 345, 353
  applications, 118
  field, 2, 99
  formulations, 160
  industries, 86
  product, 108, 109, 153
Cosmetology, 171
Cost-effective, 92, 93, 260
  bioreactor culturing system, 92
Creamy texture, 187, 350
Crop productivity, 258
Cross-linking, 89
Crude
  alginate extract solution, 101
  extracts, 62, 63, 67, 70, 193, 293, 298
Crustacea, 3
Crustaceans, 3, 30
Cryptic species, 5
Cryptococcus neoformans, 63, 65
Cultivation, 19, 28
  facilities, 23
  trials, 221
Culture media, 313, 314
Curative phytosanitary treatments, 163
Cyamopsis tetragonoloba, 256
Cyanidiophyceae, 3, 184
Cyanidiophytina, 184
Cyanobacteria, 231, 262, 263, 321
Cyanophyceae, 312, 326
Cyclooxygenase (COX), 115, 165, 295, 299, 304
Cyclooxygenase-2 (COX-2), 115
Cyprinids, 21
Cytex, 190
Cytokinins, 163, 190, 203, 206, 213, 256, 321
Cytotoxic
  activity, 60–62, 129, 249
  assay, 62, 254

D

Dictyopteris
  delicatula, 110
  polypodioides, 53, 61, 70, 166, 169
Dictyota, 51, 52, 56, 58, 62, 68, 69, 88, 90, 164, 218, 242, 249
  bartayesiana, 242
  bartayresiana, 88, 242
  fasciola, 51, 56, 58, 62, 69, 73
  indica, 249
  menstrualis, 164
Digenea simplex, 54, 64, 71
  extracts, 64
Dysdercus, 262
  cingulatus, 219, 220, 262
Dasycladophyceae, 48, 313
De-alginated, 128
Death receptor pathways, 125
Decision-making processes, 25
Defense mechanisms, 83, 189, 203, 207, 209, 218, 232, 258
Degree of,
  polymerization, 120, 127
  sulfation, 127, 233
Delisea pulchra, 238, 314
Demonstration plot, 36–38, 40, 358–361
  mass breeding development, 37, 40
Dentifrices, 109
Deoxygenation, 325
Deoxyparguerol, 58
  7-acetate, 58
Dermis, 127, 291
Desertification, 202, 203, 325
Desiccation, 216
Dessert gels, 313, 314
Destabilization, 254
Diabetes, 107, 164, 250, 252, 314, 325
Dibenzodioxin linkage, 84, 237
Dichloracetic acid, 250
Dichotomaria obtusata, 315
Dictyosiphon foeniculaceus, 110, 114
Diester-type processes, 167
Dietary
  fiber compartment, 107
  nutrients, 354
  supplement, 90
Diforina citri, 220
Dilophus fasciola, 58, 62
Dimethylsulphopropionate, 264
Dioxinodehydroeckol, 165
Diphlorethohydroxycarmalol (DPHC), 252, 294, 299, 300, 303, 304

Discrimination, 25
Disease resistance, 90, 209, 218
Dissemination, 36–40
  laboratory scale breeding technique, 37
Diterpenes, 59, 186
Diterpenoids, 186, 264
Divalent
  cations, 100, 105
  monovalent ions, 106
Diversification, 18, 36, 37, 359, 361
D-mannitol group, 120
D-mannuronic acid, 102
DNA
  fragmentation, 129
  sequencing, 75
Docosahexaenoic acid (DHA), 157, 162, 350, 362
Docosane, 264
Domestication, 16, 18, 29–31, 36, 37, 39, 358–361
  technique, 36, 37, 39
    development, 36, 39
Doxorubicin, 60–63
Dried
  seaweed, 29, 34, 122, 204, 352
  thalli weight, 98
Drug
  delivery systems, 88, 104, 107
  development, 238
  molecules, 105
Dry weight (DW), 28, 29, 84, 118, 119, 123, 156, 157, 186, 214, 233–235, 244, 245, 355
Durvillaea, 190
  antarctica, 30, 157, 161, 260
  potatorum, 89, 110, 161, 168, 210, 258

E

Ecklonia, 30, 86–88, 100, 109, 119, 122, 161, 162, 165, 168, 190, 210, 216, 234, 237, 244, 249–251, 256, 263, 294, 302, 303
  arborea, 244
  cava, 30, 86–88, 115, 165, 234, 237, 263, 294, 302, 303
  kurome, 86
  maxima, 87, 119, 122, 161, 168, 190, 210, 249, 256
  radiata, 109, 119, 122
  stolonifera, 250, 251, 302
  tolonifera, 165
Ellisolandia elongata, 57, 68, 74
Escherichia coli, 63, 66, 68, 69, 126, 166, 239, 315
Eucheuma, 6, 8, 17, 18, 28, 30, 154, 157, 160, 194, 217, 233, 245, 313
  denticulatum, 6, 157, 160, 217, 233, 245
E405, propane-1,2-diol alginate, 188
Echinodermata, 3
Eckols, 84, 188, 237
Ecological, 3, 48, 168, 191, 202, 219, 325
Economic
  capacity, 20
  funding, 92
  growth, 17, 26, 27, 344
Ecosystem
  degradation, 343
  destruction, 337
Ectocarpales, 30
Ectocarpus, 185
Edible seaweed, 30, 158, 172, 304, 349, 350, 358–361
Educational
  institutions, 36–39, 358–361
  research institutions, 36–39, 358–361
Effective dose (ED_{50}), 59, 60
Egyptian seaweeds, 47
Eicosapentaenoic acid (EPA), 156, 157, 162, 235, 350, 362
Eisenia, 30, 86, 100, 118, 122, 237, 238
  bicyclis, 30, 86, 122, 237, 238
Electricity supply, 20
Electrocuting, 351
Electronics, 90
Empowerment, 36, 358
Emulsifier, 107, 108, 187, 324, 350
Endophytic, 231
Endoplasmic reticulum, 185, 254, 293, 294
Endothelial dysfunction, 299
Endothelialization, 117
Endozoic, 231
Energy
  crops, 31
  scarcity, 336, 338, 362
  storage, 98
Enterococcus, 166

faecalis, 166, 315
Enteromorpha, 321
clathrata, 263, 313
compressa, 245, 249
flexuosa, 157
intestinalis, 61, 259, 263, 322
Environment, 24, 25, 32, 33, 89, 90, 93, 157, 168, 195, 230, 235, 255, 291, 326, 337, 341, 343, 345, 349, 354
conditions, 97, 186, 234, 237, 291
consequences, 232
degradation, 19, 23, 336–338, 342, 362
eutrophication, 32
friendly, 89, 90, 93, 230, 255
longevity, 232
management, 337
pollution, 92, 219, 293
problems, 340
sustainability, 21, 338, 344
toxins, 246
uses of seaweeds, 266
Enzymatic activity, 217
Enzyme-assisted extraction (EAE), 84, 89, 92
Epidemiological, 249, 289, 291, 298
Epidermis, 291, 293, 298
Epidermophyton floccosum, 239
Epinephelus species, 124
Epiphytes, 74, 110, 231
Epithelialization, 106
Equity, 16, 22, 336, 344, 346
Erlich ascites carcinoma cells, 62
Erythrocytes, 106
Esters, 65, 187
Ethanolic
extracts, 66, 72, 166
pre-extraction process, 110
Ethyl acetate, 62, 67, 73, 217
Eudesma-4(15),7-diene-5,11-diol, 250
Eukaryotic
algae, 184
hosts, 185
lineage, 5
organisms, 184, 231
Eurhodophytina, 184
European
Commission (EC), 34, 85, 202, 215, 313
Food Safety Authority (EFSA), 107, 133
Union (EU), 107, 108, 117, 189, 195, 202
Eutrophication problems, 33
Exploitation, 25, 91, 100, 153–155, 158, 167, 202, 229, 265
Extracellular
cellular matrix, 105
matrix components (ECM), 301
signal-regulated kinases (ERK), 299
tissue matrix (ETM), 167
Extraction, 47, 48, 66, 67, 81, 82, 84, 89, 91–93, 97–102, 110, 111, 113, 115, 119–123, 128, 130, 154, 168, 191–193, 205–207, 230, 242, 349, 356

F

Fucus, 30, 87, 88, 100, 110, 111, 114, 118, 119, 121, 122, 125, 126, 155, 190, 192, 203, 215, 221, 236, 239, 255, 257, 259, 260, 264, 315, 321
distichus, 110, 119, 122
evanescens, 110, 113, 114, 125
serratus, 88, 110, 113, 114, 119, 121, 122, 216, 236, 259
spiralis, 30, 110, 215, 239, 257, 260, 321
vesiculosus, 30, 87, 110, 111, 113, 114, 116, 117, 119, 155, 192, 203, 221, 236, 264, 321
Fusarium
oxysporum, 65, 68, 260
solani, 217, 260, 261
Fair-trade agreement, 26
Fasciola gigantica, 61
Fat accumulation, 252
Fatty acids, 62, 69, 70, 82, 99, 157, 162, 189, 203, 213, 217, 218, 221, 230, 234, 235, 252, 292, 315, 350
Feed
conversion ratios (FCR), 21, 354
supplements, 161, 354
Feedstock conservation, 110
Fermentation process, 167, 356
Fertilizer, 10, 19, 20, 31, 32, 82, 86, 132, 153, 154, 162, 171, 183, 188–192, 194, 195, 201–205, 209, 213, 214, 220, 221, 230, 255–257, 292, 311, 321, 341, 343, 344, 353
industries, 10
Fertirrigation solution, 193

Filamentous
shell phase, 28
turfs, 48
Financial
donors institutions, 36–38, 358–361
limitations, 19
Finfish production, 22
Fish
consuming country, 24
waste management, 21
Fisher organizations, 25
Fisheries, 16, 20–27, 348
contribution, 20
dependent households, 26
governance politics, 26
production, 23
Flavonoids, 66, 213, 218, 220, 236, 238, 245, 257, 314, 315
Flooding, 341, 343, 348
Florideophyceae, 3, 6, 7, 11, 184
Flucose-6-phosphatase, 314
Folic acid, 84, 157, 355
Foliose phase, 28
Food, 40, 93, 94, 107, 133, 194, 335–337, 339
Agriculture Organization (FAO), 8, 25, 40, 154, 155, 158, 160, 170, 171, 313, 314
Drug Administration (FDA), 93, 94, 107, 133
production requirements, 32
security, 1, 11, 21, 132, 202, 221, 335–339, 347, 349, 361, 362
Fossil energy, 344
Fourier transform infra-red (FTIR), 85, 94
Fractionation, 60, 62, 64, 103
Freshwater, 31, 32, 81, 82, 167, 184, 185, 262, 339–343, 345, 353
cyanobacteria, 262
ecosystems, 262
Fucaceae, 237
Fucales, 30, 100, 109, 113, 264
Fucoidan, 83, 86–91, 93, 94, 97–99, 109–118, 125, 126, 132, 133, 163, 186, 188, 195, 203, 213, 233, 234, 239, 242, 246, 249, 250, 253, 295, 302, 303, 312, 315, 355
applications, 113
biosynthesis, 93
chemical structure, 112
extraction-isolation methods, 110
general overview, 109
Fucoidanases, 111
Fuconazole, 64
Fucophlorethols, 237
Fucose, 60, 83, 112, 114, 128, 129, 131, 188, 234
Fucosterol, 71, 85, 244, 251, 300, 302, 303
Fucoxanthin, 84, 86, 165, 185, 232, 235, 236, 249, 295, 312, 321, 325, 326, 355
Fucoxanthinol, 86
Fuhalols, 84, 237
Functional group substitutions, 88
Fungicidal
activity, 239, 261
effect, 64, 261
in vitro methods, 217
Fungicides, 215, 255, 314
Fungistatic activity, 217
Furcellaria, 160
Future food insecurity, 33

G

Galaxaura
cylindrica, 52, 62, 69
elongata, 54, 67
Ganonema farinosum, 54, 55, 63, 67, 70, 71
Gelidium, 8, 56, 159, 170, 217, 263, 324
pulchellum, 263
pusillum, 217
species, 159
Gracilaria, 6, 8, 18, 19, 28, 30, 31, 51, 154, 156, 157, 159, 162, 168, 171, 194, 213, 217, 239, 251, 254, 256, 258–260, 263, 313, 315, 320, 323–325, 351
caudata, 251, 254
changii, 157
chouae, 323
confervoides, 217
corticata, 30, 239, 260, 320, 323
edulis, 30, 213, 256, 258, 260
lemaneiformis, 263, 323
tenuistipitata, 263
tikvahiae, 323
vermiculophylla, 323, 325
Galactose, 83, 131, 159, 160, 187, 188

Gas chromatography (GC), 64, 85, 94
  mass spectroscopic (GC-MS), 64
Gastric protective effects, 188
Gel filtration, 84, 112
  chromatography, 112
Gelatinases, 301
Gelation, 100, 105, 187
Gelidiaceae, 324
Gelidiella, 30, 218, 324
  acerosa, 30, 218
Gender, 16, 17, 22, 25, 26, 345
  equality, 17, 25
Gene
  expression, 259, 295, 300, 303
  transcriptions, 297
General freshwater fishes, 22
Genetic, 22
  diversity, 5, 10, 48, 348
  resources, 326, 348
Gentamycin, 64
Geosphere-biosphere processes, 325
Geotricum candidum, 64
Germination, 163, 189, 192, 193, 209, 212, 214, 217, 255, 256, 259, 261
  assay, 192, 193
Gibberellins, 163, 190, 206, 213, 221, 255, 321
Gigartina, 31, 160, 170, 233
  chondracanthus, 170
Global
  climate change, 8, 255
  poverty, 27
  protein consumption, 22
  scale, 2, 5, 155, 161
  warming, 8, 10, 336, 344, 348, 362
Gloiopeltis furcata, 166
Glucose, 83, 98, 112, 119, 120, 188, 252, 253, 326
Glucuronan, 261
Glucuronic, 83, 117, 128, 131, 234
  acid, 83, 117, 131, 234
    units, 83
Glutathione (GSH), 71, 73, 114, 244
  peroxidase (GPx), 73, 114, 244
Glycerol
  1,3-dioleyl, 71
  1-oleoyl-3-palmitoyl-2-galactoside, 71
  3-phosphate dehydrogenase activity, 253
  trilinoleate, 60
Glycine, 203, 207
  max, 259
Glycogen, 314
Glycoglycerolipid, 60
Glycuronogalactofucans, 234
Goemar, 190
Governance
  encouragement, 202
  institutions, 36–39, 358–361
  process, 27
Gracilariaceae, 324
Gracilariopsis
  lemaneiformis, 166
  longissima, 217
Gracillaria
  chilensis, 261
  edulis, 239
Gram-negative bacteria, 63, 64, 66–69
Gram-positive bacteria, 63, 68, 215, 315
Granulated
  floridean starch, 185
  sugar, 35
Grapestone, 30
Grateloupia elliptica, 250, 252
Greaseless creams, 109
Green
  algae, 3, 18, 30, 47, 50–52, 81, 82, 156, 165, 168, 184, 232, 261, 312, 349
    starch, 185
  brown particles, 191
  house gas (GHG), 21, 40, 348
  macroalgae, 18, 98
  products, 203
  revolution, 338
  seaweeds, 5, 6, 9, 75, 82, 157, 166, 186, 191, 194, 203, 207–209, 234, 235, 245
Groundwater, 341
  depletion, 343
Growth
  inhibition, 68, 250, 262, 263
  regulators, 189
Guluronic acid, 103
  blocks, 103
Gutweed, 30
Gymnema sylvestre, 260
Gynmogongrus, 160

H

Hydroclathrus clathratus, 53–55, 60, 65, 67, 261
Hypnea musciformis, 52, 62, 68, 73, 260
Halimeda tuna, 51, 54, 62, 69
H5N1 cells, 68
Habitat
complexity, 2, 50
condition, 2
destruction, 337
diversity, 6
Halopteris scoparia, 169
Haplotypes, 5
Harmful algal blooms (HABs), 230, 231, 262–264
Harvested, 10, 28, 32, 33, 113, 118, 119, 189, 191, 313, 357
prematurely, 20
Health
nursery grounds, 3
promotion, 164
Heavy metals, 107, 168, 257, 297, 311, 324, 326
Hectares, 28, 340
Hemorrhage, 117
Heparanase activity, 127
Hepatitis A virus (HAV), 64, 75
Hepatocellular carcinoma apoptosis, 249
Hepatoprotective
activity, 315
effects, 236
HepG-2 cell, 62
in vitro, 61
lines, 60, 63
Herbicides, 31, 32
Hereditary genome mutations, 246
Herpes simplex virus, 64, 116
virus 1 (HSV1), 64, 69, 116, 163
virus type-1 (HSV-Type 1), 64, 116, 242
Heterogenic, 236
Heterokontophyta, 185
Heteropolymers, 188
Heterosigma akashiwo, 262, 263
Heterotroph, 19, 32
Hevea brasiliensis, 261
Hexadecanoic acid, 264
Hexane, 62, 69, 70, 84
High cytotoxicity, 60
High population growth, 336, 362
High-density lipoprotein (HDL), 70
cholesterol, 70
Higher
mechanical resistance., 105
radical scavenging property, 85
High-level organisms, 164
High-performance liquid chromatography (HPLC), 66, 85, 94
High-quality agar, 6
High-tech biotechnology, 100
High-temperature-tolerance (HTT), 325, 326
Himanthalia elongata, 30, 111, 155, 158
Hiromi, 30
Histidine, 156
Hiziki, 30
Hizikia
fusiforme, 88, 250, 252
extracts, 252
fusiformis, 86, 115
Holistic-integral approach, 16, 17, 39, 40, 335, 337, 361
Homogenizers, 108
Hormones, 163, 194, 203, 206, 246, 321
Hormophysa cuneiformis, 53, 67, 261
Horsepower, 34
Horticulture industries, 203
HT-29 colon cancer cells, 86
Human
consumption, 1, 10, 117, 132, 153–155, 186, 194
cytomegalovirus (HCMV), 163
dermal fibroblasts (HDF), 300–302
digestive system, 108
ecological environment, 18
enterocyte-like HT-29-Luc cells, 126
environmental welfare, 311
epidermis keratinocytes, 127
hepatocellular carcinoma, 59, 60, 165
cell line, 59, 165
immunodeficiency virus (HIV), 126, 163, 164, 214, 242, 244
myeloid leukemia, 129
protein consumption, 16
rights-based approach (HRBA), 25
skin fibroblast cells, 124
Humanitarian crisis, 339

Humid microenvironment, 105
Hurghada, 50, 51, 53
Hyaluronic acid application, 108
Hydration, 98
Hydro gelling, 127
Hydrocarbon, 65
  fractions, 63
Hydrochloric acid, 101, 102, 128, 130
Hydrocolloid, 82, 86, 153, 158, 292, 313, 324, 326, 350, 356
  gums, 324
  industries, 82
Hydrodistillation technique, 65
Hydrogels, 105–107, 127
Hydrogen peroxide (H_2O_2), 315
Hydrophilic polymers, 104
Hydroponic systems, 193
Hydropuntia edulis, 213
Hydrotherapy, 320
Hydrothermal
  extraction, 89
  liquefaction, 90
Hyphal specific genes, 64
Hypnea, 30, 52, 62, 160, 260
Hypolipidemic, 53, 58, 69, 75
  activity, 58, 69, 75

I

Ishige okamurae, 250, 252, 294, 299
Ice creams, 108, 160
Immigration, 5
Immune system, 114, 115, 124, 129, 218, 246, 352
Immunomodulators, 218
Immunomodulatory, 86, 87, 89, 99, 115, 124, 129, 218, 221, 246
Immunostimulant, 115
Immunostimulatory, 123
  molecule, 130
Immunosuppressive activity, 115
In vitro cytotoxicity, 60
In vivo anti-metastatic activity, 129
Indian Major Carps, 22
Indiffusible, 128
Indole
  3-acetic acid, 326
  3-acidic acid (IAA), 321
  acetic acid, 256
Indomethacin, 71
Indonesian
  farmer paddy fields, 340
  seaweed export, 20
Industrial
  application, 119
  facilities, 290
  raw materials, 213, 292
  waste, 341
Inefficiency, 341
Inflammation, 71, 124, 130, 164, 297, 298, 301, 302, 304, 314
  mediators, 165
Infrastructure, 19, 26, 27, 34, 36, 39, 40, 342, 344, 358–361
  development, 39, 40
  management, 36, 358
Inhibition
  concentration (IC50), 59–63, 70, 125, 238, 239, 245, 246, 249–252, 254
  percentage, 62
  zone (IZ), 63–69
Insecticidal, 63, 219, 220, 260, 262
Insulin sensitivity, 253
Integrated
  Management of Coastal Zones of the Kingdom of Morocco (ICZM), 171
  multitrophic aquaculture (IMTA), 32
Intensive aquaculture activities, 23
Intercellular
  matrix, 99
  regions, 104
Interdisciplinary, 94
Interleukin 6 (IL-6), 71, 130, 165, 298–300, 302, 321
Interleukin-1β, 165
Intermediate technology, 19, 20
Internalization, 242
Intestinal flora, 187
Intrinsic factors, 92
Invasion, 29, 246, 249, 253
  species, 10
Invertebrates, 3, 109, 219
In-vitro
  antifungal activities, 67
  hypolipidemic effect, 70
Iodine, 29, 84, 154, 232, 312, 326, 347, 349, 350, 352

Ion-exchange
chromatography, 84
procedure, 101
reactions, 101
Iridaea, 29, 31, 160
Irish moss, 30, 350
Iron, 84, 157, 158, 169, 207, 232, 347, 349
Irradiation, 120, 124
Isoflavones, 213
Isoparguerene, 58
Isoparguerol, 58
16-acetate, 58
7,16-diacetate 3, 58
7,16-diacetate, 58
derivatives, 58
Isopentenyl adenine (iP), 163
Isoprenoids, 186
Isopropyl alcohol, 66, 67
Issues, 22, 24–26, 311, 347
equity and gender, 25
global warming climate change, 348
hunger, 347
livelihood, 24
poverty, 26
sustainable development, 22
IκB kinase complex (IKK), 300

J

Jania rubens, 51, 53, 56–58, 61, 62, 68, 73, 74
Jun amino-terminal kinases (JNKs), 299

K

Kappaphycus, 5, 8, 17, 18, 28, 154, 160, 194, 210, 213, 233, 245, 249, 256, 258, 313, 320, 350
alvarezii, 5, 17, 154, 160, 194, 210, 213, 233, 245, 249, 256, 258, 313, 320
Klebsiella, 67, 166, 239, 315
pneumoniae, 67–69, 166, 315
Kahalalide F, 165
Kappa, 160, 187
Kelp, 30, 210
Keratinocytes, 293, 294, 297–300, 302–304
Kerosene, 35
Ketoconazole, 65
Ketones, 65
Kingdom *plantae*, 48
Kjellmanianone, 60
Kombu, 30, 185, 350

L

Laminaria, 18, 29, 86–89, 100, 101, 110, 114, 118, 119, 121, 122, 125–127, 154, 155, 158, 161, 162, 165, 185, 190, 215, 236, 237, 250, 252, 255, 259, 261, 313, 322, 323, 352, 356
cloustoni, 121
digitata, 30, 87, 110, 114, 119, 121, 127, 155, 161, 215, 236, 259, 261, 352
hyperborea, 87, 88, 121, 126, 155, 161, 322
hyperborean, 87, 88, 322
japonica, 18, 28, 29, 86, 88, 89, 110, 114, 122, 154, 158, 161, 165, 185, 237, 250, 252, 313, 321
E. cottonii, 28
saccharina, 121, 122, 161
Laurencia, 51–53, 57, 62, 70, 220, 239, 242, 250, 251, 254, 263–265, 315
dendroidea, 251, 254
iliformis, 239
johnstonii, 220
obtusa, 242
obtuse, 250
okamurai, 239
papillosa, 52, 53, 62, 64, 315
pinnatifida, 57, 70, 263
translucida, 264
Laboratory scale
breeding technique, 37, 40
manual breeding techniques, 37
Labor-intensive, 29
Lactobacillus, 126
population, 126
Lactosoils pH, 211
Lactuca sativa, 193, 214, 258
Lambda-carrageenan, 187
Laminaran, 83, 91, 123
Laminaran sulfate, 127
Laminariaceae, 118
Laminariales, 30, 100, 109
Laminarin, 60, 83, 87, 89, 90, 97–99, 118–127, 131–133, 163, 185, 189, 203, 209, 213, 233, 257, 261, 312, 322, 355
applications, 123
chemical structure, 119

extraction methods, 121
fermentation, 125
general overview, 118
Land
food production, 342, 343
management, 31, 32
production, 342
use change (LUC), 21
L-ascorbic acid, 157
Laurepoxyene, 239
Laverbread, 30
Lead, 17, 26, 32, 48, 84, 91, 110, 114, 115, 120, 124, 131, 259, 297, 299, 324, 326, 336, 337, 347, 348
Leishmania
amazonensis, 254
parasites, 254
Leishmaniasis, 250
Lepidium sativum, 256
Less polar solvents, 84
Lessonia, 87, 100, 161, 168, 194, 258, 259, 261
flavicans, 168, 258
nigrescens, 87, 161, 194, 258, 259
trabeculata, 161, 194, 261
Leucine, 156
Leukotrienes, 165
L-guluronic acid, 102, 103, 161
Liagora farinosa, 54, 55, 63
Light, 28, 82, 185, 231, 236, 237, 298
Lignans, 213
Limited
domesticated species, 16, 18
financial capacity, 20
food resources, 336, 338, 362
Limu kala, 30
Lipid
accumulation, 251–253
peroxidation, 73, 74, 124, 293, 294
Listeria monocytogenes, 126
Lithophytic, 231
Lithothamnion, 210–212
corallioides, 211
Livelihood, 16, 22, 336, 345, 346, 362
Livestock
diets, 162
production, 22
Lobophora, 5
Loliolide, 60
Low cost green seaweed dryers, 38, 40
Low molecular weight, 118
fucoidan (LMWF), 89, 115, 302
Low product quality, 18, 20
Low technology input, 16, 18
Low-density lipoprotein (LDL), 70
cholesterol, 70
Low-end service sectors, 27
Low-income food-deficit countries (LIFDC), 24
Low-productivity jobs, 27
Lung surface metastatic nodules, 129
Lutein, 157, 235
Lycopene, 236
Lymnaea natalensis snails, 61
L-α-terpinol, 250

M

Macroalgae, 1–3, 8, 18, 30, 47, 48, 50, 51, 62, 67–69, 82, 83, 86, 90, 92, 97, 111, 113–116, 118, 120, 121, 123, 124, 133, 153, 155–158, 161–163, 166–168, 171, 184, 186, 195, 202, 203, 216–218, 221, 229–231, 233, 235, 237, 238, 242, 246, 262, 265, 351, 356
bioactive molecules, 166
biodiversity, 203
diet, 162
domestication, 18
species, 70, 155, 156
Macrocystis, 29, 87, 88, 90, 100, 101, 109, 155, 161, 162, 185, 210, 257, 320
pyrifera, 87, 88, 90, 101, 109, 155, 161, 210, 257, 320
Macronutrients, 256, 257
Macroorganisms, 231, 264
Macrophages, 124, 254, 295, 299
Macrophomina faseolina, 217
Macrophytic marine algae family, 311
Macroscopic, 231
Magnesium (Mg), 84, 100, 158, 193, 212–214, 232, 349, 350
Magnetic
resonance imaging (MRI), 169
storage media, 90
Mahorone, 264
Malnourished people, 339

Malnutrition, 338, 339, 347, 348
Malondialdehyde MDA, 71–73, 86, 125, 250
Maluku, 9
Malus
domestica, 261
hupehensis, 258
Manganese, 84, 203, 207, 324
Mangrove, 3
ecosystem, 51
Mannitol, 29, 60, 322
Mannose, 83, 112, 131, 188
Mannuronic, 83, 103, 128, 163
Marginalization, 25
Marine
biodiversity, 21
cereals, 22
climates, 2
cosmetic products, 320
derived polysaccharides (MDPs), 90
ecosystems, 230
environment, 1, 2, 7, 82, 153, 231, 232, 242, 246, 264
organism, 2, 3, 22, 48, 98, 153, 186, 265, 312, 326
resources, 17, 91, 230
sediment, 211
therapeutic plants, 326
Marmalade, 313
Mass breeding technique dissemination, 37, 40
Mastocarpus stellatus, 30
Matrilysins, 301
Matrix metalloproteinases (MMPs), 289, 291, 301–303
Maxicrop, 190, 205, 210
Maximum resistance, 189
MCF-7
breast cancer cells, 165
cells, 62
Meatballs, 313
Mechanical strength, 90, 99
Mechanization, 29
Mediterranean Sea coast of Egypt, 75
Melanoma cancer cell lines, 59
Membrane
filtration techniques, 84
functionality, 74
type (MT), 16, 22, 29, 301, 348
Mercury, 84
Meristotheca, 160
Meroditerpenoid, 264
Metabolic
diversity, 3
enhancers, 188, 194, 213
pathways, 93
wastes, 32
Metastasis, 246, 249
Methane (CH_4), 21, 167
Methanogenic, 90
Methanol, 62, 67–69, 73, 84, 191, 215, 239, 263, 264
Methicillin-resistant *S. aureus* (MRSA), 63, 238
Methionine, 156
Methoxy, 59, 65, 187
Metric tons, 16, 22
Microalgae, 30, 90, 168, 262–264, 311, 323
Microbial
activities, 21
cell, 69
Microcystis aeruginosa, 262–264
Microenvironment, 105
Micronutrients, 208, 212, 258
Microorganism, 326
Microwave, 122
assisted extraction (MAE), 84, 89, 90, 92, 94, 111, 133, 191
Millennium development goals (MDGs), 17, 40, 344, 346
Minerals, 82, 84, 91, 99, 155, 157, 158, 161, 186, 193, 195, 203, 206, 207, 213, 214, 218, 232, 292, 311, 312, 315, 321, 349, 352–355
Minimum inhibitory concentration (MIC), 64, 67, 68, 239, 266
Miscellaneous compounds, 65
Mitigation, 31, 124, 206, 262, 335, 337, 361, 362
Mitochondrial
function preservation, 117
swelling, 293
Mitochondrion, 254
Mitogen-activated protein kinases (MAPK), 299, 303
Moisture-retaining surface films, 108
Molecular
approaches, 5
characterization, 109

docking, 60
phylogeographic approaches, 5
weight, 83, 84, 89, 91, 93, 98, 112–115, 117, 120, 121, 123, 124, 162, 233, 234, 237, 244
Molluscicidal, 61
Mollusks, 3, 33, 111
Monilinia laxa, 217
Monocultures, 29
Monocyte chemotactic and activating factor (MCAF), 298
Monomeric composition, 103
Monosaccharide, 83, 93, 114, 188, 233
Monounsaturated, 234, 235, 350
Moroccan algal resources, 172
Morphological, 3, 5, 47, 48, 52, 254, 312
Mosquitocidal effect, 254
Mosquitoes, 253, 349
Mozuku, 30
Mucopolysaccharide, 293, 320
Mucor indicus, 320
Multicellular, 1, 3, 82, 185, 231, 349
Multifunctional properties, 353
Multitrophic
aquaculture, 23
cultivation, 16, 19, 23, 24, 26, 27, 32–34, 37, 38, 40
development, 16, 37, 38
system, 19
photosynthetic organisms, 183
Murine macrophage, 115
Mutation, 5
Mycelium growth, 217
Mycosporine-like amino acid (MAAs), 82, 166, 292, 304, 320
Myriophylloides, 215
Myristic acid, 71

N

Nanoparticles (NPs), 88, 90, 92, 169, 172, 239, 251, 253, 254, 266
Nanotechnology, 169
National
Agency for Development of Aquaculture (NADA), 171
economic development, 20
Natural
algal resources, 171
antioxidants, 48, 164
anti-viral substances, 260
killer (NK), 115, 129
macroalgal biopolymers, 88
metabolites, 230
phytohormones, 256
products, 30, 48, 114, 166, 186, 215, 239, 244, 259, 312, 314
resources, 195, 230, 344
Nebraskensis, 215
Necrotic lesion, 261
Neoxanthin, 157
Nereocystis, 30, 100, 155
luetkeana, 30, 155
Neurodegeneration, 116
diseases, 244
Neuroprotection, 116, 117, 236
activity, 116
Neutralization, 128, 212
Neutrophil polynuclears chemotaxy, 127
New floating cultivation system, 31
Newcastle sense virus (NDV), 67, 68, 75
N-hexadecanoic acid, 64
Nickel, 84, 207, 324
Nitric oxide (NO), 72, 73, 115, 116, 129, 165, 290, 315
Nitrogen (N), 21, 23, 29, 33, 52–56, 58, 122, 129, 164, 165, 206, 210, 212, 253, 257, 311, 323, 354
fertilizer, 29
Nitrous oxide (N_2O), 21
Nizamuddinia zanardinii, 89
Non-combustion sources, 290
Non-degradable compound, 107
Non-discriminatory, 25
Non-essential amino acids, 73, 84
Non-flowering plant-like organisms, 231
Non-food
applications, 153
products, 17
Non-polar compounds, 84
Non-specific
inflammations, 127
plant cultivars, 205
Nontoxic
chemotherapeutic drugs, 246
residues, 232
Non-woven dressing, 105
Nori, 9, 30, 156, 185, 313, 350

Normal algal diversity, 8
Nothogenia fastigiata, 163
Novantron antitumor drug, 62
Novel
agrochemicals, 220
bioactive secondary metabolites, 293
N-palmitoyl-2-amino 1,3,4,5-tetyrahy-droxyoctadecane, 164
Nuclear
factor erythroid 2-related factor 2 (Nrf2), 295, 297
factor kappa B (NF-κB), 73, 75, 295, 298, 300, 302, 303
transcription, 300
magnetic resonance (NMR), 85, 112
translocation, 297
Nutraceutical, 10, 81, 110, 113, 117, 206
Nutrient
reclamation, 31
uptake, 212, 255
Nutrition, 22, 186, 195, 202, 347, 348, 352, 355
requirements, 93

O

Osmundea pinnatifida, 57, 70
Oarweed, 30
Obesity, 251, 252
induced insulin-resistant adipocytes, 253
Ocean
acidification, 8, 325, 345
farming, 345
Ochrophyta, 7, 9, 30, 155, 184, 185, 195, 203
Octacosanoic acid, 60
Odonus niger, 218
Ointment bases, 109
Okinawa cuisine, 351
Oleic acid, 60, 71
Oligo-fucoidan, 249
Oligoglucuronans, 261
Omega-6, 157
Operational costs, 20, 35
Optimal antiadipogenesis, 252
Organ transplants, 117
Organelle, 184, 254
Organic
compounds biodegradation, 220
materials, 20
matter, 20, 23, 33, 183, 189, 202, 204, 257, 321
Organoleptic, 92, 109
Oryza sativa, 213, 256, 258
Osmotic adjustment, 83
Osteoporosis, 320
Outdoor
activities, 291
air pollution, 290
Overexploitation, 229
Oviposition, 220
Oxidation, 88, 107, 109, 113, 117, 119, 120, 124, 244
stress, 113, 164, 252, 258, 289, 293–295, 303, 304
Oximes, 250
Oxygen permeability, 89

P

Padina, 51, 53–55, 59, 62, 88, 90, 110, 127, 166, 219, 260, 261, 314
boergesenii, 314
boryana, 110, 115
commersonii, 110
gymnospora, 54, 62, 88, 90, 166, 260, 261
pavonica, 53, 55, 59, 166, 219
tetrastromatica, 88, 127
Palisada perforata, 52, 53, 62, 64, 69, 73
Phacelocarpus tristichus, 54, 62, 69
Polycladia myrica, 54–56, 64–68
Porphyra, 194
umbilicalis, 30, 156, 158, 185, 217, 320
yezoensis, 263, 313
Pseudomonas
aeruginosa, 64–67, 166, 238, 239, 315
putida, 63, 65
syringe, 216
Pterocladiella, 53, 59, 324
capillacea, 53, 56, 57, 59, 61, 68, 70, 74
Pyropia
tenera, 156, 185
yezoensis, 156, 185
Palmaria palmata, 30, 156, 158, 320, 323
Palmitic acid, 71
Palmitoleic acid, 71
Palythine, 320
Pantothenic acid, 84, 355

Paradoxical conditions, 338
Parguerene, 58
Parguerol
16-acetate, 58
7,16-diacetate, 58
derivatives, 58
Partial degradation, 88
Particulate matter (PM), 289–291, 293–304
exposed keratinocytes, 293, 298
induced
health complications, 291
inflammation, 298, 301
inflammatory responses, 299–301
oxidative stress, 290, 293, 295, 297
skin damages, 289–291, 293, 297, 301, 303, 304
skin dysfunctions, 303
Pathogen-associated cellular changes, 261
Pathogenesis, 261, 293, 300
fungi, 67, 219, 260
Pathogens, 30, 66, 67, 215, 217, 218, 220, 256, 260, 261, 291, 349
Pectobacterium carotovora, 260, 261
Pedal wheels, 34
Pelvetia canaliculata, 30, 119, 122
Penicillin-binding protein 2a (PBP2a), 238
Penicillium, 260
digitatum, 217
expansum, 261
Peptides, 9, 82, 99, 206, 233, 235, 292
Peritoneal macrophages, 254
Peroxidase, 217, 218, 258
Peroxisome proliferator-activated receptor gamma (PPARγ), 73, 251–253
Pesticides, 31, 32, 163, 189, 202, 215, 255, 259, 341, 343–345, 353
Petroleum ether, 67, 69, 191
Phaeophyceae, 3, 6, 7, 10, 11, 18, 47, 82, 84, 86, 91, 94, 97–99, 155, 185, 190, 203, 209, 232–234, 312, 327
Phaeophyta, 30, 50, 81, 85, 118, 230
Pharmaceutical, 2, 36, 37, 47, 81, 82, 86, 93, 98–100, 104, 107, 123, 127, 130, 132, 153, 163, 169, 171, 183, 187, 191, 209, 312, 313, 315, 324, 325
Pharmacological
activities, 230, 233, 326
potentialities, 266
properties, 237
Phaseolus vulgaris, 212
Phenol, 91, 220
derivatives, 65
2-methoxy-4-(2-propenyl), 64
Phenolic, 66, 73, 114, 118, 213, 218, 220, 230, 236–238, 244, 245, 257, 258, 261
compound, 66, 114, 118, 218, 220, 230, 236–238, 261
Phenylalanine ammonia-lyase, 218
Phlorethols, 84, 188, 237
Phlorofucofuroeckol A, 165, 238, 250, 251
Phloroglucinol, 84, 165, 188, 237
Phlorotannins, 84, 86, 87, 94, 165, 167, 185, 188, 190, 206, 232, 233, 236–239, 249, 251, 252, 256, 264, 292, 294, 301, 302, 314, 355
Phoenix dactylifera L., 257
Pholorotannins, 244
Phormidium sp, 263
Phosphate fertilizers, 189
Phosphatidylinositol 3-kinase, 253
Phosphor-JNK, 300
Phosphorus, 33, 84, 193, 257, 342
Phosphorylation, 113, 252, 300–303
Photo-aging, 303
Photo-damaged skins, 301
Photoprotective, 236
Photosynthetic
pigments, 74, 184, 312
process, 231, 235
properties, 3
Phycobilisomes, 184
Phycocolloid, 158, 159, 172, 185, 194, 209, 350
industry, 169
production, 31, 158, 194
Phycocyanin, 184, 232, 235, 312
Phycoerythrin, 184, 232, 235, 312
P-hydroxybenzoic acid, 71
Phyllophora, 160
Phylogenetic lineages, 3
Phylogeographic, 5
Phymatolithon calcareum, 211
Physical
isolations, 27
water scarcity, 342
Physicochemical, 83, 89, 90, 99, 103, 113, 115
biofunctional properties, 83

properties, 89, 90, 103, 115
Physiological, 3, 8, 31, 48, 105, 218, 232, 257, 259, 301, 355
Phytoalexines, 218
Phytopathogens, 259
Phytopharmaceuticals, 203, 206, 216, 220
Phytophthora
cinnamomi, 261
palmivora, 261
Phytopigments, 231
Pichia angophorae, 322
Pigmentation, 3, 82, 184, 203, 230, 354
Pigments, 82, 84, 110, 184, 185, 206, 213, 230, 232, 233, 256, 292, 312, 327, 354, 355
Plant
biostimulants, 183, 192, 195, 203, 208, 213
genome, 190
growth-promoting bacteria, 257
hormone biosynthesis, 256
respiration, 343
Plasma membrane, 69, 74, 254
Plasmodium
berghei, 253
falciparum, 63, 253, 254
3D7 strain, 253
malariae, 253
parasites, 253
vivax, 253
Plasmopara viticola, 261
Plocamium telfairiae, 250, 252
Pneumonia, 166
Polar compound extraction, 111
Political
instability, 336
institutions, 36, 39, 358–361
Polyamines, 256
Polyculture, 23
Polycyclic aromatic hydrocarbons (PAHs), 297, 304
Polyethylene
glycol, 100
pockets, 31
Polymannuronic acid monomers, 104
Polymerization, 188, 237
Polyphenol (PPs), 82, 84, 164, 167, 187, 190, 206, 213, 216, 232, 237, 244, 246, 256, 303, 325, 354–356
oxidase, 218
Polyphyletic, 3, 4, 231, 232
Polysaccharides, 31, 60, 61, 64, 68, 70, 73, 82–84, 86, 90, 91, 97–99, 110–118, 120, 121, 128, 130–132, 153, 159, 162, 163, 167, 186–190, 203, 209, 213, 216–218, 221, 230, 232–234, 242, 244, 246, 249, 251, 254, 256, 257, 259–261, 292, 301, 303, 311, 315, 325, 349, 350, 354–356
enriched extracts (PEEs), 214, 221
Polysiphonia, 216, 256
lanosa, 216
Polyunsaturated fatty acids (PUFA), 73, 82, 156–158, 165, 187, 206, 232, 234, 235, 239, 301, 355
Pomades, 109
Pond cultivation, 33
Population growth, 338, 341, 342
Porphyra, 8, 9, 28–30, 154, 156, 158, 185, 194, 203, 217, 263, 313, 320, 321, 323, 350
harvest, 9
laciniata, 30, 156
life cycle, 28
perforata, 203, 321
sp. cultivation, 28
tenera, 156, 263
thallus, 28
Porphyridiophyceae, 3, 184
Postelsia palmaeformis, 30
Post-farm, 21
Post-harvest processing, 20, 39, 40
Post-oviposition, 220
Potassium, 84, 100, 101, 107, 108, 131, 188, 193, 203, 207, 214, 257, 259
Potent
anthelmintic activity, 58
anti-inflammatory activity, 71
inhibition activity, 264
Potential, 242
alternative method, 168
application of seaweeds, 314
cosmeceutical applications, 320
impact of seaweeds in agriculture, 321
industrial uses, 324
pharmacological activities, 314
seaweed as a biofuels, 322
seaweed utilization as a bio-filter, 322
seaweed utilization in addressing environmental issues, 325

industrial applications, 186
utilization, 10
Potentiation, 218
Poultry, 162, 313, 351, 352
Poverty, 16, 17, 22, 25–27, 344, 346–348
Prasinophyceae, 48
Preservative effect, 89
Primary producers, 3, 231, 312
Pro-apoptotic effect, 249
Pro-collagen
inhibition, 303
production, 302
Product diversification, 36, 358
Production
capacity, 32, 33
employment, 17
Pro-inflammatory cytokine, 115, 298
synthesis, 115
Propylene glycol alginate (PGA), 108, 109
Prorocentrum
donghaiense, 263
micans, 263
Prostaglandin E2 (PGE2), 115, 165, 294, 295, 299, 302
Prostate cancer cell line (PC3), 62, 63
Proteins, 32, 84, 91, 106, 110, 116, 130, 153, 155, 156, 162, 188, 213, 218, 230, 232, 233, 235, 239, 244, 257, 258, 260, 261, 292, 295, 299, 300, 354, 355
content, 235
production, 31, 351
quality, 156
Proteorhodophytina, 184
Proteus, 166
mirabilis, 69
Prunus, 216
Psoriasis, 298
Pterocladia, 324
Pulmonary diseases, 164
Purification methodologies, 109
Pyrite content, 20
Pyrolysis, 90, 167
Pyropia, 30, 156, 185, 194, 203
perforate, 221
Pythium aphanidermatum, 261

Q

Quercetin, 213
Quseir, 51, 54, 66, 69

R

Rhizoctonia solani, 217, 219, 260
Ralstonia solanacearum, 260
Ramifications, 234
Reactive oxygen species (ROS), 48, 113, 116, 123, 124, 129, 133, 244, 249, 258, 291, 293–295, 304
Reclassification proposition, 5
Recombination, 5
Recommended dietary allowances (RDA), 350, 362
Red
algae, 3, 18, 29, 30, 47–51, 61, 64, 66, 81, 82, 98, 156, 157, 164, 168, 184, 185, 232, 236, 261, 349, 350
seaweeds, 5, 6, 71, 75, 166, 186, 190, 194, 203, 207, 209, 217, 233, 246, 260, 323
tide microalgae, 263
Re-jellification, 105
Renewable
energy, 336, 339, 349, 355, 357, 360–362
production, 336, 355, 360, 361
resources, 312, 353
Replication, 116, 163, 164, 242
Research institutions, 36–39, 358–361
Resource
conservation, 25
depletion, 337
Respiratory
infection deaths, 290
syncytial virus (RSV), 163, 242
Rheumatism, 320
Rhinitis, 298
Rhizophagus irregularis, 257
Rhizosphere
microbes, 203, 257
microorganisms, 257
nutrients availability, 257
Rhodellophyceae, 3, 6, 184
Rhodophyceae, 11, 47, 82, 232, 233, 235, 312, 327
Rhodophyta, 6, 7, 9, 18, 30, 49, 81, 98, 156, 160, 184, 195, 203, 209, 211, 230, 313
Ribavirin, 244
Riboflavin, 157, 355
Rice, 35, 213, 214, 219, 256, 258–260, 339, 340, 350, 351, 359, 361
Rocky shores, 2, 184
Rural communities, 21

S

Saccharina, 30, 87, 88, 90, 109, 110, 114, 118, 119, 122, 125, 155, 185, 194, 203, 221, 313, 322
japonica, 30, 88, 117, 125, 155, 185, 194, 313
latissima, 30, 88, 90, 114, 119, 155, 322
longicruris, 122, 127, 155, 203, 221
Saccharomyces cerevisiae, 65, 320
Saccorhiza polyschides, 192, 193
Sacharina cichorioides, 125
Safaga, 51, 54
Salmonella sp., 166
typhimurium, 66, 68, 126
Sarcodiotheca furcata, 56, 68
Sargassum, 5, 8–11, 30, 51, 53, 54, 56, 59, 63, 65, 67, 68, 89, 100, 110, 114, 116, 118, 119, 121, 122, 130, 131, 154, 162, 166, 168, 190, 192, 193, 210, 214, 217, 220, 237–239, 242, 245, 249–251, 253, 255, 257, 258, 260–262, 264, 295, 302, 303, 313–315, 320–324
asperifolium, 51, 53, 63
cinereum, 249
confusum, 250, 253
crispum, 51
fluitans, 242
fusiforme, 30, 53, 67, 114, 115, 122, 154, 166, 313
hemiphyllum, 249
horneri, 295, 300, 301
horridum, 220
latifolium, 51, 89, 261
muticum extract, 193, 214
muticum, 119, 192, 193, 214, 258
oligocystum, 315
platycarpum, 54, 59, 65
polycystum, 118, 245
species, 10
swartzi, 245, 257, 258, 260, 262
swartzii, 30, 257, 258, 260, 262
tenerrimum, 114, 116, 166, 220
thunbergii, 237, 250, 251, 262
vulgare, 30, 53, 67, 166, 210, 242, 249, 261
wightii, 190, 217, 239, 249, 251, 253, 260, 314, 315, 320–322
Serratia marcescens, 64, 66
Staphylococcus aureus, 63, 66–68, 74, 126, 166, 238, 239, 315
Salinity, 28, 74, 82, 120, 184, 189, 206, 208, 213, 232, 237, 258, 259, 343
Salmonids, 22
Sandy coastal areas, 2
Sanitation, 17, 341, 344
Sargaço, 183, 191, 192
Sargassaceae, 237
Sargassan, 83, 130, 132
Sargassum, 132
Saringosterols, 60
Sausages, 313
Scientific
infrastructures, 8
innovation, 100
Sea
cucumbers, 28, 32, 33, 109, 239
lettuce, 30, 156
palm, 30
whip, 30
Seabed profile, 3
Seacrop, 190
Seagrapes, 30
Seagrass ecosystems, 3
Seasonal hydrography, 3
Seaspray, 190
Seawater, 23
agitation, 99
temperature patterns, 2
Seaweed, 1–10, 15–21, 23, 27–40, 47, 48, 48, 51, 52, 56, 61–64, 66, 67, 69–75, 81, 82, 84, 85, 87, 88, 90–93, 97–101, 103, 104, 109, 110, 113–119, 121, 122, 125, 126, 128, 130, 132, 153–158, 162–172, 183–195, 201–221, 229–236, 238–240, 242–244, 247, 249–251, 254–266, 289–293, 295–304, 311–316, 320–327, 335–337, 344, 345, 349–361
aquaculture, 206
as plant protection factor, 214
antibacterial, 215
antifungal, 216
immunomodulatory effects, 218
insects protection, 219
bioactive metabolites, 232
phenolic compounds and phlorotannins, 236

seaweed lipids and fatty acids, 234
seaweed pigments, 235
seaweed polysaccharides, 233
seaweed proteins, peptides, and amino acids, 235
biodiversity, 1
biostimulants, 163
communities, 8
cultivation, 16–19, 21, 28, 31, 32, 34–40, 345
development, 17
cultivators, 35
development, 17, 208
extracts, 190, 195, 213, 255, 257, 260, 262, 264
extracts (plants foliar biostimulants), 190
scientific actualization (portuguese seaweed tradition), 191
biodiversity, 1, 2, 10, 265, 266
biostimulants, 230, 255, 256
distribution, 3, 8, 51
diversification, 5
extracts, 64, 266
farming, 35, 154, 170, 171, 357
fertilizer, 205, 211, 256, 258
industry processing, 20
inventory, 7
liquid fertilizer (SLF), 322, 327
mediterranean red sea coast of egypt, 48
mediterranean coast of egypt, 48
red sea coast of egypt, 50
metabolites, 250, 251, 289, 290, 295, 297, 299, 301, 303
multitrophic cultivation, 19
polymers, 98
processing industry, 154
production, 17, 154, 313, 353, 354
phylogeographic studies, 5
physical characteristics, 3
species, 74, 213, 292, 326
Secondary
endosymbiosis event, 185
metabolites, 82, 116, 153, 186, 219, 232, 236, 292, 302, 304, 311, 325, 354
Seed germination, 162, 192, 193, 208, 214, 258, 259, 321, 322
Seirococcus axillaris, 109
Selenium, 84, 349
Semi-closed system, 32
Semi-intensive ponds, 20
Semliki forest virus (SFV), 164
Sensory characteristics, 92
Sequential extraction, 91
Sequestration rates, 33
Sesquiterpenes, 65, 239
Sesquiterpenoids, 186, 250, 263
Seven brominated diterpenes, 58
Shabu-shabu broth, 350
Shelf-life time, 117
Shigella flexneri, 68, 315
Shinorine, 320
Shipping disruptions, 336
Shrimp, 19, 20, 22, 27, 32–34, 90, 162
culture, 19
juvenile, 19, 20
Signaling functions, 98
Silver nanoparticles (AgNPs), 125, 133, 169, 239, 260, 261
Siphoncladophyceae, 313
Sirophysalis trinodis, 51, 55, 60, 66
Size exclusion chromatography, 84
Skeletonema costatum, 263
Skill improvement, 38–40
Skin
damage-related mechanisms, 291
lightening products, 321
Small
scale
fishers, 27
production, 10
sized fish, 20
Smooth muscle multiplication, 117
Social institutions, 25
Socialization, 36, 358
Socio-economic
aspects, 36
improvement, 21
spheres, 346
Socio-techno-economic
aspects, 16, 38
improvement, 39
Sodium (Na), 73, 84, 100–103, 105, 107, 108, 121, 126, 128, 130, 131, 158, 188, 203, 207, 213, 355
carbonate, 101, 102, 121, 130
chloride solution, 128

Soil
aeration, 163, 257
amelioration, 322
compaction, 342
degraded samples, 257
erosion, 342, 343
fertilization, 191, 201, 205
moisture, 163, 220, 343
pathogenic organisms, 217
Solanum
lycopersicum, 192, 216
melongena, 261
Solar
energy conversion, 90
powered dryers, 34
Solieria
chordalis, 320
filiformis, 244, 251, 254
Soybean, 212, 256, 259, 352, 355
Spatial, 10, 120
Spawning grounds, 3
Species competition, 3
Specific
bioactive compounds, 292
poverty manifestations, 346
Spinacia oleracea, 256
Spiral wrack, 30
Splenocytes, 295
Spodopetra
frugiperda, 63
littoralis, 63
Spot correction, 320
Spyridia hypnoides, 260
Stabilizer, 23, 100, 107, 108, 159, 187, 324
properties, 100
Stabilizing agents, 104
Standardized drying technology, 34
Standard-paracetamol, 315
Starvation tolerance, 354
Stearic acid, 71
Steroids, 82
Sterol regulatory
element-binding protein 1 (SREBP1), 251
element-binding protein 1c (SREBP-1c), 252
Stocheospermum marginatum, 242, 263, 322
Strategic environmental conditions, 17, 336, 337
Stratum corneum (SC), 293, 297
Streptococcus, 166
agalactiae, 64
lactus, 65
pyogenes, 64, 239
sanguis, 64
Streptozotocin-induced diabetic, 252
Stresses tolerance, 230, 255
Stromelysins, 301
Strontium (Sr), 100
Structural
activity relationship, 120
characterization, 125
modification, 127
Stylonematophyceae, 3, 184
Subcritical water extraction (SCWE), 89, 111
Subcutaneous administration, 70
Sub-cutis, 291
Subject object method (S-O-M), 16, 17, 39, 336, 337, 361
Substantial
bargaining power, 26
freshwater pollution, 341
Substructure management, 36, 358
Sugar
composition, 113
kelp, 30, 155
Sulfated
polysaccharides (PLS), 83, 97, 98, 133, 163–165, 206, 233, 234, 242, 244, 254, 302, 314, 356
xylomannan, 163
Sulfuric acid, 117, 123
Sulphation, 113, 114, 119–121
Supernatant, 121, 122, 128, 130, 131
Superoxide dismutase (SOD), 71, 73, 114, 244, 258, 261
Support system technology, 16, 34, 38–40
Supra intensive
ponds, 34
shrimp cultivation, 19
practices, 19
Suprastructure management, 36
Surgical scars, 127
Sustainability, 24, 25, 170
aquaculture, 16, 21
development, 16, 15, 17, 21, 22, 31, 36, 40, 171, 202, 221, 229, 336, 344, 345, 347, 348, 362
agents, 345
goals (SDG), 25, 202, 221

energy, 17, 355
management, 10, 215
production systems, 22
Symbiotic
association, 231
relationships, 93
Synergistic activity, 190
Synthetic
compounds, 164, 202, 219
insecticides, 219

T

Taonia atomaria, 58, 62, 69, 73
Treptacantha barbata, 54, 66
Tricleocarpa
cylindrica, 52, 62, 69, 73
jejuensis, 263
Turbinaria, 51
conoides, 166, 315
decurrens, 242
ornata, 55, 60, 63, 66–68, 70–73, 315
Tannins, 220, 236, 245, 315
Taurine, 206
Taxonomic, 1, 10, 47, 52
biased, 5
classification, 48
groups, 97, 207, 209
revision, 5
Technical
assistance, 25
innovation, 21
Temperate
climates, 23
regions, 5
Temperature, 28, 48, 82, 98, 111, 120–122, 130, 184, 187, 191, 193, 206, 208, 231, 259, 315, 343, 348, 357
Temporal variability, 10
Terpenes, 164, 218, 220, 232, 246, 314
Terpenoid compounds, 73
Terrestrial
microorganisms, 264
systems, 29
Tetradecanoylphorbol-13-acetate (TPA), 315
Teuhetenone, 250
Textile industry, 161
Texturizers, 171
Thalassotherapy, 320
Thallophyta, 311, 327
Thallophyta division, 311
Thallus, 231, 312
Therapeutic
applications, 114
approaches, 165
botanical applications, 213
potential, 117, 123
skin preparations, 127
Thermal expansion, 343
Thiamine, 157
Thickening agents, 107, 161
Thongweed, 30
Threonine, 203, 207
Tilapias, 22
Tissue
engineering applications, 104
hydration, 83
inflammation, 115
Tobacco mosaic virus, 260
Tocopherols, 157, 164, 187, 259
Tomato, 190, 192, 212, 214–217, 256, 258, 260, 261, 321
Total aquaculture emissions, 22
Toxic sulfhydryl compounds, 204
Trace elements, 190, 311, 312, 325, 352, 355
Traditional
chinese medicine, 165
identification, 5
pond farmers, 34
sea vegetables, 10
techniques, 100
treatment, 312
Transcription, 73, 124, 217, 242, 295, 297, 300
Transforming growth factor-alpha (TGF-α), 298
Transmembrane-spanning G-protein coupled receptors, 71
Transmission electron microscopy, 67
Transparent marketing transactions, 26
Transportation facilities, 9
Trichloroacetic acid, 130
Trichophyton rubrum, 239
Trichoplusia ni, 262
Triticum aestivum, 216, 322
Tropical
climate, 2, 50

coastal diversity, 8
regions, 3, 5, 6, 10
seascape, 2
seaweed, 1–3, 5, 17
beds, 3
Tumor necrosis factor-alpha (TNF-α), 129, 130, 298, 300, 304
Tyrosine-specific protein kinase, 60

U

Ulva, 30, 51–53, 56–58, 61, 62, 71, 156, 157, 162, 164, 166, 168, 203, 209, 218, 236, 239, 249, 256, 259, 262–264, 315, 320, 322, 323, 325, 356
fasciata, 56–58, 62, 63, 164, 262, 323
intestinalis, 30, 53, 56, 57, 61, 70, 156
lactuca, 51, 52, 54–58, 61–65, 67–71, 73, 74, 156, 157, 239, 249, 259, 314, 320, 323
linza, 263
pertusa, 166, 262, 263
prolifera, 166, 263
reticulata, 218, 239, 320
Undaria, 18, 86, 87, 110, 118, 119, 123, 126, 154, 155, 158, 185, 194, 217, 234, 237, 246, 250, 253, 261, 313
pinnatifida, 18, 30, 86, 87, 110, 115–117, 119, 123, 126, 154, 155, 158, 185, 194, 217, 234, 237, 246, 250, 253, 261, 313
Ultra performance liquid chromatography quadruple time-of-flight (UPLC Q-TOF), 85
Ultrasound, 122
assisted extraction, 84, 89, 92, 94
Ultrastructural, 48, 231
Ultraviolet (UV), 48, 118, 169, 232, 237, 320
Ultraviolet A (UVA), 320
Ulvaceae, 234, 320
family, 234
Ulvans, 234
Ulvophyceae, 3, 6, 7, 9, 11, 48, 51, 313
Umibudo, 351
Uncoating, 242
Uncultivated coastal area, 2, 9
Underground aquifers, 341
Underutilized
bio-resource, 292
macroalgae, 11
seaweeds, 1, 2, 10
United Nations (UN), 25, 202, 338
Unsaturated fatty acid (UFAs), 85, 157, 263, 355
Untreated domestic wastewater, 341
Upregulated production, 293, 299, 301
Urban growth, 24
Urbanization, 342
Urchins, 109
Uronic acid, 102
Uses of seaweeds, 155
applications of seaweed (agriculture), 162
biomedical applications (seaweeds), 163
macroalgae (source for cosmetic compounds), 166
phycocolloids industry, 158
seaweeds
animal feed, 161
human food, 155
potential source of bioenergy, 167
nanotechnology, 169
water detoxification, 168
Ustilaginoidea virens, 261
Usujirene, 320

V

Valoniopsis pachynema, 251, 254
CdS NPs, 254
Vertebrata lanosa, 216
Value-added
compounds, 89
process, 20, 39
Vegetative fragments, 28
Verticillium dahliae, 215, 261
Veterinary importance, 242
Vibrio
alginolyticus, 166
fluvialis, 68
parahaemolyticus, 126
Vicia faba, 74
Vidalols A, 165
Vigna unguiculata, 257, 258
Violaxanthin, 157, 235
Viral attachment, 164, 242
Virus associated protein (VAP), 164
Viscosity, 10, 88, 100, 102

Vitamin, 84, 91, 99, 155, 157, 162, 163, 186, 189, 190, 203, 206, 213, 232, 256, 292, 311, 312, 315, 325, 349, 352, 353–355
  vitamin C, 157, 244

W

Wakame, 30, 155, 185, 350
Warmer winter temperatures, 343
Wastewater treatment, 168, 323
Water
  availability, 17, 206, 340, 341
  consumption, 341
  fraction, 68
  in-oil emulsions, 108
  insoluble molecules, 108
  pumps, 20
  quality, 21, 32, 33, 345
  retain inherent capacity, 108
  retaining capacity, 104
  retention, 162, 257, 350
  scarcity, 336, 338, 340, 362
  soluble
    polysaccharide fraction, 131
    sulfated heteropolysaccharide, 131
  vapor, 89
Wave undulation, 184
Well-integrated system, 31
WiDr cells, 86
Wildlife extinction, 337
Women empower, 17, 25
Wool production, 352
Wound
  granulation tissue, 127
  healing, 88, 99, 104, 105, 124, 315

X

Xanthomonas, 216
  arboricol, 216
  axonopodis, 260
  oryzae, 260
Xanthophylls, 84, 184, 195, 235
Xylans, 233, 355
Xylose, 83, 112, 128, 131, 188, 234
  side chains, 83

Z

Zeaxanthin, 157
Zinc, 84, 169, 207, 232, 253, 301, 324, 347, 349
  oxide nanoparticles, 253
Ziziphus mauritiana, 190, 322